Materials Science

Nanotechnology and Applications

Materials Science

Nanotechnology and Applications

H.D. Kumar
Ex Dean, Faculty of Science and Coordinator
Biotechnology Programme
Banaras Hindu University
and
Honorary Emeritus Professor
Indo-Soviet College of Pharmacy
Moga, Punjab

I.K. International Publishing House Pvt. Ltd.
NEW DELHI

Published by

I.K. International Publishing House Pvt. Ltd.
S-25, Green Park Extension
Uphaar Cinema Market
New Delhi–110 016 (India)
E-mail: info@ikinternational.com
Website: www.ikbooks.com

ISBN 978-93-81141-17-5

Reprint 2017

Published by Krishan Makhijani for I.K. International Publishing House Pvt. Ltd., S-25, Green Park Extension, Uphaar Cinema Market, New Delhi–110 016 and Printed by Rekha Printers Pvt. Ltd., Okhla Industrial Area, Phase II, New Delhi–110 020.

Preface

Human well-being rests on the foundation of three indispensable, interdependent and equally important pillars of conditions and processes related to the economic, socio-political, and environmental systems. As the planetary natural resources are limited while human needs are unbounded, we need to develop foundational technologies that can make it easier to satisfy human needs. New technologies strongly depend on appropriate materials and are equally strongly limited by materials. Judicious use of a diversity of materials and application of modern techniques of nanotechnology is making it possible, step-by-step, to engineer living and non-living systems to manipulate information, construct materials, process chemicals, produce energy, provide food, and help maintain or even enhance human health and the natural environment. The way we use energy in future will determine whether or not we can achieve many of our environmental objectives. According to J.P. Holdren, ex President of the American Association for the Advancement of Science, "energy is the hardest part of the environment problem; environment is the hardest part of the energy problem; and resolving the energy-economy-environment dilemma is the hardest part of the challenge of sustainable well-being for industrial and developing countries". Recent advances in understanding metals, minerals and other materials, coupled with continuing developments in nanoscience and nanotechnology can help in resolving the challenges facing global economic, socio-political and environmental systems.

The composition and structure of real materials are very complex. Exciting research possibilities hover around complex materials science, as exemplified by multicomponent, technologically important alloys.

Exciting discoveries from basic scientific researches in physics, chemistry, biology and engineering have expanded the boundaries of human knowledge and provide the foundation for advances in health care, computation, communication, food production, defence, energy production and use, transportation, and much more. One such discovery is related to superconductors that can carry electricity without resistance, and so are better than copper wires. But, to attain the superconducting state, these materials must be cooled below a critical temperature. Exploring the effect of intense pressure on the superconducting state could potentially lead to superconductors at ambient conditions.

The adoption of new methods for synthesizing nanomaterials and the availability of new tools and techniques for their manipulation have catalyzed explosive growth of nanoscience, nanotechnology and nanobiotechnology. Some products that were considered science fiction two decades ago, are now materializing. Effective interfacing and integration of nanotechnology and materials science is sure to lead to the creation of novel gadgets and products.

Besides the above, a few other notable aspects of materials science and nanotechnology that are briefly outlined in this monograph are molecular electronics, electrets, tiny transistors, critical minerals, nanosized catalysts, adaptive composites, graphene, graphane, nanogenerators and more.

I am indebted to all those whose work enhanced our knowledge of materials and prompted this modest attempt at writing this monograph. Sincere thanks are due to: (1) the Indian Academy of Sciences, Bangalore and the Indian National Science Academy, New Delhi, for providing recent literature; (2) the United Nations Environment Programme, Nairobi, for giving me recurring opportunities from 1989 through 2009 to visit many countries and interact with leading scientists, engineers and technologists; and (3) to the authors and editors who very kindly permitted me to reproduce their figures or tables.

I am much obliged to Sh. Parveen Garg, Chairman, ISF College of Pharmacy, Moga and the staff of the College for their sustained friendship and kindness. Special thanks are due to Prof. S.P. Vyas, Vinay Sharma, O.N. Srivastava, T.R.K. Reddy, Ashok Kumar, Er. T.S. Misra and the staff of the Department of Biosciences and Biotechnology, Banasthali Vidyapeeth, for their continuing support and academic interactions.

Several drafts of the manuscript were very ably and patiently composed by Rang Nath Singh.

H.D. Kumar

Units

ENERGY UNITS, ABBREVIATIONS AND CONVERSION FACTORS UNITS

SI (Systeme Internationale) Units

Physical Quantity	Name of Unit	Symbol
length	metre	M
mass	kilogram	Kg
time	second	S
thermodynamic temperature	kelvin	K
amount of substance	mole	Mol

Special Names and Symbols for Certain SI-Derived Units

Physical Quantity	Name of SI Unit	Symbol for SI unit	Definition of Unit
force	newton	N	Kg m s^{-2}
pressure	pascal	Pa	Kg m^{-1} s^{-2} (= N m^{-2})
energy	joule	J	Kg m^2 s^{-2}
power	watt	W	Kg m^2 s^{-3} (= J s^{-1})
frequency	hertz	Hz	s^{-1} (cycles per second)

Decimal Fractions and Multiples of SI Units having Special Names

Physical Quantity	Name of SI Unit	Symbol for SI Unit	Definition of Unit
length	Ångstrom	Å	10^{-10} m = 10^{-8} cm
length	micron	µm	10^{-6} m
area	hectare	ha	10^4 m^2
force	dyne	dyn	10^{-5} N
pressure	bar	bar	10^5 N m^{-2} = 105 Pa
pressure	millibar	mb	10^2N m^{-2} = 1 hPa
mass	tonne	T	10^3 kg
mass	gram	g	10^{-3} kg
column density	Dobson units	DU	2.68×10^{16} molecules cm^{-2}
streamfunction	Sverdrup	Sv	10^6 m^3 s^{-1}

Certain units are not part of the International System of Units, but nevertheless are important and widely used. Consistent with the recommendations of the International Committee for Weights and Measures, the units in this category that are accepted for use with the SI are given below.

Units outside the SI that are accepted for use with the SI.

Name	Symbol	Value in SI units
minute (time)	min	1 min = 60 s
hour	h	1 h = 60 min = 3600 s
day	d	1 d = 24 h = 86400 s
degree (angle)	°	1° = (π/180) rad
minute (angle)	′	1′ = (1/60)° = (π/10800) rad
second (angle)	″	1″ (1/60)′ = (π/648000) rad
liter	L	1 L = 1 dm^3 = 10^{-3} m^3
metric ton[(a)]	t	1 t = 10^3 kg
neper	Np	1 Np = 1
bel[(b)]	B	1 B = (1/2) In 10 Np[(c)]
electronvolt[(d)]	eV	1 eV = 1.60218 × 10^{-19} J, approxi.
unified atomic mass unit[(e)]	U	1 u = 1.66054 × 10^{-27} kg, approxi.
astronomical unit[(f)]	ua	1 ua = 1.49598 × 10^{11} m, approxi.

[(a)] In many countries, this unit is called "tonne".
[(b)] The bel is most commonly used with the SI prefix deci: 1 dB = 0.1 B.
[(c)] Although the neper is coherent with SI units, it has not been adopted by the General Conference on Weights and Measures (CGPM), and is thus not an SI unit.
[(d)] The electronvolt is the kinetic energy acquired by an electron passing through a potential difference of 1 V in vacuum. The value must be obtained by experiment, and is therefore not known exactly.
[(e)] The unified atomic mass unit is equal to 1/12 of the mass of an unbound atom of the nuclide ^{12}C, at rest and in its ground state. The value must be obtained by experiment, and is therefore not known exactly.
[(f)] The astronomical unit is a unit of length. Its value is such that, when used to describe the motion of bodies in the solar system, the helicocentric gravitation constant is $(0.01720209895)^2$ $ua^3.d^{-2}$. The value must be obtained by experiment, and is therefore not known exactly.

The following definitions of the SI base units are taken from NIST Special Publication 330 (SP 330), The International System of Units (SI).

Unit of:	Unit	Description
Electric current	**ampere**	That constant current which, if maintained in two straight parallel conductors of infinite length, of negligible circular cross-section, and placed 1 metre apart in vacuum, would produce between these conductors a force equal to 2 × 10^{-7} newton per meter of length
Thermodynamic temperature	**kelvin**	The fraction 1/273.16 of the thermodynamic temperature of the triple point of water.
Amount of substance	**mole**	**1.** The amount of substance of a system which contains as many elementary entities as there are atoms in 0.012 kilogram of carbon 12; symbol is "mol". **2.** When mole is used, the elementary entities must be specified and may be atoms, molecules, ions, electrons, other particles, or specified groups of such particles.

Units of Measurement

Engineers must acquire familiarity with various units of measurement, especially the International System of Units. In the United States, environmental quantities are measured and reported in both the U.S. Customary System (USCS) and the International System of Units (SI), and so it is important to be familiar with both. The latter system is usually preferred. The following Table gives conversion factors between the SI and USCS systems for some more commonly used units.

Some basic units and conversion factors.

Quantity	SI units	SI symbol x	Conversion factor =	USCS units
Length	metre	m	3.2808	ft
Mass	kilogram	kg	2.2046	lb
Temperature	Celsius	°C	1.8 (°C) + 32	°F
Area	square metre	m^2	10.7639	Ft^2
Volume	cubic metre	m^3	35.3147	ft^3
Energy	kilojoule	kJ	0.9478	Btu
Power	watt	W	3.4121	Btu/hr
Velocity	meter/second	m/s	2.2369	mi/hr
Flow rate	$meter^3$/second	m^3/s	35.3147	ft^3/s
Density	kilogram/$meter^3$	kg/m^3	0.06243	lb/ft^3

Some Energy Units (values are given for one unit of each)

joule (J) = the force exerted by a current of 1 amp per second flowing through a resistance of 1 ohm

watt (W) = 1 joule (J) per second.

kilowatt-hour (kWh) = 1 thousand (10^3) watts exerted for 1 hour

megawatt (MW) = 1 million (10^6) watts.

gigawatt (GW) = 1 billion (10^9) watts

petajoule (PJ) = 1 quadrillion (10^{15}) joules

British thermal unit (Btu) = energy to heat 1 lb of water 1°F

PJ = 947 billion Btu, or 278 million kWh

standard barrel (bbl) of oil = 42 gal (160 liters) or 5.8 million Btu

metric ton of standard coal = 27.8 million Btu or 4.8 bbl oil

Contents

CHAPTER 1

General Introduction

NANOSCIENCE AND NANOTECHNOLOGY

Nanoscience is the science of nanomolecules or matter having nanometer (nm) dimensions (1 nm = 10^{-7} cm = 10^{-9} m). Nanoscience is widely viewed as one of the most exciting technologies of today and tomorrow, for diverse kinds of applications ranging from advanced materials research, catalysis, energy storage, microelectronics to biotechnology. It is fully in tune with the concept of green sustainable chemistry. Nanoscience research has demonstrated that materials could be synthesized by using ecofriendly technologies that minimize, for example, toxic wastes as a by-product. Sastry *et al.* (2003) showed that besides the commonly used bacteria and yeasts for the preparation of nanomaterials, one could also use fungal broths for obtaining silver nanoparticles, and nanosized gold can be synthesized from an actinomycete. Indeed, it has become possible to introduce significant variations in chemical composition, size and shape of metal nanoparticles.

Nanotechnology has been hailed as the one outstanding enabling technology that is destined to redesign the future of several technologies, products and markets. It has already become a most important sector that is attracting keen interest as it can replace most existing technologies in use today. Its multi-disciplinary nature implies that its impact on society could be profound and all-pervasive. The basic unit of nanotechnology is the nanometer (nm) which is one thousandth of a micron, or one millionth of a millimeter or 10^{-9} of a meter, roughly the length occupied by five to ten atoms stacked in a straight line. For comparison, the hydrogen atom is about 0.1 nm, and a virus may be about 100 nm and a red blood cell about 10,000 nm in diameter.

In the 1970s nanotechnology was defined as the science of manipulating atoms and single molecules. Its scope has expanded greatly and it now embraces all technologies capable of building structures at the scale of a billionth of a metre. This shift may have been caused by big business and politics (see Dusastre, 2008). In fact, this broader definition has also modified the early goals of nanotechnology research.

In the early 1980s, the invention of the scanning tunneling microscope enabled imaging and manipulating single atoms and facilitated designing of nanoscale machines from individual atoms and molecules. We have come a long way from single molecules to today's molecular machines, such as organic molecules that can move and perform calculations.

Until the 1990s, atomic manipulations were developed with sustainability in mind and it was hoped that building machines from the bottom up, atom by atom, rather than etching them from larger blocks, would minimize the energy and materials spent in manufacturing.

The field of nanotechnology now covers a whole range of tiny technologies ranging from metallurgy and material sciences where new materials for various different applications are tested; to electronics where memories, computers, components and semi-conductors are explored; to biotechnology where diagnostics and new drug delivery systems are developed.

CMP Cientifica (2002) defined nanotechnology as the discipline which deals with materials in the range of 0.1 to 100 nm. Earlier NNI (2000) had defined it to mean the construction and utilization of functional structures with at least one characteristic dimension measured in nanometrers.

Nanotechnology deals with materials and systems whose structures and components show novel physical, chemical, and biological properties – properties that facilitate the exploitation of novel phenomena and processes arising from nanoscale size. Some scientific features of nanotechnology are listed in Table 1.1.

The scope of nanotechnology is not restricted to non-living materials and applications but extends to life sciences. It has become possible to replicate living organisms to perform engineered tasks. Current progress in information technology, biotechnology, semiconductors, and other technologies, is also greatly impacted by and, is in turn, being impacted, by developments in nanotechnology (Bhat, 2003).

Table 1.1 Salient properties/aspects of nanotechnology (after Bhat, 2003, with permission from *Current Science*).

- Metals become harder at the nanoscale
- Ceramics become softer
- Composites and alloys of a large variety may be made
- Thin polymers become less permeable
- Nanoscale materials are stronger, more heat resistant, transparent.
- Increased chemical resistance, reduced weight
- Different interactions with light/other radiation
- Novel electrical properties
- Novel biological properties

More recently, nanotechnology has grown even further – it is now projected as an important way of tackling the global energy crisis and water shortage. Top-down production is woefully wasteful of materials and energy. Nanotechnology will eventually affect how people work, what they eat, how they communicate, and how long they live. It will change their medical care, energy sources, water and environment. Its considerable benefits and possible risks need to be seriously and rigorously debated. Strong oversight mechanisms that can deal with the complex convergence of nanotechnology, biotechnology, and information and cognitive science, and that are up to the task of 'taming the genie' are badly needed.

The drive to miniaturize has been progressing long and is best exemplified by today's microelectronics. However, when applied to integrated electronics and electromechanical systems, lithographic techniques are approaching a physical limit. Also, below tens of nanometers, serious problems such as interconnection and quantum effects appear.

Nanotechnology has now converged with biology. According to Joachim and Pelvert (2008), we need not worry that our DNA may be modified by nanobots finding their way into cells, or about nanotechnology's association with genetically modified organisms. On the other hand, current research on building machines from self-assembly and supramolecular chemistry is likely to prove useful and could attract important future applications.

Nanotechnology attracts several applications in nanomaterials, nanoelectronics, nanobiotechnology and other areas as illustrated in Fig. 1.1.

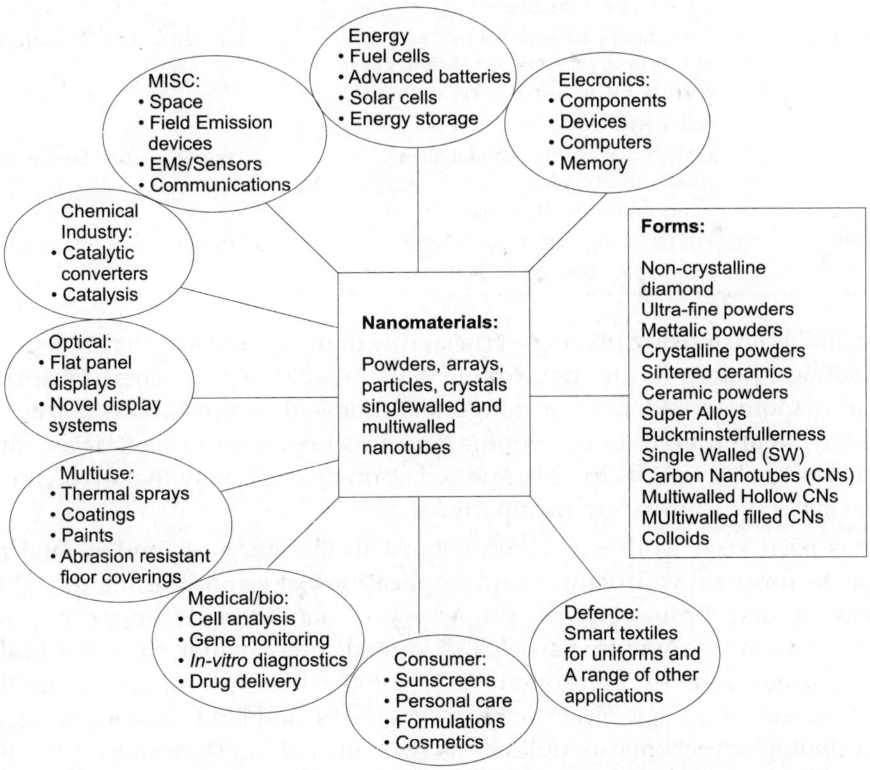

Fig. 1.1 Nanotechnology applications (after Bhat, 2003, with permission from *Current Science*).

A selected list of various innovative products created by exploiting new developments in nanotechnology is given in Table 1.2.

Table 1.2 Illustrative list of innovative products made by using new developments in nanotechnology (after Bhat, 2003, with permission from *Current Science*).

BASF	Clear and clean sunscreen based on ZCOTE nanodispersed zinc oxide	Material source: Nanophase, Technologies, Illinois, US
Babolat, France	VS nanotube power racquets for tennis using high modulus graphite with carbon nanotubes	Material source: Nanoledge, France
Wilson	Double core tennis balls with A or D-Fense nanocomposite products	Material source: InMat, New Jersey, US
Hornberg, Germany	Easy cleaning tile based on nanoparticle coating	Material source: Nanogate, Germany
Eddie Bauer	Stain and wrinkle-free clothes, jeans using molecular structures added to cotton fibers	Material source: Nano-Tex, California, US
Maui Jim	Eye glasses with protective and anti-reflective scratch-resistant glass coating made using novel chemical self assembly	Material source: Nanofilm, Ohio US
GM	GMC Safari's 2002 model uses thermoplastic olefin nanocomposite step assist	Material source: Southern Clay, Texas US
Samsung	HDTV using carbon nanotubes for display leading to better resolution	Material source: Samsung

Nanomaterials are already playing a crucial role in many futuristic technologies. One key aspect of nanotechnology is the development of reliable experimental protocols for the synthesis of nanomaterials over a range of chemical compositions, sizes and high monodispersity. With a view to developing green technologies in materials synthesis, this aspect of nanotechnology needs to be supported strongly, especially that involving the use of microorganisms in the synthesis of nanoparticles.

There has been keen worldwide keen interest in the study of fundamental properties, organization to form superstructures and applications of nanoscale matter. The unusual physicochemical and optoelectronic properties of nanoparticles appear chiefly from confinement of electrons within particles of dimensions smaller than the bulk electron delocalization length. The exotic properties of nanoparticles have attracted applications in optoelectronics, catalysis, single-electron transistors (SETs) and light emitters, nonlinear optical devices and photoelectrochemical applications (Colvin *et al.*, 1994; Schmid, 1992; Klein *et al.*, 1997; Weller, 1998; Yoffe, 1993). Magnetic nanoparticles have interested researchers for superparamagnetism and for applications such as magnetic memory storage devices, magnetic resonance image enhancement and magnetic refrigeration (Symonds, 1995; Shull *et al.*, 1993). The fact that the optical absorption/emission properties of semiconductor nanoparticles (quantum dots) can be tuned by variation in nanoparticle size is particularly attractive in the band-gap engineering of materials and the growth of quantum dot lasers (Fendler and Meldrum, 1995). Nanoscale matter has also been explored for potential application in nanocomputers, synthesis of advanced materials, energy storage devices, electronic and optical displays, chemical and biosensors, and biomedical devices (see Sastry *et al.*, 2003).

The need to develop environmentally benign nanoparticle synthesis processes which do not use toxic chemicals in the synthesis protocol has grown in recent years. Many researchers engaged in nanoparticle synthesis and assembly have turned to biological systems for inspiration – unicellular and multicellular organisms produce inorganic materials either intra-or extracellularly. Some selected examples of living organisms synthesizing inorganic materials include magnetotactic bacteria (synthesize magnetite nanoparticles), diatoms (synthesize siliceous materials, Kroger *et al.*, 1999), and S-layer bacteria (produce gypsum and calcium carbonate layers, Pum and Sleytr, 1999). Uncovering the secrets hidden in living organisms has catalyzed biomimetic approaches for the growth of advanced nanomaterials (Sastry *et al.*, 2003). Many biotechnological applications such as remediation of toxic metals make use of microorganisms such as bacteria and yeast. Materials scientists have shown keen interest in such microorganisms as possible eco-friendly nanofactories (Southam and Beveridge, 1996). Fortin and Beveridge (2000) reported that gold particles of nanoscale dimensions can be precipitated within bacterial cells by incubation of the cells with Au^{3+} ions. Klaus-Joerger *et al.* (2001) showed that when the bacterial strain *Pseudomonas stutzeri* AG259 isolated from a silver mine is placed in a concentrated aqueous solution of $AgNO_3$, the Ag^+ ions are reduced and silver nanoparticles of well-defined size and distinct morphology are formed within the periplasmic space of the bacterial cells. They also showed that biocomposites of nanocrystalline silver and the bacteria can be heated to yield a carbonaceous (cermet) nanomaterial with interesting optical properties for potential application in functional thin-film coatings (Joerger *et al.*, 2000).

Kowshik *et al.* (2002) reported that the yeast, *Torulopsis* sp. can achieve intracellular synthesis of nanoscale PbS crystallites when exposed to aqueous Pb^{2+} ions. The PbS nanoparticles can be extracted from the biomass by freeze-thawing. Kowshik *et al.* (2002a) also showed that CdS quantum dots synthesized intracellularly in *Schizosaccharomyces pombe* yeast cells show ideal diode characteristics. They used biogenic CdS nanoparticles (1-1.5 nm) for the fabrication of a heterojunction with poly (*p*-phenylenevinylene). This diode generated 75 mA/cm^2 current in the forward bias mode at 10 V, and breakdown occurred at 15 V in the reverse direction.

Eukaryotic fungi have also been used to grow nanoparticles of different chemical compositions and sizes. Indeed, fungi are suitable candidates for the synthesis of metal and metal sulphide nanoparticles. According to Sastry *et al.*, two taxa of fungi, *Verticillium* sp. and *Fusarium oxysporum*, when exposed to aqueous gold and silver ions, reduce the metal ions rapidly. In *Verticillium* sp., reduction of the metal ions occurred intracellularly leading to the formation of gold and silver nanoparticles in the size range of 2-20 nm. *F. oxysporum* behaved differently – the reduction of the metal ions occurred extracellularly and highly stable gold and silver nanoparticles (2-50 nm) were formed.

Ahmad *et al.* (2002, 2003) challenged the acidophilic fungus, *Verticillium* sp. with Ag^+ and $AuCl_4^-$ ions and noted that this leads to their reduction and accumulation as silver and gold nanoparticles within the fungal biomass. When the fungal biomass was reacted with or exposed to aqueous $AgNO_3$ solution for three days, a dark brown colour could be seen in the biomass, pointing to the reduction of the metal ions and formation of silver nanoparticles in the biomass. Gold and silver nanoparticles look coloured (pink to blue for gold, and light yellow to brown for silver) due to excitation of surface plasmon vibration in the particles (Henglein, 1993), and thus facilitate visual identification of their presence in the biomass. The growth of silver nanoparticles occurred only within the fungal biomass, not extracellularly.

It is known that in bacteria, most metal ions are toxic and hence, reduction of the ions or formation of water insoluble complexes represent a defence mechanism in the bacteria to overcome such toxicity. Fungi are not normally exposed to large concentrations of metal ions, so the observed reduction of silver ions seems surprising. Further, it was observed that the fungal cells when exposed to small concentrations of silver ions, multiply and almost completely cover the agar plate in about a week. Thus, we can exploit the utility of micro-organisms in the synthesis of nanoscale materials if their ability to multiply (and cover large surface areas) is not affected on exposure to metal ions, as demonstrated by Sastry *et al.* (2003).

While it is possible to release the intracellular silver and gold nanoparticles by ultrasound treatment of the Ag nano-*Verticillium* biomass or by reaction with suitable detergents, it is much more convenient if the metal ions exposed to the fungus could be reduced *outside* the biomass, leading to the formation of metal nanoparticles in solution (Ahmad *et al.*, 2003a). Upon exposure to metal ions, *F. oxysporum* produces enzymes which reduce the metal ions to yield highly stable nanoparticles in solution. The monodispersity of the nanoparticles produced either intra- or extracellularly was not very high (~40-50%) and is therefore very inferior to that obtained by conventional chemical methods (5-15%). Sastry *et al.* (2003) noted that the extremophilic actinomycete, *Thermomonospora*, when exposed to gold ions reduced the metal ions extracellularly, yielding gold nanoparticles with a much improved polydispersity. Extremophiles are microorganisms that thrive under very harsh conditions such as extremes of temperature, high pressure, non-aqueous environment containing 100% organic solvents; or excess heavy metal concentration. Actinomycetes share important characteristics of fungi and prokaryotes such as bacteria. Although they are normally classified as prokaryotes due to their close affinity with mycobacteria and the coryneforms, they were originally designated as 'ray fungi'. Actinomycetes have come into limelight by their phenomenal ability to produce secondary metabolites such as antibiotics (Sasaki *et al.*, 1988).

Ahmad *et al.* and Sastry *et al.* isolated a novel alkalothermophilic actinomycete, *Thermomonospora* sp. showing optimum growth at pH 9 and 50°C, from self-heating compost in Barabanki, Uttar Pradesh. After isolation in culture, the harvested biomass of this actinomycete was resuspended in 10^{-3} M aqueous $AuCl_4$ solution in 250-ml Erlenmeyer flasks at pH 9 and incubated in the dark. The bioreduction of the $AuCl_4^-$ ions in solution was monitored by periodic sampling of aliquots (2 ml) of the aqueous component and measuring the UV-visible spectra of the solution.

The gold surface plasmon band occurred at ~520 nm and steadily increased in intensity as a function of time of reaction. Complete reduction of the $AuCl_4^-$ ions takes place after nearly 120 h of reaction, showing that it is an extremely slow process. Observation of an intense red colour after suitable staining pointed to the formation of gold nanoparticles. The gold nanoparticles do not appear to aggregate into clumps. Also, the nanoparticle solution remained stable for more than six months. The reduction of the gold ions takes place extracellularly (Sastry *et al.*, 2003).

Frankel and Whitesides (2009) discussed the recently emerged structural and conceptual underpinnings at the small scale, and life and engineering. They pointed out nanotechnology's potential risks and prospects and made the revelations of nanoscience accessible. It is now abundantly clear that shape and size matter immensely to nanoscale function and that it is the layering of these simple concepts that generates the complexity of life.

Whereas the big-science communities of particle physics and genomics, for instance, are highly organized when disseminating knowledge and lobbying on behalf of their fields, nanoscience practitioners have not been able to spread its messages across wider society.

THE NANO-MICRO INTERFACE (PHYSICS VS. BIOLOGY)

Between the nano- and micrometer scales, startling properties manifest in matter and are driven by its collective behaviour. A specific example is that the pressure of a gas is a cooperative property of large numbers of particles but is not anticipated from the behaviour of one particle alone. Newton's laws of motion do account for it, but it was much later that the statistical description of atoms necessary for understanding pressure was developed (Coleman, 2007).

Quantum matter can potentially develop emergent properties. Atoms of niobium and gold are individually similar, but can join to make crystals that, if kept cold, have very different properties. Electrons move free across gold crystals and form the conducting fluid which imparts gold its lustrous metallic properties. Up to about 30 nm, gold and niobium do not differ significantly. But beyond this point the electrons in niobium couple together into what are known as 'Cooper pairs'. At the micrometer scale, pairs congregate to form a single quantum state and the crystal is transformed into new metallic state (a superconductor) which conducts without resistance, excludes magnetic fields, and can levitate magnets (Coleman, 2007).

But superconductivity is only the beginning. When softer, organic molecules assemble, a tenth of a micrometer is big enough for the emergence of life. Self-sustaining microbes around 200 nm in size have been discovered. Whereas physicists understand the principles that govern the superconductor, biologists have still to grasp those involved in the emergence of life on roughly the same spatial scale!

Condensed-matter physicists, inspired by organic evolution, believe that a key to understanding more complex forms of collective behaviour in matter is the competition not between species, but between different forms of order; high-temperature superconductors are materials that show superconductivity at liquid-nitrogen temperatures; they form in the presence of a competition between insulating magnetic behaviour and conducting metallic behaviour. Similarly multi-ferroic materials that couple magnetic with electric polarization, can develop when magnetism competes with lattice-distorting instabilities.

NANOSCALE LIGHT-MATTER INTERACTIONS

The interplay between light and matter lies at the core of several fundamental processes and various applications (Weiner and Ho, 2003). Harnessing light-matter interactions may enable operation of solid state devices under new physical principles: for instance, the a.c. optical Stark effect (OSE) has made possible coherent quantum control schemes of spins in semiconductors; these have the potential for realizing quantum devices based on spin qubits (Press *et al.*, 2008). But when the dimension of semiconductors is reduced, light-matter coupling weakens, limiting applications at the nanoscale. Khitrova *et al.* (2006) achieved significant enhancement of nanoscale light-matter interactions, though with the need for a high-finesse cavity. These interactions ultimately prevented device down-scaling and integration. Zhang *et al.* (2010) have now reported that a significant OSE can be achieved at substantial energy detuning in a cavity-free colloidal metal-semiconductor core-shell hetero-nanostructure, in which the metal surface plasmon is tuned to resonate spectrally with a semiconductor exciton transition. This resonantly enhanced OSE shows polarization dependence and can be a viable mechanism for coherent ultrafast spin manipulation within colloidal nanostructures. The plasmon-exciton resonant nature further makes possible tailoring of both OSE and spin manipulation by tuning plasmon resonance intensity and frequency. These results may pave the way for tailoring light-matter-spin interactions through plasmon-exciton resonant coupling in a carefully engineered nanostructure; they can be a basis for future applications in quantum information processing at the nanoscale. More generally, integrated nanostructures with resonantly enhanced light-matter interactions could act as a test ground for other emerging fields, including nano-biophotonics and nano-energy (Dubowski and Tanev, 2006).

Such a fine tailoring of resonant coupling may well facilitate the optimization of nanoscale light-matter-spin interactions, which could lead to development of room temperature devices (Pryor and Flatte, 2006).

SYNTHESIS AND APPLICATIONS OF NANOCRYSTALS

Several different methods have been used for preparing nanocrystals of metals such as Au and Ag and of semiconductors such as CdS and CdSe (Duff *et al.*, 1993; Rockenberger *et al.*, 1999; Fendler, 1998; Rao *et al.*, 2002). Films of nanocrystals such as Au, Ag and Pd, semiconducting sulphides such as CdS, ZnS and CoS, and oxides such as Fe_2O_3 and CuO have been prepared by performing reactions at liquid-liquid (organic-aqueous) interfaces. In this method, a suitable organic derivative of the metal taken in the organic layer reacts at the interface with the appropriate reagent present in the organic layer (Rao *et al.*, 2003). Metal nanocrystals are usually made by the controlled reduction of metal salts. Nanocrystals of metal chalcogenides and oxides are prepared by controlled precipitation or thermal decomposition (Pileni, 2003). While some control over the size distribution of nanocrystals is achieved, it has been difficult to create robust films of nanocrystals through self-assembly. Producing nanocrystal films by molecular beam epitaxy and other physical methods is quite tedious and requires expensive equipment. Rao *et al.* (2003) exploited the interface between two immiscible liquids to prepare nanocrystals and thin films of metals, semiconductors and metal oxides. In this method, the organic layer containing the metal precursor is brought into contact with the aqueous layer containing a suitable reagent for reduction, hydrolysis or sulphidation. The reaction takes place at the interface where the nanocrystals become distributed. The yield and the size distribution depend on the reaction parameters such as concentration, on how reagents are dispersed, mechanical agitation, temperature, and viscosity of the medium. By choosing appropriate parameters, it is possible to obtain either isolated nanocrystals with a narrow size distribution or their well-formed films.

By using high concentration of metal precursors, lustrous films of gold, of varying thickness, can be prepared. These can be readily transferred to a solid substrate by raising the substrate gently across the interface. Indeed, the nanocrystals at the interface can be transferred to either the aqueous or the organic phase by using appropriate capping agents. By adding dodecanethiol to the organic layer, the Au nanocrystals could be transferred entirely to the organic layer, while the addition of mercaptoundecanoic acid successfully transferred the nanocrystals to the aqueous layer. There is little or no change in the average diameter of the nanocrystals after transfer to the organic or the aqueous medium from the interface (Rao *et al.*, 2003). Nanocrystals of silver and palladium have also been observed at the liquid-liquid interface.

Nanocrystals of γ-Fe_2O_3 with a diameter of ~2.6 nm could be obtained through the reaction of iron cupferronate in the toluene layer with NaOH in the water layer. The γ-Fe_2O_3 particles happen to be superparamagnetic. Films of γ-Fe_2O_3 nanocrystals of different thicknesses were deposited on solid substrates by successive addition of dispersions of the nanocrystals in toluene or in water. Films of citrate-capped γ-Fe_2O_3 prepared on Si(100) substrates were quite smooth and extended to several microns (Rao *et al.*, 2003). These films are composed of ~12 nm diameter nanoparticles. Unlike the nanoparticles which were only superparamagnetic, the films were magnetic and showed hysteresis.

The nanocrystals of gold dispersed in water can be used for nanolithography by the dip-pen process (Demers and Mirkin, 2001). In this method, Au nanocrystals deposited on the AFM cantilever dissolved in the water meniscus between the tip and the substrate (Fig. 1.2). The dissolved nanocrystals can be transferred to the substrate when the tip is slowly scanned over the surface. Aggregates of the nanocrystals appear on the substrate when the water layer evaporates.

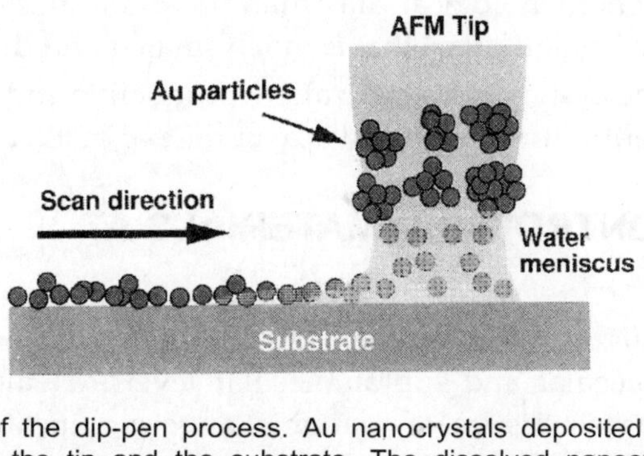

Fig. 1.2 Schematic sketch of the dip-pen process. Au nanocrystals deposited on the cantilever dissolve in the water meniscus between the tip and the substrate. The dissolved nanocrystals are transferred to the substrate when the tip is slowly scanned across the surface. Aggregates of nanocrystals are formed on the substrate upon evaporation of the water layer (after Rao *et al.*, with permission from *Current Science* and Dr. C.N.R. Rao).

The liquid-liquid interface shows high potential as a universal medium for the synthesis of nanocrystals and thin films of varying types of materials.

NANOTECHNOLOGY AND MATERIALS SCIENCE

Nanotechnology is undoubtedly the key technology for the 21st century, with far-reaching impacts on such areas as materials, electronics, information technology, energy, environment, pharmacy, and medicine.

Nanoscale science, engineering, and technology enable working at the atomic, molecular and supramolecular levels for creating large structures with fundamentally novel properties and functions. Their merger into nanotechnology can lead to greatly enhanced understanding and better control over the basic building blocks of all natural and man-made things. Already, scientists have begun synthesizing amazingly novel materials by monitoring and controlling nanoscale structures of materials.

At the nanoscale, even very small fluctuations of a single atom can produce a drastic change in the entire system. This is why nanotechnology (NT) can exert a much greater impact on technology, economy, society, and culture in the coming years as compared to the conventional technologies which affected our parents and grandparents.

NT involves an effective interdisciplinary integration among physics, chemistry, biotechnology (BT), material sciences and engineering with the objective of critically examining assemblies of atoms and molecules. Applications of NT will cover many sectors and spheres of life (e.g., communication, health, labour, mobility, housing, relaxation, energy, and food) and accordingly will be accompanied by marked changes in social, economic, ethical and ecological dimensions (Maynard, 2007).

The US National Nanotechnology Initiative (NNI) defined NT as the understanding and control of matter of roughly 1 to 100 nanometers where unique phenomena enable novel applications. One nanometer is one-billionth of a meter, or the length of ten hydrogen atoms placed in a line.

Two other definitions, proposed by the UK Royal Society and Royal Academy of Engineering, are as follows.

Nanoscience: A phenomenological and manipulation study of materials at a scale (atomic, molecular and macromolecular) that is much smaller and different than larger scales.

Nanotechnology: The design, characterization, production and application of structures, devices and systems by controlling shape and size at the nanometer scale.

DESIGNING AND CONTROLLING MATERIALS AT THE ATOMIC LEVEL

NT involves the manipulation and/or creation of material structures at the nanoscale, involving the atomic, molecular and supramolecular levels. At the nanoscale (particularly under 10-20 nm), the characteristics of matter change significantly. Nanoscale material structures of the same chemical elements change their mechanical, optical, magnetic and electronic properties, and chemical reactivity so much that surprising and unpredictable effects are produced.

Some of the most marked changes while going from the macroscale to the nanoscale are seen in metals and metal oxides. Gold, for instance, loses its golden colour when separated into nanometer-diameter particles, and becomes red. It also becomes highly reactive or corrosive at nanoscale. Likewise, silver, in nanoscale is highly resistant to bacteria. Another example is titanium dioxide becoming opaque at the macroscale, as opposed to being transparent at the nanoscale.

Eigler and Schweizer (1990) demonstrated how to manipulate individual atoms accurately. They could write I-B-M by using 35 individual xenon atoms. This feat was performed by using a low temperature ultra high vacuum scanning tunneling microscope (STM).

In a way, nanotechnologists are modern alchemists, and some day they might well control atoms and molecules as wished at room temperature to make any structures and mixtures by using any substances. When NT is fully developed, what happens in millennia would be achieved in the laboratory in a few seconds by manipulating carbons with an STM probe. In principle, any precious material such as diamond can be made out of cheap abundant materials like coal. However, in reality, it is very difficult to manipulate atoms or molecules one by one (like assembling Lego blocks) due to the thermal fluctuation of atoms at room temperature. We need at least a dozen atoms to be grouped to generate chemically stable materials. It is virtually impossible to produce on a large scale any structures or parts by manipulating atoms one by one. This means we should learn from nature which demonstrates a self-assembly (bottom-up) technology based on trial and error during hundreds of millions of years (Lee and Jhon, 2008).

Self-assembly describes processes in which a disordered system of pre-existing components generates an organized structure of pattern as a consequence of specific, local interactions among the various components without external disturbance. For example, a living thing produces proteins with various functions by making chains of amino acids at nanoscale, based on the information contained in DNA. The concept of "self-assembly" is very broad, ranging from growth in a crystal to reproduction of life (Lee and Jhon, 2008).

NT aims to grow crystals by controlling them at nanoscale and also to produce them in large quantity at once and at low price. In fact, simple structures such as wrinkle-resisting suits, perfumes, and partial semiconductor manufacturing processes (atomic layer deposition) can be generated. Some of these attempts will face strong challenges (see Hemminger, 2007). Two of the five grand challenges listed by Hemminger are (a) how to design and perfect atom- and energy-efficient synthesis of revolutionary new forms of matter with tailored properties? And (b) how do interesting or novel properties of matter emerge from complex correlations of the atomic or electronic constituents and how best may we control such properties?

Nature nurtured itself well through self-assembly with little energy loss or wastage of resources. Our dreams outlined in science fiction may materialize when we develop the self-assembly technology. If we could handle atoms and molecules freely, an electric oven may be designed that can produce any parts we want with several different elements from the Periodic Table. In future, not only novel material mixtures will be made, but also the material properties could be changed to suit the needs. Nanobiotechnology will facilitate replacing defective or damaged human body organs with new ones, and functional enhancements would be achieved by inserting new bones made by nano-engineering. Some day, even DNA and protein might be made by gathering atoms and even a living being could be created! A time can come when garbage will be regarded as a treasure house of resources.

Currently, some 70 countries are seriously considering the scope of NT which can revolutionize science, human society, and culture in the 21st century. A large market may be created by increasing the investment for NT.

According to Roco and Bainbridge (2005), $1 trillion (T) worth of products worldwide will incorporate NT in key functional components by the year 2015. The corresponding industries would require about two million workers in NT plus many more jobs will need to be created in supporting activities.

The total global investment in NT R&D reported by U.S. government organizations increased approximately 9-fold in the last 8 years, from $432M in 1997 (Siegel *et al.*, 1999) to about $4.1 B in 2005.

The Republic of Korea is a prominent country with deep interest in NT. It launched its first NT Initiative as a 10-year rolling plan (2001-2010). Its latest revision for the period 2006-2015 aims to join the ranks of the top three nations in global technological competitiveness by 2015. (The United States leads the world with Japan right behind it, followed by Germany and Korea).

In Korea, more than 270 large and new venture industrial enterprises were engaged in NT in 2006. The large companies have mostly been concentrating on the improvement of their core products and the creation of new business exploiting NT while most of the new ventures are specializing in nanomaterials. Samsung and LG are exploring various basic technologies related to organic/inorganic nanomaterials, nanodevices, and nanoprocessing for next generation semiconductor, data and energy storage. As for consumer products, many companies have developed nanosilver-coated home appliances which show superior cleaning and sterilization properties. Small and medium companies are marketing products such as cosmetics, toothpaste, soap, textiles, paint ceramics, filters and deordrants.

A highly successful national programme named Tera-Level Nanodevices (TND) was established in Korea in April 2000 by Jo-Won Lee. This centre was established around the same time as the United States' NNI, having 3 international advisors (J.W. Lee, M.S. Jhon and M.C. Roco) to strengthen the national competitiveness in nanoelectronics area and to cross the technological hurdles imposed upon upcoming semiconductor technologies. The TND program is a 10-year program consisting of three phases. The first phase was operated as a versatile basic cell development for tera-level nanodevices. The second phase saw major efforts in integrating nano-scale devices. The third phase focuses on the development of tera-level integrated array of nanodevices (Table 1.3).

The Korean Govt. proposes to launch 22 new future growth engines that will contribute to the country's economy after semiconductors, shipbuilding and cars over the next ten years. Of these, the engines related to NT include new nanofusion materials, solar cells with pollution-free coal energy, fuel cell development systems, green cars, displays, and LED lighting

(Table 1.3). Key sectors in the future growth engines are energy and environment, including pollution-free coal energy that can produce synthetic fiber using low-grade coal; seaweed-based marine biofuels that substitute coal energy; and solar cells. Eight industries are directly related to the government's "low carbon, green growth" vision.

Table 1.3 Projects operated by TND in 2006 (after Jo-Won Lee and M.S. Jhon, 2008).

Area	Title of the project
Tera-level nanoelectronics	Tera-bit-level single electron memory
	Nano CMOS
	SET logic and RF SET
	Tera Hz-level IC
Spintronics	MRAM integration process
	MR material and single cell process for MRAM
	Spin injection devices
Molecular electronics	Tera bit-level carbon nanotube devices
	Tera bit-level organic devices
Core technologies	Nanodeposition
	Nanoanalysis
	Tera-level optical interconnection

FABRICATION OF NANOSCALE FEATURES IN POLYMER THIN FILMS

It is possible to fabricate nanoscale features in polymer thin films by room-temperature forging or embossing. In this process, the material experiences inelastic flow from below a die to make an impression. The thinness of the layer below the die is limited by friction, the shear strength of the material, and the pressure generated by material building up outside the die, as well as by the magnitude of force which the die itself can tolerate before it deforms. Cross *et al.* (see *Nano Lett.* 7: 10.1021/n10624566, 2007) showed that when a small-amplitude shearing motion (approximately 10 nm) is applied in the plane of the film, surface flow is activated. The shear motion enables formation of much thinner layers and also minimizes the defects that often limit the yield of structures with the necessary critical dimensions for a given application.

Block Co-polymers

The nanotechnology revolution has been boosted by the possibility of molecular assembly, i.e., using molecular or chemical forces to self-assemble smart or functional structures which might be integrated into electronic or optical devices.

Kim *et al.* (2007) took a diblock polymer that self-assembles into micelles (known as poly(styrene-b-4-vinylpyridine). They forced it to form microdomains by tuning the micelle structure through solvent exchange, creating hexagonally ordered templates with periods of about 45 nm.

These polymer templates were then transferred by reactive ion etching to aluminium surfaces and regular pores were fabricated by anodic oxidation at 4°C. The resulting hexagonal, close-packed pores are just 12 nm across, with nearest-neighbour spacings that should be tunable over a range of about 10-50 nm. The ordering extends over large area (several micrometres square).

The seamless combination of colloid chemistry with more conventional 'top-down' processing is particularly appealing. These wet chemical methods should be cheaper and more scalable as compared to traditional cleanroom-based techniques. The structure can be tuned to provide a range of surface topologies and a wide variety of potential templates. These could greatly simplify the fabrication of periodic, sub-wavelength structures for plasmonics-based applications such as optical circuitry (Mulvaney, 2007).

TAMING THE NANOSCALE – MOLECULAR MACHINES – THERMAL NANOMOTOR

The first thermal nanomotor has been designed by European researchers. Adrian Bachtold *et al.* started with multi-walled nanotubes made by arc discharge evaporation. A gold 'cargo' was placed on them by electron beam lithography. The remaining outer nanotubes were removed by electrical breakdown, leaving a nanotube sleeve loaded with cargo which can rotate around or translate along the inner nanotube track.

The assembly is suspended between two metal electrodes. The sleeve moves when a current is applied. It was first thought that the effect was due to electrons shuttling the cargo around, but when the motion showed no dependence on the current direction, the researchers were surprised and somewhat confused. Simulations showed that heating in the nanotube leaves the center at some 1300°C. In a perfect reversal of frictional heat dissipation, phonon waves along the temperature gradient shuffle the cargo along toward cooler regions.

The mutual chirality of the inner and outer nanotubes may define a track for the nanomotor, providing a custom mix of rotation or translation. And the cargo could be anything that can go into a nanotube. Fullerenes inside nanotubes might possibly respond in the same way to temperature gradients. Further work could potentially offer new ways of taming the nanoscale.

NANOGENERATOR

Nanogenerators are tiny devices that convert movements into electricity. Nanodevices fabricated with one-dimensional nanostructures (such as nanowires, nanotubes, and nano-belts) can be generated by using very low power (Chen *et al.*, 2005; Patolsky *et al.*, 2006). Such a current is provided by an external battery which has to be replaced or recharged frequently. The dependence on an external power source significantly limits the nanodevices. Although they cannot power cities, they could possibly power arrays of invisible sensors and hand-held electronics. In a future home run by a smart computer system, for instance, an array of invisible sensors would perform many tasks ranging from tracking your movements to monitoring whether your plants should be watered (Service, 2010).

Sensor networks have been contemplated for years. One hurdle in using them has been figuring out how to power the devices. Several energy scavenging approaches have been developed with applications in wireless electronics, such as thermoelectric, piezoelectric thin-film and vibrational cantilevers (see Wang *et al.*, 2007). Wang and Song (2006) converted nanoscale mechanical energy into electric energy by piezoelectric zinc oxide (ZnO) nanowire (NW) arrays. Using a conductive atomic force microscopy (AFM) tip in contact mode, they deflected the aligned NWs, thereby converting the mechanical energy created by the deflection force into electricity with the piezoelectric properties of the NWs. Improving the power generation capabilities of the system necessitates replacing the AFM tip with a simpler source of mechanical energy that can actuate all the NWs simultaneously and continuously. This problem was tackled by employing ultrasonic waves to drive the motion of the NWs; a continuous current was produced.

In effect, they developed an innovative nanowire nanogenerator which is driven by an ultrasonic wave to produce continuous direct-current output. The nanogenerator was fabricated with vertically aligned ZnO NW arrays placed beneath a silicon zigzag electrode coated with platinum, and the electrode had a small gap. The wave moves the electrode up and down to bend and/or vibrate the NWs. A piezoelectric-semiconducting coupling process converts mechanical energy into electricity. The zigzag electrode acts as an array of parallel integrated metal tips that simultaneously and continuously produce, collect, and give out electricity from all the NWs. The work of Wang and Song (2006) and Wang *et al.* (2007) led to the development of an adaptable, mobile, and cost-effective technology for harvesting energy from the environment. It may be useful for powering nanodevices and nanosystems. Powering each network with batteries or wiring to the grid is expensive and requires effective maintenance. The recent advent of a new technology to scavenge power from vibrations and other ambient sources is poised to materialize this vision of the future.

Conventional power scavengers are large devices that harness vibrations to monitor the structural health of buildings and bridges. In recent years, however, these scavengers have been progressively shrunk to nanoscale dimensions so as to power devices ranging from minuscule sensors inside the body to arrays of self-powered environmental sensors to monitor air quality or stream flows. Such a miniaturization has been aided by the steady progress of microelectronics technology, which produces sensors and computing devices so small and so frugal with their energy needs that they can be powered with only a few nanowatts or microwatts of power.

Most nanogenerators are made from piezoelectric materials that convert mechanical motion into electricity. **Piezoelectric materials**, e.g., crystals of the ceramic lead zirconate titanate (PZT), are composed of subunits that separate electrical charge. Bending a thin piezoelectric wire changes this electric polarization of the material and makes positive and negative charges to migrate to opposite faces of the material, creating an electric voltage that is used to do work. This effect can also be reversed: Applying a voltage across a piezoelectric crystal makes it move. It is this effect that is being gainfully exploited in sonar detectors, loudspeakers, autofocusing cameras, atomic force microscopes, and several other gadgets.

The most widely used piezoelectric material (PZT) efficiently converts mechanical strain in to electricity. But PZT has some drawbacks: brittle and breakable, and contains the toxic lead which prevents its use for powering medical sensors in the body.

Now, Michael McAlpine *et al.* of Princeton University have created a flexible PZT nanogenerator encapsulated inside a biocompatible plastic. Whereas large crystals of PZT are brittle, its ultrathin ribbons can bend and flex without breaking. So these researchers grew a thin layer of PZT atop a crystal of magnesium oxide (MgO) and then used a lithographic patterning technique, like that used to pattern microchips, to pattern the PZT into an array of long ribbons. The MgO substrate was then dissolved, leaving the ribbons behind, and a rubbery plastic known as PDMS was pressed on top. This transferred the PZT nanoribbons to the plastic. The transferred PZT nanoribbons retained their same piezoelectric properties. Also they were far more efficient at transferring mechanical strain to electricity than were competing nanogenerators (Service, 2010).

Another way to produce a nanogenerator is to grow zinc oxide (ZnO) nanowires between chromium and gold electodes. This nanogenerator produces a high voltage when flexed (Service, 2010). Likewise, a flex capacitor may be built by using a nanostamping technique which turns normally brittle PZT into flexible nanowires encapsulated in a rubber sheet.

Nanofinger

Thanks to nanotechnology, policemen now have a new weapon to fight crime: a nano-solution that sharpens fingerprints.

Crime investigators have traditionally sprayed suspect surfaces with a water-based gold or silver solution to detect fingerprints. The metal ions are reduced to a black precipitate along the lines of fatty deposits left by the skin ridges. However, even with the most advanced fingerprint techniques only a few prints at crime scenes are good enough to be used as evidence.

J. Almog and Daniel Mandler of Israel reported that attaching hydrocarbons to gold nanoparicles is the key. The fat-seeking hydrocarbons guide the gold to the skin impression and leave a metal trail. If this treatment is followed with the conventional solution, the gold catalyzes the precipitation of metal in solution, and the resulting fingerprints are much sharper (see *Science* 315: 1773, 2007).

The new method could potentially revolutionize the field of crime fighting.

NANOSCALE FORCES FOR BIOLOGISTS

It is now possible to exploit emerging new technologies for exploring nanoscale forces and interactions. For long, biophysicists have engaged in characterizing the small forces associated with cells and molecules by using a variety of techniques of physics, chemistry and even materials science. They are now pushing their instruments further with a view to measuring molecular forces in the sub-piconewton range. Some biologists too wish to find out the extent to which these forces have a role in biological processes. The issue is: how can one convince oneself that forces are important if one does not measure them, but if they are not important, why measure them?

Biophysicists have developed and used 'optical traps' (which hold and manipulate molecules by means of small forces). To make them more widely applicable techniques and instrumentation need to become more robust and user-friendly if they are to be adopted for biological processes.

Biologists interested in exploring the nanoscale world can benefit from an effective integration of different technologies, such as combining Foerster resonance energy transfer (FRET) with optical traps or carbon nanotubes with atomic force microscopy (AFM) or even AFM with confocal microscopy. Several different tools are being integrated so as to enhance the understanding of highly complex molecular interactions, while potentially providing novel insights into the world of the single molecule (Blow, 2008).

Atomic Force Microscopy (AFM)

AFM, developed over 2 decades ago, has a resolution of a fraction of a nanometer – 1000 times better than the optical diffraction limit. It achieves this resolution by scanning a sample with a cantilevered tip and measuring deflection of the cantilever with a laser to give topological information such as height, length and shape. AFM differs from optical microscopy: its 'touch scanning' approach images a surface and measures its associated forces at the same time.

AFM was first used in materials science and physics to study surface properties of semiconductors for process-control applications. Many cell biologists now explore its potential in the role and influence of mechanical properties.

AFM is based on Hooke's law of elasticity – a simple equation expressing the force generated within a spring when compressed in terms of a 'spring constant' and the length of the spring. Developers view springs as cantilevers used in AFM.

Cantilevers are essentially springs, typically 50-200 micrometres long. When they are manufactured, there is some uncertainty in their dimensions. As the spring constant depends on these dimensions, such uncertainty can give erroneous measurements. This makes spring calibration a crucial procedure for AFM users. What we need is a way of getting an independent measure of the stiffness of the cantilever.

AFM cantilevers can also be used to measure forces in liquid samples. In a 'tapping mode', the cantilever oscillates across the sample while amplitude is monitored to keep the cantilever at a steady level compared with the surface. This has helped AFM to attract different biological applications. It is a gentle method of exploring soft matter such as cells and single molecules – much gentler than the 'contact mode' in which the tip is directly moved across the surface.

Continuing cantilever developments can potentially solve the issue of speed that faces AFM. Imaging cells with AFM is so slow that it can take 5 minutes or more to form the image. Speeding up this process would prompt more biologists to use AFM.

If only the size of the current cantilevers could be reduced, it would lead to faster motion and scan times.

Another part of Hooke's law relates to length or, in the case of AFM, the distance from the cantilever tip to the surface. Developers have evinced keen interest in using piezoelectric materials to move AFM tips. Piezos involve turning a knob, changing the potential by a volt – and one can move an angstrom. When the potential across the piezoelectric material is changed, it will move an angstrom – a phenomenon known as creep. Researchers have tackled this problem and a related problem – hysteresis – in various ways: some AFM developers place sensors around the piezoelectric material to measure the creep directly. These sensors provide feedback as the tip moves over the surface so that this form of drift can be largely eliminated.

A major focus of AFM developers is how to move the technology into the hands of biologists. Veeco and Asylum combine optical and force microscopes in a single system. Veeco's BioScope II is compatible with advanced light microscopy modalities when the AFM is placed on the top of an inverted microscope.

With suitable registration between the optical image and the AFM tip, a researcher can select regions of interest to correlate fluorescence signals in cells with high-resolution topography. Besides Asylum and Veeco, Park Systems in Santa Clara and Novascan Technologies in Ames (Iowa) are contemplating combining various optical microscopy techniques with AFM (Blow, 2008).

Single-molecule Interactions

One great merit of AFM is that it enables measurement of a structure's dynamics and function at the same time.

The molecular recognition of ligands and receptors is now being explored by combining AFM with antibodies to simultaneously obtain cell topography and recognition images in a method called TREC – basically a counterpart to immunofluorescence in which researchers try to localize receptors. The AFM tip is functionalized by attaching antibodies. It is then brought into contact with the cell membrane. The binding between ligand and receptor can be measured, while also obtaining topographical information simultaneously. The advantage of AFM for TREC, compared with other optical techniques lies in its higher lateral resolution, which can be combined with the ability to obtain mechanical parameters of the cell surface. As high-speed AFM is improved, we would see more work following physiological processes in living cells by using methods such as TREC.

Whereas some researchers use AFM to measure interactions between individual molecules, other teams attempt to visualize such interactions. Some groups are investigating carbon nanotubes (CNTs) as sensors. The interest in CNTs for biological applications is mainly as optical diagnostics. CNTs not only inform us about the chemical environment of a very small region of space but also have optical properties that make them unique for sensor applications.

CNTs emit light in the near infrared, which avoids issues of background fluorescence as this is a part of the spectrum in which very few biological systems autofluoresce. Further, unlike dyes, or even quantum dots, which can photobleach over time, CNTs are extremely photostable and so allow measurement of fluorescence changes to obtain molecular-level information.

One interesting aspect of CNTs is: how to get them into living cells with a view to understanding how and why, once inside, they move around. Use of single-particle tracking methods has revealed that the CNTs can either diffuse around the cell in a confined pattern or be actively transported.

CNTs also arouse interest in their application as optical sensors to detect single-molecule binding events. The binding of biologically important molecules into CNTs at the single-molecule level can in fact be detected in real time. The one-dimensional electronic property of CNTs implies high electron density at a very narrow region of energy; consequently when a

CNT absorbs a particular photon, the absorbance shows a very sharp maximum, allowing a single CNT to be visualized with spectroscopy. It also enables researchers to see if a nanotube responds to a biomolecule in a single-molecule association event.

The exciting area of single-molecule manipulation dates back to 1991 with experiments using gravity and the weight of little beads to extend the DNA. In more recent years a more sophisticated technique called 'optical tweezers' has been developed by Stephen Chu and Arthur Ashkin. By using this technique, one can trap and manipulate molecules with diameters of nanometers.

Optical tweezers exploit a focused laser beam to trap and hold dielectric objects. There is keen interest to prompt biophysicists to study things at a smaller scale – requiring more stability, less drift, more spatial resolution and improved temporal resolution from optical tweezers.

For high-resolution optical tweezers to work, the key issue of drift was sorted out by attaching a molecule to two beads and to use two optical tweezers, neither of which is physically attached to the lab environment, to hold the molecule.

Some companies are keenly interested in combining optical tweezers with single-molecule FRET to study the enzymes involved in transcription and translation – a combination that may be necessary as single-molecule research moves towards studying more complex properties of enzymes and proteins. The fundamental reaction steps of the enzymes can now be measured by using optical traps to obtain single base pair resolution; even more importantly, using FRET allows conformational changes of the enzymes themselves to be monitored at the same time. This combination allowed researchers to directly measure via fluorescence conformational changes as a function of applied force to dissect the landscape of the Holliday junction – a mobile junction between four strands of DNA (Hohng *et al.*, 2007).

The exciting prospect of using FRET with optical tweezers could well allow us to follow a particular biological process in a multi-dimensional manner.

NANOTECHNOLOGY: SOCIAL DIMENSIONS

Several countries have taken active interest in starting a National Nanotechnology Initiative (NNI) with a view to promoting nanoscale science and technology in a manner that benefits their citizens in particular and humanity in general. The U S NNI has successfully organized and funded several research projects not only on the environmental, health, and safety (EHS) aspects of nanomaterials (including associated regulatory capacity and best research and workplace practices) but also on education and outreach involving workforce preparation, educating the public about nanotechnology, and encouraging public acceptance of nanotechnology (see Sandler and Bosso, 2007). Yet, predicting and responding to broader social and ethical issues that can arise when nanotechnologies become widely disseminated has not attracted sufficient focus. Not only are some serious social and ethical issues associated with nanotechnology, but also the NNI provides an opportunity to prospectively consider the broad societal effects of a potentially transformative technology under the current nanotechnology revolution during regulatory and policy design. Grabbing this welcome opportunity can enable the NNI to effectively promote both technological and social progress (Sandler and Bosso, 2007).

Nanotechnology, like all technologies, is a patently social phenomenon. Technologies emanate from society and are not only enabled but also encouraged by society. Being implemented in and disseminated through society, they in turn exert notable impacts on society. This means that while identifying and addressing social and ethical issues associated with nanotechnology, the features of both nanotechnology and society must be carefully considered.

When the significance of the social context of nanotechnology is appreciated, social and ethical issues associated with nanotechnology materialize clearly. The lesson for any NNI's responsible development programme is that in order to be comprehensive it must tackle the issues relating to the relevant social contexts into which nanotechnology is emerging.

According to Sandler and Bosso (2007), some important social issues associated with nanotechnology are unequal access to resources and opportunities, institutionalized and non-institutionalized discrimination, differential social and political power, corporate influence and lack of accountability, inadequate governmental capacity for regulation, challenges to individual rights and autonomy, marginalization of noneconomic values, technology control and oversight, the importance of technology in creating and solving problems, and more generally, such other aspects of society and institutions that do not meet reasonable standards of justice. Any responsible development of nanotechnology must therefore consider how this development will interact with the current social contexts.

APPLICATIONS LITERATURE

Kostoff *et al.* (2006, 2006a, 2007, 2007a) applied an extensive nanotechnology/nanoscience-focused query (300 + terms) to the SCI/SSCI (Science Citation Index/Social Sci. Cit. Index) database. The nanotechnology/nanoscience research literature technical structure (taxonomy) was obtained by using computational linguistics, especially document clustering.

The applications literature associated with nanoscience and nanotechnology research was carefully examined. Through visual inspection of 60,000 of the abstract phrases of the same downloaded 2005 records, all the diverse non-medical applications were identified and their associated documents categorized in a hierarchical taxonomy. For medical applications, a fuzzy clustering algorithm in which a record could be assigned to multiple clusters was applied to the downloaded 2005 records.

The total retrieved nanotechnology database for 2005 was examined from four perspectives to identify pervasive thematic thrusts: document clustering, autocorrelation mapping, factor analysis and cross-correlation mapping. Each perspective yielded insights into the nanotechnology literature structure, but only document clustering results were presented in the papers published by Kostoff *et al.*

Document clustering: The database was divided into 256 thematic clusters by the clustering algorithm. USA produced most papers in 169 thrusts, China led in 70, Japan in 15, and India, South Korea and Spain each led in one.

When a hierarchical taxonomy was built from these 256 elemental clusters, of the 16 fourth-level categories in taxonomy, China was the leader in six, i.e. Properties of thin films; Diamond films; Applications of carbon nanotubes; Multi-walled nanotubes; Nanomaterials and nanoparticles, and Polymers, composites and metal complexes (Fig. 1.3; solid shading denotes publication lead by China, and those with vertical lines and shading refers publication lead by Japan. Light shading means category leader has 100-125% of the USA publications; medium shading 125-150%; dark shading >150%). Essentially, China led in the materials and nanostructures component of the database, whereas USA led in the physical science phenomena and biomedical components (Kostoff *et al.*, 2007a).

LEVEL 1	LEVEL 2	LEVEL 3	LEVEL 4
Quantum phenomena, Optics, Electronics, Magnetism, Tribology, and Films (32, 983 records)	Quantum phenomena, Optics, Electronics, Magnetism, and Tribology (26, 077 records)	Quantum phenomena (3326 records)	Quantum dots (2028 records) Quantum wells, Wires, and States (1298 records)
		Optics, Electronics, Magnetism, and Tribology (22, 75 records)	Optics, Electronics, (16, 432 records) Magnetism, and Tribology (6319 records) Properties of thin films (2251 records)
	Films (6906 records)	Thin films (4760 records)	Application of thin films (2509 records)
		Deposition of films (2146 records)	Deposition of thin films (1752 records) Diamond films (394 records)
Nanotubes, Nanomaterials Nanoparticles, Polymers, Composites Metal complexes and Bionanotechnology (31, 742 recoed)	Nanotubes (3211 record)	Multi-walled nanotubes (2350 records)	Application of carbon nanotubes (474 records) Multi-walled nanotubes (1876 records)
		Single-walled nanotubes (861 records)	Single- and double-walled nanotube (447 records) Single-walled nanotubes (414 records)
	Nanomaterials, Nanoparticles, Polymers, Composites, Metal complexes and Bionanotechnology (28, 531 records)	Nanomaterials, Nanoparticles, Polymers, Composites, and Metal complexs (22, 686 records)	Nanomaterials and Nanoparitcles (14, 263 records) Polymers, Composites, and Metal Complexes (8423 records) DNA (775 records)
		Bionanotechnology (5845 records)	Proteins and Cellular components (5070 records)

Fig. 1.3 Four-level hierarchical taxonomy. Numbers denote the papers published as per the database examined (after Kostoff *et al.*, 2007a, permission from *Current Science*).

Researchers in nanoscience and nanotechnology use a wide variety of instruments among which some more important instruments/techniques are X-ray diffraction (XRD), electron microscope variants, atomic force microscopy, scanning tunnelling microscopy and spectroscopy variants.

- If we consider only the use of nanotech instruments/techniques, China has produced ~25% more papers than the U.S.A. but when research in all areas of nanotech is considered, U S A produced ~25% more papers than China.
- Whereas most of China's over-production was in the XRD-related categories, the US dominance has been in atomic force microscopy (AFM).

The computer database study conducted by Kostoff *et al.* also identified the chief nanotechnology applications, both medical and non-medical, as well as the related science and infrastructure. These relationships can prove helpful to the potential user-communities to become involved with the applications-related science and performers at the earliest stages.

Non-medical applications: Applications thrust areas factor analyses were performed to show the thematic areas in non-medical applications. A six-factor analysis showed the following themes:

- Factor 1: Optoelectronics
- Factor 2: Tribology
- Factor 3: Lithography
- Factor 4: Control systems
- Factor 5: Devices
- Factor 6: Microsystems
- USA emerged as the world leader in total non-medical applications publications and in some of the themes in high-tech research areas such as devices, sensors and lithography. China was the leader in publications in catalysis, tribology and electrochemistry.
- In total non-medical applications, two of the top three institutions were Chinese, but USA was well represented by the University of California and University of Illinois.
- The journal *Applied Physics Letters* appeared in the top layer in several of the themes and was the leader in total non-medical applications publications. *Journal of Physical Chemistry B* appeared in four themes, as also *Journal of Applied Physics*.

Medical applications: A medical applications categorization constructed from visual inspection of the clustering categories revealed five broad thematic categories:

- Cancer treatment
- Sensing and detection
- Cells
- Protein
- DNA

Analysis of 19 thematic categories obtained from clustering of the total 2005 nanotechnology database revealed the following in respect of medical applications:

- USA was the publication leader in total health types, as well as in all the thematic areas. China was the second most prolific in seven thematic areas, Japan in six, Germany in four and England in two.
- The University of California system led in five clusters, the Chinese Academy of Science led in four, and the National University of Singapore led in three. The Univ. California and the Chinese Acad. Science were the most prolific in the non-medical applications as well, although their orders were reversed. The National Univ. Singapore was a prolific contributor in pharmaceuticals and biomaterials.

- The journal *Langmuir* contained the most nanotechnology articles in total health, and figured in the top layer of ten of nineteen themes. The only journals in common in the top layers of applications and health were *Langmuir* and *Journal of Physical Chemistry B*.

NANOSCIENCE AND NANOTECHNOLOGY: RISKS, UNCERTAINTIES AND CONCERNS

All new technologies including nanotechnology carry some attendant risks with them. Just three years ago, the word 'nano' used to be an advertising buzz (see *Nature* **440**: 262, 2006). Its gloss is now fading.

Some German consumers experienced respiratory problems in March 2006 after using the glass sealant Magic Nano, so the product was withdrawn from the market. Later tests showed that, despite its name, the product contained no nanoparticles. The incident, however, heightened public awareness that nano-related products might carry unforeseen risks.

Considering the lessons learnt from the marketing of genetically modified food, the nanotech sector is deeply interested in addressing any real or perceived risks associated with its products. Some consumer advocates have questioned whether nanoparticles might behave differently in the body from larger particles if ingested or inhaled. So far the greatest concern has focused on nanoparticles being used in consumer products such as cosmetics or mineral supplements. These fall outside the existing purview of such agencies as the Environmental Protection Agency (EPA) or the Food and Drug Administration (FDA) (Gewin, 2006).

Many investors are tempted by the nanotech sector, but some of them prefer to support nano-sized improvements to existing products rather than novel nanoparticles.

When the development of nanoproducts moves beyond films and coatings to more novel nanostructures, more focus and money will need to be spent on assessing the risks of nanoproducts. The EPA approved its first nanoproduct, the manufacture of a specific carbon nanotube, under a clause of the Toxic Substances Control Act. Future products also will have to be subject not only to this Act but also to other current regulations from different agencies – as well as any likely future nano-specific amendments (Gewin, 2006).

Nanomaterials are entering the market on the basis of claimed benefits. Their chemical identity is tied to existing legislation regardless of some unique characteristics and possible toxicity. Borm and Berube (2008) discussed three examples of nanomaterials (NMs) being considered for wide applications. These examples illustrate the opportunities, uncertainties, and risks for NMs.

The market for nano-based products includes medical products (e.g. heart valves, drug-delivery systems, and imaging techniques), electronic components, scratch-free paint, sports equipment, wrinkle- and stain-resistant fabrics, sunscreens, and other cosmetics. The European market for such products is currently ~€2.5 billion but could rise to €1 trillion within 2-3 decades. Nanotechnology could possibly enable:

- A cleaner environment;
- Fabrication shift from top down to bottom up.
- Existing technologies and materials may be improved so much that efficiencies increase and properties can be controlled.

- Some NPs might effectively remediate some of our worst toxic wastes.

  Titanium dioxide (TiO_2), amorphous silica (SiO_2), and iron oxides are bulk nano products being marketed and used in many consumer products, e.g. food additives, pigments, paints, and cosmetics.
- Newer generation of engineered nanoparticles (ENPs) need to be discriminated between the good, the bad, and the ugly. Methods need to be developed that can screen out bad and ugly ENPs in early phase development.
- The same is true for regulation: there are doubts that current approaches and tests in toxicology can suffice to screen all nanomaterials. Maynard *et al.* (2006) and Nel *et al.* (2006) discussed the potential impact of NMs on society and the need to resolve current uncertainties.
- Early concern was about ultrafine particles (primary particles and aggregates <100 nm), such as those occurring in ambient air pollution. These particles mostly consist of combustion-derived nanoparticles (CDNPs), such as diesel exhaust. They affect cardiovascular rhythm, peripheral blood flow, and blood coagulation, accelerate atherosclerosis in patients with existing cardiovascular disease.
- *In vitro* and *in vivo* studies with CNTs show similar actions of CNTs on blood coagulation as ambient airborne particulate matter and exert effects such as granuloma formation and inflammation in the lung after inhalation. The pulmonary effects can be explained by the normal paradigms of particle toxicology including surface dose, aspect ratio of fibrous particulates, and dose rate. Radomski (2005) tested ambient particles and ENPs in blood coagulation and found similar effects for CNTs and ambient PM.
- New studies were based on the different body distributions of nanoparticles compared with fine (<2.5 μm) particles (Kreyling *et al.*, 2002; Oberdörster *et al.*, 2005). The observation that ENPs can cross the blood-brain barrier, reach the olfactory bulb in the brain, and pass the placental barrier has aroused concern about the effects on these organs, also about the potential effects of translocated ENPs on neurodegenerative diseases such as Alzheimer's and Parkinson's. The problem is further complicated by the uncertainty and paucity of information about the environmental distribution and effects of NMs.
- Some ENPs (e.g. TiO_2, fullerenes) show photocatalytic effects and generate reactive oxygen species. Because of this, TiO_2 nanoparticles have been studied for the degradation of organic pollutants, water decontamination, air purification, and as photocatalytic coatings. But these properties, desirable in a product or application, may become hazardous if they are active in the wrong place: for example, in the aquatic environment, ENPs react with cations, anions, and natural organic matter (NOM), leading to surface modifications, aggregation of ENPs, or aggregation with particles, flocs, and microorganisms (Borm and Berube, 2008).

Computer-chip companies are selling chips packed with electronic circuitry patterned down to the nanoscale. Computer hard drives, LED-based traffic signals, CD players, and low-friction coatings are also being marketed. But some negative aspects of nanoscience should not be overlooked. Concern has been voiced that water laced with all-carbon nanoparticles called buckyballs could damage cell membranes in the brains of fish (see Service, 2004).

Regulatory agencies and others must critically investigate how nanoscale materials affect human health and the environment. Any news about environmental dangers from one form of nanomatter can potentially spark a public backlash against the whole nanotechnology enterprise. In fact, the property that makes nanoparticles so promising – their sharply different behaviour from bulk forms of the same material – also makes their potential health and environmental effects very difficult to predict.

Nanotechnology is not so much a discipline, like chemistry or physics; it is a toolkit for manipulating matter at its finest scale. The core feature of nanoscience is that the properties of materials can change drastically as their size is reduced from the bulk material to small clusters of atoms. Gold, for example, is inert in bulk but it becomes highly reactive and catalytic at the nanoscale; conceivably, electrical, optical, thermal, and other properties of materials may also undergo similar shifts. Such changes typically arise in two ways:

1. The small size of a nanoparticle means that most of its atoms are lying on the surface, and the behaviour of these surface atoms dominates the particle's properties.
2. When the atom's electrons are squeezed into smaller-than-typical spaces, properties such as the colour of the light they emit and a nanocluster's chemical reactivity, change.

It is emerging that the capability to image, manipulate, and observe all materials at the atomic level potentially touches every human aspect in the world around us. By any measure, nanotechnology is one of the hottest areas of science around. In just less than a decade, it has climbed up from being a specialty of a few physicists and chemists to a worldwide scientific and industrial enterprise. Over 750 companies and organizations in the United States alone are engaged in nanotechnology and include such giants as General Electric, Lucent, Philips, Matsushita, Intel and Advanced Micro Devices.

The rapidly accelerating race for the spoils of the nanoworld has naturally aroused concerns that the field may suffer from unintended consequences such as those that shadowed the introduction of genetically modified foods (see Kumar, 2004) and other industrial and technological revolutions. One scenario popularized by Sun Microsystems cofounder Bill Joy and others (*Science*, 24 November 2000, p. 1526), referred to specialized self-replicating robots devouring the planet, even though it is widely felt that making such destructive devices would be virtually impossible. A somewhat more plausible threat may be that nanoparticles released from coatings or other products may generate a new type of chemical pollution. Although carbon nanotubes, quantum dots, and nanoparticles do have enticing electrical and optical properties, toxicologists worry that they might harm organisms.

The properties of nanoparticles vary with their chemical composition, sizes, and interactions. We know very little about the health and environmental impacts of nanomaterials or about their synergistic impacts. Some toxicology reports have suggested that nanoparticles pose a unique risk to all living organisms. Carbon nanotubes, when washed in a suspension into the lung tissue of rats, can agglomerate, causing tissue damage and even death. Nanocrystals of buckyballs dissolved in water at a concentration of 1.5 parts per million can kill one-half of the *Escherichia coli* bacteria in the water, making them an extremely effective antibiotic.

Buckyballs are all-carbon molecules with chemical formula C_{60}: they are strong electron sponges that readily swipe loosely bound electrons from nearby molecules. This is why they are highly sought-after for use in solar cells, where they can steer electrons into a circuit. But if they find their way inside cells, the same ability could possibly convert oxygen and other molecules into highly reactive radicals that would damage cell components. Exposure of human fibroblast cells to nanocrystalline C_{60} revealed that the cell membranes had degraded and that the cells increased their production of glutathione, a small protein antioxidant that snuffs out free radicals (Service, 2004).

Nanoparticles other than C_{60} nanocrystals are also causing some concern. Exposing human keratinocytes and bronchial epithelial cells to a mixture of carbon nanotubes and nanosized iron particles increases levels of cell damage and apoptosis, or programmed cell death. Mice that breathed these particles suffered significantly more damage to their lung tissue than controls that breathed in silicon particles (Service, 2004).

Although it may be too early to start regulating nanoparticles, some caution is warranted. The best course may be to apply the precautionary principle to consider nanomaterials hazardous until proven otherwise. Some have pointed out that most of the research money is currently aimed at using nanotechnology to address environmental problems, such as remediating pollution sites. There is need for allocating more funds to research on nanotoxicology.

Regulators typically assess the safety of new compounds on the basis of their chemical composition. But nanomaterials often change their properties when their size changes, even when their composition remains constant. Deciding how to test nanoparticles for toxicity is not easy. Should regulators investigate 1-nanometer, 10-nm, or 50-nm particles? And what about everything in between? The problem is certainly unwieldy.

Although novel technologies based on nanotechnology might someday greatly lower the cost of generating electricity from solar power, purify water, and clean up past environmental contamination, researchers need to be very careful.

POSSIBLE HEALTH RISKS OF NANOTECHNOLOGY

The advent of nanotechnology also generated some fears about its safety aspects and some experts were quite worried that nanotechnology could possibly even destroy the planet Earth. It was argued that self-assembling nanobots might somehow escape from control, replicate *ad infinitum*, and could possibly spread across the landscape as a gray goo of devastation, consuming the earth and its biota.

However, today it appears that the above planet-dooming scenario was far-fetched. Our most serious worries now hover around the more immediate problems of global warming and disease. On the other hand, nanotechnology, if incorporated into improved solar panels, wind turbines or drug delivery mechanisms could even become an important tool to fight the two threats (global warming, and diseases).

In fact, like any new technology, nanomaterials have potential both for good and for bad. The strongest worries relate not to a gray goo creature but rather the more likely possibility that some of these novel materials may harm our health or the environment. It is now established that ordinary materials show unique properties at the nanoscale. So, the nanometer-size bits of an apparently benign material might turn out to be noxious. Even naturally occurring nanoparticulates can exert some deleterious effects on the human body. If natural nanoparticulates can harm us, wisdom warrants that we critically examine the possible actions of engineered nanomaterials. The size of nanoparticles is such that they can more readily escape into the environment and infiltrate deep into such internal organs as the lungs and liver. What is more, each nanomaterial happens to be unique. The most urgent research need is to gather quantitative data and determine what materials and in what concentrations nanomaterials can affect the body over what timescale.

We urgently need a robust database that can provide the hard facts needed to understand the possible health risks that specific nanomaterials might pose. In the United States alone, over 1,000 consumer products containing nanomaterials are already being marketed and the number is growing.

There are genuine concerns that with so many nanomaterials already on the market and so little public awareness of nanotechnology, even one safety scare might send a wrong signal or perception to the consumers that all nanotechnology is dangerous.

In most Western countries as well as in China, Brazil and India, industrial use of nanoscale products has been increasing over the past few years. These are commonly used in weaving, dyeing, cosmetics, medicine, and in waste-water treatments. The rapid increase in their use and applications has highlighted the need to develop suitable analysis protocols to assess nanoparticle toxicities.

Zhou *et al.* (2008) reviewed the measures which China has taken to address the issue. Several national conferences and symposia have been held in China on the safety of nanomaterials.

There is a strong need for the concerned international scientific organizations to collaborate and design reliable tests to address safety and toxicity issues for different industrial nanomaterials (see Bi *et al.*, 2009).

Real Life Examples: Nanosilver

Nanosilver coatings have antimicrobial action and so are used in wound dressings, urinary catheters, self-cleaning clothes, toothpastes, and washing machines. It may find a major use in many different coatings e.g., food packaging to wall/roof coverings.

Although the use is based on the slow formation and release of Ag ions, nothing is known about the influence of wear on these products and the release of Ag NPs or ions from coatings into the surrounding media. The toxicology of Ag ions is well understood but is mostly based on acute spills and high concentrations. From an environmental and health perspective, nanosilver could be dangerous – general exposure to Ag may damage some sensitive aquatic environments.

There is very little data on the effects of chronic exposure to low levels of Ag ions or on Ag NPs, which could act as a carrier system for local release. One argument for the introduction of

Ag is that bacteria cannot develop resistance to it as they usually do against antibiotics. But some feel that Ag resistance can be accomplished quite fast – comparable to susceptibility effects to antibiotics. It is claimed that Ag resistance is most easily developed in bacteria with already documented resistance mechanisms to antibiotics, such as methicillin-resistant *Staphylococcus aureus* (MRSA) and others. There are Ag-resistant bacteria in our mouths (Davis *et al.*, 2005), possibly related to the Ag in amalgam fillings. Ag-resistant bacteria are also found in nature, food, intestinal from different geographic locations, and hospitals (Gupta *et al.*, 1999).

Cerium Oxide Nanoparticles

Cerium oxide (CeO_2) NPs are added to diesel as a catalyst to reduce toxic exhaust gases and particulate emissions from diesel vehicles. Envirox™, a fuel-borne CeO_2 catalyst that reduces fuel consumption and particulate emissions, has been developed by Oxonica.

Titanium Dioxide and Zinc Oxide

TiO_2 and ZnO NPs are widely used in different products, especially sunscreens. Uptake of NPs from these products through the skin would cause a considerable burden in the body. There is need to determine whether nanoparticles remain on the skin or can reach systemic circulation and target organs. There are some concerns about their entry via damaged skin, and evidence of oxidative stress and damage to naked DNA *in vitro* involving inorganic sunscreen ingredients. Recent work suggests that, once in the bloodstream, these particles might be taxing on the liver, placenta, and brain (see Borm and Berube, 2008).

TiO_2 NPs are employed in coatings for saltwater vessels to control antifouling and reduce corrosion. Since they need to be reapplied, wear or sloughing occurs. The effect of TiO_2 nanoparticles on marine organisms is not known. TiO_2 is also a part of some water-treatment technologies. It degrades volatile organic compounds and kills bacteria in the presence of light.

About one-third of sunscreens probably contain ZnO nanoparticles, yet have received much less attention. As ZnO protects against ultraviolet (UV) and stays on the skin longer, its effects warrant urgent study.

For risk assessment and future regulation of ENPs, the following crucial issues need urgent consideration.

- Effects may be specific to ENPs and not seen in the same or other materials of larger size or agglomerates. In this case, ENP effects may be quantitatively different and regulation may need to be adapted by changing values and/or metrics;
- Effects may be qualitatively different, based on size, surface chemistry, or other specific interactions. In this case, existing standards can be used since the critical effect is different for fine analogs. However, this implies that the same material at different sizes may have different standards, based on different effects;
- Effects will be substantiated by studies of all kinds, but given the rapid pace of commercialization, we should focus on some NPs more than others, to study some characteristics of NPs more than others.

Nanotechnologies pose a classic dilemma for modern society: use its potential and go full speed ahead, or perform the required risk and technology assessments first. The difference with previous technologies is that nanotech can affect all aspects of human life, and an error of this magnitude might be irreversible.

Most nanotechnology experts agree that more money needs to be spent on understanding potential health and environmental dangers of exposure to materials engineered on the scale of small groups of atoms.

How nanotechnology is regulated is certainly an extremely important matter. Over 200 nanotechnology products are already on the market. These include sunscreens and cosmetics, light-weight bicycle frames, and car wax. According to a market survey conducted by Lux Research (New York City), by 2014, about $2.6 trillion worth of manufactured goods will incorporate nanotechnology. Some Fortune 500 companies have already backed out of nanotechnology research because of real and perceived risks of nanomaterials and uncertainties over how they would be regulated. For the same reason, Venture-capital funders and insurers have withdrawn services for some clients.

To stop this trend, several experts argue that nanotoxicology research funding needs to be increased dramatically. But there is no such agreement on how that money should be spent and coordinated. In general, nanotech (environmental health and safety) is being done in an ad hoc fashion depending on individual priorities. That approach has left strong gaps between what the agencies are pursuing and what is needed to tune regulations to products already on the market. One example is that while carbon-based nanomaterials are incorporated into only about one-third of nanotech products, the vast majority of nanotoxicology studies focus on those materials and ignore broad classes of other materials already on the market. Such anomalies may be avoided if a more centralized top-down research approach is adopted.

The need to find the best way to ensure that the up-and-coming industry remains safe for both people and the environment cannot be overemphasized.

NANOTECH PATENTING

With the advent of nanoscience and nanotechnology two decades ago, there has been a rush to patent an area that is a fledgling in its technology life cycle. It implies that carbon nanotubes, semiconducting nanocrystals, light emitting nanocrystals, metal oxide nanorods, atomic force microscopes, methods of making self-assembling nanolayers and of making nanotubes by chemical vapour deposition are being patented. Somehow, the buckminsterfullerene was left out of the patent system. The term "patents" in nanotechnology has been used ambiguously as it also consists of rods of carbon, or nanotubes and bucky balls, all of which have been patented 250 times in different forms. The areas of patenting include composition of matter, devices, apparatus, systems and control of nanomaterial and devices, and methods. Nanotechnology has also been divided into nanomaterials and nanodevices. In nanotechnology, changing the size of magnetic nanoparticles by only a few nanometers and changes in the magnetic characteristic can change from ferromagnetic to diamagnetic to super paramagnetic. Properties of nanoparticles are strongly size-dependent.

Mark A. Lemley (Standford Law School) discussed the cross-industry and multi-disciplinary nature of nanopatents as follows:

> *'Unlike other new industries in which the patentees are largely actual or at least potential participants in the market, a significant number of corporate nanotechnology patentees will own rights not just in the industry in which they participate, but in other industries as well' (see Krishna, 2008).*

For patenting issues, prior art forms a critical element of an examiner's decision on novelty and non-obviousness. As scale is an important consideration in nanotechnology, it is possible that the same structure might have been already patented though at a larger, e.g. micrometres (larger scale/size). The rule of inherency commonly followed in the USA directs a patent examiner to consider whether the earlier patent at a larger (e.g. μm) scale was actually put in use/practised. Only if it was not, is novelty deemed to be valid for the nano patent. This reasoning flows from the view that the older patent on a larger scale has inherent in it the nanoscale and anticipates the nanoscale patent in advance. This warrants some caution in favour of nano-patents: 'Mere similarity in scale between the claimed invention and the prior art may not suffice to support an argument that the claimed invention and the prior art are the same (see Krishna, 2008). The patent examiner must also determine if the claims in the patent application overlap with those in other patent documents. It is important that during infringement litigations, both the patentee and infringer pay attention to the claims and understand their scope. In most cases, claims relate either to the physical structure or to the physical properties of the nanomaterial or object. Besides the structural claims there are claims on particle size, in combination with other structural elements, e.g., presence of coating on the particles. Claims related to properties are not usually as strong as those relating to structures. Physical structural claims must be broad enough to also cover the physical properties.

THE NANOWORLD

While projecting the future prospects and potential of a new **nano era**, the technology aspect is usually more emphasized that the underlying science. This may be due to the hope that we might reap the benefits of a new technology without necessarily investing a great deal of time and money on understanding the underpinning nanoscience. Drexler (1981) had stated: 'Engineers (in contrast to scientists) need not seek to understand all proteins but only enough to produce useful systems in a reasonable number of attempts.... Thus, the difficulties encountered in predicting the conformations of natural proteins do not seem insurmountable obstacles to protein engineering.' More than two decades since 1981, the obstacles to protein engineering in the production of technologically-useful devices, have still to be crossed. Yet, the term nanotechnology has become a very popular term, possibly in the light of its following attributes: manufactured, products are made from atoms; their properties arise from how those atoms are arranged; by rearranging the atoms in coal, diamond can be made. If the atoms in sand are rearranged (and a few other trace elements are added), computer chips can be made.

The electron microscopes of today are so powerful that we can see the nanoworld. Consequently, nanoscientists can accurately describe their observations of 'nanotubes'. The fullerene (carbon) nanotubes are minute hollow carbon pipes which have attracted great attention. Nanotubes have been observed in unique situations: for instance, Perutz *et al.* (2002) assertively titled their paper 'Amyloid fibrils are water filled nanotubes'.

2

CHAPTER

Useful Novel Materials

INTRODUCTION

Some promising new materials are the following:

High performance polymers (engineering plastics; polymer separation membranes; photosensitive resins).

Fine ceramics (silicon nitride and silicon carbonate; high purity alumina used in engine materials, artificial bones and ceramic condensers, etc.).

New metallic materials (shape memory alloys such as Ni-Ti which can return to a given form at a selected temperature (used for pipe joints and brassiers; hydrogen storage alloy which stores and releases hydrogen and desorbs and absorbs heat at certain times – used in heat storage and transport systems; rare earth magnets, used in headphones and small speaker motors).

Composite materials (carbon fibre reinforced plastics – materials in which carbon fibres are used to reinforce plastics; used in sporting goods, aircraft body materials).

Anthony Hoffman (Princeton University in New Jersey) and his associates have engineered the first multilayer semiconductor that is capable of bending incoming light in the opposite direction to all known naturally occurring materials. The material was assembled by superimposing many alternating ultra-thin layers of two types of semiconductor. When an infrared beam was shone into this at any angle, it was negatively refracted through the material.

Certain other metamaterials also comprise alternating layers of metals and semiconductors, but they can be grown to only limited thickness.

The new material is fairly easy to produce and could improve infrared laser devices such as those used in high-speed communication and medical diagnostics. It may also lead to improved flat lenses to reduce distortion in telescopes and microscopes.

Super-black Material

A 'super-black' material has been created recently (see *Nano Lett.* Doi: 10:1021/nl072369t, 2008). It reflects only 0.045% of all the light of a certain wavelength which makes it the darkest synthetic material ever.

This material is made of row upon row of vertically aligned carbon nanotubes of different lengths, most of which are multi-walled. This arrangement imparts the material a rough surface with no repeating topology, and makes it scatter light in random directions, so that the material looks dark black.

The material appears dark across the rest of the visible spectrum, with a reflectance ranging from 0.045% to 0.07% as the wavelength changes from 633 nm (red) to 457 nm (blue). These properties may be gainfully exploited in devices to capture solar energy.

Hydrogels

Some interesting soft materials, such as hydrogels, are made from polymers that swell or shrink with changing exposure to a solvent. These materials respond rapidly but there is a limit to the complexity in their shape change, as well as to the degree of stress that can be exerted on these materials.

Two architectures have been in which silicon nanocolumns are either freely embedded in a polymer hydrogel or attached to the underlying substrate. (see *Science* **315**: 487, 2007). When exposed to water vapour the hydrogel changes dimension and the nanocolumns lift up and stand up off the surface. This actuation occurs very fast and is reversible. The hydrogel can also be templated so that the motion of the silicon nanocolumns can take place in a specific pattern.

GLASS SURFACES

In many materials, atoms or molecules on surfaces are more mobile than those in the interior. The diffusion of an atom along a single crystal surface is usually much faster than that inside the crystal. The layer-by-layer growth of crystalline materials for semiconductor lasers and other devices can be controlled. Some technologies also depend on noncrystalline materials such as glass (a solid without the regular packing of a crystal); Fakhraai and Forrest (2008) proved that a polymer glass surface can be much more mobile than the interior. Their results cannot be accounted for by atoms or molecules skittering across a surface. Rather, it was found that a liquid-like layer, at least several nanometers thick, exists at the surface of a polymer glass.

Surface mobility can be measured by preparing polystyrene films with well-defined nanoindentations on the surface. Annealing the polymer films at a suitable temperature allows the gold spheres to sink a few nm into the surface. At room temperature, mercury dissolves away the gold spheres, leaving behind hemispherical nanoindentations about 5 nm deep. Surfaces prepared in this way are then annealed at low temperature for various periods of time, and atomic force microscopy can be used to image the filling of the holes. Filling of the nanoindentations on the polystyrene glass surface was virtually completed quite rapidly. The observation that nanoscale indentations on a polymer glass relax rapidly indicates the presence of a highly mobile liquid surface layer.

The wide application of polymer glasses has aroused interest in understanding the dynamics of these surfaces (Roth and Dutcher, 2004). Whereas earlier studies measured the relaxation of a free surface upon which a pattern has been templated, Fakhraai and Forrest combined a reproducible deformation with a simple geometry, a penetration depth of only a few nms needed for probing only the surface, and a fairly mild method for removing the template from the sample. Previous studies had shown evidence for enhanced mobility at the surface of polymer glasses (Ellison and Torkelson, 2003); now the surface relaxation time has been quantitatively characterized over a wide temperature range. The important point is that it highlights a critical difference between mobility at crystal surfaces and at glass surfaces. The polymer segments in these experiments were connected to each other in long chains extending 20 nm or more into the glass and therefore the nanoindentations cannot be filled by polymer segments skittering across the surface; each segment at the surface remains chemically anchored to other segments lying below the surface. Instead, mobility at the surface of a glass can be considered as the motion of a thin liquid-like layer which responds to surface tension and fills the nanoindentations to minimize surface area. It appears that the range of enhanced surface mobility is at least several nms (perhaps more than 10 nm); for comparison, the polymer segments are roughly 1 nm. A liquid-like layer of several nms seems reasonable in view of the fact that molecular motions are highly cooperative. That is, the packing is so tight that one polymer segment can move only when a large number of neighbours together adjust their positions (Weeks *et al.*, 2000). The size of this group of segments appears to be several nms, and it is likely that the presence of a surface could perturb motions over a length scale at least this large.

The surface mobility of polymer glasses is important for understanding adhesion, friction, and instabilities in thin polymer films and nanostructures, used in advanced lithography (Yoshimoto *et al.*, 2005). By devising new strategies to decrease surface mobility, photoresists could well be used to transfer patterns on even shorter scales. Surface mobility may also be critical in a wide variety of nonpolymeric.

METAL TURNED TO GLASS

Glasses are frozen liquids made of closely packed, randomly ordered assemblages of molecules that no longer flow on any reasonable timescale; these properties make glasses solids for all practical purposes – solids which lack the long-range order typical of that phase of matter that gives glasses highly useful physical properties. One interesting member of this ubiquitous category of materials described recently by Bhat *et al.* (*Nature* 448, p. 787, 2007) is made by cooling a pure, 'monatomic' metal.

It is possible to make glasses from several different materials, irrespective of the type of interaction prevalent among their constituent atoms or molecules – whether covalent, ionic, hydrogen, van der Waals or metallic bonds (see Tarjus, 2007). Commonly, however, glasses are formed by cooling a liquid, with the prerequisite that crystallization into a regular solid does not occur first along the way. It was discovered half a century ago by D. Turnbull that molten metals can be supercooled to below their melting point and still retain their liquid structure. But such systems tend to crystallize and metallic glasses have so far been obtained only by mixing several specially chosen components into an alloy (Tarjus, 2007).

Although finding a one-component metallic glass is important, it is probably less notable in terms of direct technological applications than for the progress it might catalyze in our understanding of glass formation. To quote David Turnbull: "The most convincing evidence of the universality of the glass state would be the demonstration that pure monatomic substances… can be put into the glass form". These substances should be characterized by simple inter-atomic forces that do not involve strong directional bonding: liquid selenium, for instance, is a non-metal in which the atoms form covalently bonded, polymer-like chains, and is easily vitrified.

It is crucial to determine the factors that allow crystallization to be bypassed as a substance is cooled past its melting point. It has been known for long that the rate at which the liquid is cooled must be faster than the time taken for crystals to form. So we must find ways of slowing the crystallization down. In the case of metallic materials, widely applicable protocols have been devised by which the thermodynamic tendency towards crystallization is repressed, and the temperature of crystal formation is reduced.

Indeed, with a view to forming a glass by cooling a liquid, the normal process of solid crystallization must be bypassed. Achieving that for a pure metal had seemed impossible – until pressure was applied to liquid germanium (Tarjus, 2007).

MATERIAL FOR OPTICAL STORAGE DEVICES

In Japan an exciting new material has been found that could be used to make a low-price super disc with data storage capacity thousands of times greater than a DVD. The material changes from a black-colour metal state that can conduct electricity into a brown semiconductor when hit by light, according to Shin-ichi Ohkoshi, of the University of Tokyo.

The material is a new crystal form of titanium oxide. It can switch back and forth between the metal and semiconductor states at room temperature when exposed to light, **creating an effective on-off function for data storage**. It is highly promising as a material for a next-generation optical storage device.

A material that changes colour with light can be used in storage devices – colours reflect light differently to contain different information. The Japanese researchers created the material in particles measuring five-to-20 nanometers in diameter.

If we use the smallest particle, the futuristic disc could hold more than 1,000 times as much information as a Blu-ray disc, provided that matching data-writing and reading equipment could be developed.

A single-layer Blu-ray disc can contain five times as much data as a conventional DVD. Titanium oxide's market price is about one-hundredth of the rare element – germanium-antimony-tellurium – that is currently used in rewritable Blu-ray discs and DVD.

Being cheap and safe, titanium oxide is already being used in many products ranging from face powder to white paint.

QUANTUM DOT OPTOELECTRONICS

Architectures are being developed that allow the quantum state of one part of a system to be controlled by the quantum state of another part. The ability to manipulate and control the optical properties and/or the charge and spin states of a single quantum dot (q.d.) have marked them as ideal candidates for further development. Controlled interaction has been demonstrated between a pair of self-assembled quantum dots, whereby the optical response of one of the dots is conditioned on the state of its neighbour (see *Science* 320: 772, 2008). A single q.d. coupled to a photonic crystal cavity can be used to induce some optical phase shift down at the single photon level (see *Science* 320: 769, 2008). The ability to control and manipulate the optoelectronic properties of q.ds. is an essential ingredient for quantum logic applications.

Organic Light

Light-emitting transistors made from organic materials could become the next generation of low-cost display technologies. The **organic light-emitting transistor** (OLET), made by Muccini and Capelli (2010) is 100 times more efficient at converting electricity into light than the equivalent organic light-emitting diode (OLED), the latest technology for thin, lightweight displays.

The OLET comprises a luminescent matrix sandwiched between two semiconductor layers, all stacked on a substrate. The architecture of these layers decreases or prevents the loss of photons and quenching of excitons – both particles responsible for the light emission – that are common in OLEDs.

SOFT COLLOIDS FOR STRONG GLASSES

Glass formation in colloidal suspensions has many parallels with glass formation in molecular materials (Mason and Weitz, 1995; Kegel and van Blaaderen, 2000). For hard-sphere colloids, which interact only as a result of excluded volume, phase behaviour is controlled by volume fraction. An increase in Ø moves the system towards its glassy state, analogously to a decrease in temperature, *T*, in molecular systems. When $Ø^* \approx 0.53$, the viscosity starts increasing significantly, and the system eventually moves out of equilibrium at the glass transition, $Ø_g \approx 0.58$, where particle crowding greatly restricts structural relaxation (Weeks *et al.*, 2000). The large particle size facilitates to study of both structure and dynamics with light scattering and imaging; colloidal suspensions have therefore provided valuable insight into the glass transition. However, hard-sphere colloidal suspensions do not show the same diversity of behaviour as molecular glasses. This is highlighted by the wide variation in behaviour observed for the viscosity or structural relaxation time when the glassy state is approached in supercooled molecular liquids. This variation is characterized by the unifying concept of fragility, which has spurred the search for a 'universal' description of dynamic arrest in glass-forming liquids (Angell *et al.*, 2000). For 'fragile' liquids, the structural relaxation time is highly sensitive to changes in *T*, whereas non-fragile, or 'strong', liquids show a much lower *T* sensitivity. In contrast, hard-sphere colloidal suspensions are restricted to fragile behaviour, as determined by their Ø dependence; this limits their utility in the study of the glass transition.

Mattsson *et al.* (2009) showed that deformable colloidal particles, when studied through their concentration dependence at fixed temperature, do show the same variation in fragility as that observed in the *T* dependence of molecular liquids at fixed volume. Their fragility is governed by elastic properties on the scale of individual colloidal particles. Also, an equivalent effect was found in molecular systems, where elasticity directly reflects fragility. Colloidal suspensions may thus provide new insight into glass formation in molecular systems (Mattsson *et al.*, 2009).

Structural relaxation in molecular glass-formers is described by thermal activation within a complex energy landscape, which reflects the possible configurations of the system. Fragility is reflected in the detailed topology of this energy landscape (Sastry *et al.*, 1998). Fragility could probably also have a parallel in the temperature dependence of purely thermodynamic properties, such as entropy (Martinez and Angell, 2001).

In molecular glass-formers, strong dynamic behaviour commonly reflects some directional bonding that is typical of network-forming glass-formers such as silica. By analogy, it is speculated that the origin of the behaviour of colloidal suspensions could be similar: softest particles are more easily deformed, and as the particles shrink, their shapes will increasingly deviate from the spherical, which in turn leads to an increased directionality of interparticle interactions (Wyart, 2005), corresponding to strong behaviour. By contrast, hard particles retain their spherical shape and their interactions are more isotropic, corresponding to fragile behaviour.

The work of Mattson *et al.* demonstrates striking similarities between the behaviours of colloidal and molecular glass-formers, and suggest that an enhanced understanding of glass formation could possibly come from exploring the properties of other soft colloidal systems, such as solutions of star polymers and block-copolymer micelles (Xia and Hu, 2004).

SOFT MATERIALS

Soft materials traditionally cover polymers, liquid crystals, and aggregating systems such as micelles and vesicles. Their science dimensions face several analytical challenges but also offer technological opportunities. The challenges come from the fact that these materials are often solution-based, have low inherent contrast between the different phases, or are fragile and easily damaged by physical probes or energy beams. The opportunities emanate because soft materials are governed by weak forces, so small changes in chemistry or external conditions can lead to strong changes in the system properties (Clery and Lavine, 2002).

If we consider soft surfaces and interfaces, the analytical challenges are greater, because the amount of material being probed is small. Happily, however, the possibilities for interesting chemistry are magnified, because surface properties can be tweaked through post-processing or by designing the chemistry to be responsive to external stimuli.

Russell (2002) investigated the behaviour of molecules that are attached to a surface when the surface area changes. As these molecules must migrate into or out of the surface, their surface properties can change. Hydrophobic materials tend to become hydrophilic, and tacky materials can lose their stickiness, with only mild changes in temperature or humidity.

These surface properties are critical for implantable sensors. These devices are quite limited by the response of the body's immune system, which recognizes the material as a foreign body and thus attacks or immobilizes it.

Soft interfaces are particularly crucial in biology. Vesicles exemplify transport capsules used by cells. Discher and Eisenberg (2002) reviewed the world of polymer vesicles, where advances in synthetic chemistry have enhanced the potential for creating these bilayer sacs of weakly associating molecules, which can possibly serve as drug-delivery vehicles. In the areas of optics and tribology, the surface of a material often impacts the bulk properties.

Polymer Vesicles

Vesicles are microscopic sacs which enclose a volume with a molecularly thin membrane. Vesicles and biomembranes perform crucial roles in compartmentalization functions as varied as nutrient transport and DNA protection. While phospholipids exemplify the natural amphiphiles of cell membranes, vesicle-forming materials used in products ranging from cosmetics to anticancer agents can be synthetic as well as biological. The membranes are generally self-directed assemblies of amphiphilic molecules with a dual hydrophilic-hydrophobic character (Discher and Eisenberg, 2002). Biological amphiphiles produce vesicles critical for cell function; there are mainly lipids of molecular weight less than 1 kilodalton. Block copolymers that mimic lipid amphiphilicity can self-assemble into vesicles in dilute solution, but polymer molecular weights are often orders of magnitude greater than those of lipids. Structural features of vesicles, as well as properties including stability, fluidity, and intermembrane dynamics, are strongly affected by characteristics of the polymers. Future applications of polymer vesicles will depend on utilizing the unique property-performance relations. However, it is already clear that biologically derived vesicles are only a small subset of what is physically and chemically achievable (Discher and Eisenberg, 2002).

Soft materials science has an important role in the coatings industry, even in cases where the final product turns out to be harder than the substrate it is applied to. This industry can generate tremendous added value by the addition of a very small amount of material. Matheson (2002) discussed how soft coatings have always slowly evolved with the discovery of new materials, but the industry today faces a number of new challenges, as commonly used solvents and heavy metals have been forced out of existing formulations because of environmental and toxicity concerns (Clery and Lavine, 2002).

Whether in drug-delivery packaging, value-added coatings, or futuristic medical probes, many industries are driving these technologies.

Surface-responsive Materials

Synthetic polymers offer many opportunities to design responsive materials triggered by external stimuli. A well-known example of a responsive material is an elastic band which, within some limit, can be stretched rapidly, and yet can respond to the applied deformation. The polymer chains are able to stretch and align, and this molecular mobility is typical of materials that we consider soft or responsive. In contrast, if you hammer a sheet of glass, it will shatter, so it cannot be termed a responsive material. The slow dynamics of the glass have affected its ability to respond. Changing the length, chemical composition, architecture, and topology of the chains allows response mechanisms and rates to be easily manipulated; and devices based on the entropy of the chains, surface energies, and specific segmental interactions can be made (Russell, 2002).

SILKWORM SILK

Spider dragline silk, aptly hailed as a 'super-fibre', seems to be stronger and more extensible than commercial silkworm silk (Kaplan *et al.*, 1994; Lazaris *et al.*, 2002). But Shao and Vollrath (2002) showed that the mechanical properties of silkworm silks can approach those of spider dragline silk when reeled under controlled conditions. According to them, silkworms might be able to produce threads that compare well with spider silk by changing their spinning habits, rather than by having their silk genes altered.

Typical commercial silkworm silk from *Bombyx mori* cocoons has a tensile strength of about 0.5 gigapascals (GPa), a breaking elongation of 15%, and a breaking energy (toughness) of 6×10^4 J kg^{-1} (Shao and Vollrath, 2002). On the other hand, *Nephila* spider dragline silk typically has a strength of 1.3 GPa, a breaking elongation of 40% and a toughness of 16×10^4 J kg^{-1}. These mechanical measures vary considerably in both types of silk. In spider silk, the variability arises from spinning conditions that are affected by the spider's body temperature and the speed of drawing. Silkworm silk comes from a natural cocoon that is spun by the moving silkworm, which accelerates and decelerates its head in arcs that are attached at points that correspond to each change of direction (see Shao and Vollrath, 2002).

Artificial reeling of silk from immobilized silkworms under steady, controlled conditions generates fibres that are superior to naturally spun ones. The silkworm, like the spider, produces stronger, more brittle fibres at faster spinning speeds. Slower speeds lead to weaker, more extensible fibres.

Force-drawn silkworm fibres compare well with *Nephila* dragline silk—slow spinning at 4 mm s^{-1} imparts a breaking elongation for *Bombyx* silk that is comparable to that of spider silk (37% and 35%, respectively). When spinning is faster (13 mm s^{-1}), breaking energy approaches that of *Nephila* silk (12×10^4 J kg^{-1} for *Bombyx;* versus 16×10^4 J kg^{-1} for *Nephila*), although the breaking strength remains lower (0:7 GPa compared with 1.3 GPa). At still higher spinning speeds, silk toughness decreases because of loss of extensibility.

Other differences between silkworm and spider silk originate from specific differences in the composition of the principal silk molecules, as well as to their arrangement during and after spinning. Depending on reeling speed, *Bombyx* silk becomes strong or stretchable, whereas spider dragline silk typically combines the two at the speeds that are characteristic of web construction (10-20 mm s^{-1}) (see Shao and Vollrath, 2002).

Spider silk fibres are studied in their native state. Silkworm silk first needs to be washed to remove most of the sericin gum which coats the fibroin filaments. Washing extraction is necessary for the thread to be unravelled from the cocoon and probably weakens the thread. Spinning conditions affect the silkworm silk's material properties more than washing. More than anything else, washing increases the elasticity of fibroin filaments.

The mechanical properties of *Bombyx* silk, like those of spider silk, depend on spinning conditions – silkworm silks can be made stronger, stiffer and more extensible merely by adjusting the harvesting parameters. If it were possible to reel silk straight from the silkworm, as from spiders, then the silkworm would produce highly valuable fibres of natural and artificially spun spider silks; silk fibres produced by artificial reeling are considerably superior to those that are spun naturally.

SELF-CLEANING CLOTHES

Fabrics can be made in such a way that they clean themselves. The fibres are coated with nanocrystals of titanium dioxide, a photocatalyst that decomposes dirt and stains when exposed to light. Walid Daoud, a chemist at Monash University (Australia) and others at the Hong Kong Polytechnic University have developed a treatment for fabric fibres that chemically bonds the TiO_2 nanocrystals to the fibres. Sunlight produces the best results but the self-cleaning also works under any light source, even while the clothes are being worn. The crystals also inhibit the growth of odour-causing bacteria.

The aim is to reduce the consumption of water, detergent, and energy used in laundering and dry cleaning. Such self-cleaning fabrics are expected on the market very soon (see *Chem. of Materials* 26 Feb. 2008).

ELECTRONIC TEXTILES

Computer engineers are poised to set fashion trends in the coming years—electronic textiles that change colour on a computer command are on the horizon. Besides revolutionizing the fashion world, computer, electronics, and textile researchers are also engaged in revolutionizing the fields of communication, medicine, advertising, and even warfare. Plans are afoot to electrify textiles ranging from jackets and shirts to advertising displays and carpets with electronic sensors, processing chips, and displays (Service, 2003).

Snowboarding jackets that play MP3 music files with the help of controls stitched onto the sleeve have already reached store shelves. A Georgia company in Atlanta has designed a shirt that monitors a patient's vital signs and alerts a doctor through a wireless signal in case of trouble. The U.S. military is examining fabric-based sensor nets that could forewarn it against the approach of enemy tanks and other vehicles. Textile-based sensor nets could enable army units to trinagulate the location of opposing forces. A German computer chip company is designing carpets that can detect intruders and fire.

Perhaps the most exciting of the potential applications of e-textiles is as effective monitors of patients' medical conditions. But many experts also caution that addressing some teething troubles associated with the diverse applications of e-textiles will take some time because ways have to be found to integrate flexible wires into clothing; link them to electronics that can withstand bending, twisting, and stretching; and power the whole system.

Rudimentary electric blankets have been prescribed by doctors for several decades to encourage tuberculosis patients to sleep outside in fresh air. These blankets had a resistive heating coil stitched between twin sheets of fabric. Modern e-textiles, by contrast, weave conductive threads right into fabrics themselves, which makes them much more like traditional fabrics.

A British inventor, Robert Rix has created carbon-based textiles (called Gorix), that are being incorporated in, for instance, scuba diving suits and car seat covers; these carry a small electric current throughout the fabric where the power is converted into heat. SOFT switch company in West Yorkshire (UK) is manufacturing a foldable touch-sensitive fabric keyboard and mouse that can connect to a **personal digital assistant** (PDA) and cell phone.

Sundaresan Jayaraman, a Park and Georgia Tech textile engineer launched valuable medical textiles in 1996 by incorporating current-carrying fibres into fabrics to power electronic sensors that monitor the wearer's breathing, temperature, and heartbeat. This system included a wearable motherboard, a wired shirt into which off-the-shelf monitors can be plugged in and later unplugged—when the shirt needs washing. Georgia Tech has already licensed the technology to Sensatex for commercialization. This company plans to develop a medical monitoring **SmartShirt**, which can relay information wirelessly to doctors. Although low cost wristwatch-style heart rate monitors are already available, shirt-based monitors may prove to be much more versatile and capable of working with a large variety of bio-sensors to track body temperature, oxygen, and perhaps even blood glucose levels. Indeed, medical monitoring shirts could become lifesavers particularly in combat. Some military labs are toying with the idea of adapting e-textiles for remote sensing. Work is underway to develop an acoustic sensor fabric that can help to pinpoint the location of enemy vehicles. Proper use of e-textiles could even allow the military to incorporate fixed-sensor networks into everything from tents to parachutes (Service, 2003).

Chameleon Cloth

Colour-changing electronic textiles have been created that consist of thin metal wires wound into yarn coated with thermal chromic inks that change colour when heated. When the juice is turned on, these novel fabric swatches generate a rainbow of colour combinations, changing from black to red or white to blue. According to Margaret Orth of Cambridge, Mass., (see Service, 2003) such chameleon cloths have tremendous potential for interior designers and advertisers as they make it possible to express art and craft on everything from walls to carpets.

The major hurdle for electronic textiles is power. While carpets and other fixed fabrics could easily plug into conventional outlets, wired clothing and other mobile e-textiles will have to rely on batteries. But batteries pose a serious problem in wearable electronics. To address this hurdle lighter power sources are being contemplated—a slimmed-down version of the lithium batteries that now power everything from laptops to cell phones. Infinite Power Solutions (IPS) in Golden, Colorado, is designing lithium nano batteries, a few micrometers thick, that are essentially painted on metal foils. Such batteries may conceivably be incorporated into the lining of garments to power sensors and logic chips as well as on the surface of backpacks and other fabrics used in everyday life. Upon returning home, the wearer may one day plug in his backpacks or jacket in the same way as he does with cell phones today.

Other researchers are engaged in developing even sleekier batteries that could be incorporated right into textiles themselves. One company is designing fibre-based batteries, each of which layers the traditional anode, cathode, and electrolyte right on top of a 100-

micrometer-wide fibre that's incorporated into a textile (see Service, 2003). It is being hoped that textilemakers should be able to weave fibre batteries into garments in the same way that they add different coloured threads today, without significantly making the fabrics heavier.

SMART SUPRAMOLECULAR MATERIALS

Biological systems have inspired new possibilities through which designer building blocks can be made to self-associate in a predictable manner to create cutting edge materials. Haldar *et al.* (2007) designed peptides which self-assemble by intermolecular hydrogen bonding, van der Waals and π-π stacking interactions to form spherical vesicles in both nano and macro scales. The **nanospheres** (diameter 20 nm) tend to interact with dye and drug molecules and adsorb on their surfaces. **Nanoporous materials** having unique bulk, surface and structural properties are in the limelight of current research in view of their potential importance in several fields like catalysis, sensor, ion exchange, separation, purifications, biomolecular isolation and delivery vehicles. Attempts are underway to develop peptides that self-assemble to form **porous nanomaterials** (pores size 1-10 nm) and act as *in situ* reducing agent in solid state. The pore sizes can be adjusted by controlling the self-assembly process. **Peptide nanofibers** may possibly be decorated with metal nanoparticles and chromophores to tune their optical properties and develop artificial light-harvesting systems (see Haldar, 2008).

DISORDERED MATERIALS WITH PHOTONIC BAND GAPS

Photonic band gap (PBG) materials were first introduced in 1987 (John, 1987). Their unusual properties have attracted several applications, including efficient radiation sources, sensors, and optical computer chips (Chutinan *et al.*, 2003; Altug *et al.*, 2006; El-Kady *et al.*, 2006). To date, however, the only known large-scale dielectric heterostructures with sizeable, complete band gaps have been periodic; this limits the rotational symmetry and defect properties crucial for controlling the flow of light in applications. Florescu *et al.* (2009) showed that we can design 2D, isotropic, translationally disordered photonic materials of arbitrary size with large, complete PBGs, blocking all directions and polarization. The designs could be created through a method used to build various disordered, hyperuniform structures in two or more dimensions, which are distinguished by their suppressed density fluctuations on long length scales and could possibly act as templates for designer materials with several other novel physical characteristics including electronic, phononic, elastic, and transport behaviour.

The designs reported by Florescu *et al.* with the largest band gaps can be created by a constrained optimization method that starts from a hyperuniform disordered point pattern, an array of points whose number variance within a spherical sampling window grows more slowly than the volume. Florescu *et al.* focused on adapting the method for fabricating materials with optimal photonic properties because of their valuable applications and because it was thought feasible to manufacture the dielectric heterostructure designs by employing existing techniques. Although their objective was to produce designs for isotropic, disordered heterostructures, they also reported how the same procedure might be gainfully used to obtain photonic quasicrystals with complete PBGs (see Florescu *et al.*, 2009a).

The design protocol adopted by Florescu *et al.* includes two free parameters which are varied to find the largest possible band gap in the constrained subspace of structures. The optimization requires only a modest computational cost. For computing the PBGs for the various disordered structures, Florescu *et al.* used a supercell approximation in which the disordered structure is treated as if it repeats periodically. Systematic convergence tests ensured that results converge as the supercell size increases.

According to Florescu *et al.*, hyperuniformity, combined with uniform local topology and short-range geometric order, can explain how complete photonic band gaps are possible without long-range translational order. They also pointed out the ramifications for electronic and phononic band gaps in disordered materials.

The work done by Florescu *et al.* explains how to design isotropic, disordered, photonic materials of arbitrary size with complete PBGs. Although photonic crystals have larger complete band gaps, disordered hyperuniform heterostructures with substantial, complete PBGs are well-suited for several applications. Disordered heterostructures are isotropic, a positive point for use as highly efficient isotropic thermal radiation sources and waveguides with arbitrary bending angle (Miyazaki *et al.*, 2003). The properties of defects and channels that enable controlling the flow of light are different for disordered structures. Whereas crystals have a unique, reproducible band structure, the band gaps for the disordered structures show some random variation for different point distributions. Also, light of frequencies above or below the band edges are propagating modes that are transmitted through photonic crystals but are localized modes in the case of 2D hyperuniform disordered patterns, which give the former advantages in such applications as light sources or radiation harvesting materials (Florescu *et al.*, 2009, 2009a). On the other hand, in view of their compatibility with general boundary constraints, PBG structures based on disordered hyperuniform patterns offer a flexible optical insulator platform for planar optical circuits. It also seems that eventual flaws that might strongly degrade the optical characteristics of photonic crystals and perhaps quasicrystals are expected to have only a weak effect on disordered hyperuniform structures, therefore relaxing fabrication constraints.

MAGNETIC MATERIALS

Magnetic materials represent a class of functionally important materials. Magneto-structural phase transitions are critical for the behaviour of materials exhibiting fresh magnetoresistance, giant magnetocaloric effect or giant magneto-elasticity. Certain features associated with magneto-structural phase transitions are quite general to different types of materials. The key features associated with such magneto-structural phase transitions are phase coexistence and metastability; these features are actually universal characteristics of a disorder-influenced first order phase transition. The phase-coexistence phenomenon is also related to the so called vortex solid melting phenomenon.

Kaul (2005) reviewed the magnetic behaviour of various systems depending on the nature of magnetic inhomogeneities. These systems cover a large variety including ferromagnets, antiferromagnets, soft magnetic alloys, and nanostructures. Unfortunately, some attempts to understand the origin of magnetic inhomogeneities have not been successful but it appears that the infinite ferromagnetic spin clusters model (originally proposed by Kaul) may provide a working basis for understanding the nature and origin of magnetic inhomogeneities, especially in amorphous Fe-Zr alloys.

People's economic and environmental needs today strongly depend on renewable and sustainable sources of energy in which water serves as the primary carbon-neutral hydrogen source, and light provides an energy input. This is why there is considerable interest in research devoted to the synthesis of new polynucleating ligands and their transition metal compounds (see Khanra *et al.*, 2007, 2008, 2009). Understanding the factors that regulate a four-electron-four-proton reaction is extremely difficult. Khanra *et al.* are exploring the development of viable systems for the oxidation of water to oxygen, especially non-biological approaches to oxygen-oxygen bond formation. Cluster compounds can have potential applications towards quantum derived properties in physical applications. Small molecule activation and atom transfer to functionalize important organic precursor molecules need to be critically investigated. The coordination chemistry of a range of transition metals, as well as the relatively unexplored low-valent early transition metals chemistry also require study. Spectroscopic methods such as electron paramagnetic resonance, Mössbauer and electronic absorption spectroscopy along with magnetochemical investigations could be gainfully exploited for this purpose. Any materials that can be synthesized are bound to attract many applications.

Magnetic and Electrical Transport Properties of Intermetallic Alloys and Spectroscopic Properties of Inorganic and Organic Materials

Alloy samples of transition metal and rare earth have been developed by using the arc-melt method (see Banik *et al.*, 2006; Ranjan *et al.*, 2006; Mukhopadhyaya *et al.*, 2006). Magnetic and electrical characterization of these materials can facilitate investigation of the role of d- and f-electrons on eliciting their physical properties. Uday Kumar and his associates have also developed polycrystalline diluted magnetic semi-conducting materials and studied their magneto-optical coupling. These researchers are specially interested in understanding the effect of magnetic and optical degrees of freedom on the physical properties of these materials in view of their importance in the field of spintronics (Kumar and Deshpande, 2009).

ROOM-TEMPERATURE FERROMAGNETISM IN ULTRATHIN MANGANITE LAYERS

Strong magnetoresistance, metal-insulator transitions, charge/orbital ordering, and half-metal ferromagnetism exemplify the puzzling processes that operate in manganites and have boosted interest in the family of perovskite manganese oxides (ABO_3 perovskite structure, B site occupied by Mn) (Tokura, 2000). With a Curie temperature (T_c) of ~370 K, the half-metal $La_{0.7}Sr_{0.3}MnO_3$ (LSMO) has attracted much interest for spintronics applications. Complete spin polarization in LSMO has been inferred from photoemission measurements and a record tunnelling magnetoresistance (TMR) ratio of 1,800% was obtained at low temperature in tunnel junctions with manganite electrodes separated by a thin insulating layer of $SrTiO_3$ (STO) (see

Kourkoutis *et al.*, 2010). But the TMR decreased rapidly as the temperature increased and diminished while still far below T_c. The reduced TMR may conceivably originate by the reduction of ferromagnetic ordering at the interface between the manganite and the STO. These interfacial effects probably become dominant as the LSMO layer thickness decreases, causing the magnetization and T_c to degrade and the resistivity to increase, creating films that are insulating at all temperatures when the layer thickness decreases below a reported critical value of 7-13 unit cells (Kavich *et al.*, 2007; Kim *et al.*, 2010). This critical thickness for sustaining conductivity below T_c appears to be due to an inherent "dead layer" at the interface between LSMO and STO (Yamada *et al.*, 2004).

$La_{0.7}Sr_{0.3}MnO_3$ happens to be a conducting ferromagnet at room temperature. Combined with thin $SrTiO_3$ layers, the resulting heterostructures have been used as highly spin-polarized magnetic-tunnel-junction memories. However, when shrunk to dimensions below some critical thickness, the structures become insulating and ferromagnetic ordering is suppressed. Interface spin and charge modulations probably create an interfacial dead layer; this limits the use of this material in atomic-scale devices. Kourkoutis *et al.* (2010) used atomic-resolution electron spectroscopy to show that the degradation of the magnetic and transport properties of $La_{0.7}Sr_{0.3}MnO_3/SrTiO_3$ multilayers correlates well with atomic intermixing at the interfaces and the presence of extended 2-D cation defects in the $La_{0.7}Sr_{0.3}MnO_3$ layers (in contrast to 3D precipitates in thick films). Eliminating these extrinsic defects stabilizes metallic ferromagnetism at room temperature in five-unit-cell-thick manganite layers in superlattices, placing the upper limit for any intrinsic dead layer at two unit cells per interface (Kourkoutis *et al.*, 2010).

The work done by Kourkoutis *et al.* suggests that LSMO/STO multilayers with manganite layer thicknesses of ~2 nm show ferromagnetism with T_c above room temperature and remain metallic below T_c, if the structure is optimized by tuning not only the oxygen partial pressure and growth temperature but also the laser fluence at the target. When spot sizes are smaller and fluences larger, the quality of the multi-layer deteriorates, resulting in less abrupt interfaces, and an LSMO A-site/B-site cation ratio greater than one. Introduction of extended defects reduces the magnetization and increases the resistivity such that the multilayers remain insulating at all temperatures. The laser fluence corresponding to the smaller spot size is typical for previous PLD-grown LSMO/STO multilayers (Huijben *et al.*, 2008); this seems to account for their inferior properties.

ORGANIC ELECTRONIC MATERIALS

Conjugated oligomers and polymers are potentially promising materials for applications in electrochromics, organic photovoltaic and light emitting devices. Considerable work has been done on the design, synthesis and characterization of a broad family of π-conjugated systems for redox and optoelectronic applications (see Zade *et al.*, 2007, 2008, 2009). Work is being done on developing the π-conjugated systems that can be useful for certain absorptive/transmissive and reflective electrochromic devices, polymer-based solar cells, light emitting devices and field effect transistors. Special focus is on designing π-conjugated systems that comprise heavier chalcogens. Quantum mechanical studies can be useful tools in the hand of chemists who wish to design 'tailor-made' organic electronic materials and to predict their properties. There is need to create synergism of the combination of experimental (synthesis and study of physical properties) and theoretical quantum mechanical studies in order to intensify productive research.

METAL COMPLEXES IN MOLECULAR MAGNETISM AND BIOCATALYSIS

Single molecules of high nuclearity contain a definite number of interacting spins that provide welcome opportunities to understand basic concepts of magnetism. Quite a few such molecules carry the potential to act as single molecule magnets. Mukherjee *et al.* (2007), Mukherjee and Sadler (2009) and Mulyana *et al.* (2009) have studied MOFs (Metal Organic Frameworks) and polynuclear complexes of multidentate ligands and their magneto-structural properties. They have also been engaged in modulation of the active site of metalloproteins in view of their potential to develop bio-friendly catalysts and sensors. In this context, they examined the structure-activity relationships of native and mutant bacterial (ferric-ion binding protein; FbpA) and mammalian transferrins and ferritins, as also the kinetics of metal binding and design protein catalysts by mutating the active site. They have also designed metal complexes to produce novel anticancer agents and enhance their target specificity to cope up with the emerging drug resistance in different forms of cancer. They are particularly focusing on synthesis of metal complexes bearing ligands and tethers having some potential as photoactivable moiety, telomere binding ability, cytotoxicity against cancer cells, being lipophilic, or having entities for which receptors are over-expressed in cancer cells.

DYE-SENSITIZED SOLAR CELLS

Dye-sensitized solar cells based on nanocrystalline TiO_2 or ZnO provide a cheaper alternative to conventional silicon based solar cells. The current efficiency of these solar cells is ~10% and in need of some enhancement. Electronic structure calculations of ruthenium polypyridyl complexes used as dyes in dye-sensitized solar cells are being attempted to understand the factors affecting their photo-conversion efficiency. These calculations are being made on dyes adsorbed on cluster models of TiO_2 surface to understand the effect of adsorption on the electronic structure of dyes. This might enable researchers to correlate the electronic structure of dyes with their observed photoconversion efficiency. Work is also underway on synthesis of new dyes and fabrication of dye-sensitized solar cells with these dyes.

MATERIALS FOR BIOTECHNOLOGY AND PHARMACEUTICAL INDUSTRIES

Technical inventions have been changing our lives continuously in the past, and current developments are going to change our lives in the future.

The future of pharmaceutical production cannot be projected only on the basis of technical developments – the regulatory factors are equally critical and can be both inhibiting and encouraging. The pharma industry is not fully exploiting existing technical capabilities for designing plants. The chemical industry is far more aggressive in the application of new technologies.

In this context, the "discipline of the future", as biotechnology is often proclaimed, is relatively innovation-averse. Indeed, the "discipline of the future" tends to follow proven past practices. One reason is associated with reducing the risk of regulatory disapproval; another seems to be the interdisciplinary nature of biotechnological process development. Even when a single technical innovation is introduced, its effect needs to be studied in detail and tested exhaustively from amongst different perspectives.

Besides process technology and regulatory factors, business matters also act as crucial drivers of the future: the expiry of patents is as crucial as the measures needed to recruit and retain qualified human resources.

Intelligent Process Control

One important factor is the refinement of process control sensors and tools. Materials and mechanics are not static; their developments are often understood only by specialists and so remain too obscure to attract the attention of a wider audience. Typical examples include enhancements in sterile and hygienic design of fermenters. In this context, progress in **compact component engineering** can be just as valuable in the production environment.

It seems likely that the future of automation in the coming years will be based on the distribution of intelligence to in-process components on the one hand and on the systemic integration of these highly sophisticated elements on the other. Mettler Toledo is a leading developer of **intelligent sensors**. This new generation of intelligent devices will be used for plant maintenance as these sensors are able to monitor themselves. This increased transparency of data quality will also be gainfully exploited for process control purposes. The improved precision and reliability of data will impact the validation and dependability of the processes. Additionally, such sensors will serve as springboards for greatly improved control strategies as an enhanced understanding of processes unfolds.

For fermenter specialists, the challenge will be to first address the researchers' individual needs and then offer tailored solutions while also reliably integrating new developments as and when they become available. Indeed, some manufacturers have already initiated slightly disparate communications protocols and data formats. Any systems engineering company that is reluctant to be a niche player must be able to reconcile these market-strategy-driven variances at the technical level (Niklaus, 2009).

Validation and Automation Strategies – Process Analytical Technology (PAT)

The U.S. Federal Drug Administration (FDA) in 2004 proposed that as long as critical process parameters measured in the course of production remain within a design range, the product of the process is automatically considered validated. In this approach, a product is not tested subsequently (and sometimes rejected) - an intelligent process control system equalizes potential deviations before they actually occur. Hence, the important point is to develop a control strategy under which deviations - ideally - are no longer possible. In this case, the process data (critical parameters) can be used to validate the product, thereby significantly improving the effectiveness of production processes.

Unfortunately, in the past, pharmaceutical companies were often disinclined to accept an obligation to grant FDA inspectors deeper insights into process data. This may be why PAT is not *de facto* being used as a validation strategy. This does not point to any stagnation in the evolution of PAT as an efficiency improvement strategy. In fact intelligent automation systems are destined to gradually establish themselves in pharma production as well. In the chemical and semiconductor industries, for instance, such a process design approach has long been widely accepted. In view of the long development times, intermediate products become highly cost-intensive and each improvement of the process exerts a major impact. Cost pressures and shorter times to market will drive this development. Accordingly, PAT will find broader acceptance in the "discipline of the future" even without any initiative of the FDA.

One-way Fermenters versus Control-disposable Fermenters

There is considerable current interest in disposable fermenters. As disposables have long been used in med-tech applications, the pharma sector adopted them and, at least in laboratory operations, has been using disposables for about two decades. Since the development of stirred one-way fermenters, larger systems of this kind are now being contemplated for the production environment. Some difficulties are expected with so-called leachables and spontaneous flaking of product-contacting material surfaces but may be solved soon. Some benefits related to time savings are also undisputed: as compared with a production plant based on fixed piping, one can save engineering time. Not only are in-service cleaning times eliminated but in multi-purpose plants, the operator benefits from fairly short reconfiguration times. However, the price advantage is not much. In view of the high cost of developing drugs and prolonged pharmaceutical planning horizons, the sums calculated are not very relevant.

Disposable fermenters are usually exposed to stronger constraints in sensors than steel fermenters, because the options are either to integrate the sensors in the disposable fermenter or to insert them just before use. However, high-end technology is not just thrown away after one use, and the subsequent insertion of an expensive sensor that is reused involves serious sterility problems. Even if this dilemma is solved with technical advances, the availability of newly developed sensor technologies for disposables will suffer from delays.

The crucial issue is: can high-end sensors be made available in the long term for the comparatively small market of the pharmaceutical industry? The future market share that disposables can capture in competition with steel fermenters depends on how much emphasis is placed on intelligent process control and comprehensive process monitoring. To some degree, the effectiveness of process control argues against the deployment of one-way fermenters.

Another serious disadvantage of disposable fermenters is the so called simplified validation - disposable fermenters are deemed validated after they have been exposed to gamma radiation. But validation refers to processes, not to equipment. And because the use of disposable fermenters requires personnel for bag handling and transportation, the human factor – the people who necessarily need to be present in the clean room poses the greatest risk for the process. For disposables, the validation process is thrown back to the era of old-fashioned cleaning validation where specifications determined how exactly a worker had to move a sponge. The operation of disposable fermenters can be justified if the following activities are also validated: storage and transport of bags, inspection of the surface on which a bag is placed, the assembly and disassembly of sterile unions and the transport of filled bags (Niklaus, 2009).

Conversely, the operators themselves also can be endangered. The constant presence of supervisors of disposable technologies poses a threat particularly where highly active substances are manufactured. Each manoeuvre with a filled bag turns into a safety risk for personnel. This implies that disposable fermenters will be deployed predominantly in Asian countries where labour is "less expensive".

The implication of these serious downsides is that technical considerations alone discourage major suppliers to the pharma industry from depending solely on the delivery of disposable technologies even though this would not be mandated on the basis of their product lines. Such decisions are motivated largely by commercial aspects. The market for disposable fermenters is growing briskly, and business considerations will prompt suppliers to position themselves in such a way that they can leverage their pricing power with increasing market shares.

MATERIALS FOR ORGANOMETALLIC CATALYSIS AND NANOBIOLOGY

Commendable work is being done to design and synthesize materials for catalytic and biological applications (see Mandal *et al.*, 2006, 2007, 2007a; Nembenna *et al.*, 2006). The major objectives of this research are: (1) development of useful catalysts using low cost main group green metals for a variety of industrially and pharmaceutically important processes, (2) development of nanocatalysts, (3) synthesis of metal nanoparticles coated with chemically functionalized single walled carbon nanotubes and their application in catalytic organic transformation, and (4) use of metal nanoparticles as anticancer and anti-bacterial agents. A simple, mild method has been developed to create palladium nanoparticles of nearly uniform size. These nanoparticles have been used as catalysts for the Ullmann coupling reaction (a new C-C bond forming reaction) of several aryl halide substrates at an excellent conversion rate into the symmetrical biaryls under fairly mild reaction conditions. Also, copper nanoparticles (CuNPs) stabilized with sodium citrate, have shown marked cytotoxic effect towards bacterial and cancer cells. It was noted that DNA degradation occurred in dose dependent manner both *in vivo* and *in vitro* after treating the isolated genomic DNA with CuNPs (Mandal *et al.*, 2007).

GRAPHENE

Graphene is a two-dimensional giant, flat sheet of carbon atoms. The sheet is just one atom thick. Its planar, hexagonal arrangement of carbon atoms is a single layer that is peeled off from the three-dimensional graphite crystal. The single-layered honeycomb structure of graphene makes it by far the most important of all carbon-based systems – the graphite in our pencils is just a stack of graphene layers; CNTs are nothing but rolled up sheets of graphene; and fullerenes or buckyballs are nanometer-size spheres of wrapped-up graphene.

Graphite is an allotrope of carbon. Graphene is a new nanocarbon possessing, single-, bi- or a few to ten layers of carbon atoms that form six-membered rings. Various types of graphene have been examined by X-ray diffraction, atomic force microscopy, transmission electron microscopy, scanning tunnelling microscopy and Raman spectroscopy. The extraordinary electronic properties of single- and two-layered graphenes are unique. Such other properties of graphene as gas adsorption, magnetic and electrochemical properties and the effects of doping by electrons and holes are equally fascinating. Graphene shows high electrical conductance: electrons glide much faster through graphene than other materials such as silicon; this makes graphene attractive for use in microchips.

Lin *et al.* (2010) have developed a field-effect transistor – a key component of microchips – in which the channel, the part through which electrons flow, is made of graphene. The graphene allows this transistor to switch at speeds of up to 100 gigahertz, more than twice as fast as current silicon transistors.

Molecular charge-transfer markedly affects the electronic structure and characteristics of graphene. There is need to explore other aspects of graphene by synthetic strategies that can yield sufficient quantities of graphene with the desired number of layers.

Being the **strongest material** known, graphene should make tomorrow's mix of power-hungry consumer gadgets lighter and more efficient. The problem is that graphene is extremely difficult to manufacture in large amounts.

Stride *et al.* (2008, *Nature Nanotech.* Doi: 10.1038/nnano.2008.365) first reacted ethanol and sodium, then broke down a solid intermediate of the reaction by heating. This created a fused bundle of graphene sheets which could then be broken apart with gentle sonication.

The crystal structure of graphite is shown in Fig. 2.1. This structure is made of layers of hexagonally arranged carbon atoms in which each carbon atom is bonded to three coplanar neighbouring atoms by strong covalent bonds. The length of the bonds is 1.42 Å and the distance between planes is 3.35 Å. The interlayer bonds are quite weak – they are held together by van der Waals type of forces (Pierson, 1993).

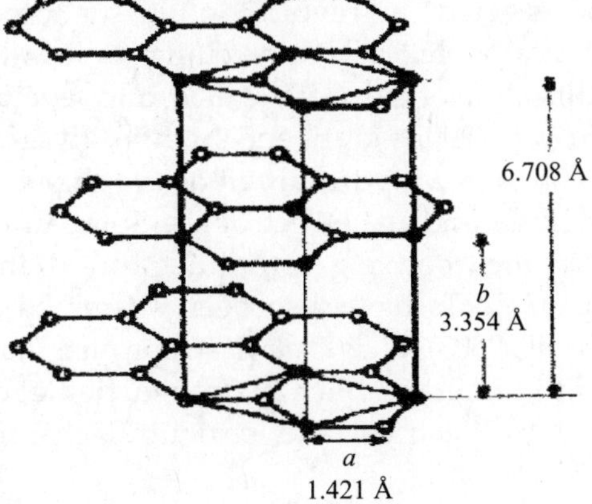

Fig. 2.1 Crystal structure of graphite (after Srinivasan and Saraswathi, 2009, with permission from *Current Science*).

Graphene was first prepared by Novoselov *et al.* (2004) who just peeled off the top layer of a highly oriented pyrolytic graphite with a scotch tape (Fig. 2.2). Graphene is a 2D nanomaterial that shows several distinct properties (Srinivasan, 2007; Srinivasan and Saraswathi, 2009). It is considered to be a viable alternative to silicon. The unique properties of graphene have already attracted applications in electronics where its exceptionally high electrical conductance is a big advantage. Another use for this wonder material has also been reported by researchers at Pohang University in South Korea who used a composite material made from reduced graphene oxide (RGO) and magnetite to effectively remove arsenic from drinking water (*ACS Nano*, 16 June 2010; doi: 10.1021/nnl008897).

Fig. 2.2 Extracting graphene from graphite (after Srinivasan and Saraswathi, 2009, with permission from *Current Science*).

Graphene also exists as an oxide. RGO is a state of graphene that has gained electrons. The researchers dispersed a magnetite-RGO composite in water containing arsenic. The RGO composite soaked up the arsenic from water and could be removed by means of a permanent magnet.

The researchers at Pohang made a new type of magnetite composite based on RGO. The hybrid material is super paramagnetic at room temperature and can remove over 99.9% of arsenic in a sample. The composite more effectively removes arsenic than bare magnetite because the presence of the graphene flakes among the magnetite particles increases the number of arsenic adsorption sites. The RGO also enhances the stability of magnetite so that it can be used in continuous-flow systems for fairly long periods.

Much work has been done to design simple and effective methods to prepare and process graphene on a large scale and to use it for a variety of applications (Katsnelson, 2007; Sungjin and Ruoff, 2009). Some of its applications include gas sensors, energy storage and nonelectronics. An end-to-end rolled graphene sheet generates the CNT, whose discovery by Iijima (1991) has already started a revolution in nanotechnology. The chemical functionalization of CNTs has attracted several exciting applications (Sun *et al.*, 2002). It is hoped that the chemical modification of graphene will lead to several new materials with novel properties. According to Sofo *et al.* (2007), graphene might be hydrogenated to yield a stable 2-D hydrocarbon. They termed this new hydrocarbon as **graphane** and discussed its potential applications in hydrogen storage and nanoelectronics. Some attempts have been made to hydrogenate certain forms of carbon e.g. graphite and diamond. Physisorption and chemisorption of hydrogen on CNTs have also been examined in the context of possible hydrogen storage (Stojkovik *et al.*, 2003). Elias *et al.* (2009) proved that a single layer of graphene can indeed be hydrogenated to graphane. Such hydrogenation could be achieved by using a cold hydrogen plasma for 2 h. Electrical and structural characterization confirmed the formation of graphane. Unlike graphene, graphane showed an insulating behaviour such that the resistivity grew by two orders of magnitude when the temperature fell from 300 to 4 K. Graphane remained stable at room temperature for several days and could be reconverted into graphene by annealing at 450°C in Ar atmosphere for 24 h.

Examination of the electron-diffraction pattern of graphane indicated that graphene hydrogenation neither changed the hexagonal symmetry nor crystallinity, but led to a strong decrease of lattice constant by about 5%.

The successful creation of high-quality crystals of graphane from graphene constitutes a new way of designing many other ultra-thin crystalline materials based on graphene. A very striking aspect is that the semi-metal graphene is changed to an insulator, graphane. This welcome modification in the electronic behaviour of graphene following addition of hydrogen has prompted the idea of fine tuning its electronic properties by binding of various atoms and groups to its scaffolding. In fact, the long-cherished dream of fabricating nanoscale transistors from graphene wafer, and one in which all interconnects are made from highly conductive, pristine graphene, and its other parts are modified chemically to produce semiconductors, may soon materialize. If so, the transistor will contain only graphitic materials, thus providing a new excitement after the imminent end of the silicon era. Another welcome prospect is that as compared to graphene, graphane may be a much better, even an ideal, material for hydrogen storage (Srinivasan and Saraswathi, 2009).

High-pressure Cylinders and Nanoextruders

The covalent bond between two carbon atoms in graphite being extremely strong, a graphene plane consisting of a hexagonal arrangement of carbon atoms serves as a tearproof net at the nanoscale. CNTs consisting of cylindrically closed graphene sheet therefore show exceptionally remarkable mechanical properties (Lu, 1997; Zhang *et al.*, 2005) and have attracted several applications (Dalton *et al.*, 2003; Barber *et al.*, 2003). While the rupture of nanotubes under axial tensile load has been well studied (see Motta *et al.*, 2005), their strength against internal pressure has not been critically explored. As nanotubes can be filled with various materials (Maniwa *et al.*, 2002; Gogotsi *et al.*, 2002), it is possible to test the pressure-holding ability of nanotubes by studying the behaviour of encapsulated compressible materials upon tube contraction (Sun *et al.*, 2006).

Nanotubes can be contracted by knocking carbon atoms out from the tube lattice and allowing the atomic network to rebuild. Such contraction has been studied in spherical carbon onions under electron irradiation (Banhart and Ajayan, 1996). Closed-shell carbon nanostructures, such as carbon onions, act as self-contracting high-pressure cells under electron irradiation.

Nanotubes have previously been shown to be self-healing structures under irradiation (Li and Banhart, 2004; Banhart *et al.*, 2005). Vaccines remain mobile at sufficiently high temperature but can coalesce to form divacancies, which ultimately mend by reconstruction of the lattice (Krasheninnikov *et al.*, 2006), reduce surface area and generate local tension in the graphene sheet. But although such collapse of hollow nanotubes under electron irradiation has been repeatedly observed (see Sun *et al.*, 2006), the pressure effects produced inside the cores during collapse have not been studied. Sun *et al.* observed the dynamics and compressibility of encapsulated nanowires inside collapsing nanotubes and reported that even hard materials such as iron carbide or cobalt can be extruded under the pressures prevailing inside nanotubes. They showed that controlled irradiation of multiwalled carbon nanotubes can cause large pressure buildup within the nanotube cores that can plastically deform, extrude, and break solid materials encapsulated inside the core. Atomistic simulations suggested that the internal pressure inside nanotubes often reaches values higher than 40 gigapascals (Sun *et al.*, 2006). Sun *et al.* did this work by preparing multiwalled carbon nanotubes (MWNTs) encapsulating Fe, Fe_3C, or Co nanowires by means of a modified ethanol-metallocene-based chemical vapour deposition process.

The results reported by Sun *et al.* clearly demonstrate the great strength of CNTs against internal pressure. Interestingly, the ultimate internal pressure (~40 GPa) that nanotubes can resist is only an order of magnitude below the pressure in the centre of Earth (~360 GPa) or the highest pressure that has been achieved in diamond anvil cells (~400 GPa). The merit of their experimental protocol is that individual nanometer-sized crystals can be studied *in situ* and deformed under high pressure, although irradiation effects in the crystals might occur and need to be considered. The behaviour of the encapsulated crystals could be monitored directly with high spatial and temporal resolution. CNTs thus provide a useful template in compression/extrusion cells for studying pressure-induced phase transformations and deformations of various solid nanomaterials.

Sequencing with Carbon

Postma (2010) described a device that could rapidly sequence a single strand of DNA passing through a gap in a piece of graphene.

This device exploits graphene's conducting ability. The graphene would act as the electrodes to measure the conductance of DNA as it moved through the gap. Each of the four bases that make up DNA shows a unique conductance, which allows the DNA sequence to be read.

Although some other nanopores have already been devised for DNA sequencing, graphene's innate conductance and strength makes it more attractive.

GRAPHITE OXIDE

Graphene has usually been obtained by mechanical methods which have very low productivity (Novoselov *et al.*, 2005). Large cohesive energy between the layers in graphite neither allows easy separation nor makes a suspension of graphene stable in liquids. In contrast, chemical methods allow formation of stable suspension of sheets of graphite oxide (GO; in high concentration) in water and these can be reduced to graphene (Stankovich *et al.*, 2006, 2007; Niyogi *et al.*, 2006). Hydroxy, epoxy and carboxyl groups in GO make individual sheets water soluble. GO has enabled fabrication of paper-like films with excellent mechanical properties, electrically conductive polymer, and transparent electrode (see Stankovich *et al.*, 2006a, Dikin *et al.*, 2007; Becerril *et al.*, 2008). The electrical conductivity of reduced GO and its composite formulations is significantly enhanced. Thermal treatment is a good reduction method for conversion of GO to a black graphitic material, suggesting that a single graphene sheet might be obtained by thermal reduction of a single sheet of GO deposited on a suitable substrate. Ali *et al.* (2009) reported temperature dependent electron paramagnetic resonance (EPR), thermogravimetric and chemical analysis which has thrown some light on the mechanism involved in the reduction of GO by thermal treatment. Ali *et al.* prepared GO by a method of Hirata *et al.* (2004) and showed it to be EPR-active with ~10^{16} electron spins/g. Heating GO increased the free radical concentration at ~140°C, together with the elimination of CO_2 and water; black, diamagnetic graphitic material was finally produced at high temperature. Significant loss of carbon in the thermal process implies that the chemical route to graphene may be less attractive.

Commercially available microcrystalline graphite powder (< 25 μm) was oxidized to produce water soluble GO. Graphite was treated with potassium permanganate and sodium nitrate in H_2SO_4 for five days, and processed further to produce yellow coloured GO. GO contains a number of oxyfunctional groups, e.g. epoxy, carboxylic and hydroxyl. X-ray diffraction (XRD) of GO powder revealed that the interlayer spacing in GO had increased to 9.33 Å in graphite. When sonicated, single sheets of GO can be stabilized in the suspension, which can be dropcast on a substrate such as mica. Average thickness of GO sheet is about 1 nm (Ali *et al.*, 2009).

Thermal treatment of the yellowish GO powder yielded a black graphitic powder. Its thermogravimetric analysis (TGA) indicated a weight loss of ~55%: weight loss of ~15% at < 110°C, a sharp weight loss at ~140°C, followed by a further weight loss up to 300°C. Simultaneous thermogravimetric and mass spectrometric (TG-MS) analysis suggested that weight loss of water occurred throughout whereas simultaneous loss of CO_2 and water occurred only at ~140°C. According to Ali *et al.*, elimination of CO_2 at ~140°C increased hydrophobicity, leading to co-elimination of water bound to hydrophilic sites.

GO turned out to be paramagnetic. The EPR spectrum of GO powder showed a single line with *g*-value of 2.0028. The number of spins per gram of GO were calculated to be about 10^{16}. An increase in the EPR intensity at ~140°C could be due to the formation of free radicals accompanying elimination of CO_2 and H_2O. At ~150°C and above, the EPR signal was completely lost and the compound turned irreversibly black.

The free radicals present initially were destroyed upon heating. The black colour of the final product and the absence of the peaks of hydroxyl and carboxyl groups in the FT-IR spectrum point to the possibility that graphene regions had developed in GO.

The results reported by Ali *et al.* suggest that graphite oxide prepared by Hirata *et al.*'s method is EPR-active. Thermal treatment leads to co-elimination of CO_2 and H_2O and there is a substantial enhancement in transient free radicals, finally leading to EPR-inactive graphitic material. Conceivably, single sheets of GO deposited on mica substrates could probably be thermally converted into graphene-like molecule. However, there is a significant loss of carbon in the thermal process which makes this route towards pure graphene not so attractive (Ali *et al.*, 2009).

PATTERNED GRAPHENE TRANSPORT

Graphene, consisting of single sheets of graphite, or unzipped carbon nanotubes, shows interesting transport properties which depend on geometry and crystallographic orientation, in much the same way the electronic properties of CNTs are dependent on their chirality, or how they are rolled up. Han *et al.* (2007, *Phys. Rev. Lett.* 98: 206805) studied single sheets of graphene extracted from bulk graphite and patterned into strips of widths ranging from 10 to 100 nm and a selection of crystallographic orientations. The ensuing transport properties were then critically examined. It was observed that the energy band gap increased with decreasing width, as expected from theory, but there was no systematic variation with orientation. These findings suggest a method to engineer the band gap of graphene nanostructures with potential applications for various devices.

Today, many devices use indium tin oxide. A much better option may be to use a touch-screen with transparent electrodes made from graphene. These electrodes could be thinner, cheaper and longer lasting than the commonly used devices of today. Unfortunately, it has been very difficult to manufacture these transparent films efficiently on a large scale. Hong *et al.* (2010) have succeeded in making such films by using a scalable industrial manufacturing process. They have also incorporated them into a touch screen. They were able to make 30-inch graphene films by depositing carbon atoms onto copper, which is later etched away.

THERMAL TRANSPORT PROPERTIES OF GRAPHENE

It has been known for sometime that unsupported graphene sheets have exceptional thermal transport properties, but it was not clear whether these properties are also expressed when a graphene sheet is in contact with a substrate. Seol *et al.* (2010) measured the thermal conductivity of graphene supported on SiO_2. They noted that while the conductivity is significantly lower than that of free-standing graphene, it is still higher than that of copper. According to them, the out-of-plane flexing vibrations of the graphene have a crucial role in thermal transport. This means that graphene may attract interest in applications such as conducting heat away from electronic circuits.

In order for graphene sheets to be used in sensors, transistors and other devices, reactive molecules need to be bound to them so that the sheets can be chemically bonded to other materials. Yan *et al.* (2010) described a simple way to do this. They created three variants of perfluorophenylazide. This compound reacts with graphene's carbon-carbon bonds when heated or illuminated with a mercury lamp. Each variant carried a different molecular group. By mixing each variant individually with a solution of graphene flakes, the researchers could attach a specific molecule to the graphene, preparing the carbon sheet for further chemical bonding.

The efficiency of many electronic and optoelectronic devices depends on temperature. Therefore, these devices need to be made of materials with high thermal conductivities so that the heat generated spreads locally in the devices (Mahajan *et al.*, 2006). Bulk copper, being widely used as heat spreader in computers, shows a thermal conductivity of ~400 W m^{-1} K^{-1} at room temperature. In contrast, copper thin films used as electrical interconnects show lower thermal conductivity (below 250 W m^{-1} K^{-1}). This is why researchers are seeking materials with thermal conductivities higher than that of copper. Seol *et al.* (2010) reported that single monolayers of graphite (graphene) in contact with silicon dioxide (SiO_2) have a thermal conductivity of ~600 W m^{-1} K^{-1}).

The reason why pure-carbon materials (e.g., diamond, graphite, and carbon nanotubes) have very high thermal conductivities is because the strong covalent bonding between carbon atoms results in a large phonon (lattice vibration) contribution to the thermal conductivity. According to Geim (2009), graphene shows very high intrinsic charge carrier mobility, and Yan *et al.* (2010) described the fabrication of 100-GHz graphene-based transistors.

For future applications, it is extremely important to understand heat generation and dissipation from graphene-based devices. Yu *et al.* (2005) and Balandin *et al.* (2008) had made measurements on single monolayers of SWCNTs and graphene in suspended form where the thermal conductivity was quite high because of the strong covalent bonds between the carbon atoms. The thermal conductivity of suspended graphene is 5000 W m^{-1} K^{-1} at room temperature, about 2.5 times as large as that of diamond. Diamond has the highest thermal conductivity among natural materials (Prasher, 2010).

Because in some practical applications, e.g., as composites for thermal interface materials (Prasher, 2006, 2010) or heat spreaders in devices, single-layer materials are bound to a much thicker bulk material, such as SiO_2 or a polymer through weak van der Waals forces, the question arises as to how such weak contacts can affect thermal transport in the monolayers. Seol *et al.* (2010) attempted to address this question by first attaching a graphene layer to a SiO_2 surface through a process called **mechanical exfoliation** (Novoselov *et al.*, 2004) and then measuring the thermal conductance of the combined structure. The graphene layer was then etched away and the thermal conductance of only the SiO_2 layer was measured. The difference between the two values represented the conductance of the graphene layer, from which the thermal conductivity of the graphene layer could be extracted. Whereas at room temperature the thermal conductivity of the supported graphene layer is ~600 W m^{-1} K^{-1}, an order of magnitude lower than that of suspended graphene, this is still higher than the thermal conductivities of bulk or thin-film copper.

The work of Seol *et al.*, by combining thermal, structural, and mechanical phenomena into one theoretical framework, represents a crucial milestone towards understanding the thermal conductivity of supported graphene. However, some questions still remain; we know that the large thermal contact resistance of CNTs has greatly limited their use as a high-thermal-conductivity material. Can graphene overcome this limitation because of its flat structure?

CLUSTERS

The term **cluster** means a group or bunch. Physicists and chemists consider this term to mean a group of atoms or molecules formed by interactions ranging from very weak van der Waals contacts to strong ionic bonds. Studies of clusters in mass spectrometer ion sources and later in molecular beams started in the 1950s. The advent of the laser vaporization technique allowed researchers to produce clusters of virtually any element in the periodic table, and much work ensued on clusters of various compositions, beyond systems of volatile materials, which had been the focus in the 1960s and 1970s (see Becker, 1986; Dietz *et al.*, 1981). Clusters are now viewed as a new form of matter that is intermediate between atoms and their bulk counterpart. Clusters not only bridge phases but also disciplines (Castleman and Jena, 2006). They bridge the disciplines of environment, materials science, and biomedicine.

Clusters are not only an intermediate form of matter but their properties also can be changed by adding a single atom or electron (Jena *et al.*, 1987; Haberland, 1994; Khanna and Castleman, 2003). Some clusters show high stability and unique properties, and may be considered as super atoms; they serve as building blocks for a new class of materials with tailored properties (Khanna and Jena, 1995).

How does a cluster differ from a molecule? Both are groups of atoms. Molecules found in nature are stable under ambient conditions, whereas clusters tailored in a laboratory often are metastable. Whereas molecules have specific stoichiometry, the composition of a cluster can usually be changed by a change of production conditions. Clusters tend to coalesce when brought close together, and they usually react with ambient gases. Thus, although all molecules can in principle be identified as clusters, very few clusters can be called molecules, the exceptions being fullerenes and Met-Cars (Kroto *et al.*, 1985; Guo *et al.*, 1992).

Beyond a certain size limit, the properties of a cluster vary smoothly, eventually reaching the bulk limit. The size at which the transition to bulk happens depends on the kind of atoms or molecules that form the cluster and on the property being studied. This makes a cluster intrinsically an "artificial molecule" that is metastable and whose properties depend on the exact number and kind of atoms that compose it. A cluster can sometime be stabilized by passivation or ligation (Meiwes-Broer, 2000), and these stable clusters can be used as the building blocks of new materials (Castleman and Jena, 2006, 2006a).

Clusters of volatile constituents can usually be made by supersonic expansion through nozzles or, in the case of ion clusters, by sequential growth in low-pressure gases available in ion sources or flow-tube reactors (Castleman and Wei, 1994). The intensity of the mass spectra contains information on the relative stability of the clusters – this stability is studied by reacting the clusters with various reagent gases and reexamining the intensity of their mass peaks. The electronic, optical, and magnetic properties of mass isolated clusters are studied by using such techniques as resonant ionization and electric and magnetic deflection methods. The ionization potential, electron affinity, dissociation energies, gap between **highest occupied molecular orbitals** (HOMO) and **lowest unoccupied molecular orbitals** (LUMO), and reactivity provide valuable information that can be gainfully utilized to characterize a cluster and study the evolution of its properties with size and composition; for this, photoelectron spectroscopy is generally used (Ganteför, 2003).

One notable development now hovering on the near-term horizon is a strong bridging of ideas beyond a focus on the physical sciences towards new and emerging problems in the life sciences, such as the identification and destruction of cancer cells, and to learn about processes involving ion transport, proton transfer, and even mutagenesis.

Clusters in Environmental/Atmospheric Science

The formation of finite particles and dust grains in interstellar media, heterogeneous reactions occurring in the stratosphere, and unidentified infrared bands in interstellar gas clouds may have a cluster origin.

Fine particles are ubiquitous throughout the universe; they are present in dense clouds where ion clustering reactions are possibly involved in the formation of molecules. Some theories of planetary evolution are based on clusters to explain the origin of planets themselves. Fine particles greatly interest environmental scientists – they affect visibility and health (Rossi, 2003; Castleman and Jena, 2006a). Clustering and related nucleation phenomena are at least partly responsible for their formation (Whitten, 1982).

Critical study of the clustering of neutral molecules onto ions is a good approach to understanding the behaviour of acid clusters. Cluster ions in the atmosphere reveal the role of acids. Strong acids become preferentially clustered to negative ions, which act as bases. Under normal atmospheric conditions, ambient ions of the type $HSO_4^-(H_2SO_4)_x(HNO_3)_y$ or $NO_3^-(HNO_3)_n$ (the latter ion is probably hydrated in the lower troposphere) are present throughout the lower atmosphere (Huey, 1996). Some work has been done on the thermodynamic stability of $HO_3^-(HNO_3)_n$ clusters and the reactivity of nitric acid and sulfuric acid with negative ions (Huey *et al.*, 1995). These studies have enhanced our knowledge of the pathways involved in the production of the acid clusters and their relationship to the gas-phase acid loading of the atmosphere.

Clusters in Materials Science

The building blocks of natural materials are atoms or molecules. Depending on the chemistry of the atoms, the ambient pressure, and temperature, the atoms arrange themselves in distinctive arrays to form different crystal lattices. The properties of these materials depend on how the atoms are arranged. Molecular crystals, on the other hand, show unique properties because molecules – not atoms – form the building blocks.

Khanna and Jena (1992) floated the idea that by varying the size and composition, one atom at a time, it is possible to design atomic clusters to mimic the chemistry of atoms in the periodic table and so create a 3D periodic table where clusters make up the third dimension. Should these clusters be unusually stable, they could be used as building blocks of a novel class of cluster-assembled materials. Analogous to molecular crystals, size-selected cluster-assembled materials may possibly reveal unique novel properties.

The core issue is how to design and synthesize stable clusters. In clusters of simple metals, the stability is governed by the electronic shell structure. In covalently bonded clusters such as those of C and Si, the atomic arrangements are crucial for stability. Stable compound clusters may also be generated by using appropriate alloying techniques.

Covalently Bonded Clusters

It was the discovery of C_{60} in the gas phase by Kroto *et al.* (1985) as an unusually stable cluster and its chemical synthesis in macroscopic quantities (Kratschmer *et al.*, 1990) that created the new field of fullerene chemistry. The ability to introduce metal atoms inside C_{60} generated some hope that novel cluster-assembled materials with unique properties may be synthesized. Applications of these materials range from superconductivity to medicine.

Although both Si and C belong to the same column in the periodic table, Si clusters behave very differently from those of C. Repeated attempts to synthesize Si_{60} in the fullerene form have failed, the only way to confine Si_{60} in a cage structure may be to encapsulate it in a metal such as $Al_{12}Si$.

Metallocarbohedrynes (Met-Cars)

The first member of the Met-Car class of clusters to be discovered was made of 8 titanium atoms bound with 12 carbons (Ti_8C_{12}) and assembled into a cage-like structure (Guo *et al.*, 1992). A wide range of Met-Cars form from metals such as Ti, Zr, Hf, V, Nb, Cr and Mo, and possibly Fe (Pilgrim and Duncan, 1993). Binary Met-Cars can be made from the above metals. Some other metals that do not apparently form the pure Met-Car can be substituted into at least the titanium member of the class; these include Y, Ta, W, Mn, Th, and even Si, which interestingly replace a metal rather than a carbon position (Castleman and Jena, 2006). It appears that the Met-Car assembles from species comprised of metal-C_2 units, which ultimately undergo carbon loss at the point of cage closure to the 8-12 structure. These species, plus the Buckminster-fullerene family, offer the prospect that molecular clusters as well as cluster-assembled materials, may possibly exert significant impact in materials science.

May *et al.* (1995) showed that Met-Cars display unique behaviour after laser photoexcitation. Besides delayed ionization, the building blocks of Met-Cars tend to undergo delayed atomic ion emission.

Biology represents the ultimate expression of molecular engineering where nature has perfected the science and technology of organic and metalloorganic complex matter at the **nanoscale**. Research on the properties of inorganic nanomaterials bridged biology and physical sciences, leading to the emergence of biomaterials (Stupp *et al.*, 2005) designed and synthesized to imitate, monitor, and repair biology and to learn from it in making new materials. Clusters play a basic role in this effort and facilitate understanding of the principles governing the success of biomaterials. There are several examples where clusters can be designed and synthesized with specific optical and magnetic properties for noninvasive treatment of tumours, also examples where clusters mimic natural enzymes in being highly efficient, low-temperature catalysts (see Castleman and Jena, 2006, 2006a).

A major goal in cancer therapy has been to seek ways by which cancer cells can be selectively destroyed without damaging healthy cells. Nanoparticles that have the same length scales as those of tumours can help to achieve this cherished goal – the idea is based on creating three-component nanoparticles that target, image, and destroy tumours. Some experiments have already proved promising in laboratory tests in animals and test tubes. Hirsch *et al.* (2003) targeted tumours with gold-coated silica nanoparticles, which absorb infrared light and generate enough heat to potentially kill the cancer cells. Sun *et al.* (2004) later showed that small silica clusters containing just three molecular units absorb infrared light when coated with one or two gold atoms. The gold atoms attach to the dangling bonds of silicon atoms in the silica nanoparticle, so modifying the highest occupied molecular orbital – lowest unoccupied molecular orbital (HOMO-LUMO) gap in such a way that even a tiny cluster can absorb infrared light. The merit of using a small cluster with the same functionality as of the large nanoparticles is that it can penetrate the crowded environments such as biological milieu of cells and live tissues for effective drug delivery. Unfortunately, the above technique is unsuitable for tumours such as anaplastic astrocytomas or glioblastomas that are buried in the human body. For these buried tumours, magnetic fluid hyperthermia (Pankhurst *et al.*, 2003) is a better option – a magnetic nanoparticle can be heated by absorbing radio frequency, and the resulting heat can destroy the tumour. Further, it is possible to manipulate the magnetic nanoparticle that are widely used in experiments and animal testing. These are functionalized by coating with gold and are then targeted to attack specific cells (see Kinoshita *et al.*, 2005).

Sublimed Clusters

It is possible to make inorganic thin films under vacuum by transferring material from one surface as a vapour and depositing it on a nearby substrate. But creating the vapour phase usually involves such harsh conditions as strong heating or bombardment with high-energy ions or photons – to desorb atomic species or a distribution of clusters. Consequently, preserving intricate synthetic structures during the transfer is quite difficult. Chao *et al*. (2007, *Nature Nanotechnol.* 2: 486) showed that silicon nanocrystals (with mean diameters ~2 nm) that are capped with undecyl hydrocarbon chains suffer thermal desorption at fairly low temperature (200°C) and can be transferred from one surface to another for various types of substrates. Composition of the transferred particles has been confirmed by x-ray photoemission spectroscopy, and structures have been examined with different microscopy techniques. The nanocrystals at first grow as two-dimensional islands but can also form three-dimensional clusters.

Clusters in Liquid Metals and Glasses

Clusters with icosahedral short-range order constitute basic structural elements of liquid metals and glasses and their presence has been demonstrated experimentally (Sheng *et al.*, 2006; Wochner *et al.*, 2009) as well as *ab initio* molecular dynamics simulations (MDS) (Jakse and Pasturel, 2003). However, the structure of liquids cannot be defined easily at solid-liquid interfaces nor can the influence of interfaces on ordering and supercooling predicted. The extent to which a liquid can be supercooled depends on the substrate in contact with it, and on thermal history. Supercooling in metals (the preservation of a disordered, fluid phase in a metastable state well below the melting point) has aroused speculation that local atomic structure configurations of dense, symmetric, but non-periodic packing act as the main barrier to crystal nucleation (Torquato and Jiao, 2009). For liquids in contact with solids, crystalline surfaces induce layering of the adjacent atoms in the liquid (Oh *et al.*, 2005) and may prevent or lower supercooling. This seed effect probably depends on the local lateral order adopted in the last atomic layers of the liquid in contact with the crystal. Schülli *et al.* (2010) reported supercooling in gold-silicon (AuSi) eutectic droplets, enhanced by a Au-induced (6 × 6) reconstruction of the Si(111) substrate. *In situ* X-ray scattering and *ab initio* molecular dynamics showed that pentagonal atomic arrangements of Au atoms at this interface favour a lateral-ordering stabilization process of the liquid phase. Such an interface-enhanced stabilization of the liquid state highlights the importance of the solid-liquid interaction for the structure of the adjacent liquid layers. These processes are crucial not only for present but also future technologies, as fluidity and crystallization have a key role in soldering and casting, as well as in processing and controlling chemical reactions for microfluidic devices or during the vapour-liquid-solid growth of semiconductor nanowires (Schülli *et al.*, 2010).

Protection of Gold and Silver Clusters

Sub-nanometer clusters of gold and silver can be protected on monolayers. Such clusters are molecule-like with well-defined absorption and emission features, spanning across the visible-near infrared window. In a few clusters, the emission can be quite intense. In general, several clusters are prepared together by reduction of metal ions in the presence of appropriate ligands, followed by size separation by polyacrylamide gel electrophoresis (PAGE). Suitable methods have been developed for fast and efficient synthesis of clusters. All of these clusters are stable, highly water-soluble and can be made in the solid state. The clusters are characterized by electrospray mass spectrometry which gives multiple charged species. Isotope patterns and theoretical mass spectra are used to characterize the molecular composition. Pradeep (2009) separated several distinct clusters such as Au8, Au22, Au23 and Ag4, and showed that fluorescence resonance energy transfer can be achieved in these clusters by linking appropriate fluorescent molecules to the cluster core. Interfacial routes for the direct synthesis of some of these clusters have also been reported. The synthesis of a variety of clusters, their chemical stability and intense luminescence may attract several applications in areas such as energy transfer, sensors, bio-labelling and drug delivery (Pradeep, 2009).

Nanophase Clusters

There exists a class of unique structures known as clusters between isolated atoms or molecules and bulk materials. This class consists of a few to hundreds of atoms or molecules. Within this "nanophase" range, many physical and chemical properties of the materials depend on cluster size, and materials may exhibit novel properties due to quantum confinement effects. Clusters not only serve as controllable model systems for revealing the complexity of condensed-phase and biological structures, but also play crucial role in defining nanoscience and nanotechnology (Cheng *et al.*, 2006).

Real-time dynamics of clusters have been explored in the past three decades by studying time-resolved spectroscopy, with picosecond (ps) and then femtosecond (fs) resolution in molecular jets and beams (see Bernstein, 1996; Felker and Zewail, 1983; Zewail, 1994). Early results (Phillips, 2003) paved the way to many applications as they brought out the influence of microsolvation and stepwise solvation effects on reactivity and photophysical processes in clusters. The ultimate goal was not only to narrow the gap between the isolated-molecule behaviour and that in the bulk phase, but also to understand the influence of solvent on reactions at a microscopic level (Cheng *et al.*, 2006).

Some phenomena that attracted research interest included the influence of one-atom solvation and caging (Wan *et al.*, 1997) and stepwise solvation of acid-based and isomerization reactions in clusters of up to 30 solvent molecules (Cheng *et al.*, 2006). Electron and proton transfer processes were also investigated. With mass selection, Paik *et al.* (2003, 2004) investigated ionic clusters, the premier being solvated electrons and dioxygen anions, both of which are involved in biological activities.

Some other major subjects and systems that were studied from 1983 onwards included biomolecular reaction, vibrational-energy redistribution and vibrational predissociation, microscopic solvation, caging and recombination, and biological complexes (see Cheng *et al.*, 2006).Throughout these studies, the main focus was on understanding the microscopic dynamics, and on bridging the isolated system behaviour to that of the bulk.

Some notable new concepts that emerged from such studies include those describing the mechanism of fragment caging in solvent shells (hard sphere interaction vs. vibrational energy loss), local sphere solvation (as opposed to continuum model solvation), ultrafast back electron transfer (which explains phenomena of chemical and biological nature), critical solvent-size effect in proton and electron transfer reactions, and the critical role of collision complex lifetimes in controlling the outcome of reactions. All these concepts are crucial to the description of phenomena on the nanometer scale, because they emerged from studies on such a scale, with size selection, and not in the bulk phases. For biological dynamics and function also these concepts are relevant.

One particularly striking finding in ultrafast dynamics of complex systems has been the ability of such systems to direct energy through electronic and nuclear couplings to enhance the efficiency and useful work. This became clear from early studies of femtochemistry and similarly in femtobiology (Zewail, 2000). Chemical and biological systems efficiently direct the energy by limiting the phase space to a few coordinates, so avoiding dissipation or wasting of energy.

Studies of isolated systems add another dimension to our understanding of the role of complex environments, such as proteins, in the process of molecular evolution. Lee *et al.* (2006) successfully probed the dynamics of the isolated negative anion chromophore of PYP (photoactive yellow protein) and compared with its dynamics in the protein environment. This protein reduces the phase space and prepares the ground state in <1 ps, without creating the radical species that is detected in the isolated chromophore. It achieves this by using a few nuclear motions (bond rotation/twisting) and a conical intersection. Studies with stepwise hydration have revealed the nature of the forces involved and purpose of biological complexity (Cheng *et al.*, 2006).

3

CHAPTER

Nanomaterials and Nanoparticles

NANOMATERIALS

Nanomaterials (NMs) are materials whose sizes are less than 100 nm at least in one dimension. Nanoparticles (nps), nanointermediates and nanocomposites (ncs) exemplify NMs. When nps are made out of larger materials, their reactivity increases; gold is chemically inert at normal scales, but acquires catalytic ability at nanoscales. Properties of nps depend on their dimensions; unique spectroscopic, electronic and chemical properties are manifested by their small size and high surface/volume ratios. Nps can be stabilized with organic molecules to modify their solubility, stability and luminescence. Colloidal semiconductor nanocrystals or quantum dots (qds) can be used as light emitting devices and lasers. Magnetic nps (mnps) have attracted several applications in biology and medicine. Mnps (size, <20 nm) are strongly magnetic. Supermagnetic molecules are attracted to a magnetic field but do not retain residual magnetism after the magnetic field is removed (Zhong *et al.*, 2003).

Among mnps, nps of λ-Fe_2O_3 and its derivatives are specifically important in biomedical application. Spinel ferrite (AB_2O_4, where A sites have tetrahedral coordination by oxygen while B sites are octahedrally coordinated) is a suitable crystal system for understanding magnetic properties of nps (Fig. 3.1). Spinel ferrites, MFe_2O_4 (M = Mn, Fe, Co, Zn, and Mg) are ferromagnetic. Magnetic properties of substituted iron oxide γ-$Mn_xFe_{2-x}O_3$ or $M_xFe_{3-x}O_4$, where M = Al, Zn, Mn or Co can be further improved beyond those of parent oxide (Prasad *et al.*, 2007). Kijima *et al.* (2007) reported that α-Fe_2O_3 nps work as a rechargeable electrode material. Sun and Zeng (2002) and Zhang *et al.* (2005) described hexagonal Fe_3O_4 nps and size controlled synthesis of mnps (dia. 3-20 nm).

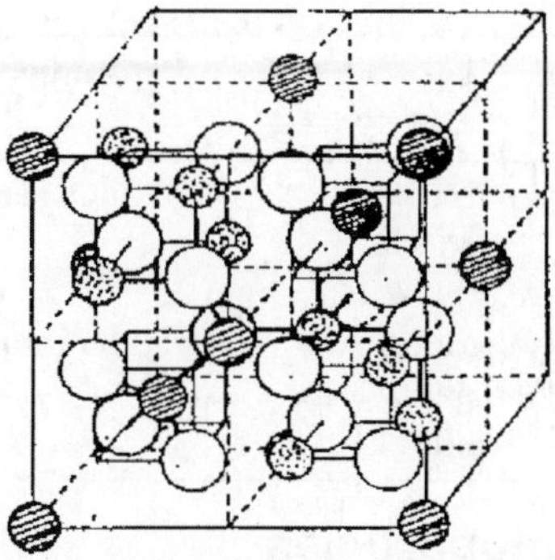

Fig. 3.1 Spinel structure AB_2O_4 (after Das and Ansari, 2009, with permission from *J. Sci. Ind. Res.*).

Synthesis

NMs of different geometries are exemplified by (i) metal nps; (ii) metal oxide nps; (iii) carbon NMs; (iv) polymer NMs; and (v) ncs. Some more advanced nps are dots, rods, spindle, tetrapod wires, tubes, ribbons, nanobelts, nanocables (Zhu *et al.*, 2003; Das and Ansari, 2009), nanosheets and nanobuds (Table 3.1). NMs can be synthesized by means of sol-process, micelles, sol-gel sol-process, chemical precipitation, hydrothermal synthesis, pyrolysis and vapour deposition.

Table 3.1 Some NMs of different geometries (after Das and Ansari, 2009, with permission from *J. Sci. Ind. Res.*).

Nanostructures (Dimensions)	Shape	Examples
Zero-D	Symmetric, isotropic, spherical, cubicle, tetradecahedron and tetrahedral	Spherical semiconductor metal oxides and metal nanocrystals.
One-D	Rods, wires, papers	Group II-IV semiconductors (CdS, CdSe, CdTe, ZnSe); Group III-V semiconductor (Ge, PbS, PbSe, MnS); transition metal oxides (ZnO, TiO_2, $W_{18}O_{49}$, Mn_3O_4, V_2O_5, $BaTiO_4$ and Fe_3O_4); carbon nanotube and semiconductor nanowires.
Two-D	Disks, polygonal plates	Transition metal sulphides Cu_2S, NiS, metal nanodisks, Gd_2O_3 nanodisks, Fe_2O_3 trigonal prisms(D_3b) CdS, CdSe, CdTe, MnS, ZnSe

Contd...

Table 3.1 *Contd...*

Nanostructures (Dimensions)	Shape	Examples
Three-D	Flower-like morphology consisting of uniform nanorods (diam 100-400 nm)	ZnO nanostructures
Multipod structures of semiconductors	Bipods (C_2v), tripods (C_{3v}), and tetrapods (Tá), star shaped dendritic	PbS (0h), CdSe, CdTe

Metal and Metal Oxide Nanoparticles

Reproducible synthesis of NM yields nanostructures of a specific size. Yang *et al.* (2006) described antioxidative copper nps. Song *et al.* (2004) reported that seeding and autocatalytic reduction of platinum salts in an aqueous surfactant solution containing ascorbic acid (aa) as reductant produces dendritic NMs. Photocatalytic reduction of platinum salts by SnP can be achieved by exposing to light and an electron donor aa.

Metal nps can be prepared by introducing metal salts in reversed micelles. Strong reducing agents such as $NaBH_4$, N_2H_4 and H_2 gas can be used. Pt, Rh, Pd, Ir, Ag, Au, Cu, Co, Ni and bimetallics FeNi, Cu_3A_u, CoNi have been successfully synthesized in this way. Normal micelles (oil droplets in water) serve as a polymer that controls np size. Use of micelles yields Cu nps. Small pored Ni and Au wires have been made from 300-nm channel alumina templates and 140-nm silica spheres (Li *et al.*, 2003). Porous wires are made from anodized alumina or polycarbonate membranes containing spheres of silica or polystyrene. Metal nanocrystals can be prepared by photochemical, electrochemical, biochemical or thermochemical methods (Zhou *et al.*, 2006).

Variously shaped (monopod, bipod, tripod and tetrapod) gold nps have been prepared (Chen *et al.*, 2003). Shin *et al.* (2003) prepared gold nps by citrate reduction. Palladium (Pd) nps could be electrodeposited from aqueous solution of $PdCl_2$ in 0.1M HCl (Bera *et al.*, 2004). Light has an important role in shaping nps. Whereas gold nanorods can be made spherical by using femtosecond laser pulses, spherical silver-silver nps have been transformed into silver nanorods.

Versatile semiconductor zinc oxide with a wide direct-band gap of 3.37 ev has attracted applications in sensors, optoelectronic devices, solar cells and field emitters. Li *et al.* (2007) synthesized flower-like ZnO microsctructures with several different sizes and shapes via a simple aqueous solution route by using $ZnCl_2$ and NaOH as reactants and triethanolamine as modifying agent. Zhong *et al.* (2008) synthesized porous plates of ZnO. Porous single crystalline ZnO nanodisks have also been synthesized (Jayan *et al.*, 2008).

Thermal evaporation and hydrothermal, solvothermal, sol-gel and electrochemical deposition methods have been used to synthesize one-D and 2D ZnO NMs. In thermal evaporation method, use of high temperature (>900°C) and a gold catalyst is needed for making 3D ZnO NMs. Mridha *et al.* (2008) developed an activated carbon assisted method to synthesize ZnO nps having a higher surface area than those synthesized without activated carbon.

Zhou *et al.* (2005) synthesized 3D flower-like ZnO NMs by a hydrothermal method in which KNO_3 is used as inorganic mineralizer. Preparation of ultrafine particles of Fe_3O_4 could be accomplished by means of a hydrothermal process using single crystal and ceramic powder. Size controlled Fe_3O_4 (magnetite) nps (diam, 4 nm) were made by reaction of iron (III) acetylacetonate [Fe(acac)$_3$] in phenyl ether in presence of alcohol, oleic acid and oleylamine (Sun *et al.*, 2002).

Metal nps can also be formed inside reverse micelles by hydrolysis where metal alkoxide dissolved in oil reacts with water within the droplets. Nanowires of $BaCO_3$ and $BaSO_4$ have been made by using reverse micelles.

Carbon Nanomaterials

Carbon nanowires and nanotubes made of organic or inorganic materials, show highly valuable electronic, mechanical and structural properties. Carbon nps of several different geometries such as cylindrical nanotubes and fullerenes attract applications in medicinal chemistry and material science. The archetype of fullerene family is carbon 60 (C_{60}; diam, 1 nm); fullerenes are closed cage carbon molecules with three coordinate carbon atoms; they have potential use as lubricants or catalysts and in drug delivery (Das and Ansari, 2009).

Carbon nanotubes (CNTs) are hollow cylinders (diam. 100 nm) of carbon atoms consisting of graphite sheets or cylinders. Depending on the arrangement of hexagon rings along the tubular surface, CNTs can be either metallic or semiconducting. The best applications of CNTs are as field effect transistors, nanotube interconnects and nanosensors. Depending on growth process, single walled nano tubes (SWNTs) or multiple walled nano tubes (MWNTs) may be formed SWNTs usually arrange in bundles of a few scores of tubes in a hexagonal lattice, whereas MWNTs are concentric arrangements of several tens of tubes (inner tube distances, ~ 3.41 A°). Unique properties of SWNTs have prompted exploration of their potential application in electrostatic discharge, structural reinforcement and thermal dissipation. A SWNT is nothing but a one atom-thick sheet of graphite resembling chicken wire rolled seamlessly into a tube. Semiconducting nanotubes can be electrically switched on or off as field effect transistors (Ciuparu *et al.*, 2004). Fabrication strategies make use of arc-discharge, laser ablation, chemical vapour deposition, templated synthesis and/or high pressure experiments (Beckyarova *et al.*, 2003).

A stable carbon nanobuds structure (Fig. 3.2) is an interesting material that combines allotropes of carbon: CNTs and fullerenes; fullerene-like buds are covalently bonded to outer sidewalls of underlying CNT. The nanobud combines useful properties of both fullerenes and CNTs, and is an excellent field emitter.

Fig. 3.2 Nanobud (after Das and Ansari, 2009, with permission from *J. Sci. Ind. Res.*).

Bimetallic Nanocrystals

Bimetallic nps of $Fe_{58}Pt_{42}$ were made by reduction of platinum acetylacetonate [$Pt(acac)_2$] and decomposition of $Fe(CO)_5$ in octyl ether solvent under air-free conditions. $Pt(acac)_2$, 1, 2 hexadecanediol and dioctyl ether can be mixed followed by heating to 100°C. Oleic acid, oleylamine and $Fe(CO)_5$ are added and heated to reflux at 297°C for 30 min. Good quality $Zn_xCd_{1-x}Se$ nps have been made by Kijima *et al.* (2007) at high temperature by incorporating appropriate quantities of Zn and Se into pre-prepared CdSe nps. CdTe nps show a cubic zinc blend structure that reorganizes into crystalline nanowires (Tang *et al.*, 2002).

Sulphide Nanoparticles

Samokhvalov *et al.* (2003) synthesized CdS quantum particles assembled at predetermined distance from a gold substrate. Two kinds of CdS particulate film had been generated *in situ* by Pan *et al.* (1997) by exposing stearic acid Langmuir monolayer at air aqueous $CdCl_2$ interface to hydrogen sulphide gas: particulate films consisting of oriented rod-like nps and of dot-like nps were made and formed a stripe-like domain with straight edges aligned with 6-fold symmetry. Hu *et al.* (2004) synthesized ZnS nanobelts through H_2 assisted thermal evaporation. These nanobelts were grown on a Si substrate and assembled spontaneously into a network structure. Zhu *et al.* (2003) synthesized nanocable aligned ZnS tetrapod nps; they mixed ZnS and carbon powders (precursors) and placed them in an inner graphite tube enclosed within an outer quartz tube. Precursors were heated at 900°C for 1 h in a N_2 atmosphere to prepare ZnS-C nanocables, followed by heating at 1100°C for 1 h to fabricate nanocable aligned tetrapods.

Shchukin *et al.* (2003) synthesized nanosized magnetic ferrite particles within hollow polyelectrolyte capsules. Different ferrites ($CoFe_2O_4$, $ZnFe_2O_4$, $MnFe_2O_4$) and magnetite could be generated from corresponding salts exclusively inside poly(styrenesulphonate)/ poly(allylanine hydrochloride) polyelectrolyte capsules of diam. 10 μm. Polyelectrolyte capsules with synthesized ferrite (magnetite) particles show high enough magnetic activity to be easily manipulated in water solution by an external magnetic field.

Silica Nanoparticles

Suzuki *et al.* (2004) made mesoporous silica nps by means of a cationic surfactant as a templating agent and a nonionic surfactant as a suppressant of grain growth. Hydrolysis of tetraethoxysilane with HCl followed by assembly of cationic surfactant micelles and anionic silicate at a basic condition arranged mesopores (20 nm) hexagonally in nps.

Polymer Nanomaterials

Nanoscale p-conjugated organic molecules and polymers are being used in biosensors, electrochemical devices, single electron transistors, and as nanotops in field emission displays. Zhang *et al.* (2006) studied wire, ribbons and sphere-like nanostructures of polypyrrole obtained by solution chemistry methods in presence of surfactants with oxidizing agents.

Silver/polypyrrole ncs were developed by Dallas *et al.* (2007) through interfacial polymerization.

Nanocrystals of coordination polymers. Coordination polymers are metal-ligand compounds usually prepared as:

$$\text{Metal cation + bridging ligands} \xrightarrow{\text{self-assembly}} \underset{\text{(2-100 nm)}}{\text{Nanocrystals}} \xrightarrow{\text{(inorganic parts)}} \text{bulk crystals}$$

The characteristics of coordination polymers (2-100 nm) differ from those of bulk coordination polymers.

Dendrimers. Dendrimers are repeatedly branched materials now widely used in nanomedicine as nanoscale catalysts and reaction vessels, chemical sensors, and for delivering drugs into cells – multiple molecular hooks on their surface facilitate attachment to cell-identification tags, fluorescent dyes, enzymes and other molecules. Two structural types of dendrimers are: globular, with a central core from which branches radiate; and (ii) those lacking central core; these consist simply of a series of highly branched polymers.

Nanocomposites

These are multiphase materials containing two or more very dissimilar components mixed at nanometer scale. The relevant phase may be inorganic-inorganic, inorganic-organic, or organic-organic; the resulting material can be amorphous, crystalline or semi-crystalline (Das and Ansari, 2009). Ncs provide an innovative route to synthesis of complex advanced materials. They are also used in sporting equipment such as racing cars and bicycles; and in cutting tools, laser mirrors and aerospare components. Nanotube-polymer ncs have potential application in aerospace science. They have been derived from polyethylene oxide (PEO) and polyethyleneimine (PEI). Ncs derived from the nylon family, polyester, polypropylene or polystyrene have found commercial applications as automobile parts and coatings for packaging and fire retarding materials. Some notable CNT epoxy composites are hyrdrocarbon polymer composites, conjugated polymer composites, polyacetonitrite composites, composites of CNTs with polycarbonates, fluoropolymers, polyvinyl alcohol, polyethylene glycol, silicon polymers, polyelectrolytes, polyesters, polyamides, polyvinyl carbazole, poly (p-phenylene benzobisoxazole), phenoxy resin and natural rubber.

Bionanoparticles

BioNps are naturally produced nanoscale entities. Rising demands for nanoparticulate products (viruses, plasmids, protein np and drug delivery systems) have highlighted the requirement for predictable and controllable processes. Living organisms are built of cells typically 10 µm across). Proteins (5 nm) compare well with smallest man-made nps. This implies the potential of using nps as ultrasmall probes that would allow us to critically examine the cellular machinery without disturbing it significantly (see Taton, 2002). Indeed, a thorough understanding of biological phenomena at the nanoscale is a major driving force behind the explosive development of nanotechnology (Whitesides, 2003). Liposomes are lipid-based nps being used extensively in pharmaceutical and cosmetic industries because of their capacity for releasing their cargo of drugs etc. inside cells (see Vyas and Kumar, 2010).

Characterization

NMs are characterized by using a variety of techniques such as scanning electron microscopy (SEM), atomic force microscopy (AFM), dynamic light scattering (DLS), X-ray photoelectron spectroscopy (XPS), X-ray diffractometry (XRD), fourier transform infrared spectroscopy (FTIR), matrix assisted laser desorption time of high mass spectroscopy (MALDI-TOF), UV visible spectroscopy, ^{1}H and ^{13}C NMR, high resolution mass spectrometry, and superconducting quantum interface device (SQUID) for reading out signal from detector. Np tracking analysis (NTA) makes it possible to directly track Brownian motion. This method allows sizing of individual nps in solution. Scanning probe techniques are simple, inexpensive and capable of direct patterning on a surface. AFM has proved a robust tool for imaging and lithography of nps. Grobenly *et al.* (2006) used scanning tunnelling microscopy (STM) for imaging on gold surface.

SEM and TEM images of ZnO revealed that flower-like ZnO bundles consisted of prism-like or petal-like branches (Li *et al.*, 2007). Scanning transmission electron microscopy (STEM), high resolution transmission electron microscopy (HRTEM) and selection area electron diffraction (SAED) confirmed that nanodiscs of ZnO are single crystalline and porous (Jayan *et al.*, 2008). MnO_2 nanosheets have been characterized by X-ray absorption spectroscopy.

The diameter of nanosized silica particles can be measured by using photon correlation spectroscopy (PCS), AFM and TEM (Costa *et al.*, 2003).

Photoinduced aggregation of polymer nps in dilute nanoaqueous dispersion revealed that there exist two limiting regimes for aggregation: diffusion limited cluster-cluster aggregation (DLCA) and reaction limited cluster-cluster aggregation (RLCA). Whereas DLCA produces aggregates with fractal dimension D approximately 1.7-1.8, RLCA generates more uniform aggregates with a higher fractal dimension (2.0-2.5). Some MgO_2 nanostructures such as those of fish bone fractal nanostructures induced by Co have been synthesized (see Das and Ansari, 2009).

LANTHANIDE-DOPED LUMINESCENT NANOMATERIALS

There is considerable interest in developing novel lanthanide-doped luminescent nanomaterials showing high photostability and luminescent quantum efficiency. Various different synthetic routes, e.g., sol-gel, solid state, hydrothermal, thermal decomposition, and microwave could possibly be adopted to synthesize these nanomaterials both in powder and colloidal forms (see Venkataramanan *et al.*, 2007, 2008, 2008a). The structure-property relationships of such nanomaterials should be investigated by applying several microscopic and spectroscopic techniques. These nanomaterials with appropriate Ln^{3+} doping acquire the ability to convert near infrared (NIR) radiation into high energy visible or ultraviolet radiation through a process known as upconversion. Following certain chemical modifications of their surfaces, these nanomaterials could possibly be used as luminescent tags to address issues in biomedicine.

Furthermore, there is interest in developing template-assisted synthetic routes to construct new **hollow nanomaterials** and **nanocomposites** with multifunctional capabilities such as manifesting both magnetic and luminescent property (see Venkataramanan *et al.*, 2009, 2009a).

PROBING NANOMATERIALS AT ULTRASHORT TIME SCALES

Pal *et al.* (2007, 2008, 2009) have studied exciton and spin dynamics in semiconductor nanostructures by employing modern laser spectroscopy in the subpicosecond to millisecond time scales. Some of the fundamental issues addressed by Pal *et al.* have been Mott transition for coherent and incoherent excitons, intersubband relaxation of n = 2 excitons (a Fano resonant state), coherent effects in resonant reflectivity measurements, dynamics of type-II excitons in core-multishell nanowires, and electron and nuclear spin dynamics in quantum dots. These researches have thrown useful light on physical understanding of these phenomena. For example, it has been clarified that for high density optical excitation of excitons in quantum wells, the phase space filling effect is more important than the Coulomb screening effect. This is why the excitons density saturates below the theoretical Mott density. Pal's work revealed a very long spin lifetime of about 1 ms of doped electrons in InP quantum dots. It also suggested the possible spin relaxation mechanism involved. Influence of nuclear spins on electron spin dynamics was also elucidated. He is currently investigating the optical properties of transition-metal-doped ZnO nanocrystals - dilute magnetic semiconductors which express ferromagnetism at room temperature. This research has some potential towards achieving effective integration of electronics, photonics and **spintronics**.

FUNCTIONAL PROPERTIES OF NANOMATERIALS

Gold Nanoparticles

Metal particles in colloidal state have attracted people since times immemorial, largely owing to their medicinal value and fascinating colours (Mishra, 2004; Freestone *et al.*, 2007). Colloidal gold has an envious therapeutic history rooted in the Indian subcontinent. The beautiful colours of metal nps have adorned glass windows in cathedrals in Europe.

The light absorption in metal nps occurs from localized surface plasmon resonance. According to the Drude-Lorentz model (see Thomas, 2009), the atoms in metals exist in a plasma state in which a core of positively charged nuclei is surrounded by a pool of negatively charged electrons and termed 'plasma electrons'. In the presence of an electromagnetic radiation, the electric vector displaces the free electrons and the electrostatic attraction of the nuclei restores the electrons to the original position. Because of the oscillating nature of the electric field of light, the electron oscillates over the surface with a characteristic resonance frequency (Kelly *et al.*, 2003). When the frequency of this oscillation matches with that of incident radiation, a resonance is created and results in the intense absorption termed 'surface plasmon (SP) absorption'. A good example is that of spherical gold nps which show a single surface plasmon band at around 520 nm, attributed to the collective dipolar oscillation of the electron cloud (Fig. 3.3).

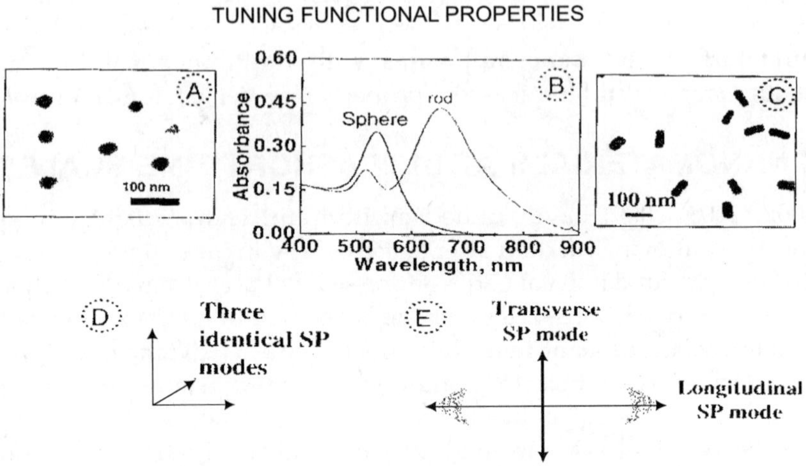

Fig. 3.3 (A) TEM images of gold nps (obtained by reducing $HAuCl_4$ using gallic acid at room temperature); (B) absorption spectrum of Au np and nanorod; (C) TEM images of gold nanorod (obtained by photochemical reduction); (D) surface plasmon oscillation of spherical gold nps and (E) the surface plasmon oscillation of gold nanorods in the transverse and longitudinal axes (after Thomas, 2009, with permission from *Curr. Trends in Sci.*, Plat. Jub.).

Depending on the dimensionality of the nanostructured materials, various different boundary conditions can be imposed on the electron plasma. Unlike the bulk metal, the electron cloud in nps is confined to a finite volume that is smaller than the wavelength of light. Hence, the frequency of oscillation of metal nps is governed chiefly by four parameters: number density of electrons, effective mass of the electron and the size and shape distribution of the charge. This makes it possible to tune the optical properties of noble metal nps by varying their size, shape and dielectric environment. According to Mie (1908), the total cross-section consists of scattering and absorption (often termed extinction) and given as summation over all electric and magnetic oscillations. The contribution of absorption and scattering depends on the size and shape of the particles.

Nanostructured metallic systems of noble metals e.g. silver, gold and copper, show localized surface plasmons and resonate at the visible range of the electromagnetic radiation. Typically, the surface plasmon (SP) band for spherical Au nps having diameter of ~20 nm is around 530 nm. The surface plasmon absorption of Au nps obtained by reducing $HAuCl_4$ using gallic acid at room temperature and gold nanorods prepared by photochemical reduction and their corresponding TEM images are shown in Fig. 3.3. A bathochromic shift in the λmax occurred from 530 to 550 nm on increasing the diameter of the nanosphere from 20 to 80 nm – attributed to the electromagnetic retardation in larger nps. With changing shape of the np, the surface electron density and hence the electric field on the surface varies, causing strong variations in the oscillation frequency of the electrons and generating different optical cross-sections (including absorption and scattering). Nanorods of gold and silver split the dipolar resonance into two surface plasmon bands in which the induced dipole oscillates along the transverse and longitudinal axes (Figs. 3.3B and E) (Link *et al.*, 2005). While the longitudinal surface plasmon band shifts to longer wavelengths with increase in aspect ratio, the status of transverse surface plasmon band is not significantly affected. These results suggest that the dimensionality has critical role in determining the plasmon resonance of metal nps. Au nps in the size range of 2-5 nm display interesting quantum size effects and can act as conductor, semiconductor or insulator depending on its dimension (Thomas, 2009). The surface plasmon band of Au np, a property of its metallic nature, broadens and dampens when its size falls below 5.0 nm (Templeton *et al.*, 2000). A sharp decrease in the intensity of surface plasmon band occurs in nps of diameter below 3.2 nm due to the transition from metallic to semiconductor/insulator behaviour. No surface plasmon band is seen in Au nps of core diameter less than 2 nm. Gold nps <5 nm in diameter are catalytically active for several reactions, such as the oxidation of CO (an automobile exhaust) at room temperature (Janssens *et al.*, 2007). In fact, the unique features of Au nps/nanorods create excellent opportunities for their use in biomedical field for diagnosis and imaging (Jain *et al.*, 2008). The efficient conversion of absorbed light in Au nps/nanorods into localized heat has been exploited for the laser photothermal destruction of cancer cells – Au nps/nanorods can be targeted to tumour site by conjugating them with bioactive molecules (Thomas, 2009).

Besides gold, several recent developments in the classical wet chemistry methods have made possible the synthesis of various anisotropic nanostructures of other elements possessing well defined shapes (for example, ellipsoids, rods, cubes, disks, tetrahedra, cylinders, pyramids, triangular prisms, and multipods) (Wiley *et al.*, 2005). The functional properties of various NMs have been well correlated with their size and shape.

Semiconductor Quantum Dots

When a bulk semiconductor is photoexcited, an electron is transferred from valence band to conduction band and an electron-hole pair, called 'exciton', is created. The lowest energy needed to generate such an exciton is termed the band gap energy. Excitons can be considered a hydrogen-like system in which the spatial separation between the charge carrier pair is termed exciton Bohr radius (a_B). This can be deduced from the Bohr approximation. When the physical dimensions of matter are comparable to or less than the exciton Bohr diameter ($2a_B$), a semiconductor becomes functionalally sensitive to the size and shape due to confinement of excitons. This phenomenon is described as quantum size or **quantum confinement effect**. For example, CdSe has a Bohr exciton radius of ~56Å; the electron and hole cannot achieve their desired distance when the diameter of nanocrystal is smaller than 112 Å (Nirmal and Brus, 1999). This situation can be considered a 'particle-in-a-box' system.

The confinement of electrons in semiconductor qds affects their electronic structure in two ways. The energy gap of semiconductor qds increases sharply with decrease in radius. Also the continuum observed in the conduction band and valence band in the case of bulk materials is replaced with discrete atomic like energy levels as the particle size decreases (Thomas, 2009).

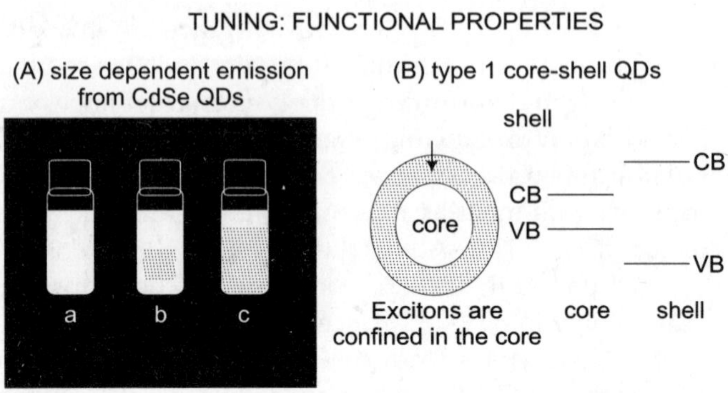

Fig. 3.4 (A) The size dependent emission from CdSe quantum dots having radius of (a) 2.6 nm (b) 3.7 nm and (c) 4.7 nm; (B) the energy levels of the valence and conduction bands of core and shell in type I heterostructure qds (after Thomas, 2009, with permission from *Curr. Trends in Sci.*, Plat. Jub.).

Quantum dots (qds; qdots) are fluorescent semiconductor nanocrystals that have attracted several applications in cell and animal biology (Michalet *et al.*, 2005). They serve as effective multimodal contrast agents in bioimaging (Fig. 3.5).

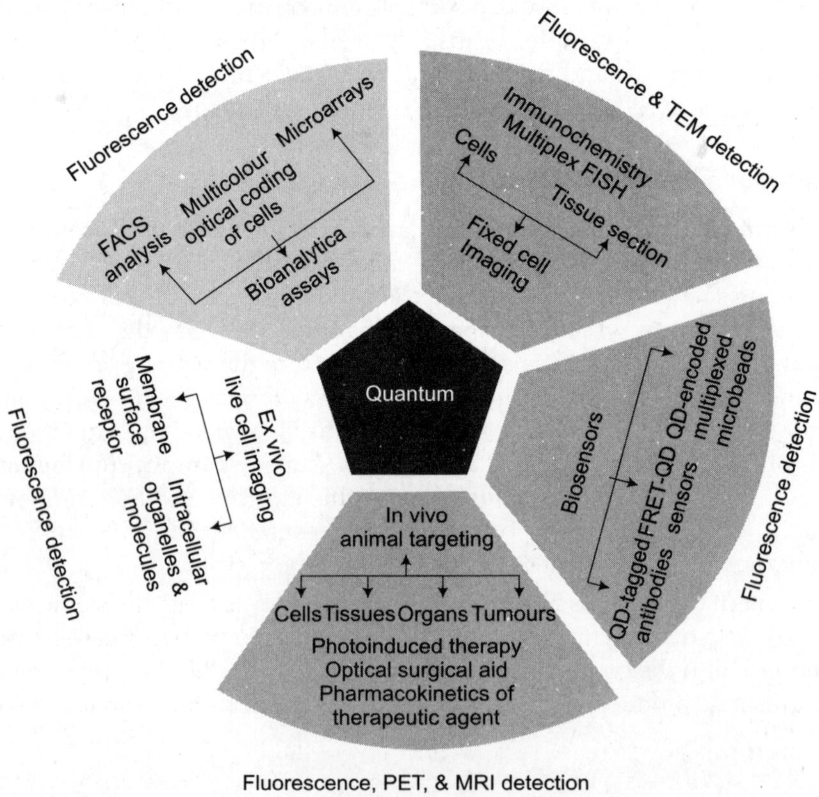

Fig. 3.5 Applications of quantum dots as multimodal contrast agents in bioimaging. (From Michalet, X. *et al. Science* 307: 538-544, 2005. Reprinted with permission from AAAS).

One good strategy for conferring photostability and improving the photoluminescence efficiency of qds is to overcoat them with an inorganic shell material of similar lattice parameters and higher band gap energy (E_g). Such core-shell systems in which the core is overcoated with large band gap shell material are described as type I heterostructure qds (Fig. 3.4). In these systems, the conduction band of the shell has higher energy than that of the core whereas the valence band of the shell has lower energy than that of the core. Consequently, both electrons and holes (exciton) are confined in the core. Some shell materials being used for overcoating CdSe qds include zinc sulphide (ZnS), zinc selenide (ZnSe) and cadmium sulphide (CdS) (Bleuse *et al.*, 2004). Among these, ZnS is the most widely used in view of its higher band gap energy, allowing efficient confinement of excitons in the core. Overcoating prevents photoinduced charge transfer in bare qds and also prevents the leakage of core materials thereby avoiding any toxicity inside the cell when used in biological applications. However, core-shell qds of greater shell thickness are less effective for biological applications in view of the decrease in luminescence quantum yield.

Carbon Nanotubes

Carbon nanotubes (CNTs) represent an allotrope of carbon. An ideal CNT is a hexagonal network of carbon atoms rolled into a seamless cylinder, with a diameter of just a few nanometers and a tube length of thousands of nanometers. CNTs are quasi one-dimensional nanostructures that show unique electrical properties suitable for fabrication of nanodevices (Rao and Govindraj, 2002; Dresselhaus *et al.*, 2007). CNTs can be of two types: single-walled (SWCNTs) and multi-walled (MWCNTs). Although, in principle, the electronic properties of 'ideal' CNTs depend on their diameter and chirality, certain impurities such as pentagons, heptagons, vacancies or dopant can drastically modify the electronic properties (Charlier, 2002).

The welcome electronic properties of CNTs emanate from the quantum confinement of electrons normal to the nanotube axis. In the radial direction, electrons become confined by the monolayer thickness of the graphene sheet and periodic boundary conditions prevail around the circumference of the nanotube. Electrons can only propagate along the nanotube axis because quantum confinement and their wave vectors point in this direction. The resulting number of 1-D conduction and valence bands depend on the standing waves around the circumference of the nanotube (Thomas, 2009).

CNTs have been successfully grown by three methods: (i) arc discharge, (ii) laser furnace and (iii) chemical vapour deposition (CVD). Commercially available nanotubes contain a mixture of metallic and semiconductor tubes, one-third being metallic and the rest semiconducting. CNTs constitute versatile building blocks for the design of optoelectronic devices, but their availability in a scalable form having (i) uniform diameter and bandgap, and (ii) metallic/semiconducting character, is a major limitation.

HYBRID NANOMATERIALS AND THEIR FUNCTIONAL PROPERTIES

The functional properties of materials in the nanometer dimension originate chiefly from the quantum confinement effect or the so-called '**particle-in-a-box**' model. Dimensional confinement of charge carriers in nanoscale materials introduces strong modifications in their **density of states** (DOS) giving rise to novel size and shape dependent properties. Based on dimensional confinement of electrons, NMs have been generally classified as 2D, 1D or 0D. The confinement of electrons occurs only along one axis in the case of 2D NMs (e.g., quantum wells). In 1-D NMs, electrons become confined along two dimensions and can freely move only in one dimension; the best examples are semiconductor nanowires and CNTs. When the materials show confinement of electrons along all the three dimensions, they are termed quantum dots (qds). The dramatic blue shift observed in the absorption and emission spectra of semiconductor qds, with decrease in size, is one of the direct observations of quantum size effect (Thomas, 2009).

It has become possible to tune the functional properties of materials without changing their chemical constituents. This has opened up exciting possibilities in physics, chemistry, biology, medicine and engineering. The major focus of nanoscience today is to design materials with novel properties and utilize them for the miniaturization of devices with better performance. Among the various nanoscale building blocks, metal and semiconductor nps and carbon nanotubes have received sustained attention.

The functional properties of NMs can be fine-tuned by a stepwise integration of nanoscale building blocks into organic-inorganic and inorganic-inorganic hybrid systems. In the hybrid NMs, the functional properties of the individual nanoscale building blocks can integrate with each other, generating newer properties fundamentally different from those of isolated components. The novel optical properties of NMs generate excellent possibilities in biomedicine for diagnosis, imaging and therapeutics. Hybrid NMs with diverse functionality are also facilitating the design of a new generation of nanophotonic and optoelectronic materials (Thomas, 2009).

Exciting technological innovations in the past few years played a crucial role in the controlled organization of materials with micrometer scale precision leading to the miniaturization of devices. Designing of materials in the nanometer scale has enabled the exploration of a new size regime and one in which materials show properties quite different from that of the bulk. In this regime, size and shape have a greater role in tuning their properties than those of the constituent elements and their chemical composition (Kreibig and Vollmer, 1995; Daniel and Astruc, 2004; Burda *et al.*, 2005). Some better known examples of nanostructured materials are metal and semiconductor nps and nanostructured carbon based systems such as fullerenes, SW and MWNTs. The noble metal np based systems in the nanoscale exhibit strong optical absorption from the localized surface plasmon resonance (yellow, red and blue colour for Ag, Au and Cu, respectively). Another unique feature of nanostructured materials, which make them a suitable candidate in catalytic processes, is the larger surface to volume ratio (Rao *et al.*, 2002).

Recent methods of characterization based on electron and probe microscopy have enhanced critical understanding of the crystallographic properties of NMs. Spectroscopic methods have revealed details of the optical and electronic properties of these materials. Among various NMs, gold nps and nanorods, semiconductor qds and SWCNTs have been widely used as components for the design of hybrid nanostructures.

NMs can be functionally tuned by stepwise integration of nanoscale building blocks (nps, nanorods, nanotubes, etc.) into hybrid NMs When integrated as hybrid NMs, the functionalities of nanoscale building blocks augment each other to create novel properties. Thus, hybrid NMs may either possess new properties different from those of isolated components, or show complementary properties useful for performing specific functions. Some strategies for designing hybrid nanostructures and representative systems include organic-inorganic and inorganic-inorganic hybrid systems. Organic-inorganic hybrid systems can be designed by functionalizing photo or electroactive molecules onto metal or semiconductor NMs wherein electron/energy transfer processes may operate. Inorganic-inorganic hybrid NMs are produced through the hierarchical integration of metal and semiconductor building blocks into higher order assemblies. Properties of these hybrid nanostructures can be tuned by varying the length of the spacer group. Hybrid NMs with heterojunction have been designed by bringing together (i) metal and semiconductor nps, (ii) dissimilar metal nps and (iii) dissimilar semiconductor nps. Such hybrid NMs show properties quite different from those of isolated components. In fact, they often possess more promising functional properties than individual building blocks.

Stable metal as well as semiconductor nps have been produced by capping with a monolayer of organic molecules. The ability of functional groups such as quaternary ammonium halides, amines, or thiols to bind to the surface of nps has been gainfully utilized for organizing organic molecules around them (Fig. 3.6). The commonest method for the functionalization of organic molecules onto the surface of metal nps is based on a two-phase synthesis reported by Brust *et al.* (1994) and Templeton *et al.* (2000). Such monolayer-protected clusters (MPCs) are seen as 3-D assemblies having a core-shell structure (metal core covered with ligand shell; (Figs. 3.6 A-C). Solubility of MPCs can be tuned from nonpolar to polar medium by varying the nature of the stabilizer layer. Likewise, their optoelectronic properties can be adjusted by linking photoactive molecules on to the surface of nps. Appropriate techniques are available for incorporating a desired number of chromophores onto the np. For instance, metal nps having mixed monolayers are synthesized by means of ligand (place) exchange reactions reported by Song and Murray (2002). The addition of pre-synthesized nps possessing thiol bearing ligands to a solution of another thiolate molecule showing diverse functionality leads to the partial substitution of the protecting shell on the np surface (Fig. 3.5D). It is this strategy that allowed the functionalization of a desired number of photoactive or electroactive molecules around the metal np (Fig. 3.6E-G).

Interesting physical processes emerge when photoactive or electroactive molecules are linked on to metal nps (Thomas and Kamat, 2003). The most remarkable are electron/energy transfer processes, which can be further modulated by varying the size, shape and chemical constituents of NMs. Indeed, Au nps in the size range of 2-5 nm tend to behave as conductor, semiconductor or insulator depending on their dimensions due to quantum size effects (Kamat, 2002). Optoelectronic properties of hybrid NMs are tuned by anchoring chromophores on to Au nps in this size range. A variety of chromophores have already been functionalized on the surface of Au nps (Fig. 3.6E-G) and proposed as active components in light-harvesting systems. Fig. 3.7 summarizes the significant photophysical events occurring in these hybrid NMs.

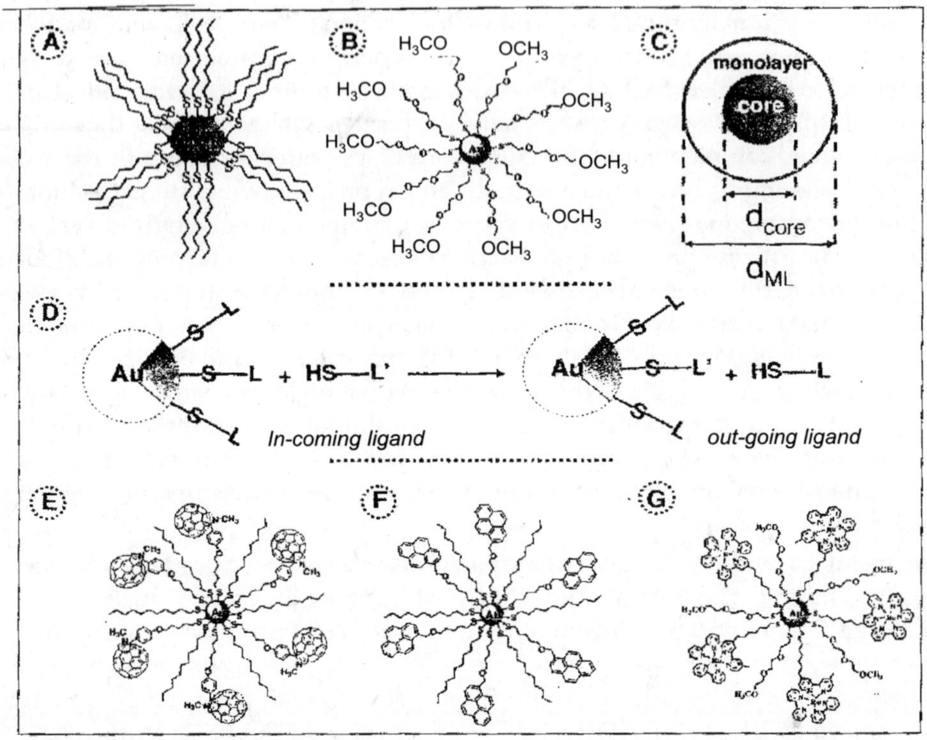

Fig. 3.6 Monolayer-protected clusters (MPCs) having metal core covered with (A) alkyl thiol and (B) monothiol derivative of triethylene glycol as ligand shell; (C) Schematic sketch of a core-shell hybrid nanostructure; (D) sketch of ligand (place) exchange reaction for preparing mixed monolayers; (E-G) gold nps having mixed monolayers of (E) fullerenethiol and dodecanethiol; (F) pyrenethiol and dodecanethiol; (G) ruthenium trisbipyridine thiol and monothiol derivative of triethylene glycol (after Thomas, 2009, with permission from *Curr. Trends in Sci.,* Plat. Jub.).

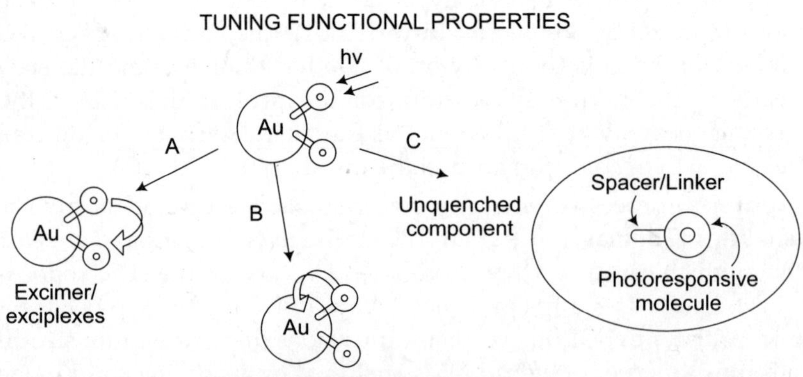

Fig. 3.7 Excited-state deactivation processes in fluorophore-metal hybrid NMs (after Thomas, 2009, with permission from *Curr. Trends in Sci.,* Plat. Jub.).

Semiconductor qds linked with photoactive molecules are used as hybrid NMs for fluorescence resonance energy transfer (FRET) analysis. Medintz *et al.* (2005) and Sapsford *et al.* (2006) used qd based hybrid NMs for the detection of proteins and DNA. As compared to organic fluorophores, qds have several unique photophysical properties that make them attractive as biolabels: (i) broad absorption spectra and large Stokes shifts; (ii) molar extinction coefficients that are much higher than organic dyes; (iii) size-tunable photoluminescent emission with fairly high quantum yield and lifetime; and (iv) high resistance to photobleaching and chemical degradation. The native organic ligands used as a monolayer for protecting the qds can be partially replaced with a bioactive/photoactive molecule. The emission properties of qds can be size-tuned to achieve better spectral overlap with a specific acceptor dye. The FRET efficiency can be increased by loading a central qd donor with multiple fluorophores. Unique photophysical properties of qds generate newer opportunities for the design of FRET-based biological assays, providing much better distance resolution than achievable with traditional donor acceptor FRET systems. The only cause of concern with qds may be their possible toxicity, especially for biomedical application. Newer generation qds such as InP are now emerging as alternatives for this purpose.

Plasmon Coupling in Hybrid Nanomaterials

The transport of optical energy using materials significantly smaller than the wavelength of light is a most challenging issue in the miniaturization of photonic components. However, nanostructures can convert photons into surface plasmons that address this limitation. Within the propagation length, the surface plasmon modes can be decoupled to light, offering tremendous opportunities in the design of nanoscale optical and photonic devices such as metal-np based plasmon waveguides. This goal can be realized by designing higher order hybrid NMs, for example, 1-D arrays of noble metal nps with defined particle spacing. Electron beam lithography is used for the construction of higher order nanostructure (Girard and Dujardin, 2006). These authors showed that the optical properties of metallic nps can be fine tuned by their controlled organization into periodic arrays.

Closely packed 1-D arrays of Au nps can, in principle, act as (i) guides of electromagnetic radiation (**waveguides**) allowing miniaturization of devices below the diffraction limit, and (ii) **interconnectors** in optical and photonic devices. Unfortunately, the isotropic nature of spherical Au nps prevents the selective binding of molecules on surfaces and limits the possibility of designing 1-D array of NMs by chemical functionalization methods. In contrast, the anisotropic features of Au nanorods allow their assembly in several different orientations. Pramod and Thomas (2008) attempted to organize Au nanorods by using electrostatic/ supramolecular/ covalent approaches. Recent work has included the (i) linear organization of Au nanorods using biotin-streptavidine connectors and lateral organization through electrostatic interactions by varying the pH of the medium (Murphy *et al.*, 2005), (ii) longitudinal assembly through covalent functionalization by using dithiols (Joseph *et al.*, 2006; Prasad and Thomas, 2008), (iii) cooperative intermolecular hydrogen bonding by using 3-mercaptopropionic acid and electrostatic interaction by using cysteine and glutathione (Fig. 3.8) (Sudeep *et al.*, 2005), and (iv) end-to-end electrostatic assembly of Au nanorods on MWCNTs (Correa-Duarte *et al.*, 2005).

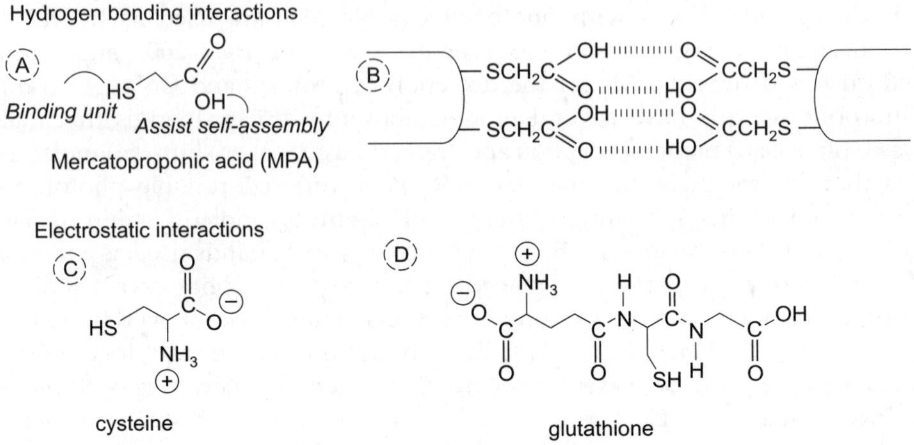

Fig. 3.8 Molecules used for organizing Au nanorods using hydrogen bonding and electrostatic approaches (Thomas, 2009, with permission from *Curr. Trends in Sci.*, Plat. Jub.).

Recent work has focused on the design of Au nanorod dimers and critical study of plasmon coupling in the above systems as a function of their distance and orientation. The plasmon coupling in Au nanorod dimers linked through rigid molecules (e.g., 1, 2-phenylenedimethanethiol) turned out to be quite pronounced, confirming that the nature of the linker group plays a critical role in plasmon coupling of metal nps. Such hybrid metal nps are promising components in the design of nanoscale devices, a good example is using stable Au nanorod chains with effective dipolar overlapping as elements in the design of plasmonic wave guides. Light waves can be channelled through hybrid NMs depending on the assembly of Au nanorods. Rigid Au nanorod dimers also have potential as nanoscale interconnectors in future molecular electronics devices – the distance and orientation between the nanorods can be varied by choosing proper linker groups.

Heterojunctions in Hybrid Nanomaterials

Hybrid NMs containing heterojunctions have potential applications in optoelectronic devices as they exhibit novel properties fundamentally different from those of nanoscale components (Liz-Marzan, 2006). Heterojunctions in hybrid NMs can be created by bringing together NMs of (i) dissimilar metals and (ii) dissimilar semiconductors and (iii) semiconductor and metal.

Heterojunctions of dissimilar metal nps tend to form alloys showing novel optical and electronic properties. Interestingly when two dissimilar semiconductor NMs come in contact, the heterojunction created assists the spatial separation of charge carriers (excitons) formed upon photoexcitation. This is done by tuning the energy levels of the valence and conduction bands of semiconductors. If the valence and conduction band position of the semiconductor 'S_1' is lower (or higher) than semiconductor 'S_2', then one carrier is mostly confined in 'S_1', and the other in 'S_2' (Fig. 3.9). Core-shell nanostructures of this type are termed type-II nanostructures. These core-shell heterostructures are synthesized by **molecular beam epitaxy** (e.g., GaSb/GaAs as well as chemical methods (e.g., CdTe/CdSe and CdSe/ZnTe) (see Yu *et al.*,

2005). Photoexcitation of CdTe/CdSe Qds leads to the spatial separation of charge carriers (excitons). This hole is mostly confined to the CdTe core while the electron is mostly in the CdSe shell. The situation is reversed in CdSe/ZnTe qds, because the energy levels of the valence and conduction bands of the shell are higher than that of the core. In CdSe/ZnTe Qds, the electron resides mostly in the CdSe core, whereas the hole is mostly in the ZnTe shell. Emission in type-II nanostructures comes from the radiative recombination of the electron-hole pair across the core-shell interface. Type-II qds emit at energies that are smaller than the band gap of either materials; the photoemission from CdTe/CdSe and CdSe/ZnTe qds occurs at longer wavelengths than from the corresponding core and shell components (Thomas, 2009). The emission properties of type-II qds can be fine-tuned by altering the shell thickness and core size. The emission spectra from CdTe/CdSe qds e.g., can be tuned in the range from 700 nm to over 1000 nm by changing the core size and shell thickness. Heterojunction hybrid NMs of dissimilar semiconductors are potentially promising materials for photoemission band and longer lifetimes.

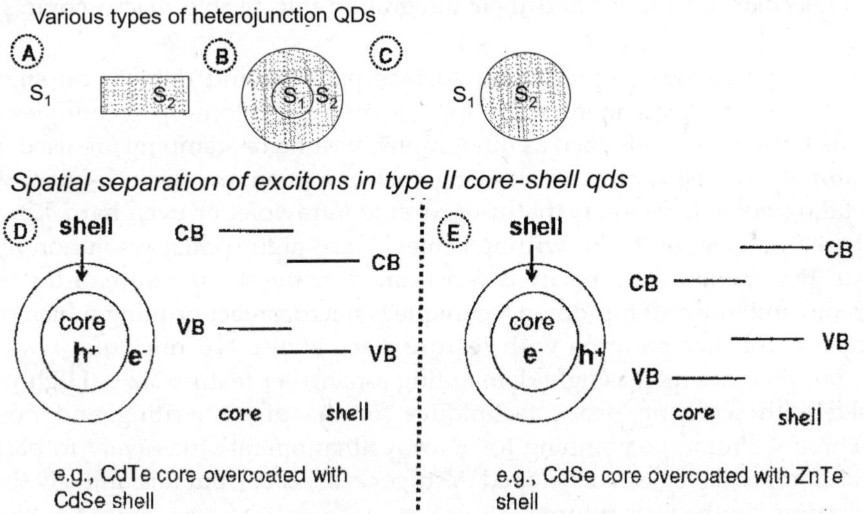

Fig. 3.9 (A-C) Various types of heterojunction hybrid NMs of dissimilar semiconductors and (D-E) schematic representation of the spatial separation of excitons in type II core-shell qds and the corresponding energy levels of the valence and conduction bands of the core and shell (after Thomas, 2009, with permission from *Curr. Trends in Sci.*, Plat. Jub.).

ENGINEERING OF NANOSTRUCTURES

Although the microelectronics industry has been producing smaller and smaller devices, it will soon reach their fundamental limits. One likely alternative route to even smaller functional systems with nanometer dimensions involves the autonomous ordering and assembly of atoms and molecules on atomically well-defined surfaces. This strategy combines ease of fabrication with precise control over the shape, composition and organization of the surface structures formed. Attempts are underway to understand the mechanisms controlling the self-ordering phenomena so that the self-assembly and growth processes can be tailored to make a wide range of surface nanostructures from metallic, semiconducting and molecular materials.

The invention of the scanning tunnelling microscope has made it possible to image and manipulate individual molecules and atoms (see Eigler and Schweizer, 1990). Several nanostructured materials with extraordinary functional properties, such as semiconductor qds and CNTs, have become available in recent years (see Dekker, 1999) and our understanding of how structural features regulate the function of such small systems has grown significantly (Barth *et al.*, 2005).

All these welcome developments represent different aspects of nanotechnology which focuses on the nanometer scale and can potentially meet the emerging needs of industries which thrived on continued miniaturization but are now facing serious difficulty in maintaining the trend, particularly in microelectronics and magnetic data storage (Thomson and Best, 2000; Bohr, 2002). Unfortunately, development of nanosystems and nanodevices with desirable performance characteristics cannot meet the emerging needs unless nanodevices are affordable in respect of costs and are economically viable. What we badly need is to develop effective methods for the controlled large scale fabrication of reasonably priced functional atomic or molecular assemblies and their integration into usable macroscopic systems and devices.

Two basic approaches to producing surface patterns and devices on substrates in a controlled and repeatable manner are the 'top-down' and 'bottom-up' techniques (Gates *et al.*, 2005). For the former, methods such as lithography, writing or stamping are used to define the desired features. The relevant tools allowing such precision are electron-beam writing, and advanced lithographic techniques that use extreme ultraviolet or even hard X-ray radiation. Methods based on electron-beam writing achieve very high spatial resolution at reasonable capital costs, but operational capacity is constrained by the serial nature of the process. One versatile, rapid and low-cost top-down technique is microcontact printing which uses soft and hard stamps to transfer patterns with feature sizes above 100 nm onto a wide range of substrates, but becomes increasingly demanding for smaller feature sizes. High precision can be achieved with scanning probe techniques for the direct writing and positioning of individual atoms. Prototype scanning force arrays that operate massively in parallel and so multiply throughput have been developed (Vettiger *et al.*, 2002), but it is unlikely that these will find industrial use in the near future.

The bottom-up techniques exploit self-processes for ordering of supramolecular or solid-state architectures from the atomic to the mesosopic scale. All bottom-up strategies for fabrication of nanostructures at surfaces are essentially based on growth phenomena. Atoms, molecules, or both are deposited on the substrate (in vacuum, ambient atmosphere or solution) and nm-scale structures arise as a result of a multitude of atomistic processes. This is inherently a non-equilibrium phenomenon and any growth scenario is governed by the competition between kinetics and thermodynamics. The term 'self-organized growth' describes autonomous order phenomena mediated by mesoscale force fields or kinetic limitations in growth processes, whereas 'self-assembly' refers to the spontaneous association of a supramolecular architecture from its molecular constituents (Philp and Stoddart, 1996). The term 'self-organization', in contrast to self-organized growth and self-assembly, relates to dissipative structure formation in systems far from thermodynamic equilibrium and the initial emergence of biological macromolecules (see Nicolis and Prigogine, 1977).

Whereas top-down methods 'impose' a structure or pattern on the substrate being processed, bottom-up methods direct the assembly of atomic and molecular constituents into organized surface structures through processes inherent in the manipulated system. Barth *et al.* described how self-organized growth and self-assembly at well-defined surfaces (some of which may have been created by means of top-down methods) are an efficient tool for the bottom-up fabrication of functional structures and patterns on the nm scale. With a focus on atomic-level investigations, they provided informative illustrations of how this approach might lead to useful nm-scale surface structures.

Supramolecular Engineering

Only a limited range of functional nanometer-sized structures can be created from metallic or semiconducting materials through self-organized growth because design and fabrication methods have to be based on the functional and structural features of these materials. This is unlike the situation applicable to the building of molecular nanoscale structures and patterns: chemical synthesis gives access to a vast range of functionally and structurally diverse building blocks (**tectons**), which can be linked through different types of relatively weak, non-covalent interactions (mostly hydrogen bonds and metal-ligand interactions) to tailor and fashion organized supramolecular architectures (Lehn, 1995). We know much about how the chemistry of the intermolecular non-covalent bond can be tuned to create desired supramolecular crystals or supramolecular compounds in solution, but this knowledge is not amenable to direct translation to guide the assembly of adsorbed molecules into larger surface structures (Feyter and Schryver, 2003) because our understanding of the influence of the substrate atomic lattice and substrate electronic structure on non-covalent bonds is scanty. However, although it is very difficult to translate supramolecular concepts developed for crystals or solutions, they can nevertheless be used as tools for steering non-covalent interactions through the choice of templates with appropriate symmetry, surface patterns or chemical functions (Barth *et al.*, 2005).

Because metal-ligand interactions tend to be stronger than hydrogen bonds, they result in more robust entities. The incorporation of metal centres also enhances the functional properties of the nanoarchitectures so that we can use design strategies based on metal-directed assembly (Holiday and Mirkin, 2002). This renders the controlled fabrication of surface structures based on metal-organic coordination quite attractive, provided the deposition sequence and substrate temperature are very carefully controlled to minimize spurious effects and to achieve the formation of regular nanosystems.

Self-organized growth and self-assembly at surfaces can potentially act as an efficient and versatile tool for creating low-dimensional nanostructures. Not only feature size but also organization on the atomic and mesoscopic length scales can be easily controlled. These process characteristics together with the ability to produce high-density structures in a fast and parallel fashion, are essential requirements for any nanofabrication methodology that facilitates further miniaturization not only in the microelectronics industry but also elsewhere. One challenge is how to incorporate nanostructures into more complex organized architectures and how to effectively interface them to the macroscopic world. Conceivably, this might be achieved by combining bottom-up and top-down techniques – the former providing easy access to features with sizes below 10 nm, and the latter allowing for integration of these structures into larger functional systems. This general approach is likely to result in new materials and devices with use beyond the applications traditionally targeted by miniaturization efforts, particularly when it is guided by recent insights into the physics of small systems or combined with chemical (Ouyang and Awscholom, 2003) and biological bottom-up methods (Niemeyer, 2001; Seeman and Belcher, 2002; Sarikaya *et al.*, 2003).

APPLICATIONS OF NANOMATERIALS

NMs are being used in catalysis, photochemistry, sensors, tagging in optical, electronic and magnetic devices. Copper nps are used in polymers and plastic, lubricants, inks and metallic coatings (Yang *et al.*, 2006). Nanotechnology has enabled fabrication of semiconductor NMs. Multipod gold nanostructures, synthesized by simple chemical reduction at room temperature, are poised to be used in nanoelectronics and other related fields (Grass and Stark, 2006). Highly effective NM catalysts may serve as good starting materials for reactions in photochemistry, sensors and other devices (Song *et al.*, 2004).

Nanoscale engineering based on np assembly, manipulation and lithography opens up exciting possibilities for developing novel devices in several different applications. 3D nanostructured ZnO products are suitable for anode materials in Li ion batteries (Mridha *et al.*, 2008). MnO nps show ferromagnetic characteristics even though this material is antiferromagnetic in bulk form (Park *et al.*, 2004). Iron oxides and hydroxides are used in pigments, magnetic devices, anticorrosive agents, and catalysis (Zhou *et al.*, 2005). Unique electronic and mechanical properties of CNTs carry huge potential for many applications (Jung *et al.*, 2003).

Organic-inorganic nps in different combinations may be exploited for developing advanced nms. A few selected applications of NMs in biology and medicine are as: (i) fluorescent biological labels; (ii) drug and gene delivery; (iii) pathogen detection; (iv) detection of proteins; (v) probing of DNA structure; (vi) tissue engineering; (vii) tumour destruction through hyperthermia; (viii) separation and purification of biological molecules and cells; (ix) MRI contrast enhancement; and (x) phagokinetic studies (see Molday and MacKenzie, 1982; Yoshida and Kobayashi, 1999; Edelstein *et al.*, 2000; Parak *et al.*, 2002).

A mammalian bone is really a NM composed of hydroxyapatite crystallites in an organic matrix mainly made of collagen. An artificial hybrid material has been made from 15-18 nm ceramic nps and poly (methylmethacrylate) copolymers (see Das and Ansari, 2009). Hybrid bionps may also be pressed into service to create useful electronic, optoelectronics and memory devices (Yan *et al.*, 2003; Karen *et al.*, 2003). Gold nps have been used in immunohistochemistry to identify protein-protein interactions. In biosciences, nps can replace dyes in applications requiring high photostability as well as high multiplexing capabilities (Das and Ansari, 2009).

Hybrid Nanomaterials for Energy Research

The possibility of tailoring the dimensionality of nanostructured hybrid materials has aroused great interest because such tailoring could facilitate modifying their basic properties such as redox potential, conductivity and charge storage as compared to similar behaviour of their bulk analogues. A template-assisted route has been one of the widely used for shape selective synthesis of NMs as it has several clear merits over other approaches. However, even though processing within such porous templates produces uniform morphologies, this route has been successfully employed only for linear structures (Vijaymohanan, 2008). Fabricating three-terminal morphology (e.g. Y-junctions) could potentially lead to the development of molecular-scale electronic devices, but it is quite difficult to fabricate these junctions by using conventional methods because the linear structure cannot be changed in a controlled manner. Some ideas floated for post-growth modifications include electron beam welding to form molecular junctions, although these sometime introduce defects.

Several nanostructured materials of silver, gold, platinum and palladium have been produced with or without the use of templates on a variety of shapes, sizes and aspect ratio to unravel their distinct behaviour. Comparison of their properties with those of their bulk analogues points to their promise for many new applications. Subhramannia *et al.* (2008) and Kannan *et al.* (2008) have demonstrated the utility of hierarchically developed alumina templates for the formation of platinum Y-junction nanostructures for fuel cell application.

NANOPARTICLES

Nanoparticles (nps) are tiny engineered lumps of matter with unusual industrial properties. They have been used unlabelled and untested in numerous consumer products for several years, without much attention to possible toxicity problems they may pose. Recently many governments and NGOs participated in intergovernmental organizations, and 39 in the intergovernmental Forum on Chemical Safety in Dakar, Senegal and agreed that producers should label NMs in consumer products and that countries should have the right to refuse imports of nanoproducts. The Austrian government suggested a moratorium on food uses of nanotechnology. The European Commission already agrees that new labelling laws are needed for nanotechnology products.

Several different kinds of nps made from pure metals (e.g. gold, silver, silicon, cobalt, etc.), oxides of silicon, zinc, iron, aluminium and titania, and carbon (fullerenes) have attracted many applications in ceramics, polymer composites, filler materials, pigments, electronics, and catalysts (Willert *et al.*, 2001). In bacteria, nps can affect growth kinetics in both adverse and favourable ways, though the effect is stronger against bacterial growth. Barium nps inhibit the growth not only of bacteria but also fungi, mosses, and algae (IPCS, 1990). Titanium dioxide, silver, and gold nps have shown antimicrobial activities (Fu *et al.*, 2005; Duran *et al.*, 2007). Flores *et al.* (2004) reported that cobalt-ferrite nps stimulate the growth of *E. coli*.

Carbon-based nps are widely used in cosmetics, drug delivery vectors, and semiconductors. Fullerenes show various kinds of pharmacological activities and have attracted industrial and synthetic applications. The commonest, stable fullerene is Buckminster fullerene, having 60 carbon atoms. Fullerenes have proved very useful in fullerene-polymer combinations, in electrooptical devices, and in biological applications (Prato, 1999; Bosi *et al.*, 2003).

Plastic waste management is another area where nps may prove useful. Low-density polyethylene (LDPE) has been widely used for manufacturing various containers, dispensing bottles, tubing, bags, and moulded laboratory equipment. It shows high resistance against dilute and concentrated acids, alcohols, bases, and esters, and has been accumulating in the environment at a rate of several million tonnes annually (Orhan and Buyukgungor, 2000). It may possibly be biodegraded by microorganisms – land filling, recycling, and incineration pose serious environmental constraints. Satlewal *et al.* (2008) biodegraded HDPE and LDPE by a microbial consortium. One consortium showed enhanced LDPE biodegradation in the presence of nanobarium titanate particles of size 38 nm (Kapri *et al.*, 2009). In this work, the feasibility and extent of LDPE biodegradation was determined in the presence of fullerene-60 nps, which otherwise are toxic to bacteria.

Sah *et al.* (2010) studied the effects of fullerene-60 nps on LDPE biodegradation efficiency of two potential polymer-degrading consortia comprising three bacterial strains each. At a concentration of 0.01% (w/v) in minimal broth lacking dextrose, fullerene showed no adverse influence upon the consortia growth. But fullerene proved detrimental to bacterial growth at higher concentrations (viz., 0.25%, 0.5%, and 1%). Addition of 0.01% fullerene into the biodegradation assays containing 5 mg/ml LDPE reduced bacterial growth significantly, but subsequent analysis of the degraded products revealed an enhanced biodegradation. Thus, it appears that fullerene-60, which is otherwise toxic, may accelerate the polymer biodegradation process caused by bacterial consortia.

Researchers at Brigham Young Univ., Utah noted that their cerium oxide nps were displaying odd magnetic behaviour, and that impurities were to blame for the odd behaviour. B. Woodfield and S. Liu designed an easy-to-use general technique for making dozens of different types of metal oxide nps that could have a major impact on everything from catalysts to electronics.

Researchers in many countries are looking to metal oxide nps to improve everything from fuel cells to optical films. One common method to make nps uses heat to vaporize bulk-sized starting materials. As the vapour cools, its atoms condense into larger nps. Another approach precipitates nps from ions in liquids. But such techniques produce pure nps only when the chemistry and kinetics are just right.

S. Liu instead started with high-purity metal salts, which, like table salt, are a fusion of positively and negatively charged ions that come together to form a neutral compound. He ground the salt together with ammonium bicarbonate, a white powdery material used e.g., as a leavening agent in breads or as fertilizer. The grinding rearranged the chemical partners. Mixing aluminium nitrate and ammonium bicarbonate, for example, produced aluminium hydroxide and ammonium nitrate, with excess carbon and oxygen bubbling off as CO_2. The mixture was baked at 300°C for an hour. The heating drove off several additional components as gases, leaving behind a nanopowder of aluminium oxide nps. The simple process is a good method of making uniformly sized metal oxide nps. Woodfield *et al.* have already made many different metal oxide and mixed-metal oxide particles.

Although all the current particles are oxides, it is possible that the technique could be extended to combine other negatively charged ions with the metals. That should facilitate making a variety of chloride, nitride, and phosphide nps with a novel exotic optic, electronic, and catalytic properties. Indeed this powerful technique could even become a powerhouse (Service, 2008).

Chemists probing a basic mystery of magnetism in cerium oxide nps discovered this general recipe for making numerous flavours of the tiny grains. Starting with a metal salt, ammonium bicarbonate is added. The mixture is stirred, heated, and instant nps result.

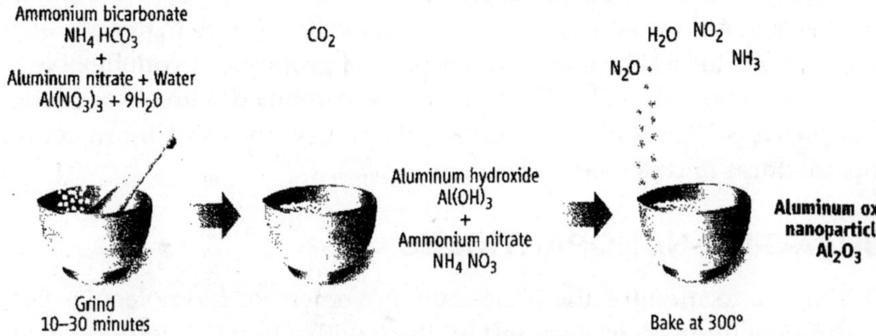

Fig. 3.10 Recipe for cerium oxide nanoparticles (From Service, R.F. *Science* 320: 1584-85, 2008. Reprinted with permission from AAAS).

Although making nps is relatively easy, assembling them into more complex structures in a predictable way is very difficult. Recently, however, gold nanorods have been so modified that they polymerize to make linear, branched and cyclical structures.

Kumacheva and Rubinstein (2010) synthesized their nanorods with 'arrowheads' on each end that could be coated with polystyrene molecules. Side-by-side aggregation could be prevented by covering the sides of the rods with another compound. The rods connected with each other end-to-end to form chains through the association of polystyrene molecules when the researchers reduced the solubility of polystyrene by adjusting the composition of the solvent. The aggregation number and architecture of the structures were also successfully controlled.

NANOPARTICLE SUPRASTRUCTURES

Light induced self-assembly (LISA) is a promising method of producing novel structures and materials having potential applications in optics, sensing, and delivery systems (Whitesides and Grzybowski, 2002; Oh and Mirkin, 2005). LISA has been successfully induced at both colloidal and macromolecular scales (Gong and Marr, 2004; Liu and Jiang, 2006), but not at the nanoscale. Nanoscopic components of dimensions significantly smaller than the wavelength of light cannot be assembled by using optical confinement techniques such as laser interference or optical trapping, on which virtually all colloidal LISA systems are based. However, Klajn *et al.* (2007) have described a system in which photoisomerization of dithiol molecules bound onto the surfaces of metal nps mediates their LISA into ordered, 3-D suprastructures: light-reversible or irreversible crystals and supraspheres of various sizes. These nps are decorated with ligands combining photo-switchable dipoles and covalent cross-linkers and are assembled by light into organized, three-dimensional suprastructures of various types and sizes. Nps covered with only a few photoactive ligands form metastable crystals that can be assembled and disassembled "on demand" by using light of different wavelengths. For higher surface concentrations, self-assembly is irreversible, and the nps organize into permanently cross-linked structures including robust supracrystals and plastic spherical aggregates.

The use of ligands which simultaneously mediate LISA and confer a desired level of permanence on the assemblies is a novel way to nanostructured materials of controllable stability (from metastable to irreversible) and mechanical properties (from flexible to extremely rugged). Conceivably, in the future, LISA might well be combined with electrostatic or entropic forces/effects to provide a basis for a "photoregulated" synthesis of more complex and/or adjustable nanostructures and devices (Klajn *et al.*, 2007).

COLOUR-CHANGING NANOPARTICLES

One popular technique to monitor the nanoscale movement of biomolecules is to tag these molecules with different fluorescent dyes and hit them with a burst of light. Because of the way the dyes absorb and re-emit each other's light, tags very close together glow a different colour from those farther apart. This technique is known as fluorescent resonance energy transfer (FRET). It works only if the dye molecules are less than 10 nm apart, and the tags wink out after less than a minute of light exposure. But now a novel molecular ruler has been developed that solves both problems at once (see Service, 2006). P. Alivisatos (Berkeley California) has described how to use pairs of gold nps to measure distances out to 70 nm and keep track of their targets indefinitely.

By extending the length of the ruler, nps (qds) allow one to gauge the proximity of molecules across stretches equivalent to large complexes of proteins. That, in turn, may make it possible to track events from the binding of DNA strands to one another to the ability of transcription factor proteins to bind with and initiate genetic transcription.

DIVALENT METAL NANOPARTICLES

Nanoparticles (nps) consisting of crystals of a large number of atoms are being used as "artificial atoms" to make supracrystals which mimic ionic solids (Shevchenko *et al.*, 2006), or to produce the nanoscale equivalent of ionic liquids (Bourlinos *et al.*, 2006). A strong challenge is how to break the interaction symmetry in these isotropic and almost spherical materials. Nps are unable to bond along specific directions (atoms and molecules can do so). Nps could become a very powerful research tool if we could place a small but known number of molecules in their ligand shell to achieve directional assembly. Such a capability can greatly expand the scope of potential applications of nps in view of the fact that anisotropic assemblies have characteristic properties as seen or produced in isotropic assemblies (Thomas *et al.*, 2004). A major stumbling block in achieving directed assembly of nps based chiefly on biomolecules or other templating agents is that researchers cannot control the number of receptors that interact with the templating agent. Recently, however, Praharaj *et al.* (2004) and Huo *et al.* (2006) have successfully developed methods to introduce single valency in nps through solid-state reactions. DeVries *et al.* (2007) suggested a method to functionalize monolayer protected metal nps at two diametrically opposed points that exist as a result of the topological nature (Jackson *et al.*, 2004) of the particles. They did this by making nps which serve as divalent building blocks ("artificial monomers"), and can react with complementary divalent molecules to produce chains which then create self-standing films. This simple method involves placing target molecules specifically at two diametrically opposed positions in the molecular coating of metal nps. The approach is based on the functionalization of the polar singularities that always form when a curved surface is coated with ordered monolayers, such as a phase-separated mixture of ligands. The molecules placed at these polar defects can be used as chemical handles to form np chains that in turn can generate self-standing films.

Silver Nanoparticles

As size- and shape-dependent silver nps and their assemblies have found application in many different fields such as plasmonics, disease prevention and control, electronics and catalysis (Elechiguerra *et al.*, 2005; Morones *et al.*, 2005), these applications may be commercialized if the nps can be produced on a large scale at reasonably low cost in bulk quantities by continuous flow processing. Rapid, room-temperature and environmental friendly processes are needed to minimize capital, design and environmental costs upon scaling up. There is an urgent need of suitable methods that use hazardous chemicals as reducing agents.

Sivaraman *et al.* (2009) described a rapid, energy-efficient, clean, and economically scalable room temperature process for the synthesis of silver nps. Tannic acid, a polyphenolic compound derived from plant extracts was used as the reducing agent. Silver nps of mean size ranging from 3.3 to 22 nm were successfully synthesized at room temperature by adding silver nitrate to tannic acid solution maintained at an alkaline pH. The mean size could be changed by varying the molar ratio of tannic acid to silver nitrate. The protocol appears to be suitable for room temperature continuous flow processing.

Not only did Sivaraman *et al.* use tannic acid as a reducing and stabilizing agent to synthesize silver nps on a timescale of seconds through a simple room temperature protocol, but they also demonstrated the suitability of this protocol for energy-efficient, continuous flow synthesis by using a simple co-axial flow reactor (see Fig. 3.11). Tannic acid was previously used as a reducing agent in the presence of additional stabilizers or as both the reducing and stabilizing agent; but the reactions were quite slow and the particle sizes reported were >15 nm. By comparison, Sivaraman *et al.* adjusted the pH of the tannic acid solution prior to the addition of metal salt. The alkaline pH increased the reducing and stabilizing capability of tannic acid and made possible room temperature synthesis of silver nps in seconds, while also enabling tuning of the mean size from 3 to 22 nm. The increase in particle size with increasing molar ratio of tannic acid to silver nitrate points to a new role for tannic acid as an organizer for facilitating nucleation. This idea of using tannic acid at alkaline pH as a reducing, organizing and stabilizing agent may well be extended to other elements known to chelate with tannic acid (Cruz *et al.*, 2000).

Fig. 3.11 Snapshot of co-axial flow pattern, at various axial locations, visualized using a dye solution as the inner fluid (after Sivaraman *et al.*, 2009, with permission from *Current Science*).

IRIDESCENT GOLD NANOSTRUCTURES

Iridescent gold nanostructures have been exploited since times immemorial in the medical, decorative, and theological arts. Jadzinsky *et al.* (2007) have reported total structure determination of a large gold-thiolate cluster – a finding that can help reveal the detailed structure of **self-assembled monolayers** (SAMs). SAMs are chemically modified gold electrodes with applications from sensing to nanolithography (Love *et al.*, 2005).

While atomic structures of larger molecules can be easily determined by x-ray crystallography, structural studies of nanoscale metal clusters or of surfaces are extremely difficult. In the case of clusters, identical cluster cannot be easily created (see, however, Mednikov *et al.*, 2007). Surface diffraction studies also face resolution-limiting problems, e.g., large contributions from subsurface metal layers and multiple scattering effects.

Molecules used to create SAMs on gold surfaces have an organic group (R) connected via a sulfur atom (S) to gold (Au) substrate atoms. Being usually highly ordered, extended SAMs on crystalline substrates allow the orientation of the R group to be determined, but the structure of the buried head group can only be inferred.

As chemical bonding is mostly a local affair, a typical approach to attacking this nanostructure problem uses gold nanocrystals as the metal substrate, assuming that the gold-thiolate bond remains constant at all scales and that a suitable sample will crystallize for total structure determination. Massive thiolate monolayer-protected gold clusters of several different sizes and containing different R groups have been made. The properties of such clusters typically resemble those of the extended SAMs: they continue to be highly stable during size separation and analysis, but till recently no total structure determination was reported (Whetten and Price, 2007).

Jadzinsky *et al.* (2007) reported the successful isolation, crystallization, and structure determination of a cluster with 102 gold atoms and 44 RS groups. They did this by using water-soluble, rigid aromatic benzene thiolates with a carboxylate group on the opposite side of the aromatic ring from the sulfur atom. Crystallization attempts with the reaction products produced large single crystals suitable for x-ray diffraction measurements. Out of the 102 gold atoms, 79 formed a truncated-decahedral cluster. This "grand core" shows gold-gold bond lengths and coordination consistent with bulk metals, supported gold nanocrystals, and a prior model for gold-thiolate clusters ~1.5 nm in diameter (Bagley, 1970). However, the 15-atom caps at each pole are rotated with respect to the ideal compact decahedral shells, yielding denser packing and placing all 40 surface atoms of the grand core on the surface of a common sphere (Whetten and Price, 2007).

The remaining gold atoms make up an outer shell which strongly interacts with the thiolate sulfur atoms.

The weak interactions of the thiolates with the surface atoms of the grand core are still not clear. A strongly polar, if not ionic, interaction could be possible by the distinctive anchoring of thiolate sulfur atoms in formally anionic complexes (Hakkinen *et al.*, 2006).

It has now become possible to rationalize known properties of nanoscale clusters in terms of atomic ordering. Should structures of this type hold generally for flatter clusters and extended SAMs, for smaller clusters with more highly curved surfaces, for non-aromatic R groups, and for diverse gold-plated and gold-alloy nanostructures, then there is virtually no published interpretation (for example, of electron transfer, capacitance, or density) that will not be in need of reinterpretation (Whetten and Price, 2007).

SPHERICAL NANOPARTICLES WITH CORE SHELL STRUCTURE

Heteroepitaxy refers to the overgrowth of one crystalline material onto a second crystalline material. It is an important fabrication method for making thin films and nps. But if the lattice mismatch between the two materials happens to be too high or messy, fractured interfaces result. Zhang *et al.* (2010) reported a synthesis strategy to obtain spherical nps with a core-shell architecture that does not depend on heteroepitaxy. They deposited silver and converted it into various semiconductors through chemical transformations to yield structurally perfect single-crystal semiconductor shells on a gold core, despite mismatches approaching 50%.

SELF-ASSEMBLY OF SEMICONDUCTOR NANOPARTICLES INTO TWISTED RIBBONS

The helical structures so widespread in natural macromolecules result from well-coordinated bonding interactions. These structures influence their physical properties strongly. The collective properties of nps are expressed by their ability to self-organize into complex microscale structures. The ribbons could be twisted into helices because they were doped with cadmium sulfide nps, which underwent surface oxidation and created localized stresses that could only be relieved by a conformational change. It was possible to control the pitch of the twisted ribbons by the intensity of illumination applied. This behaviour may affect application in the development of materials with interesting optical properties. Srivastava *et al.* (2010) produced helical nps by slowly oxidizing cadmium-tellurium under visible light which yielded ribbons of nanostructure. Slow oxidation of tellurium ions in cadmium telluride (CdTe) nps results in the assembly of 1- to 4-micrometer-long flat ribbons made of several layers of individual cadmium sulfide (CdS)/CdTe nanocrystals. Twisting of the ribbons with an equal distribution of left and right helices was induced by illumination with visible light. The pitch lengths (250 to 1500 nm) varied with illumination dose, and the twisting was associated with the relief of mechanical shear stress in assembled ribbons caused by photooxidation of CdS. Unusual shapes of multiparticle assemblies, such as ellipsoidal clouds, dog-bone agglomerates, and ribbon bunches, were observed as intermediate stages. Computer simulations showed that it is the balance between attraction and electrostatic that repulsion determines the resulting geometry and dimensionality of the np assemblies (Srivastava *et al.*, 2010).

Spirals, helicoids, helices, twisted ribbons (TRs), and other helical structures have elegant geometries of interest in the fields not only of mathematics but also biology, optics, and mechanics. The formation of helices from nps will enable exploitation of the unusual properties of helices arising from quantum confinement within nps and also widen the design space (Yang and Wang, 2006; Glotzer and Solomon, 2007). This can create a new method of controlling the pitch and/or chirality of the helical structures. Self-organization of such intricate objects can be achieved by fine tuning the overall balance of forces, including anisotropy of interactions that drives assembly of nps into larger structures. Control of these processes will become possible when many-body interactions are discovered at the nanoscale, and our understanding of their dynamics and capabilities of transitioning from nano- to microscale improves.

In general, the observed multiparticle behaviour and transition from packing into chains, ribbons, and sheets may be understood as a competition between the face-face attraction and the charge-charge repulsion. For low charge and strong face attraction, particles pack very densely and would, if constrained to the same packing structure form infinite two-dimensional sheets. If the amount of charge is increased, an infinite sheet becomes energetically unfavourable in view of the long-range electrostatic repulsion, and nps assemble as ribbons. Further increase of particle charge can eventually result in chains (see Zhang *et al.*, 2007).

What Srivastava *et al.* demonstrated is that light can induce microscale twisting of the matter due to marked effect on the mutual interactions of nanoscale building blocks in a multibody system. In fact, the modulation in the pitch length for the TRs under different light intensities suggests a new approach in the synthesis of nanostructures and new possibilities to form NMs with controlled circular dichroism and other optical, electronic, and mechanical properties (Srivastava *et al.*, 2010).

ULTRATHIN LEAD SULPHIDE SHEETS

For device manufacture, materials have to be deposited or assembled in specific patterns. Schliehe *et al.* (2010) achieved the oriented attachment of lead sulfide (PbS) nanocrystals into 2-D sheets. The packing is driven by the choice of suitable solvents which influence the interactions between the nanocrystals. The nanocrystals show excellent photoconductive properties and have been incorporated into a photodetector without any additional chemical processing.

As effective control of anisotropy is crucial in the generation of complex functionality in advanced materials, oriented attachment of nanocrystal building blocks (a self-assembly of particles into larger-single-crystalline objects) is a most promising approach in nanotechnology. In fact, nanoparticle arrays have moved in the limelight for low-cost device fabrication, like photodetectors (Talapin and Murray, 2005; Szendrei *et al.*, 2009). The 2-D oriented attachment of PbS nanocrystals into ultrathin single-crystal sheets with dimensions in the micrometer scale. This process is initiated by cosolvents, which change nucleation and growth rates during the primary nanocrystal formation, and is finally driven by dense packing of oleic acid ligands on (100) facets of PbS. The nanosheets so made can be integrated in a photodetector device without any more treatment. Typical 2-D assemblies need suitable treatments to remove the insulating organic shell capping. The lack of ligands in the in-plane dimensions of raw PbS 2-D sheets generates excellent photoconductive properties. PbS bridging two gold electrodes on a SiO_2 substrate shows low, dark conductance between ±0.1 V without substantial hysteresis. When illuminated with a 532-nm laser, however, the conductance improves greatly at an illumination intensity of 2.0 mW/cm^2.

Schliehe *et al.* showed that chlorine-containing cosolvents can direct the colloidal synthesis of PbS nanoparticles into 2-D nanosheets by oriented attachment. This is probably a consequence of the size of the nanoparticles and, hence, the presence of reactive facets that disappear as the nanoparticles grow.

COLLOIDAL NANOCRYSTALS

Colloidal nanocrystals are solution-grown, nanometer-sized, inorganic particles stabilized by a layer of surfactants attached to their surface. They are sometimes termed 'artificial atoms' because the density of their electronic states - which controls many physical properties - can be readily tuned by adjusting the crystal's composition, size and shape. The inorganic cores possess useful properties that are controlled by their composition, size and shape, and the surfactant coating ensures that these structures are easy to fabricate and process further into more complex structures. The typical combination of size- and shape-dependent physical properties, easy fabrication and processing renders colloidal nanocrystals promising building blocks for materials with designed functions (Alivisatos, 1997; El-Sayed, 2004), for example, as inorganic fluorophores in biomedical assays (Yin and Alivisatos, 2005). The ability to control the uniformity of the size, shape, composition, crystal structure and surface properties of the nanocrystals is not only of technological interest: access to defined nanoscale structures is essential for revealing their intrinsic properties which are not adversely affected by sample heterogeneity. Rigorous understanding of the properties of individual nanocrystals will enable us to exploit them gainfully and facilitate designing and building of novel electronic, magnetic and photonic devices and other functional materials based on these nanostructures. Chemists are striving to achieve more and more precise control over the composition, size, shape, crystal structure and surface properties of nanocrystals, thus setting the stage for fully exploiting the potential of these remarkable materials.

The inorganic core of colloidal nanocrystals is stabilized by a surface layer of surfactants. Nanocrystals with a semiconductor as the inorganic material - so-called quantum dots - show size-tunable band gaps and luminescence energies owing to the quantum-size effect (Alivisatos, 1996). These colloidal qds have been widely employed as targeted fluorescent labels in biomedical research applications (Michalet *et al.*, 2005). As biological labels, quantum dots are brighter than organic fluorophores and do not photobleach. They provide an easily accessible range of colours. Other applications that may emerge from the combination of low-cost processing and solid-state performance include the use of colloidal quantum dots and rods as alternatives to semiconductor polymers in light emitting diodes, lasers and solar cells (Tessler *et al.*, 2002; Kazes *et al.*, 2002; Huynh *et al.*, 2002). The exciting scope for applications has prompted critical study of the synthesis of these materials to optimize colloidal semiconductor nanocrystal fabrication, and many new ideas for controlling the size, shape and connectivity or coupling of colloidal nanocrystals have not only been developed for these materials, but a unified set of synthesis control possibilities is also being applied to other metals and metal oxides. These materials may also extend the range of applications for colloidal nanocrystals to catalysis and many other areas.

For nanocrystals of 1-100 nm diameter, the average and the dispersion of the diameter, as well as the aspect ratio can be defined. The precision with which the desired structure is synthesized is fairly similar to that realized with synthetic polymers. As with artificial polymers, researchers can control the size and shape of colloidal nanocrystals; commendable progress has been made towards the tailored synthesis of colloidal nanocrystals with well-defined structures. A wide variety can now be successfully produced by different methods, such as coprecipitation in aqueous phase, using reverse micelles as templates, hydrothermal/ solvothermal synthesis and surfactant-controlled growth in a hot organic solvent (Pileni, 2003; Cushing *et al.*, 2004).

A typical synthesis protocol for colloidal nanocrystals consists of three components: precursors, organic surfactants and solvents. In some instances, surfactants also serve as solvents. When a reaction medium is heated to a high enough temperature, the precursors chemically transform into active atomic or molecular species (monomers) which then form nanocrystals whose subsequent growth is greatly affected by the presence of surfactant molecules (Yin and Alivisatos, 2005).

One important requirement for the generation of colloidal inorganic nanocrystals is the identification of suitable precursor molecules, such as organometallic compounds. The precursors must either rapidly decompose or react at the required growth temperature to yield reactive atomic or molecular species, which then cause nanocrystal nucleation and growth. The best example of this process is the use of dimethyl cadmium and trialkyl phosphine selenide to yield cadmium selenide (CdSe), where injection of the precursors into a hot solution yields supersaturation, nucleation and subsequent growth.

PROTEIN-NANOPARTICLE INTERACTIONS

Nps offer real and novel opportunities in biomedicine, materials science and several other fields (Tenne, 2006). Because these particles are small enough to enter most areas of the body, they open a new approach to medicine, including rationally targeted intracellular delivery of genes and other therapies (Bharali *et al.*, 2005). However, there has been some concern that their use might involve some biological hazard (HSE, 2005; Service, 2006). Our understanding of the interaction of nanoscale objects with living matter, even at the level of single cells, has lagged much behind the explosive development of nanoscience in recent years. The work of Cedervall *et al.* (2007) on development of suitable methods for understanding the association of proteins to nps is a very important contribution to knowledge. Because such association is usually a first step when nps enter a biological fluid, when we talk of the interactions of nps with a living system, we really refer to protein-coated particles (see Klein, 2007). Proteins compete for the np surface, leading to the formation of an adsorbed protein layer or "corona" which defines the biological identity of the particle.

Cedervall *et al.* observed a large variation in dissociation rates for proteins on nps with the implication that different sets of proteins may be identified as part of the particle-protein corona. The concentrations of particles and biological fluid affect the outcome of identification experiments.

In spite of the fact that nanotechnology has great potential in biomedicine, very little is known about the interaction between nps and biological systems. Some serum proteins that bind to nps have been identified, but it is not clear how these proteins interact with nanoscale particles. Cedervall *et al.* (2007) used tailored copolymers in tandem with isothermal titration calorimetry and size exclusion chromatography techniques to investigate protein-np binding. They estimated the rates, affinities, and stoichiometries of this binding in a simple serum model system. It was found that particle size, hydrophobicity, and protein identity all contribute to np-protein association. Smaller particles had less surface coverage, possibly because the higher curvature interfered with binding. At equilibrium, hydrophobic particles had greater coverage but albumin, the most prevalent serum protein, showed lower exchange rates for hydrophilic particles. The substantial variation in protein interactions depending on particle composition demonstrated that many technical hurdles must be crossed before protein-np interactions can be fully characterized. However, these techniques appear to be well suited to identify which proteins from a complex mixture are presented on np surfaces; this can be helpful for predicting the biological effects of nps.

The novelty of the work done by Cedervall *et al.* is in their imaginative use of familiar methodologies to examine the issue of np-protein interactions and the resulting coating of the particles by a protein layer (so-called corona). We may expect this critical issue to attract increasing focus in the nanomedicine and, in particular, nanotoxicology, which recognizes the potential for harmful interactions between living tissues and submicrometer- or nanometer-scale objects in a manner that often differs qualitatively from the better known large-scale particles with which living organisms have evolved (Seaton and Donaldson, 2005).

NANOPARTICLES AND PROTEIN FIBRILLATION

NMs can potentially revolutionize biomedical industries including nanomedicine and nanodiagnostics (EC, 2005). They present enormous surface areas and enhance the rate of protein fibrillation by decreasing the lag time for nucleation. Protein fibrillation is involved in many human diseases, including Alzheimer's, Creutzfeld-Jacob disease, and dialysis-related amyloidosis. Fibril formation occurs by nucleation-dependent kinetics – whereby formation of a critical nucleus is the crucial rate-determining step, after which fibrillation proceeds rapidly. However, relatively little is known about any possible biological risks from nps (EC, 2005; Radomski *et al.*, 2005). There is no evidence to suggest that nanoscale objects can cause disease, except from the long-known induction of the aggressive cancer mesothelioma by nanorods of blue asbestos (Ramos-Nino *et al.*, 2003). Many nps can be small enough to enter any part of the body, including the brain (Oberdorster *et al.*, 2004). They enter into cells via receptor-mediated endocytosis, and there exists an *in vivo* link between nps and reactive oxygen production (Carter *et al.*, 2006). Linse *et al.* (2007) highlighted the potential for nps to promote protein assembly into amyloid fibrils *in vitro* by assisting the nucleation process. They showed that nps (copolymer particles, cerium oxide particles, qds, and CNTs) increase the probability of appearance of a critical nucleus for nucleation of protein fibrils from human b_2-microglobulin. The observed shorter lag (nucleation) phase depends on the amount and nature of particle surface. An exchange of protein occurs between solution and np surface, and b_2-microglobulin forms multiple layers on the particle surface, providing a locally increased protein concentration that promotes oligomer formation. This and the shortened lag phase suggest a mechanism involving surface-assisted nucleation that may increase the risk for toxic cluster and amyloid formation. It also paves the way to new routes for the controlled self-assembly of proteins and peptides into novel NMs. These findings may enhance our understanding of nucleation mechanisms. In fact, nps may well create new routes of controlling biological self-assembly for use in nanomedicine, based on a novel and general surface-mediated nucleation mechanism.

As pointed out by Gessner *et al.* (2000) and Labarre *et al.* (2005), nps are very often coated with proteins when they enter a biological fluid with consequent structural and functional perturbations of the surface-bound state of the protein (Lynch *et al.*, 2006; Linse *et al.*, 2007). Proteins can be found in either a native form or denatured form depending on their surface charge, hydrophobicity, and stability, and also depending on particle characteristics (Lynch, 2007). Nps possess enormous surface-to-volume ratios. The high concentration of proteins adsorbed at the particle surface combined with the low dimensionality of the surface tend to increase the probability of partially unfolded proteins coming into frequent contact, leading to faster clustering, or even new protein clusters. These properties influence protein self-assembly reactions, so that either crucial biological processes are perturbed or diseases involving protein misfolding and assembly are aggravated.

Amyloidoses are human diseases that exemplify and involve some protein self-assembly. Some 30 different proteins and peptides can cause human amyloid disease (see Chien *et al.*, 2004; Westermark *et al.*, 2005; Chiti and Dobson, 2006). All these diseases involve self-assembly of soluble proteins into large insoluble fibrils through nucleation-dependent assembly, usually through the production of toxic oligomeric structures (Cleary *et al.*, 2005; Baglioni *et al.*, 2006). In fact surfaces presented by lipid bilayers, collagen fibers, polysaccharides, and various interfaces can significantly promote amyloid formation (Powers and Kelly, 2001; Lu *et al.*, 2003; Knight and Miranker, 2004). It appears that interactions with different surfaces promote protein self-assembly into amyloid fibrils and enhance protein conformational changes associated with other protein misfolding diseases (Linse *et al.*, 2007).

Although the work of Linse *et al.* used a tailored set of model nps, it prompted the issue of whether the observed effects also arise with other kinds of nps, especially those with proposed imminent industrial applications. To address this question, they also conducted other experiments in tubes for b_2-microglobulin (b_2m) alone and in the presence of 16-nm hydrophilic polymer-coated quantum dots, 16-nm cerium oxide particles, or multiwalled carbon nanotubes of 6 nm diameter. All these different particles increased the rate of fibrillation by shortening the lag phase consistent with the view that nucleation is a surface-assisted process.

It seems certain that various different kinds and large amounts of new "surfaces" will emanate from the diversity of engineered nps that are now emerging for use in industrial applications. This will increase the likelihood of human exposure to nps. It is yet to be seen whether the great enhancements in nucleation rate for protein fibrillogenesis reported by Linse *et al.* also occur *in vivo* or in more complex biological fluids where competitive binding may screen the nucleation surface.

NANOPARTICLE-PROTEIN CORONA

In a biological fluid, when proteins associate with nps, the amount and presentation of the proteins on the surface of the particles generates an *in vivo* response. Proteins compete for the np "surface", leading to a protein "corona" that largely defines the biological identity of the particle. Knowledge of rates, affinities, and stoichiometries of protein association with, and dissociation from, nps is essential for understanding the nature of the particle surface seen by the functional machinery of cells. Cedervall *et al.* (2007) developed approaches to study these parameters so that they could be applied to plasma and simple model systems, albumin and fibrinogen. They used a series of copolymer nps with variation of size and composition (hydrophobicity) and showed that isothermal titration calorimetry is well suited for studying the affinity and stoichiometry of protein binding to nps. Cedervall *et al.* (2007) also determined the rates of protein association and dissociation using surface plasmon resonance technology with nps that are thiol-linked to gold, and through size exclusion chromatography of protein-np mixtures. This method is milder than centrifugation, and has been developed into a systematic protocol to isolate np-associated proteins. The kinetic and equilibrium binding properties depend on protein identity as well as particle surface characteristics and size (Cedervall *et al.*, 2007).

Nps facilitate several exciting developments in biomedicine and technology, for example in diagnostics, treatment and novel functional materials. Being small, these particles enter almost all areas of the body, including cells and organelles, leading potentially to both a new approach to medicine (NanoMedicine) and a possibility of biological hazard. Unfortunately, very little is known about the interaction of nanoscale objects with biological systems – this hampers progress in both the new fields of nanomedicine and nanotoxicology. Upon entering a biological fluid, nps become coated with proteins that sometime exert biological effects through changed protein conformation, exposure to novel epitopes or disturbed function.

The isolation and identification of np-associated proteins is an essential prerequisite for nanobiology, nanomedicine and nanotoxicology but is a very difficult task. In the context of the biological response, the more abundantly associated proteins do not necessarily exert the most profound effect. A less abundant protein that has high affinity and specificity for a particular receptor can instead be more relevant. Another challenge is how to separate free protein from protein that is bound to nps; we need nonperturbing methods that neither disrupt the protein-particle complex nor induce additional protein binding. The method of choice so far has been centrifugation, identifying the major serum proteins albumin, IgG and fibrinogen as being associated with a broad range of particles of apparently disparate molecular composition (Gessner *et al.*, 2003; Sun *et al.*, 2003; Göppert and Müller, 2005; Labarre *et al.*, 2005). In view of its high abundance, albumin is usually always found on particles and can be retrieved even though it has relatively low affinity. Other proteins associated with nps in these centrifugation assays are immunoglobulins, apolipoproteins, alpha-1-antitrypsin and several more.

Our understanding of protein-np interactions and their biological consequences will be enhanced if we could go beyond mere identification of particle-associated proteins and gather information on the binding affinities and stoichiometries for different combinations of proteins and nps followed by ranking of the affinities of proteins that coexist in specific body fluids or cellular compartments. The rates by which different proteins bind to and dissociate from nps (the time scales on which particle-associated proteins exchange with free proteins) are crucial for understanding their interaction with receptors, and their biological effects.

The lifetimes of typical protein-protein complexes range from microseconds to weeks, and protein-ligand complexes usually have lifetimes of microseconds to days. Some reported time scales of exchange of adsorbed proteins from silica, polymer and TiO_2 nps range from 100 s to many hours (Sousa *et al.*, 2004; Renner *et al.*, 2005; Lundqvist *et al.*, 2005).

A PROGRAMMABLE DNA NANOSCALE ASSEMBLY LINE

Researchers can now synthesize nps of desired shapes and composition. This ability can potentially generate novel functional materials and devices by appropriately combining them into larger structures. While self-assembly can perform this work efficiently, it suffers from some thermodynamic and kinetic limitations: e.g. reactants, intermediates and products tend to collide with each other throughout the assembly time course, thereby producing some non-target species instead of target species. These limitations can, however, be addressed by adopting an alternative approach to nanoscale assembly that uses information-containing molecules such as DNA to control interactions, thereby minimizing unwanted cross-talk between different components (Seeman and Lukeman, 2005; Gu *et al.*, 2010). In principle, this approach should make possible the stepwise and programmed construction of target products by joining individually selected nanoscale components in much the same way that an automobile is built on an assembly line. Gu *et al.* have shown that a nanoscale assembly line can be realized by the judicious combination of three known DNA-based modules: (a) a DNA **origami tile** (Rothemund, 2006) which acts as a framework track for the assembly process; (b) cassettes containing three independently controlled two-state DNA machines which act as programmable cargo-donating devices and are linked in series to the tile (Ding and Seeman, 2006); and (c) a **DNA walker** which can move on the track from device to device and collect cargo. When the walker covers the route prescribed by the origami tile track, it sequentially comes across the three DNA devices, each of which can be independently switched between an 'ON' state, allowing its cargo to be transferred to the walker, and an 'OFF' state, in which no transfer occurs. Gu *et al.* used three different types of gold np species as cargo and showed that the experimental system allows the controlled fabrication of the eight different products that can be obtained with three two-state devices.

Drexler (2001) and Smalley (2001) had discussed the possibility of creating assembly lines on the nanometer and chemical scales (1-2 Å), analogous to those used on the macroscopic scale. While the operation of such assembly lines differs from that of their macroscopic counterparts because different forces and effects dominate at the different scales, the basic idea is still that it might be possible to build products difficult to produce by more conventional techniques. According to Gu *et al.*, it is indeed possible to use nanometer-scale DNA devices to assemble, in a controlled way, several complex non-covalent constructs with acceptable yields. Indeed, the system described by Gu *et al.* adds both programmability and temporal control to DNA-assisted assembly and might therefore, with some modifications, even enable the building of novel chemical species that cannot be easily synthesized by other means.

NANO-TWINNED COPPER

Nanocrystalline metals with grain size smaller than 100 nm typically show much higher strength than their coarse-grained counterparts but have very poor ductility (Zhu and Liao, 2004; Sutton and Balluffi, 1995). Lu *et al.* (2004) and Dao *et al.* (2006) showed that introduction of coherent nano-twins, a few tens of nanometers in thickness, in ultrafine-grained copper (with grain size of several hundred nanometers) generates a striking combination of ultrahigh strength and high ductility. The plastic deformation characteristics of nanocrystalline metals have traditionally been rationalized on the basis of mechanisms such as grain boundary (GB) sliding, grain rotation and diffusional creep (Zhu *et al.* (2007). The active deformation mechanism can be effectively probed by measuring the sensitivity of flow stress to the rate of loading because both the sensitivity index m and the associated activation volume v vary greatly for different rate-limiting processes. For face-centred cubic metals such as copper, grain refinement into the nanocrystalline regime increases m by up to an order of magnitude relative to microcrystalline metals with grain size in the micrometer range, and a concomitant decrease in the activation volume v by two orders of magnitude (Asaro and Suresh, 2005). Nano-twinned copper shows the same properties of enhanced rate sensitivity and reduced activation volume as non-twinned nanocrystalline copper (Lu *et al.*, 2005) even when the average grain size is several hundred nanometers. However, it achieves very high strength without severely compromising ductility (Lu *et al.*, 2004). Thus nano-twinned copper shows a welcome combination of ultrahigh strength and high ductility, plus increased strain-rate sensitivity. A mechanistic framework has been developed for predicting the rate sensitivity and elucidating the origin of ductility in terms of the interactions of dislocations with interfaces. By using atomistic reaction pathway calculations, it was reported that slip transfer reactions mediated by twin boundary are the rate-controlling mechanisms of plastic flow. The relatively high ductility of nano-twinned copper may be attributed to the hardening of twin boundaries as they gradually lose coherence during plastic deformation. These valuable findings suggest possible means of optimizing strength and ductility through interfacial engineering (Zhu *et al.*, 2007).

The work of Zhu *et al.* throws light on the underlying mechanism of the strength-ductility trade-offs in plastically deforming materials and could well open possible avenues for optimizing strength and ductility in nanostructured metals through controlled introduction of coherent internal interfaces.

NANOFLUID FOR ENHANCED HEAT TRANSFER

Heat transfer fluids are very important for energy supply and production, transportation, electronics and aerospace applications. Thermal conductivity of a fluid can be improved by using suspension of solid particles because thermal conductivity of a solid is much greater than that of a fluid. Unfortunately, dispersion of mill- and micrometer-sized particles is prone to sedimentation and clogging in microchannels. In contrast, nanofluid is a stable colloidal suspension of low (<1%) volume fraction and ultrafine solid particles of nanoscale dimensions in a conventional heat transfer fluid. It sharply enhances thermal conductivity of fluid without any problems relating to effective dispersing of coarse particles. In the past, nanofluid used to be synthesized either by collecting nanometric alumina (Al_2O_3) and copper oxide (CuO) particles (prepared by chemical vapour deposition) or dispersing nanostructured copper, gold or carbon nanotubes in such common heat transfer fluids as water, transformer oil or ethylene glycol. Manna (2006) synthesized and characterized Al rich Al-Cu and Al-Ag binary alloy powders prepared by mechanical alloying. He also developed nanofluid by dispersing these ultrafine powders in ethylene glycol in very low volume fraction, and measured thermal conductivity of the nanofluid by means of a thermal comparator. It appears that both composition and thermal property of solid dispersoids directly influence the thermal conductivity of a nanofluid (Manna, 2006).

The variation of thermal conductivity ratio with the volume fraction of nps follows a sigmoidal course involving a steady and sharp increase in between 0.75 to 1.5 volume percentage of particles. For both the alloy powders dispersed in ethylene glycol, this increase exceeded 200% with merely 1.5 vol. % particles present in the liquid. Though conductivity ratio increased with greater amount of the particles, stability of the nanofluid was undermined due to sedimentation and inhomogeneity beyond 1.5 vol. % of particles.

Work on the effect of np size on thermal conductivity ratio of ethylene glycol containing 0.5 vol. % $Al_{70}Cu_{30}$ nanofluid revealed that thermal conductivity of the nanofluid increases with reduction in crystallite size. Such an enhancement of thermal conductivity ratio with reduction in particle size, for the same volume fraction of nps, is possibly due to the increase in specific surface area to volume ratio. The smaller the sizes of the colloid particles, the more intense are the interaction and collision among particles and fluid; the faster the particles move, the energy transport inside the liquid also becomes stronger. Below 20 nm size of dispersed nps, thermal conductivity of nanofluids tends to increase suddenly.

Thus, a nanofluid comprising a small amount (<1 vol. %) of nanocrystalline metallic particles as a stable colloidal dispersion in ethylene glycol can be prepared. Thermal conductivity measurements showed that the conductivity of this nanofluid was about 1.2 to 2.4 time greater than that of the base fluid. The increase in conductivity seems to be a function of size, volume fraction and thermal property of solid suspension. This makes the nanofluid useful in such strategic heat transfer applications as MEMS and automobiles.

NANOCRYSTALLINE SUPERPARAMAGNETIC SPINEL

Soft magnetic materials to be used in electromagnetic gadgets, memory or data storage devices, switches, actuators or sensors should have high permeability, remanence and saturation magnetization but low core loss and coercivity. Some widely used materials for these applications are oxide-based soft magnetic ferrites. The property of nano-crystallinity, especially in Mn-Zn ferrites that are useful for low frequency (10-50 kHz) power transformer cores, have received attention for possibly creating high magnetic permeability, high saturation magnetization and low core losses (i.e., superparamagnetic behaviour) below a certain crystalline/domain size.

Nanocrystalline $Mn_{1-x}Zn_x\ Fe_2O_4$ (0.2< x < 0.9) has been made by mechanical alloying of the concerned oxide precursors followed by annealing in air and Ar atmosphere, respectively. Milling and annealing (at 973 K, 15 min) in air generates Zn-ferrites ($ZnFe_2O_4$) instead of Mn-Zn ferrites because MnO not only is further oxidized at higher oxygen partial pressure but also fails to dissolve in the spinel phase; in contrast, single-phase Mn-Zn ferrite results from the identical precursor oxide blend when milling and annealing are done under controlled (Ar) atmosphere. In fact, a ferro-to-paramagnetic transition takes place at room temperature with decrease in Mn content. Alloys with < = 0.4 show magnetic splitting but lack any superparamagnetic component. The saturation magnetization values of the present nanocrystalline alloys are about the same as for identical coarse-grained ferrites. In fact saturation magnetization is quite marked and fairly useful for practical purposes for $Mn_{1-x}Zn_x\ Fe_2O_4$ with < = 0.4. The magnetization vs. temperature plots showed that T_B gradually increases with increase in Mn content. In fact the blocking temperature (T_B) exceeds room temperature for Mn-rich ferrites, implying that the coercivity of Mn-rich nanocrystalline Mn-Zn ferrites is significantly greater than that of their coarse-grained counterparts because of blocking of superparamagnetic grains. This high T_B is attributed to the high anisotropy found in the milled ferrite coupled with grain coarsening (Manna, 2006).

More work on $Mn_{0.6}Zn_{0.4}Fe_2O_4$ revealed that a different annealing cycle at a lower temperature (773 K) for extended time (1 h) can prevent grain growth and reduce the blocking temperature of the superparamagnetic grains to 310 K, which, in turn, yields a very long coercivity (<1 Oe) and hysteresis when magnetic measurements are made at 330 K, i.e. at a temperature that is only slightly higher than the concerned T_B (= 310 K). This shows that an early super-paramagnetic relaxation and a significant decrease in blocking temperature in Mn-Zn spinels can be achieved by suitable selection of process parameters/cycle (time, temperature, atmosphere) that precludes grain coarsening and ensures better soft-magnetism.

NANOPARTICLES FOR CONTROLLING BIOFOULING

Considerable biofouling occurs on submerged man-made surfaces in marine habitats. Such biofouling is caused by unwanted growth of various algae and other organisms on mobile surfaces like ship hulls and aquaculture equipment, fixed structures such as water intakes, wharf and pier pilings, and oil rigs in the sea. Biofouling reduces inflows, increases abrasion, wear, and corrosion, and decreases buoyancy. Biofouling on ship hulls increases drag, and boosts fuel consumption by as much as 40%.

The diversity of fouling organisms that grow on these structures is quite high. To tackle such high biodiversity, conventional antifouling (AF) coatings have contained a cocktail of toxic biocides, but many of these have lately been banned in view of concerns over their persistence, accumulation, and non-target toxicity in the environment. A few notable examples of banned AF biocides are organo-mercury, lead, arsenic, DDT, and tributyl tin (TBT); this TBT has been used on some 75% of the world's shipping fleet. The global ban on TBT as an antifouling biocide has resulted in a shift to cooper-based formulations as the dominant AF agents. These coatings have been estimated to contribute 15,000 tonnes of copper to the marine environment annually (Depree, 2008).

In the face of mounting concerns at the potential environmental impacts of copper and associated organic biocides, researchers are developing new technologies that may provide effective AF performance without releasing toxic biocides into the marine environment. Today, the only commercially available non-toxic AF coatings are those based on 'non-stick' technology – the surface still fouls under static conditions but, due to relatively weak adhesion, the fouling comes off when the boat moves fast. This property limits the usefulness of these non-stick coatings to high-speed vessels; it is of no use for large commercial vessels and stationary objects.

Nps can potentially provide a novel way to develop non-toxic surfaces for various kinds of marine applications. Nps differ markedly in behaviour from larger particles of the same material, because the larger surface-area-to-volume ratio enables unique properties of the 'surface atoms' to dominate over the 'bulk material'.

In New Zealand, researchers are developing and evaluating the effectiveness of photocatalytic coatings containing a semi-conducting oxide, namely titanium dioxide (TiO_2), in the form of nps. This research is inspired by the fact that TiO_2 is a semiconductor: upon its exposure to ultraviolet (UV) light, an electron moves to a higher energy level in the TiO_2 particle (the photocatalytic reaction). This process generates negatively charged electrons (e^-) and positively charged holes (h^+), which react to form powerful oxidants on the surface which break down any attached organic matter.

TiO_2 is harmless in itself. The oxidants produced at the TiO_2 surface during photocatalysis are short-lived, so pose no serious threat to the environment. It is this photocatalysis that forms the basis for existing technologies that use TiO_2 for self-cleaning windows and coatings, air purification, antibacterial surfaces, and wastewater treatment. Now the photocatalytic properties of semi-conducting oxides have been gainfully exploited for submerged AF applications, and their effectiveness is being tested against common fouling organisms such as algae, bryozoans, and barnacles. Promising results have been reported on TiO_2-treated surfaces inhibiting fouling both from diatoms and the larger bryozoan larvae.

NANOPARTICLES FOR BIODEGRADATION OF POLYETHYLENE FILMS

Neal (2008) reported that metal nps affect bacterial growth kinetics in both negative as well as positive ways, but the influence is more profound against bacterial growth. Whereas silver nps show antibacterial and antiviral properties (Oka *et al.*, 1994; Oloffs *et al.*, 1994), cobalt-ferrite nps stimulate the growth of *Escherichia coli* and *Corynebacterium xerosis* (Flores *et al.*, 2004). Silicon nps change the growth profiles of bacteria (Perez *et al.*, 2003). Williams *et al.* (2006) evaluated the growth response of *E. coli* to several inorganic nps including silica, silica/iron oxide and gold, and found no significant adverse effect.

Among various different ferrite-based nanostructures, **superparamagnetic iron oxide nps (SPION)** have attracted applications in such fields as permanent magnets, magnetic fluids, magnetic drug delivery, microwave devices, and high-density information storage. Magnetic SPION nps probably interact with the electric polarity of the bacteria, thereby affecting their growth. Barium titanate ($BaTiO_3$) is an important material for the electronic ceramics industry and has been widely used in **multilayer ceramic capacitors (MLCCs)**, piezoelectric and ferroelectric components, **embedded decoupling capacitors (EDC)**, PTC thermistors and ceramic filters. Titanium nps and nanotubes, in mixed composites like titania-nickel ferrite and titania silver have been reported to be antibacterial (see Rana and Misra, 2005; Ling *et al.*, 2005; Keskinen *et al.*, 2006).

Kapri *et al.* (2010) studied the effects of SPION and **nanobarium titanate (NBT)** nps on plastic waste, especially polyethylene (PE), a synthetic, highly hydrophobic, high molecular weight polymer. Low-density polyethylene (LDPE) is used for manufacturing various containers, dispensing bottles, wash bottles, tubing, plastic carrying bags, and some moulded laboratory equipment. It shows excellent resistance to dilute and concentrated acids, alcohols, bases and esters, and according to Orhan and Buyukgungor (2000) has been accumulating at an estimated rate of 2.5 million tonnes annually. As LDPE is non-biodegradable, its use in the production of disposal or packing materials creates serious environmental problems.

Satlewal *et al.* (2008) described an efficient bacterial consortium that can degrade LDPE and synthetic polymers like epoxy and epoxy silicone blends (ESBs) (Negi *et al.*, 2009). In fact, the participating strains have also been exploited jointly with other bacteria to degrade high-density polyethylene (HDPE) as well as non-poronized and poronized LDPE (Soni *et al.*, 2009).

Kapri *et al.* compared the effects of SPION and NBT on *in vitro* and *in situ* LDPE biodegradation efficiency of a potential polymer-degrading microbial consortium. Supplementation of 0.01% (w/v) concentration of the nps in minimal broth not only increased the bacterial growth, but also led to an early onset of the exponential phase. Simultaneous **thermogravimetric-differential thermogravimetry-differential thermal analysis (TG-DTG-DTA)** pointed to multiple-step decomposition of LDPE degraded in the presence of nps. Scanning electron micrographs revealed greater dissolution of the film surface in the presence of nps. Progressive degradation of the film was enhanced when incubated with the nps under soil conditions for 3 months. The work done by Kapri *et al.* points to the significance of bacteria-np interactions, which greatly influence biodegradation, and possibly also certain other microbial processes.

CHAPTER 4

Carbon Nanotubes and Nanostructures

BUCKYBALLS AND NANOTUBES

Fullerenes represent a kind of solid carbon that is quite distinct from diamond and graphite. Fullerenes were discovered in the 1980s by scientists at the Rice University (USA) working on graphite. The researchers recorded an abundance of carbon clusters in which each contained the equivalent of 60 atoms. This was puzzling because one could not figure out how 60 atoms could arrange themselves so stably. Intense thinking, however, enabled researchers to hit upon the solution – one carbon atom must lie at each vertex of 12 pentagons and 20 hexagons arranged like the panels of a soccer ball. This puzzling molecule was named "buckminsterfullerene", as tribute to Buckminster Fuller's similar geodesic domes. It was this discovery by R.E. Smalley and his co-workers that inspired research leading to elongated versions called carbon nanotubes, which Sumio Iijima of Japan described in a seminal paper in 1991.

The remarkable properties of fullerene nanotubes make them suitable for several applications. These properties include strength, toughness, chemical robustness, thermal conductivity and electrical conductivity. Depending on their precise molecular symmetry, some nanotubes are semiconducting while others show truly metallic conductivity. This behaviour, coupled with their nanoscale geometry, makes them ideal candidates for wires, interconnects, and even devices for true molecular electronics (Colbert and Smalley, 1999).

The isolation and detection of fullerenes as the first "free-standing" carbon-free elemental hollow cages identified by spectroscopy experiments and synthesized in the bulk, have attracted perpetual interest ever since the discovery and synthesis of the buckyball C_{60} and the higher fullerenes (Kroto *et al.*, 1985; Kratschmer *et al.*, 1990; Bulusu *et al.*, 2006). Bulusu *et al.*

presented evidence of hollow cages consisting of pure metal atoms Au_n (where n = 16 to 18). These hollow gold-cages ("bucky gold") have an average diameter >5.5Å, which can accommodate one guest atom inside. "Free-standing" inorganic cages were synthesized by Bai *et al.* (2003) but bare elemental metal cages have not previously been observed in nature or detected in the laboratory. Among metals, gold has some unique relativistic properties and aurophilic attraction. Johansson *et al.* (2004) and Gu *et al.* (2004) predicted a fullerene-like hollow cage with 32 Au atoms to be highly stable. However, at the relatively large size the overwhelming population of low-lying clusters for Au_{32} near room temperature appears to consist of only compact structures (Ji *et al.*, 2005). Larger gold clusters with cage-like local minimum structures may also exist but none has been observed experimentally. Conversely, it has been experimentally proved that the most stable anion gold clusters (Au_n^-) in the size range n = 5-13 possess planar structures, with a structural transition from planar to three-dimensional (3D) structures occurring at n = 14. Beyond n = 14, it appears that all low-lying isomers of gold clusters assume space-filling compact structures (Bulusu *et al.*, 2006). Au_{20} is the most interesting among the larger gold clusters as it has a pyramidal structure with tetrahedral symmetry just as carved out of the bulk face-centred cubic crystal (Li *et al.*, 2003).

According to Bulusu *et al.*, it is possible that the hollow golden cages studied by them can accommodate a guest atom with very little structural distortion to the host cages. In fact, an icosahedral Au_{12} cage with a central metal atom, $M@Au_{12}$, has already been verified experimentally (Li *et al.*, 2002; Zhai *et al.*, 2004). A larger gold cage with a central atom ($M@Au_{14}$) has also been predicted to be very stable (Gao *et al.*, 2005). However, bare Au_{12} and Au_{14}, as also their anions, do not have hollow-cage structures, and the endohedral cage structures $M@Au_{12}$ and $M@Au_{14}$ are chiefly stabilized through the interaction between the central impurity atom M and the outer gold cage. The current mid-sized hollow golden cages with n = 16-18 investigated by Bulusu *et al.* suggest that a novel kind of endohedral gold clusters may exist, analogous to the endohedral carbon fullerenes with a metal inside (Cai *et al.*, 1991).

The field of molecular electronics has been actively pursued with the goal of chemical synthesis of electronic devices whose elements are individual molecules. The successful materialization of this interesting idea can result in electronic devices at the ultimate size scale—the molecular realm. There exists a strong possibility that fullerene nanotubes may provide an ideal—perhaps unique—basis for a workable, robust molecular electronics.

Fullerene nanotubes were discovered by Iijima (1991) who found multi-wall carbon nanotubes in the cathode deposit of a carbon arc discharge for making fullerenes. These consisted of 4-25 concentric tubular layers of sp^2-bonded carbon. Two years later, Iijima and Ichihashi (1993) and Bethune *et al* (1993) independently discovered single-wall carbon nanotubes while attempting to produce endohedral metallofullerenes in an arc discharge (Colbert and Smalley, 1999). Their work made available fairly perfect single-wall nanotubes (SWNTs) for the first time but the arc process they used yielded only about 4-5% SWNTs in the soot produced, the remainder consisting of amorphous carbon, fullerenes and metal particles coalesced from the transition-metal catalysts required to produce SWNTs. The yield was low.

Guo *et al* (1995) developed a new synthetic route to SWNTs : a composite graphite - metal target was vaporized by a laser, with the condensing carbon-metal vapour swept down a heated flow tube, depositing on the cooler walls downstream. This procedure produced SWNTs in yields of 80-90% of all the carbon vaporized (Thess *et al.*, 1996).

These most-perfect carbon fibres (SWNTs) are now established to be the stiffest known fibres (Wong *et al.*, 1997) and may well turn out to be the strongest and toughest as well. Their thermal conductivity should be similar to diamond. The property that has aroused the greatest interest in physics, chemistry, biology and engineering, however, is their electrical conductivity, which may make possible revolutionary applications from the tiniest electrochemical probes for biology to true molecular electronics (Colbert and Smalley, 1999).

NANOTUBES IN DISCOTIC LIQUID CRYSTALS

CNTs were accidentally discovered by Iijima in 1991 and generated brisk activity in many areas of science and technology in view of their remarkable physical and chemical properties. CNTs represent the fourth allotrope of carbon, after diamond, graphite and fullerene. They are one dimensional, well ordered all-carbon hollow cylinders of graphite with a high aspect ratio (Bisoyi and Kumar, 2009). CNTs are divided in two classes on the basis of their structure and dimensions: (a) single-walled carbon nanotubes (SWNT) consist of one layer of cylindrical graphene and have diameters of 0.4-2.0 nm and lengths up to about 1000 nm; (b) multi-walled carbon nanotubes (MWNT) have many concentric graphene sheets and their diameters can be as large as 100 nm and lengths range from 1 to a few microns. Tubes with exactly two concentric carbon sheets have been synthesized, prompting a new concept of double-walled carbon nanotubes (DWNT) (Hutchison *et al.*, 2001). SWNTs can be metallic or semiconducting depending on the sheet direction about which the graphite sheet is rolled to form nanotube cylinder. CNTs can be produced by various methods such as chemical vapour deposition (CVD), carbon arc discharge, high pressure carbon monoxide (HiPCO) method and laser ablation. These methods produce random mixtures of tubes of different diameters, chirality, length distributions and different electronic properties (Dresselhaus *et al.*, 2001). As SWNTs and MWNTs are usually synthesized by using catalysts, metal nanoparticles may be present along with unwanted carbon particles, amorphous carbon and nontubular fullerenes. CNTs are chemically virtually inert and insoluble. They cluster to form crystalline ropes which in turn can aggregate in entangled and unorganized networks or bundles. It is possible to separate semiconductive CNTs from conductive CNTs for various applications (Tromp *et al.*, 2008). Katz and Willner (2004) showed how to chemically modify the side wall or tube-end by molecular or biomolecular components so as to obtain purified and soluble CNTs. Other suitable methods for functionalizing CNTs are noncovalent exohedral with polymers and surfactants and endohedral functionalization with fullerenes (Tasis *et al.*, 2006). The combination of excellent mechanical, thermal and electronic properties of SWNTs and of MWNTs make them well suited for many different applications such as conductive and high-strength composites, catalyst supports in heterogeneous catalysis, energy-storage and energy-conversion devices, field emitters, transistors, sensors, gas storage media, tips for scanning probe microscopy and molecular wires. Many biomedical applications of CNTs have also been reported (Lin *et al.*, 2004; Park *et al.*, 2006). Unfortunately, their applications as one-dimensional conductors or semiconductors have not been very successful in view of problems in aligning them in the desired direction. Well-arrayed CNTs are badly needed for use in various nanodevices, particularly where one dimensional charge migration is critical. The alignment of CNTs has crucial role in the properties of nanotube based devices (Xie *et al.*, 2005). One early method of aligning MWNTs was based on cutting thin slices (50-200 nm) of a CNT-polymer composite (Ajayan *et al.*, 1994). Other possible methods to align CNTs may involve applying magnetic and electric field, shear flow or mechanical techniques.

Liquid crystals (LCs) are interesting functionally soft materials possessing both order and mobility at the molecular and supramolecular levels. In fact, the liquid crystalline state represents an intriguing fourth state of matter and an intermediate state between the solid and the liquid. LC phases share some properties of both isotropic liquids and crystalline solids. Though the molecules in these phases show certain positional and orientational order, they also behave as fluids – they integrate the fluidity of liquids and the anisotropy of crystals. The commonest application of LCs is as liquid crystal display (LCD), but besides this application, LCs also find use in surfactants and detergents, membranes, high strength polymers, photonics thin films, semiconductors, artificial muscles and in thermal, chemical and electrical sensors (Demus *et al.*, 1998).

The dispersion of CNTs in both thermotropic and lyotropic liquid crystalline phases, their mesophase behaviour, and preparation of CNTs from discotic LCs has attracted much interest (Zakri, 2007; Zhang and Kumar, 2008). One most important application of optoelectronic devices based on LCs is liquid crystal display (LCD). Optimizing the performance of an LCD requires good understanding of physical parameters of the LC mixture, e.g., the elastic constants of deformation, dielectric and optical anisotropies, conductivity and rotational viscosity. Mixing CNTs into thermotropic LC hosts has been attempted to modify the physical properties of LCs along with CNT alignment in an LC matrix. LCs show long range orientational order which makes them anisotropic phases. Should CNTs be well dispersed in an LC matrix, they will generally align with their long axes along the LC director, minimize distortions of the LC director field and hence the free energy (Bisoyi and Kumar, 2009). Proper dispersion of CNTs in LCs provides low cost, versatile and effective method of controlling nanotube orientation on macroscopic scale with no restrictions being imposed on nanotube type. There are two advantages of thermotropic LCs in this context: (a) by using existing alignment techniques, well developed for display industry, the ground state of the LC and hence of the dispersed CNTs can be readily determined on macroscopic scale; (b) the ease and speed of modifying the director field by exposure to electric or magnetic fields can create switches based on reorientation of CNTs. The commonly used thermotropic LCs for dispersing CNTs are E7 and SCB. Lee and Chin (2001) first reported the doping of CNTs in LCs; they observed self-diffraction by gratings in nematic LCs doped with MWNTs. Lynch and Patrick (2002) achieved good CNT alignment by exploiting self-assembly properties of LCs followed by removing the LC by vacuum suction which leaves the CNTs alignment intact. Doping with nanotubes not only effectively reduces the dc driving voltage but also improves the switching behaviour of a twisted nematic-LC cell (Lee *et al.*, 2004). Voltage-dependent transmittance and capacitance under ac and dc electric field revealed that the residual dc, being related to an image sticking problem in LC displays, is much reduced due to the ion trapping by CNTs following strong charge transfer from the adjacent LC molecule (Baik *et al.*, 2005). Under applied ac voltage the electro optical properties of CNT doped TN-LCD cells can be rectified by reducing the driving voltage and rise time.

Cervini *et al.* (2008) successfully aligned MWNTs by a template assisted method; they used a homeotropically aligned LC monomer which subsequently polymerized to a rigid polymeric material conducive to retention of the orientation of both the mesogens and more importantly the nanotube. This may turn out to be a promising means of creating composite materials showing novel functionality or enhanced electrical, mechanical and/or thermal properties due to the presence of aligned CNTs.

LYOTROPIC LIQUID CRYSTALS

Carbon nanotubes are also being incorporated into lyotropic liquid crystals. Since lyotropic LCs are water based and quite common in biological systems, it might be worthwhile to use CNT-lyotropic LC mixtures for biotechnological applications. Unfortunately, recent indication to the effect that CNTs could be as harmful as asbestos warrant more critical studies before CNT-lyotropic LC mixtures can be used for *in vivo* medical application. Unlike thermotropic LC-CNT composites, the lyotropic LC-CNT composites can be prepared by two different methods: (a) by preparing an isotropic low-surfactant-concentration suspension of CNTs and then making it liquid crystalline by adding more surfactant (see Fig. 4.1); or (b) it is added to an already prepared lyotropic liquid crystal sample. The incorporation of CNTs into lyotropic liquid crystalline phases was investigated by Weiss *et al.* (2006) who observed that both the *d*-spacing and viscosity of the hexagonal columnar phase increased with increase of CNT concentration. CNTs have been aligned in the lyotropic nematic liquid crystals formed by both rod shaped and disc shaped micelles. The nanotubes were aligned along the LC director and allowed the determination of the order parameter. Jiang *et al.* (2007) dispersed CNTs in a lyotropic liquid crystal formed in room-temperature ionic liquids; here also the *d*-spacing and the viscosity of the medium increased with increase in nanotube concentration.

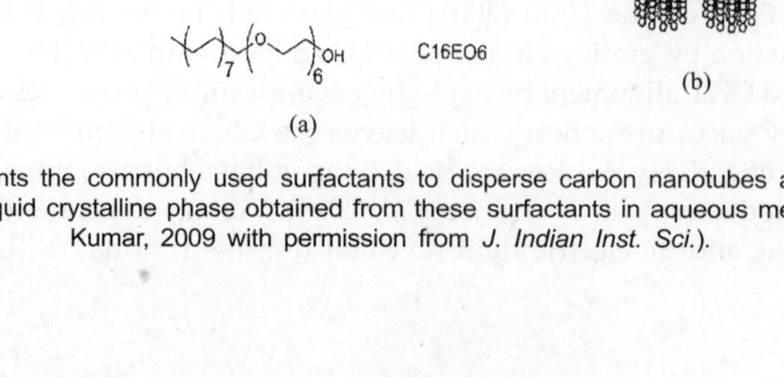

Fig. 4.1 (A) Represents the commonly used surfactants to disperse carbon nanotubes and (B) represents the columnar hexagonal liquid crystalline phase obtained from these surfactants in aqueous medium (after Bisoyi and Kumar, 2009 with permission from *J. Indian Inst. Sci.*).

LIQUID CRYSTAL PHASES OF CNTs

Carbon nanotubes are highly anisometric rigid rod-like particles. Like other anisotropic and one-dimensional molecules, they can also form lyotropic LC phase. Several methods to describe different ways of making LC phases of CNTs in solution have been described (Song and Windle, 2005, 2008). These are based on acid functionalization, acid protonation, gradual evaporation or absorption of amphiphilic molecules. Rai *et al.* (2006) achieved lyotropic liquid crystalline phases of SWNT by protonating the side walls of CNTs in strong acids. Another approach to obtain LC phase from unmodified SWNT is reported by Badaire *et al.* (2005). They dispersed CNTs by non-covalent wrapping of denatured DNA to CNTs which show nematic phase but retain the core properties of CNTs. The LC behaviour of CNT solutions has been gainfully utilized by Ko and Tsukruk (2006) to fabricate highly oriented CNT arrays for thin film transistors.

PREPARATION OF CNTs FROM DISCOTIC LIQUID CRYSTALS

CNTs are usually prepared by applying high temperature or high pressure. Being chemically complicated, the reactions cannot be controlled except with great difficulty - the catalysts cannot be easily separated from the nanotube product because typically the catalyst particles are encapsulated into the nanotubes. Also, the orientation of the graphene layers in the nanotubes is critical. One possible strategy to make ordered graphite nanostructures without metal catalysts and with desired graphene layer orientations is based on carbonization within a discotic columnar mesophase. Here the columnar superstructures of discotic liquid crystals are suitable precursors as they have a nanographene subunit as the aromatic core and also because these molecules stack one on top of another to build up columns which organize themselves in different two-dimensional lattices (Fig. 4.2). Upon carefully controlled carbonization the preorganized ordered columnar superstructures can be converted into nanotubes. Wu *et al.* (2005) and Zhi *et al.* (2005) successfully produced nanotubes from thermotropic discotic liquid crystals (Fig. 4.3). Chen *et al.* (2005) and Sousa *et al.* (2005) likewise produced nanotubes from both thermotropic and lyotropic discotic liquid crystals. Pyrolysis of well-defined discotic molecules in the bulk state enabled production of novel carbon nano- and microstructures. Here the temperatures are much less than the normally used graphitization temperatures (2000-3000°C). Discotic molecules can be converted into 1D CNTs having well-controlled graphene layer orientation by a template-directed solid state pyrolysis process. Hill *et al.* (2004) described a method of self-assembling graphitic nanotubes from an amphiphilic hexabenzocoronene (Fig. 4.3). Unlike the above nanotubes, their nanotubes have graphitic wall formed from numerous molecular graphene sheets stacked parallel to the long axis of the tube (Bisoyi and Kumar, 2009).

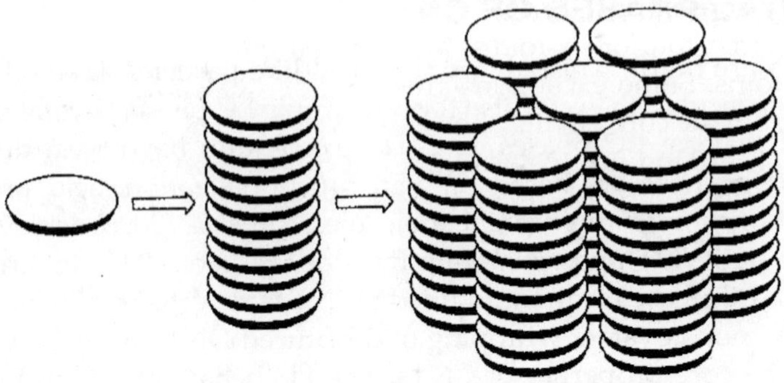

Fig. 4.2 Discotic molecules self-assemble one on top of the other to form columns and these so formed columns self-organize in two-dimensional lattices, only columnar hexagonal structure is shown (after Bisoyi and Kumar, 2009 with permission from *J. Indian Inst. Sci.*).

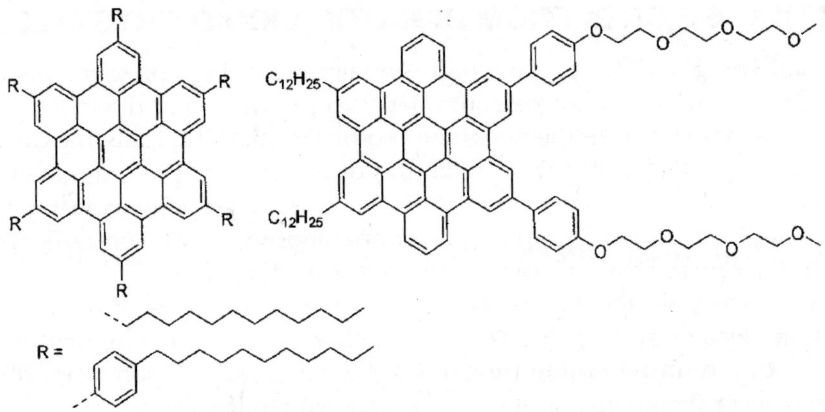

Fig. 4.3 Precursor discotic molecules from which carbon nanotubes can be obtained via pyrolysis (A) or by dissolving in a suitable solvent (B) (after Bisoyi and Kumar, 2009, with permission from *J. Indian Inst. Sci.*).

VARIOUS PHASES OF DISCOTIC LIQUID CRYSTALS

LCs formed by disc-shaped molecules have attracted great interest in view of the fundamental importance of supramolecular order of their columnar phases as models for the study of energy and charge migration in self-organized systems and also as functional materials for device applications such as one-dimensional conductors, photoconductors, light emitting diodes, photovoltaic solar cells, field effect transistors and gas sensors. The functional capabilities of these materials arise from their easy processability, spontaneous alignment between electrodes and self-healing of defects owing to their dynamic nature. In fact, some negative birefringence films formed by polymerized nematic discotic liquid crystals have already been commercialized as compensation foils to increase the viewing angle of commonly used twisted nematic liquid crystal displays (see Bisoyi and Kumar, 2009). So far the number of discotic liquid crystals derived from more than 50 different cores has reached about 3000. It is the hierarchical self-assembly of disc-shaped molecules which leads to the formation of discotic liquid crystals (DLCs).

DLCs typically consist of flat, rigid aromatic cores surrounded by flexible chains. They usually have two, three, four or six-fold symmetry and three or more peripheral chains of three or more carbon atoms. Some exceptions are materials with low symmetry, or those with a non-planar, non-aromatic core having shorter number of chains. Flexible chains are connected to the core via such linking groups as ether, ester, benzoate or alkyne. Fig. 4.4 illustrates a general molecular architecture for discotic mesogens (Bisoyi and Kumar, 2009).

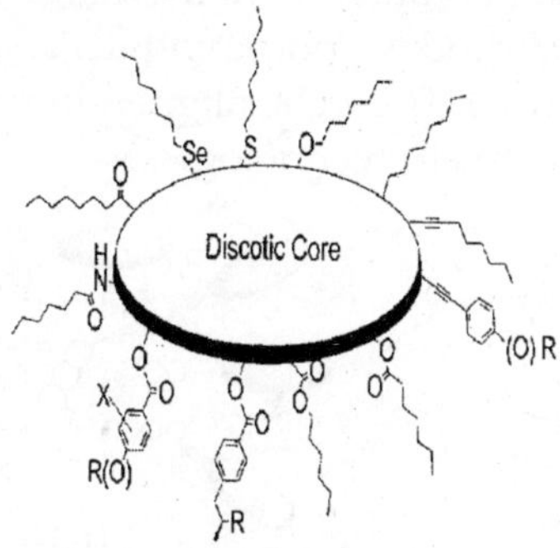

Fig. 4.4 General template of a discotic liquid crystalline molecules. The substituents around the core can vary from 3-8. The central core may be benzene, naphthalene, triphenylene or hexabenzocoronene (after Bisoyi and Kumar, 2009). (after Bisoyi and Kumar, 2009 with permission from *J. Indian Inst. Sci.*).

CNTs in the Supramolecular Order of Discotic Liquid Crystals

The dispersion of CNTs in the supramolecular order of DLC monomers and polymers can potentially lead to some novel materials useful for device applications. In contrast to isotropic and nematic phases, the ordered columnar phases offer welcome opportunity for processing a material with oriented nanotubes – a critical factor in view of the strongly anisotropic properties of CNTs. In fact, there are several similarities between CNTs and DLCs: (a) both are anisotropic materials; (b) both have self-assembling properties and often form hexagonal aggregates; and (c) the remarkable one-dimensional (1D) conductivity properties of both CNTs and DLCs. Conductivity along the columnar axis of DLCs is much greater than perpendicular to it; the same is also true for CNTs. The conductivity along the CNT axis is much higher than across the axis. This is why both the materials are considered as "molecular wires". Prompted by all such similarities and other attributes, Bisoyi and Kumar (2009) are attempting to disperse functionalized CNTs into the columnar matrix of liquid crystalline discotic monomers and polymers. They specially focused on dispersion and alignment of functionalized CNTs in the columnar matrix of discotic liquid crystal monomers and polymers having electron-rich and electron-deficient cores exhibiting mesomorphism at room temperature.

With a view to chemical functionalization of CNTs to improve their compatibility with the DLC matrix, acid-purified and carboxylic acid-functionalized CNTs were chemically functionalized at the ends and sidewalls by a hydroxyl functionalized hexabutyloxy triphenylene (H_4TP) derivative (Fig. 4.5). First the carboxylic acid groups of the CNTs were made highly reactive by converting them into acid chlorides followed by esterification with the hydroxyl functionalized H_4TP derivative to furnish discotic functionalized SWNTs (f-SWNTs)

(Fig. 4.6). Such f-SWNTs happen to be very soluble in common organic solvents such as dichloromethane and chloroform. The formation of discotic functionalized CNTs was confirmed by IR, ^{1}HNMR and ^{13}C NMR spectroscopy and thermogravimetric analysis (TGA). However, triphenylene capped SWNTs were found to be non-liquid-crystalline (Bisoyi and Kumar, 2008, 2009). According to Bisoyi and Kumar, their functionalized CNTs occupy the space between the disc columns and are oriented parallel to the columnar axis (Fig. 4.6), and such stable CNT-DLC hybrid systems could prove useful for many device applications, such as photoconductors, light-emitting diodes, photovoltaic solar cells, sensors and thin-film transistors in which the dispersed CNTs can be aligned in the desired direction by exploiting well established liquid crystal alignment technologies.

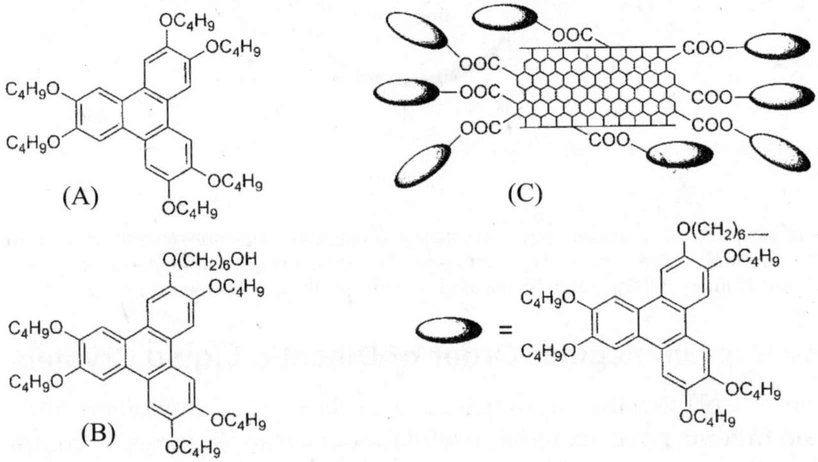

Fig. 4.5 H4TP (A) hydroxyl functionalized H4TP derivative (B) and discotic functionalized single walled carbon nanotube (C) (after Bisoyi and Kumar, 2008, 2009, with permission from *J. Indian Inst. Sci.*).

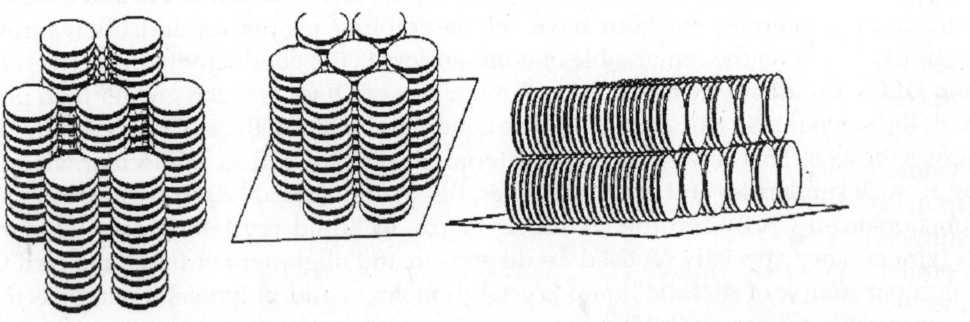

Fig. 4.6 Alignment of CNTs along the columnar axis (A), homeotropic alignment of columnar phase where the column axis is perpendicular to the substrate (B) and uniaxial planar alignment of columnar phase where the column axis lies parallel to the substrate (after Bisoyi and Kumar, 2009, with permission from *J. Indian Inst. Sci.*).

HIGH-PERFORMANCE NANOTUBE FIBRE

Development of high-performance synthetic fibres, based on polymer molecules or graphene sheets, has been strongly motivated by the high strength and stiffness of the covalent carbon-carbon bond and by the ability to achieve alignment of these bonds with the fibre axis. Considering their impressive individual properties, CNTs should form high-performance fibers. The key to producing such fibres is to maximize the number of covalently bonded carbon atoms per unit volume or mass, and so to reduce the proportion of other types of atoms or groups attached to polymer chains. The mechanical properties of pure carbon fibres accrue from the in-plane stiffness and strength of graphene sheets, without any adulterating effect of additional atoms to satisfy available carbon bonds. The problem is that the route to carbon fibres involves the alignment of precursor structures, which are then covalently bonded to each other to produce the final structure. It is this second phase of chemistry that complicates the processing operation and also generates a structure in which the basic mechanism that produces toughness in linear polymer systems (i.e., chain pullout) is not known. This is why carbon fibres are rather brittle, especially when heat-treated to maximize stiffness (Koziol *et al.*, 2007).

Critical work done by Demczyk *et al.* (2002), Natsuki *et al.* (2004) and Dumitrica *et al.* (2006) demonstrated the very high axial strength and stiffness of individual CNTs, and created the possibility of processing them directly into fibres without any need for a subsequent cross-linking – the benefits of high-performance polymeric fibres – especially directness of processing and fiber toughness can now be combined with the advantages of a fiber consisting only of carbon atoms. If we consider CNTs as extremely strong and stiff polymer molecules, it is clear that the processing routes developed till hitherto have been based on concepts from polymer fiber-processing technologies. The major approaches for production of nanotube fibres are (i) spinning from a lyotropic liquid crystalline suspension of nanotubes, in a process like that used for polymeric fibres such as aramids; (ii) spinning from multiwall nanotubes previously grown on a substrate as semi-aligned carpets; and (iii) spinning directly from an aerogel of single- and double-walled CNTs as they are formed in a chemical vapour deposition reactor (see Jiang *et al.*, 2002; Ericson *et al.*, 2004; Li *et al.*, 2004). Koziol *et al.* (2007) used this last process.

Koziol *et al.* explored the roles of nanotube length and structure, fiber density, and nanotube orientation in achieving optimum mechanical properties. They noted that CNT fibre, spun directly and continuously from gas phase as an aerogel, combines high strength and high stiffness with an energy to breakage (toughness) that is considerably greater than that of any commercial high-strength fibre. Different levels of CNT orientation, fiber density, and mechanical properties were achieved by drawing the aerogel at various winding rates. This work points to the considerable potential of CNT assemblies in the search for maximal mechanical performance (see Loiseau *et al.*, 2006). .

Chemists have created bundles of nanoscale fibres which stretch, bend, and can be packed alongside cells. Stupp *et al.* (2010) produced the noodle-like material by heating and then cooling peptide suspensions. Living cells were embedded in the stringy matrix. From these, individual strings were teased out by means of a pipette and salty solution. This method seems superior to 'electrospinning' methods of nanofibre formation, in which the high electrical or mechanical energies involved can kill cells. Stupp *et al.* also showed that stem cells elongated preferentially along the direction of the fibres, and that the noodles also allowed heart cells to transmit electrical signals just like that in a wire. The material has potential use as a scaffold in medical applications.

DOUBLE-WALLED NANOTUBES

Recent years saw a surge of interest in the development of materials for storing hydrogen which is a good alternative to hydrocarbon fuels. Single- and multi-walled carbon nanotubes (SWNTs and MWNTs, respectively) have been extensively studied for use as energy storage materials and electronic devices. Japanese researchers have explored the ability of double-walled carbon nanotubes (DWNTs) to adsorb H_2. They noted that a DWNT bundle can adsorb twice as much H_2 as compared with a close-packed SWNT bundle (Miyamoto *et al.*, 2006). A new type of high-purity DWNT buckypaper, synthesized by catalytic chemical vapour deposition (CCVD), was also compared with SWNTs, which are usually preferred in view of their greater uniformity of structure. The buckypaper, consisting of nanotubes of two graphene layers, shows extremely high purity and uniformity, and contains only trace amounts of amorphous carbon and metal impurities. The highly ordered bundled structure of DWNTs provides greater stability as well as a much better adsorption field for H_2 than SWNTs.

High-resolution transmission electron microscopy confirmed that the DWNTs pack loosely into an ordered hexagonal array. It appears that this array has interstitial pores that can adsorb H_2 molecules. Adsorption isotherms of supercritical H_2 on SWNTs and DWNTs at 77 K showed that DWNT samples adsorb much more H_2 than the SWNTs despite having a 40% smaller specific surface area. This may be accounted for not only by the accessibility of H_2 to the interstitial sites in DWNTs but is also explained by the fact that these sites have a much deeper molecular potential field, as a result of the overlapping of molecular potential from the double tube walls. This work points to a clear possibility of using these nanotubes for future hydrogen storage applications (Reinhold, 2006).

NANOFLUIDICS IN NANOTUBES

A combination of transport efficiency and selectivity have conferred on CNT membranes a high potential for the next generation of water desalination, water purification, nanofiltration, and gas separation applications. Molecularly smooth hydrophobic graphitic walls and nanoscale inner diameters of CNTs are well suited for ultra-efficient transport of water and gas through these ultra-narrow molecular pipes. Water and gas molecules move through nanotube pores much faster than through other pores of comparable size. Their water transport mechanism resembles the transport mechanisms of biological ion channels (Noy *et al.*, 2007). CNT membranes have also become a versatile and robust nanoscale platform for fundamental studies of nanofluidics.

According to the continuum fluid hypothesis, both state and dynamic variables such as density, temperature, and velocity of fluid flow vary in a continuous and differentiable manner with respect to the position in the fluid and to time. The continuum theory is independent of the nature of the molecular structure and configuration (Batchelor, 1967). Advances in nanoscience have propelled the dimensions of synthetic objects closer to the molecular scale but pose an important issue: does the continuum-based description of fluid flow work in a situation where the reduction in the system size causes those variables to vary significantly over the molecular length and relaxation time scales? Answering this question requires estimation of the smallest size of water droplet that still can be treated as a continuum object. To obtain less than 1% statistical fluctuation or resolution of a physical property we need ~10000 particles. A sphere containing 10000 equidistant water molecules has a diameter of ~27 molecules or ~6.5 nm.

Some have suggested that continuum flow theory might be used to predict bulk fluid flow in channels as small as about 2 nm if the viscosity variation near the channel wall is taken into account. It is now established that flow through CNTs with lateral dimensions of less than 2 nm would fall in the domain of the nascent field of noncontinuum fluidics, or nanofluidics (see Noy *et al.*, 2007).

The CNT has already become the iconic molecule of nanoscience (Terrones, 2003). Even though it has a very simple chemical composition and structure, a CNT can display a huge variety of unique properties. A CNT is simply a nanosized rolled-up graphene sheet that forms a perfect seamless cylinder which is capped at the ends by fullerene caps. The structure of the nanotube is characterized by its rolled-up vector, termed chirality or helicity, which defines the position of the matched carbon rings during the roll-up of the graphene sheet (Dresselhaus *et al.*, 1998). This roll-up vector defines the nanotube morphology, its diameter, and its electronic properties.

A CNT either has one (as in case of a single-walled CNT or SWNT), or several concentric graphitic shells (as in multiwalled nanotubes or MWNTs). A most remarkable property of this structure – one that makes it so attractive for transport applications – is its unique combination of extremely high aspect ratio with small dimensions: a nanotube can be several millimetres long and yet retain a diameter of only a few nanometres. Another important property critical for transport applications is the remarkable atomic scale smoothness and chemical inertness of the graphitic walls of CNT.

CNTs can be produced by several methods. In the lab, catalytic chemical vapour deposition (CVD) is preferred over arc discharge and laser ablation because it produces higher quality CNTs. CVD reactors yield individual isolated nanotubes, as well as densely packed vertically-aligned arrays. Unfortunately, the long cherished aim of CNT synthesis, producing a uniform population of nanotubes with a given chirality has not been realized. The size of the catalyst particle during growth determines the size of the CNT to less than 10% (Yamada *et al.*, 2006); yet the attempts to control the size of CNTs with greater precision have failed, and synthesizing a vertically aligned CNT array with a narrow size distribution is still extremely difficult.

It is indeed surprising that hydrophilic liquids, especially water, can enter and fill so narrow and hydrophobic CNTs. If water does enter CNTs, what influence does extreme confinement exert on the water structure and properties? How these changes in structure influence the rates, efficiency, and selectivity of the transport of liquids and gases through CNTs? Hummer *et al.* (2001) used simulations to observe the filling of a (6,6) CNT (0.81 nm in diameter and 1.34 nm in length) with water molecules and found that water fills the empty cavity of a CNT within a few tens of picoseconds; also that the filled state persists over the entire simulation time (66 ns). The water molecules confined in such a small space form a single-file configuration that is not seen in the bulk water. Maniwa *et al.* (2007) also demonstrated water filling of CNTs. Water molecules inside and outside a nanotube appear to be in thermodynamic equilibrium. One important phenomenon associated with nanofluidics system is that nanoscale confinement leads to a narrowing of the interaction energy distribution, which lowers the chemical potential; confining a liquid inside a nanotube channel actually lowers its free energy! Further simulations revealed that the filling equilibrium is very sensitive to water-nanotube interaction parameters: whereas a 40% reduction in the carbon-water interaction potential results in the emptying of the CNT cavity, only a 25% reduction results in a fluctuation between filled and empty states (bi-stable states) (Hummer *et al.*, 2001). Such a sharp transition between the two states has been recorded for other hydrophobic nanopores as well (Beckstein and Sansom, 2003).

CNTs as Biological Channel Analogs

One defining feature of water structure in CNTs is the formation of hydrogen-bonded 'water wires' oriented along the nanotube axis (Kalra *et al.*, 2003). Such one-dimensional hydrogen-bonded structures bring to mind the water wires observed in biological channels specializing in water transport, such as aquaporins (Sui *et al.*, 2001). The similarity between aquaporin and CNT channels is quite striking: like the hydrophobic interior of CNTs, the inner cavity of the aquaporins is lined with hydrophobic residues that facilitate formation of the one-dimensional hydrogen-bonded water chains. Weak interactions of water molecules with the hydrophobic walls, in conjunction with the smooth nature of the walls, enable smooth, frictionless water transport. Kalra *et al.* (2003) showed that water moves very rapidly through a nanotube channel under osmotic pressure, and friction at the channel walls in the system is so low that water transport depends largely on happenings at the nanotube entrance and exit. The calculated rates of water transport approach 5.8 water molecules per nanosecond per nanotube – comparable to water transport rates in aquaporins.

One-dimensional water wires play a critical role in proton transport in biological structures such as bacteriorhodopsins and cytochromes (Schlichting *et al.*, 2000). CNT channels also seem to facilitate proton transport along the nanotube axis. In these simulations, the calculated proton mobility exceeds proton mobility in bulk water by a factor of 40 (Noy *et al.*, 2007).

Sorin and Pande (2006) used CNTs to probe the role of water in the stability of confined proteins. They observed that the protein helix stability is low for the narrowest CNT and increases monotonically with nanotube diameter – contradicting classic polymer theory, according to which peptide helix stability should decrease in this situation. According to them, the primary effect of confinement in this system is related to changes in the solvent entropy, which in turn changes the free energy cost of protein-water hydrogen-bond formation and the relative stability of the peptide helix. Solvent entropy confinement in ultra-narrow channels could possibly affect the rules of protein stability.

Gas Transport in CNTs

The inherent smoothness of the graphene wall of CNTs seems to be a defining feature for the gas phase transport in these channels. Gas flux in CNTs reaches a value almost three orders of magnitude higher than in zeolites with equivalently sized pores (Skoulidas *et al.*, 2002). Our understanding of gas transport through CNTs that has emerged from various simulations hinges on the nature of the molecule-wall collisions inside CNTs: a CNT effectively allows gas transport since most of the gas molecules move along the tube in an almost billiard-ball-like manner.

Fabrication of CNT Membranes

Testing the peculiar predictions of fast transport through CNTs, which emerged from simulations, has necessitated the fabrication of a robust test platform. Sun and Crooks (1999) attempted to observe molecular transport through a single nanotube. But practical considerations require that experimental efforts are focused on a geometry that can allow observation of transport through multiple nanotube pores, such as a *CNT membrane* geometry. These membranes are typically made of an aligned array of CNTs encapsulated by a filler (matrix) material, with the nanotube ends open at the top and bottom. There exist many ways to produce such a structure; the best approach is one that involves growing an aligned array of CNTs, followed by infiltration of a matrix material in between the CNTs. The extremely high aspect ratio of the gaps between the nanotubes in the array (length/diameter on order of 1000 or larger) poses a strong fabrication challenge, yet researchers have developed appropriate strategies to address this issue (see Noy *et al.*, 2007).

Hinds *et al.* (2004) designed a membrane fabrication strategy based on polymer encapsulation of CNT arrays. It involves infiltrating MWNT arrays with a liquid polystyrene precursor which, after curing, produces a high-density MWNT membrane of ~7 nm pore size. As the process occurs in the liquid phase, elaborate procedures are required to ensure that the CNTs do not aggregate in bundles when the solvent evaporates.

Holt *et al.* (2006) developed a method for the encapsulation of a vertically aligned array of CNTs with low-stress Si_3N_x by a low pressure CVD process. This method has been widely adopted for several microfabrication processes as it produces an extremely conformal coating around CNTs. The membrane produced is quite strong and withstands pressure gradients exceeding 1 atm. After encapsulation, the membrane is subjected to certain etching steps to remove excess Si_3N_x from the tips of the CNTs, followed by oxygen plasma to uncap the CNTs. The double-walled CNT (DWNT) membranes so produced consist of pores less than 2 nm in diameter, consistent with diameters of as-grown nanotubes, and no nano- or microvoids are seen.

CNT Polymer Network Fabrication

Kim *et al.* (2007) described a different approach to producing an aligned CNT-polymer composite membrane. Amine-functionalized CNTs were dispersed in tetrahydrofuran and then filtered through a hydrophobic (0.2 μm) polytetrafluoroethylene (PTFE) filter, which aligned the membrane pores. Spin coating with a dilute polymer solution (Polysulfone) produced a mechanically stable thin-film structure in which CNT tips protruded from the top of the membrane. Membranes so produced show enhancements in gas transport rates. This approach is potentially more scalable and economical than direct CVD growth of CNTs on a substrate, but the nanotube density (and hence the available pore density) is much smaller than for the membranes produced by CNT array encapsulation (Noy *et al.*, 2007).

Nanofiltration Properties

The highly smooth graphitic walls of CNT composite membranes offer a welcome combination of extremely fast flow and extremely small pore size, which make them far superior to traditional membrane materials for energy-efficient, low-cost ultra-filtration and nanofiltration applications. Concentration-gradient or pressure-driven flow are used to determine the size exclusion properties of CNT composite membranes. Hinds *et al.* (2004) showed that 10 nm Au nanoparticles are fully excluded by a MWNT membrane having a pore inner diameter of ~7 nm, whereas small dyes (0.5-2 nm) diffuse with low hindrance. Aligned DWNT array membranes completely excluded 2 nm and 5 nm Au nanoparticles in pressure-driven filtration experiments. Interestingly, at the same time the membrane showed fast water permeation (Holt *et al.*, 2006).

Ion Exclusion

Nanoscale diameters and high water permeabilities make CNTs a valuable platform for ion removal from water for desalination and demineralization. Simulations have predicted that if a nanotube is uncharged, size-based exclusion of small ionic species such as Na^+, K^+, or Cl^- requires CNT diameters of ~0.4 nm, comparable to the hydrated ion size (Park *et al.*, 2006). At this scale, the ion loses part of its hydration shell to enter the CNT. Molecular simulations predict that 0.34 nm diameter CNTs, bearing negative charges along the walls, conduct K^+ ions while excluding Cl^-, whereas positively charged 0.47 nm diameter CNTs exclude K^+ ions while conducting Cl^-. Joseph *et al.* (2003) showed that the presence of charged groups on open CNT tips induces preferential ion transport for CNT diameters of 2.2 nm. These results suggest that dedicated functionalization of small-diameter CNT membranes could conceivably enable the control of ionic flow or even the exclusion of very small ions – an exciting prospect for water purification and desalination.

Membrane Functionalization and Transport

Transport through nanotube channels may be controlled by using chemical modifications of the nanotube to change the channel permeability or size exclusion characteristics. It may even be possible to construct controllable 'gate' to regulate the transport efficiency based on an external stimulus. To maintain efficient flow through a nanotube, these treatments should preserve the fundamental smoothness and hydrophobicity of the nanotube walls. This is best done by focusing on modification strategies that target only the entrance and exit of the nanotube. Most membrane fabrication strategies involve the use of an oxidation step to remove the fullerene cap from the nanotube. This typically produces carboxylic groups at the mouth of the nanotube, which then serve as the target for a variety of chemical modification approaches (Wong *et al.*, 1998).

Hinds *et al.* have gainfully used carbodiimide chemistry to attach a variety of organic and biological molecules to the mouth of CNTs. They found that for polymer-matrix-based aligned nanotube membranes, the idealized picture of the oxidation step producing only a ring of carboxylic acid groups on the nanotube end and leaving the rest of the nanotube intact misrepresents reality. Decorating nanotube surfaces with Au nanoparticles showed that oxidation produces reactive groups in regions of the nanotube up to 700 nm away from the tip (Majumder *et al.*, 2005). These workers used carbodiimide chemistry to attach aliphatic chains, charged dye molecules and polypeptides to polystyrene-based CNT membranes. These modifications have a significant effect on the flux of the two large organic cations (methyl viologen (Mv^{2+}), and $Ru(bpy)_3^{2+}$) used: for example, attaching a C_{40} alkane chain reduced the flux of Mv^{2+} by six times.

Majumdar *et al.* (2007) used charged organic dye molecules to regulate transport through CNT membranes as a function of an external applied voltage. An innovative modification procedure was used whereby a constant flow of an inert solution through the nanotube membrane helps to protect the inner walls of the nanotube and localizes the modification to the rim of the nanotube. After modification, it appears that the applied voltage can draw the grafted charged dye molecules from the pore entrance into the nanotube core (Noy *et al.*, 2007).

Future Potential

Nanoscale size and atomically frictionless surfaces make CNTs a unique system for studying molecular transport and nanofluidics. A clear understanding of the behaviour of water and ions in confined geometries is essential for many biological and geological processes. Recent advances have enabled the simulation of molecules inside CNTs. Experimental platforms have been developed that allow the study of molecular confinement inside CNTs and which can validate the predictions of simulations.

Understanding the transport mechanisms and molecular structure within CNTs will allow researchers to utilize the high permeabilities characteristic of CNTs for developing membrane-based transformational technologies for separation and purification. The fast flow and small pore size of CNT membranes can lead to substantial energy reductions in desalination and gas separations. The use of CNT as fluidic channels also paves the way to true molecular-scale transport gating, since the size of the pore not only compares well with the size of large organic molecules but also has important implications for the development of protective gear and drug delivery systems (Noy *et al.*, 2007).

CONTROLLING NANOTUBE REACTIONS

For several potential applications, CNTs need to be chemically modified, but the reactions involved are difficult to control. How to achieve the right balance between chemical reactivity and selectivity – a measure of the number of unwanted side reactions associated with a desired reaction – is a challenging task, best seen in the chemistry of CNTs. In view of their low reactivity and high stability, harsh reaction conditions are needed to make CNTs react. But when they do react, the number of chemical groups that become covalently attached to the carbon skeleton (the degree of functionalization) cannot be controlled easily. It is this tendency that is so critical for many applications of CNTs (Prato, 2010).

"Naked" CNTs cannot be easily handled as they are insoluble in most solvents. They tend to aggregate into tight bundles, which makes them difficult to process, and it becomes difficult to study the properties of individual CNTs. However, attaching suitable chemical groups to CNTs makes them more soluble in organic solvents and even in water. Further, chemical modification helps prevent tight bundles of CNTs from forming, almost completely separating individual CNTs from each other (Tasis *et al.*, 2006). A primary objective is to control the degree of functionalization, because although chemical modification helps in the processing of CNTs, too much modification can adversely affect their desirable properties. One example: strong chemical modification can alter the conductance of CNTs, which decreases considerably upon increasing functionalization (James and Tour, 2005). This effect is related to the type of functionalization and is specially important for CNT-based transistors and for electronic applications in general. Conversely, a high degree of functionalization might be needed for other potential uses of CNTs, such as drug delivery.

The first step of the process is to covalently attach alkyl groups – saturated hydrocarbons – to the side walls of single-walled CNTs (SWCNTs) as per the method of Graupner *et al.* (2006) that provides high yield of functionalized SWCNTs (f-SWCNTs). Syrgiannis *et al.* (2010) further improved Graupner's method by making the process reversible by judiciously modifying the reaction conditions. By treating the f-SWCNTs with sodium or lithium metal in liquid ammonia, the appended alkyl groups can be thinned out and the metals start donating electrons to the nanotubes; when the f-SWCNTs become negatively charged, they 'eject' the attached alkyl groups. Indeed, some of the alkyl groups could be replaced by n-propyl-amino groups merely by adding n-propylamine ($C_3H_7NH_2$) to the reaction mixture. In this way, f-SWCNTs containing two different chemical groups could be made (Prato, 2010).

The ability to attach a wide variety of chemical groups to CNTs has created new possibilities. For instance, electron-donor groups (such as porphyrins and phthalocyanines) have been covalently attached to SWCNTs. Exposure of these nanotubes to visible light enables electrons to transfer from the donors to the SWCNTs (Guldi, 2007), mimicking a crucial step in natural photosynthesis: the generation of charge separation to produce photocurrents.

Another example is that CNTs modified to become water soluble can cross cell membranes readily. In principle, if drug molecules are attached to such CNTs, then CNTs might be used as drug-delivery vehicles (Kostarelos *et al.*, 2009). This might be particularly useful for drugs showing low bioavailability i.e., failure to reach high concentrations in the body either because they are extensively metabolized or are inefficiently absorbed. The anticancer drugs doxorubicin, cisplatin and paclitaxel (Taxol) have all been attached to CNTs, for example, and clinical trials are being planned.

NANOTUBE PHOTODIODES

A single-walled carbon nanotube (SWNT) is like a rolled sheet of graphene because it generates many species of one-dimensional charge carriers. For a semiconducting SWNT, the band-gap energy is the energy required to generate an electron-hole (e-h) pair whereas higher subband energies correspond to excitations with greater effective mass.

SWNTs have quite small Fermi velocity and low dielectric constant which indicate that high-energy carriers should efficiently generate e-h pairs. Highly efficient e-h pair generation could potentially lead to highly sensitive photodetectors (Yang *et al.*, 2006), electroluminescent emitters (Chen *et al.*, 2005), and improved-efficiency photovoltaics (Lee, 2008; Tian *et al.*, 2007). Multiple electron-hole pair generation could in principle enhance the efficiency of photovoltaic solar cells beyond standard thermodynamic limits (see Gabor *et al.*, 2009). As suggested by Ellingson *et al.* (2005) and Kim *et al.* (2008), such processes do occur in semiconductor nanocrystals, but the results are difficult to interpret.

Gabor *et al.* (2009) studied electron-hole pair generation in p-n junctions consisting of individual SWNTs in a split-gate field-effect geometry as described in Bosnick *et al.* (2006). Gabor *et al.* reported highly efficient generation of electron-hole pairs due to impact excitation in single-walled carbon nanotube p-n junction photodiodes. Optical excitation into the second electronic subband led to striking photocurrent steps in the device characteristics that occur at voltage intervals of the band-gap energy. Spatially and spectrally resolved photocurrent combined with temperature-dependent studies suggested that these steps result from efficient generation of multiple electron-hole pairs. This process is not only of fundamental interest but is also relevant to applications in future ultra-efficient photovoltaic devices (Gabor *et al.*, 2009).

Potentially, the e-h pair creation process observed by Gabor *et al.* could enable better power conversion efficiency in future photovoltaic devices. The work of Gabor *et al.* shows that multiple e-h pairs can be generated and collected in a nanotube p-n junction. Although actual creation of devices exploiting e-h pair generation into photovoltaic cells will require much future effort, the process has strong implications for such technology and represents a milestone toward ultra-efficient photovoltaic devices with high power conversion efficiency.

AMBIPOLAR-TRANSPORTING COAXIAL NANOTUBES

Development of solar cells which can efficiently convert light energy into electrical outputs can significantly relieve the imminent energy problem. Silicon-based solar cells have already been commercialized but there is still heavy demand for organic solar cells as organic materials are quite flexible and easy to process (Brabec *et al.*, 2008). A suitable configuration for organic photovoltaics (PVs) is made of well connected hole- and electron-transporting layers (p/n heterojunction) formed from electron-donating (D) and electron accepting (A) molecular components, respectively, with neither charge-transfer complexation nor macroscopic D/A segregation (Thompson and Frechet, 2008). In recent years, considerable progress has been recorded by the "bulk heterojunction" approach (Yu *et al.*, 1995; Li *et al.*, 2005; Kim *et al.*, 2007). In this, electron-accepting molecules such as fullerenes are integrated into hole-transporting media composed of π-conjugated polymers. The problem with this useful approach is that the interface for the resulting p and n domains is, in principle, difficult to tailor at the molecular level. To address this limitation, molecular assembly of covalently or noncovalently connected D-A modules is being examined (Hoeben *et al.*, 2005). Dye-sensitized wet solar cells have been fabricated by a modification of electrodes through layer-by-layer assembly of oligo(p-phenylene)/naphthalenedimide and prophyrin/C_{60}, designed to make a heterojunction (Yamamoto *et al.*, 2009). Yet, despite much research, rational design of molecularly engineered D/A heterojunctions, leading to PV outputs, has not been accomplished.

Despite a large steric bulk of C_{60}, a molecular graphene with a covalently linked C_{60} pendant [hexabenzocoronene (HBC)-C_{60})], self-assembles into a coaxial nanotube whose wall is made of a graphite-like π-stacked HBC array, whereas the nanotube surface is covered by a molecular layer of clustering C_{60} (Brabec *et al.*, 2008). In view of this clear coaxial configuration, the nanotube shows an ambipolar character in the field-effect transistor output [hole mobility (μh) = 9.7×10^{-7} cm^2 V^{-1} s^{-1}; electron mobility (μe) = 1.1×10^{-5} cm^2 V^{-1} s^{-1}] and shows a photovoltaic response when illuminated with light. Successful coassembly of the above (HBC)-C_{60} and an HBC derivative without C_{60} makes it possible to tailor the p/n heterojunction in the nanotube, so that its ambipolar carrier transport property can be optimized for increasing the open-circuit voltage in the photovoltaic output (Yamamoto *et al.*, 2009). As evaluated by an electrodeless method involving flash-photolysis time-resolved microwave conductivity technique, the intratubular hole mobility (2.0 cm^2 V^{-1} s^{-1}) of a coassembled nanotube containing 10 mol % of HBC-C_{60} is as large as the intersheet mobility in graphite. The homotropic nanotube of the HBC derivative without C_{60} blended with a soluble C_{60} derivative [(6, 6)-phenyl C_{61} butyric acid methyl ester] has shown a photovoltaic response with a much different composition dependency, where the largest open-circuit voltage attained was lower than that realized by the coassembly of (HBC)-C_{60} and the HBC derivative without C_{60}.

SIGNAL MULTIPLICATION WITH Y-SHAPED NANOTUBES

Knowledge of molecular scale signal conversion and multiplication has attracted many physical and biological applications, such as molecular switches, nano-gates, biosensors, and various neural systems in recent years (Kwok and Ellenbogen, 2002; Xu, 2005; Litvinchuk *et al.*, 2007; Koenig *et al.*, 2008). However, being extremely complex, these molecular systems have prevented clear and detailed understanding of the molecular processes affected by noises arising from thermal fluctuations as well as interferences between branch signals. Carbon nanotubes, conversely, have much greater potential for applications such as nanoscale sensors, devices, and nanomachines (Ghosh *et al.*, 2003; Holt *et al.*, 2006; Gong *et al.*, 2007; Krasheninnikov and Banhart, 2007; Whitby and Quirke, 2007; Köfinger *et al.*, 2008). As water molecules confined within nanoscale channels are known to show structures and dynamics very different from bulk (Joseph and Aluru, 2008; Ball, 2008), this could provide a medium for molecular signal transmission. Given a suitable nanochannel radius, water molecules in the nanochannel can form a single hydrogen-bonded chain, with the "concerted" water orientations (i.e., water dipoles ordered cooperatively inside the carbon nanotube) (see Gong *et al.*, 2007). Hummer *et al.* (2001) estimated the time for reorientation of the dipole chain to be 2 to 3 ns for carbon nanotubes of a length of 1.34 nm. The water molecule chains in a nanotube can remain dipole-ordered up to lengths of approximately 0.1 mm, with durations up to approximately 0.1 s (Köfinger *et al.*, 2008). This means that if we could "tune" the orientation of a water molecule with a charge at one end, it might be possible to control the orientations of water molecules in various chains. This would involve a molecular-level "signal transduction", i.e., converting a charge signal at one end to water dipole orientation signal and then transmitting and multiplying it to many other ends (Tu *et al.*, 2009).

Several workers have reported fabrication of Y-shaped nanotubes by such different methods as alumina templates, chemical vapour deposition of products generated from a pyrolysis of metallocenes, nano-welding of overlapping isolated nanotubes using high-intensity electron beams, and spontaneous growth of nanotube mats using Ti-doped Fe catalysts (Satishkumar *et al.*, 2000; Deepak *et al.*, 2001; Terrones *et al.*, 2002). Those nanotubes exhibit not only electrical switching but also logic behaviour (Xu, 2005; Bandaru *et al.*, 2005). Single-file water molecules confined within a Y-shaped nano-channel can perform the signal conversion and multiplication, ignited by a single electron, due to the surprisingly strong interactions of water molecules at the Y-junction.

Tu *et al.* (2009) used molecular dynamics simulations to show that a signal at the single-electron level is converted and multiplied into 2 or more signals by water chains confined in a narrow Y-shaped nanochannel. This remarkable transduction capability is related to the fact that the concerted water orientations in the 2 branches of the Y-shaped nanotubes can be well modulated by the water orientation in the main channel. The response to the switching of the charge signal is very rapid, from a few nanoseconds to a few hundred nanoseconds. Furthermore, simulations with various water models, including TIP3P, TIP4P, and SPC/E, show that the transduction capability of the water mediated signal with Y-shaped carbon nanotubes is very robust at room temperature, while the interference between branch signals is negligible. This may possibly be the first such observation of the remarkable water-mediated signal conversion and multiplication with a Y-shaped nanotube at a molecular level (Tu *et al.*, 2009). Tu *et al.* actually used (6,6) uncapped armchair single-walled carbon nanotubes (SWNTs) in their Y-shaped carbon nanotube system.

Although these findings were from a specific size of SWNT, it is hoped that these phenomena would be replicable for other nano-channels, provided that the water molecules inside the channels are arranged in a single file. The observations reported by Tu *et al.* may have significance for future applications in molecular-scale electronic devices and biological systems.

NANOTUBE SPEAKERS

Whereas traditional loudspeakers produce sound by vibrating back and forth, new speakers made from sheets of CNTs produce music in the same way that lightning generates thunder. Physicists at the Tsinghua-Foxconn Nanotechnology Research Center in Beijing unexpectedly discovered that when an audio-frequency electric current was applied to stretchable, flexible transparent films of 10-nm thick CNTs, they could make sounds as loud as commercial speakers. It seems that the electrified nanotubes heat and expand the air near them, producing sound waves. It is possible to stretch these loudspeaker membranes up to twice their original length without breaking and with little change to sound intensity. Conceivably, these sheets might be placed over paintings, windows, video screens, or clothing, as a means to keep a person warm (see *Nano Lett.* Online Oct. 29, 2008).

INTERCALATING SEMICONDUCTOR NANOPARTICLES IN CARBON NANOTUBE BUNDLES

SWNTs have attracted several applications in nanoelectronics, pressure and flow sensors and actuators, high-strength composites, electron emitters and hydrogen storage (Baughman *et al.*, 1999; Zhou *et al.*, 2001; Rao *et al.*, 2001; Dalton *et al.*, 2003; Gundiah *et al.*, 2003). SWNTs can assemble to form a triangular lattice with an inter-tubular gap of ~3.2Å. Nath *et al.* (2003) showed that SWNT bundles can serve as templates to form nanocrystalline semiconducting systems in the van der Waals inter-tubular gaps. CdSe, CdS and ZnS quantum wires have been successfully generated in this way.

The lengths of the intercalate inside the bundles have been varied by changing the reaction conditions. Electronic absorption and photoluminescence studies from the semiconductor intercalates revealed the expected blue-shift with respect to the corresponding bulk samples in CdS and ZnS samples. The SWNT lattice expands on incorporating CdSe (Nath *et al.*, 2003).

Absorption and photoluminescence bands from intercalated CdS and ZnS nanoparticles show blue-shifts as expected from quantum size effects. X-ray diffraction measurements showed that whereas the inter-tube van der Waals gap in the nanotube bundles increases by the intercalation of CdSe, it is almost unchanged in the case of ZnS.

HYBRID CARBON NANOTUBE-CLAMPED METAL ATOMIC CHAIN

Metal atomic chain (MAC) is a one-dimensional structure having unique physical properties such as quantized conductance, colossal magnetic anisotropy, and quantized magnetoresistance. Electron transport across an MAC leads to ultimate miniaturization and novel functionality of electronic devices (Ohnishi *et al.*, 1998; Sokolov *et al.*, 2007; Smogunov *et al.*, 2008). MACs have great potential as useful components of nanoscale electronic and spintronic devices. Techniques such as scanning tunneling microscopy (STM), mechanically controlled breaking junction high resolution transmission electron microscopy (HRTEM), self-terminated electrodeposition and controlled electromigration have been developed for the production of suspended MACs (Kondo and Takayanagi, 1997; Chopra *et al.*, 2005; Bolotin *et al.*, 2006). However, as MACs are often suspended between two-macroscale metallic electrodes, significant technical challenges face the interconnection and integration of MACs. Tang *et al.* (2010) addressed such challenges and reported a CNT-clamped MAC in which CNTs play the roles of both nanoconnector and electrodes. A fabrication strategy was designed by which a metal nanorod is filled into a CNT, after which the carbon shell covering the filled metal nanorod is peeled off and the exposed metal nanorod is thinned by strong electron irradiation within a TEM. Finally, a CNT-clamped MAC is formed through further elongation and thinning by the impact of the strain in the TEM sample or by stretching when a TEM-STM holder is employed. This nanostructure can be made by *in situ* machining of a metal-filled CNT.

It includes peeling off carbon shells by spatially and elementally selective electron beam irradiation, followed by some elongation of the exposed metal nanorod. The microstructure and formation process of this CNT-clamped MAC were studied by both TEM observations and theoretical simulations. Strong covalent bonds appear to form between the CNT and MAC. Tang *et al.* also experimentally measured the electrical transport property of the CNT-clamped MAC and observed quantized conductance. The enigmatic characteristics of Fe atomic chain (ACs) such as half-metallicity and quantized conductance were observed to be retained after the combination of Fe ACs and CNTs. As the strategy proposed by Tang *et al.* seems to be quite effective in fabricating various MACs and *in situ* connecting these MACs with CNTs, it is expected to attract potentially useful applications in the assembly of nano/subnanoscale devices.

CARBON MATERIALS: ENERGY-RELATED APPLICATIONS

Carbon materials are exemplified by fullerenes, nano-sized single and multi-walled carbon tubes, graphene, and carbon foam. As they are inert and have good electrical conductivity, carbon materials can be extremely useful as electrodes in re-chargeable batteries, storage media for fuel cell for on-board hydrogen supply, fuel cell components, nano-electronic devices for computer chips, and as superconductors. Their high surface area and layered structure also make carbon materials attractive for these applications.

Carbon materials are particularly valuable for producing, transmitting and storing energy. They can facilitate development of low-cost, cleaner and more energy-efficient technologies (see Srivastava *et al.*, 2009). In its various electronic configurations, carbon can form structures such as diamond, graphite, fullerenes, CNTs and graphene. Starting with a suitable carbon material and selecting certain processing routes, several modifications of carbon (carbon fibres, carbon foam, meso-carbon-micro beads, carbon-carbon composites etc.) can be made (Srivastava *et al.*, 2001).

Energy Generation

The electrolyte used in a Leclanche battery is a mixture of carbon black, powdered graphite, manganese dioxide and an acid electrolyte. Carbon black is produced by partial combustion of acetylene or aromatic oils. The electrical conductivity of carbon black is lower than that of graphite but it retains more electrolytes. This is why graphite and carbon black are used in electrolyte, depending upon high conductivity (camera flash light) and low conductivity (torch) applications. In automobile batteries, carbon black along with short carbon fibres, enhances electrical conductivity and maximizes surface areas.

Being a renewable clean source of energy, a lithium-ion re-chargeable battery is an important component of many popular portable electronics for entertainment, computing and telecommunication equipment. In Li-ion batteries, Li^+ ions move between guest-host type arrangements acting as cathode and anode. Such a movement prevents reduction of Li^+-ion to the toxic lithium metal (Alper, 2002). The major hurdle in the development of these batteries is non-availability of a suitable electrode material that will not degrade over repeated discharge-recharge cycles. Other requirements for use as anodic materials in Li-ion battery are high capacity, excellent reversibility and high degree of lithiation. Some carbon materials (graphite, carbon fibres, meso-carbon micro-beads, pitch and coke) have been tested as anode material (Flandrois and Simon, 1999; Suzuki *et al.*, 1999). These carbons fall into three categories, viz., soft (pitch based), hard (polymer based) and graphite. Graphite is usually preferred because of its low cost, high lithiation capacity and high energy-density.

Fuel Cells

A fuel cell converts chemical energy into electrical energy in a clean and energy-efficient manner. Oxygen and hydrogen react together to produce energy and water as by-products. H_2, which powers a fuel cell, is produced either from natural gas or fossil fuels. Although fuel cell was developed half a century ago, it has not yet been commercialized. The chief problem is the unavailability of suitable H_2 storage material for continuous and consistent supply of H_2, and the availability of appropriate fuel cell components is also scarce (Steele and Heinzel, 2001). However, two promising fuel cell technologies, viz., polymer electrolyte fuel cell (PEFC) and phosphoric acid fuel cell (PAFC) have been successfully developed by using components (bipolar plates, catalyst-supports, gas-diffusion system, current-collector plate) made of carbon-based materials, that are highly effective.

Bipolar plates of fuel cells fabricated from graphite/polymer show high electrical conductivity. Being hydrophobic, gas diffusion layer made from woven carbon cloth or carbon paper is not affected by water produced during reactions. In fuel cell, platinum metal dispersed on carbon support is used as electro-catalyst. A collector plate made of graphite minimizes gas diffusion in the fuel cell.

Graphite also increases electrical conductivity. Exfoliated graphite has better electrical conductivity and performance. Membranes prepared from catalyst-filled CNTs and used in fuel cell for electrocatalyzing oxygen reduction and methanol oxidation are highly effective (Che *et al.*, 1998).

Many automobile manufacturers are developing prototype fuel cell cars. Honda has already launched a petroleum-free fuel cell car. Use of fuel cell for power generation in spacecraft also attracts keen interest (Hyder *et al.*, 2000).

Energy Storage

H_2 is a clean fuel. On combustion, it produces only energy and water vapour. The energy density (38 kWh/kg) is higher than gasoline (14 kWh/kg), making H_2 a very attractive transportation fuel. But the problem with H_2 technology is the lack of availability of a safe and practical storage device that could conveniently load and unload H_2 supply continuously and consistently to the fuel cell.

Carbon materials offer high surface area, cavities, and micro-pores for H_2 adsorption. CNTs seem good candidates for H_2 storage (Takagi *et al.*, 2004). CNTs can be produced by arc-discharge, laser ablation and chemical vapour deposition. SWNTs are made of individual graphene sheets rolled into seamless hollow cylinders (diam. 1-20 nm). MWNTs are composed of some tens of cylindrical graphitic sheets. CNTs (Gordon and Saeger, 1999) have large external surface and internal hollow cavity where gas can be stored by chemisorption or physical sorption.

H_2 can be adsorbed on the outer walls of CNTs but cannot be stably adsorbed inside the cavity (Lee and Lee, 2000). In case of MWNTs, H_2 can be adsorbed between layers of CNTs; however, presence of H_2 between tubes increases the radius of CNTs and so destabilizes them. In case of bundled CNTs, H_2 can intercalate between CNT bundles. The chief obstacle in commercializing CNT based H_2 storage technology is the exorbitant cost of CNTs (~50-600 $/g), depending upon their purity.

Carbon nano-fibres (Chambers *et al.*, 1998) have been explored for storing H_2. These are stacked graphite layers (diam. 5-100 mm; length, 5-100 mm), and can adsorb H_2 up to 50% wt. at ambient temperature and high pressure (see Srivastava *et al.*, 2009). These fibres are made by interaction of metal-catalyst nanoparticles with hydrocarbon vapour at high temperature.

Fullerene is another potential carbon material for H_2 storage. It contains only carbon atoms – unlike diamond and graphite – arranged in the form of soccer balls. Most stable fullerenes are C_{60} and C_{70}, composed of 12 pentagons, and 20 and 25 hexagons respectively. C_{60} forms compounds like $C_{60}H_{24}$, $C_{60}H_{36}$ and $C_{60}H_{48}$. Further addition of H_2 atoms to a fullerene ruptures its cage. $C_{60}H_{48}$ represent 6.3 wt% of H_2 adsorption. Chen *et al.* (1998) reported that fullerene can adsorb H_2 up to 6 wt% at 180°C and ~25 bar. But, desorption of H_2 requires temperatures of at least 225°C. If desorption temperature could be brought down, fullerenes can form a good material for H_2 storage.

Mechanical Storage

Besides H_2 storage, CNTs may also be suitable for mechanical energy storage. They act like a spring on applying 25 Kbar pressure – so this property of CNTs can be exploited for designing energy absorbing composite materials.

Energy Transmission

Depending upon their chirality (armchair or zigzag) and diameter, CNTs act as conductor or semiconductor in portable, high-speed robust electronic devices like flat panel displays, portable cell phones, or computer chips. Fullerenes make promising material for making electronic textiles, molecular transistors, and electromagnetic interference (EMI) shielding. Intermolecular versatility of CNTs makes them suitable for producing nano-circuits for electronic devices.

Carbon Nano Electronics Company (CNCL, 2008) has successfully developed a conductive textile made of surface grafting of C_{60} conjugate polymers for such electronic applications as intelligent chem/bio-electronic sensing and wearable computing system, miniature electronic gadgets, and miniature power sources. CNCL has also developed antistatic coating for electronic devices from nanoparticles of a fullerene-polyaniline conjugate. Fullerenes-polyaniline based CNTs are used as robust electromagnetic interference (EMI) shield coating for laptops, cellular phones, electro-chemical switches, remote control devices and microwave ovens.

Graphene also holds great potential for the development of single-electron transistor and composite materials etc. (see *Nature Materials* 6: 169, 2007). Graphene-based microprocessors may appear in the market in the new few years.

CARBON NANOTUBES FOR POLYMER ELECTROLYTE MEMBRANE FUEL CELLS

Different types of carbonaceous materials have been developed for various applications such as heterogeneous catalysts, field emitters, supercapacitors, Li-ion batteries, gas sensors, nanoprobes, actuators, hydrogen storage, photovoltaic cells and fuel cells (see Steen and Prinsloo, 2002; Frackiowiak and Beguin, 2001; Baughman *et al.*, 1999; Ago *et al.*, 1999; Rajalakshmi *et al.*, 2000; Zuttel *et al.*, 2002; Kymakis and Amaratunga, 2002; Luxembourg *et al.*, 2004; Penza *et al.*, 2004). Carbon is now being engineered for special purposes to satisfy application-specific criteria such as surface area, particle size, pore size distribution and electronic conductivity, so that each criterion can be adjusted for optimum performance. An example is that of partially disordered graphitic carbon which seems more suitable for reversible Li storage in batteries than other forms of carbon (Nitin and Joachim, 2009).

Since their discovery in 1991 by Iijima, carbon nanotubes have been in focus for their inertness under various chemical environments, highest Young's modulus, electrical conductivity, high surface area, light weight and easy interfacing capability with many inorganic and organic compounds. Their unique electronic structure boosts the catalytic activity of the supported metal besides providing mechanical integrity (see Kannan and Pillai, 2009).

Among the various different energy related applications of CNTs, **polymer electrolyte membrane fuel cell (PEMFC)** represents one promising case for greening the world from a renewable energy point of view. PEMFCs exemplify remarkable environment-friendly options for electricity generation, if hydrogen is produced by solar technologies without any carbon foot print. Fuel cells also show much better efficiencies over commercial combustion engines if combined heat-power generation is targeted. Recent innovations in nanostructured materials have further enhanced the future potential of fuel cells both for stationary and mobile applications, even though they still face some challenges in practical applications such as inexpensive and durable materials as electrodes, electrolytes and other components.

Carbonaceous materials have played a crucial role in the performance of PEMFC. Their use in composite electrolyte membranes is more recent, especially after their functionalization to impart enhanced proton conductivity to the composite electrolyte.

The commercialization of PEMFCs has been hampered by two key factors: (a) longevity or durability of some key components like membranes and electrodes, and (b) the high cost of fuel cell systems. Whereas the durability of a PEM fuel cell accrued mainly from the performance of the membrane electrode assembly (MEA) of the stack, the cost reduction depends on enhancing the performance of the MEAs while minimizing the Pt content (Bahr *et al.*, 2005). Failure usually occurs by the breakdown of polymer electrolyte membrane and the loss in active surface area of the Pt electrocatalyst with time, besides corrosion of the support. The life time and cost as a function of performance is strongly connected with design, materials and operation strategies.

The cost target fixed by the US Department of Energy (DOE) for PEM fuel cell stack is 30$ per kW by 2015, which is much less than the current value of 110 $ per kW. Most of this reduction must come from the Pt catalyst, bipolar plate and PEM. Different types of inexpensive carbon could accomplish some of these objectives as carbon with varying properties can be prepared to meet these technology-specific requirements. This is why carbon has been thoroughly investigated by many research groups as a support material in PEMFC (Yuan *et al.*, 2007; Lin *et al.*, 2009; Beena *et al.*, 2009).

Previous attempts to address these challenges involved use of three different forms of carbon viz., activated carbon, carbon black and graphite or graphitized materials as the primary choice of support for catalyst materials in different types of fuel cells. The commonest form of carbon used in a fuel cell electrode (including that in gas diffusion layers) has been Vulcan XC-72, a kind of activated carbon having moderate surface area (250 m^2/g) and good electrical conductivity; this is at variance with the usual choice of activated carbon with a surface area of more than 3000 m^2/g for certain other applications. The mesopores in Vulcan XC-72 result in part from the Pt nanoparticles being buried deep inside the pores and thereby becoming inaccessible to the triple phase boundary (TBP) formation, which is essential for sustaining the electrode reactions in fuel cells. Further, Vulcan XC-72 undergoes corrosion (more common under conditions of peroxide intermediate formation in fuel cell cathodes) that leads to aggregation as well as dissolution of Pt nanoparticles.

CNTs with their unique properties can effectively address some critical problems. Nitrogen or boron doped CNTs can replace Pt as an electrocatalyst. It might also be feasible to develop new **nanostructured electrocatalysts.** Kannan and Pillai (2009) reviewed recent work on the use of CNTs in polymer electrolyte fuel cells in both electrodes and electrolytes and pointed out their multifunctional role as catalyst layer, support, and reinforcing component in polymer composite membranes (see Kannan *et al.*, 2008, 2009, 2009a).

CARBON NANOTUBES AS ELECTRODE MATERIAL FOR PEMFCs

Effective utilization of Pt nanoparticles decreases the cost of fuel cell stacks – very low Pt loading (a few hundred micrograms/cm^2) without any change in performance is essential. This is achieved both by effective distribution of Pt on the supporting material with a high surface area, and higher electrical conductivity (Sun *et al.*, 2003). For this purpose, a good support material for PEMFC electrodes is one that can satisfy the following requirements: high surface area; chemical stability under oxidative/reductive conditions; mechanical robustness under both open and closed circuit conditions, and good electrical and thermal conductivity.

Many of these conditions are met very effectively by CNTs; however, CNTs are quite costly.

One chief trouble with commercial electrode formulations is the isolation of carbon particles by the use of Nafion as a binder in the catalyst layer. The insulating nature of binder tends to block the access of Pt particles associated with carbon to the external circuit due to lack of electrical network and hindering further utilization of Pt. But when Pt nanoparticles are deposited on CNTs, effective electrical contact with external circuit avoids the problems associated with the presence of a thin insulating layer of Nafion covering the carbon particles (Wang *et al.*, 2005). Careful use of CNTs can address many problems which other forms of carbon-based electrodes cannot, in power source related applications (Kannan and Pillai, 2009; Kannan *et al.*, 2009, 2009a).

Wang *et al.* (2005, 2006) investigated the use of CNTs in PEMFC electrodes, especially in the cathode, to improve Pt utilization. Their work on depositing 4-nm Pt nanoparticles on CNTs improved current and power density in all the regions (i.e., activation, ohmic and mass transport regions). Britto *et al.* (1999) had earlier reported improved charge transfer at CNT electrodes. Enhanced mass transport may also be achieved in CNT-based electrodes if ohmic loss is reduced. The problem in this area is related to water clogging in the cathode which limits the performance of PEMFC under high humidity conditions (Wang *et al.*, 2006).

Being generally hydrophobic, CNTs tend to control gas diffusion. Oriented CNTs show greater hydrophobicity than disordered CNTs. Li *et al.* (2005) oriented the CNTs by a filtration method followed by transfer to the membrane. Functional groups such as –COOH were introduced by refluxing with concentrated HNO_3/H_2SO_4 mixture followed by *in situ* chemical reduction of Pt precursor solutions for proper anchorage of the Pt nanoparticles on to the CNT surface. After the chemical reactions, the functionalized CNTs can be filtered through a hydrophobic nylon membrane with precisely controlled pore size/distribution in the oriented CNTs. Such oriented CNTs elicit better fuel cell performance as compared to that by randomly aligned CNT based electrodes and Pt/C based electrodes.

Liang *et al.* (2001) attempted to bring down the fuel cell cost by reducing the particle size and by increasing the Pt distribution by a wet chemical modification route. They stabilized Pt nanoparticles of 1-3 nm while simultaneously maintaining a uniform Pt distribution due to the anchoring groups present on the CNTs surface.

Considerable work has been done on preparing CNT-based electrodes to increase the Pt utilization by varying preparation conditions such as chemical reduction in a formaldehyde bath, and electrochemical deposition of Pt on CNTs. Another possibility involves growing CNTs directly on the carbon paper support in order to reduce the ohmic resistance, and the modification of the reduction method to prepare smaller nanoparticles with narrow size distribution (see Kakade and Pillai, 2008; Kannan *et al.*, 2009; Lin *et al.*, 2010).

In view of the highly problematic nature of the hydrogen oxidation reaction, only a few reports are available for anode supports. The real challenge in the anode is how to sustain performance by using the reformed H_2 which contains some CO. Catalyst systems (e.g., Pt-Ru) that show good activity towards methanol oxidation reaction may possibly elicit greater tolerance for CO. As stronger metal support interactions can facilitate the electron transfer between the metal and support during electrochemical reactions, Pt nanoparticles deposited on CNTs tend to show greater catalytic activity than that of unsupported metal. Notable improvements have been reported for methanol oxidation on Pt surfaces with and without the presence of CNTs: significant enhancement in the oxidation current of 50-60 mA/cm^2 for CNT/Pt electrode, as against only 6 mA/cm^2 for unsupported Pt. While this enhancement may partly be due to the increased surface area, the kinetic aspects revealed a clear improvement in the catalytic activity of Pt that is supported on CNTs. The work done by Villers *et al.* (2006) clearly points to the possibility of easy removal of the adsorbed CO at a much lower onset potential, and could potentially elicit better performance even with increased CO level in the hydrogen stream.

CNTs in Composite Electrolytes of PEMFCs

Solid polymer electrolyte membrane is a key material that affects the performance and cost of the PEMFCs as electrolyte in MEAs. A good PEM should possess high protonic conductivity, yet be electrically insulating to avoid short circuiting. It should have very low permeability values towards such fuels as hydrogen, methanol and ethanol. It should also be highly stable chemically to withstand acidic conditions prevailing in fuel cell operating environment, as well as high mechanical stability to withstand the stresses of stack fabrication. PEMs may be classified into two groups: those operating at temperatures less than 100°C often with perfluorosulphonic acid (PFSI) electrolyte, and those operating above 100°C with a variety of new thermally stable polymeric electrolytes. In the latter kind, life time data for more than ~5000 hours has rarely been reported despite the availability of many types of polymeric, blended and organo-inorganic hybrid membrane electrolytes. Consequently, these have not been commercialized. Considering Nafion as a typical ionmer, Ma *et al.* (2004) discussed some promising aspects of CNT-based polymeric composite electrolytes for PEMFC applications for temperature less than 100°C.

Successful fabrication of composite membranes of CNTs and Nafion at 1:99 wt% after ball-milling and solution casting performed as well as pure Nafion membrane in terms of proton conductivity, but with much less dimensional change for the case of CNT reinforced composite membranes. Thomassin *et al.* (2007) used melt extrusion to incorporate the CNTs on to nafion membrane and recorded reduced methanol permeability to about 60% along with an unusual increase (up to 140-160%) in the Young's modulus in comparison to that of commercial Nafion membranes.

What is the effect of electronic conductivity of CNTs? Certain types of functionalization enhance the conductivity. We should consider the risk of electrical short-circuiting even if only very low amounts of CNTs (<0.1%) are used. Proper dispersion ensures uniform behaviour and the normally reported value of percolation threshold for CNTs in Nafion is around 5-11% depending upon the CNTs used and their mode of preparation. Liu *et al.* (2006) designed a three-phase composite membrane in which a Pt-CNT-Nafion composite membrane is sandwiched between two pure Nafion membrane phases. This led to marked improvement in the performance of Nafion membranes in terms of mechanical stability and water. The addition of CNTs leads to both improved mechanical stability and chemical inertness while the Pt nanoparticles anchored on it produce water by reacting with H_2 and O_2 that penetrate the membrane; the membrane is kept wet under dry conditions (see Thomassin *et al.*, 2007).

CNT-Nafion composite membranes have shown conductivity values similar or less than those of pure Nafion membranes, especially if single- or multi-walled CNTs are used for composite fabrication. It is also possible to functionalize CNTs by attaching desired groups at their side walls by means of appropriate chemical method which, if used in a constructive manner, enhances the conductivity of Nafion membranes. Some notable merits of using functionalized CNTs for composite polymer electrolytes are outlined below.

1. Reduction in electronic conductivity; CNTs show electronic conductivity, so functionalization can effectively reduce such conductivity.

2. Tuning the interfacial structure; molecules that are structurally analogous to that of polymer backbone can be attached to have uniform properties in the composite.
3. Better dispersion and adhesion: agglomeration and phase segregation of CNTs at isolated regions can be avoided or minimized by using appropriate functionalization.
4. Easy processability: As CNTs can be dissolved or dispersed in aqueous or organic solvents by means of appropriate chemical functionalization, the composites can be easily processed in the form of films of uniform thickness and properties. However, the choice of functionalization is critical since some functional methods tend to enhance carrier density to finally yield better electronic conductivity.

Table 4.1 summarizes the data on variation of proton conductivity, domain size and yield strength with multi-walled sulfonic acid functionalized CNTs (s-MWCNTs).

Table 4.1 Variation of proton conductivity, domain size and yield strength with s-MWCNT composition and water content (after Kannan *et al.*, 2008, 2009, with permission from *J. Indian Inst. Sci*).

Membrane composite	Proton conductivity (S cm^{-1})	Hydrophilic domain size (nm)	Water content (%)	Yield strength (MPa)
Nafion 115	0.028	48	29.2	2.68
Recast Nafion	0.020	51	42.3	2.4
Naf-sMWCNT 0.01%	0.029	54	39.7	2.5
Naf-sMWCNT 0.1%	0.032	70	30.4	2.67
Naf-sMWCNT 1%	0.030	35	27.2	4.3

Graphene as Electrode Material for PEMFC

Graphene has attracted intense focus in view of its valuable properties and the ability to tune its band gap. Functionalized graphene sheets have been employed as the cathode support material for Pt electrocatalysts and showed greater electrochemical surface area and oxygen reduction activity with longer stability as compared with the commercial catalyst. Pt decorated graphene showed higher methanol oxidation activity than its commercial counterpart (Yumura *et al.*, 2009; Li *et al.*, 2009).

Limitations of CNTs in PEMFC

Even though CNTs are suitable both as electrode support and as novel composite polymer electrolytes, some hurdles need to be crossed before they can be used more widely. SWCNTs exist as a mixture of metallic and semiconducting SWCNTs which may introduce some variation in properties. Both SWCNTs and MWCNTs contain certain impurities which can undermine PEMFC performance. Even traces of Fe impurities degrade Nafion due to the presence of sulphonic acid moieties in the backbone (Kou *et al.*, 2009). Highly purified CNTs are too expensive and increase the cost of MEAs and PEMFC stacks. There is also an acute paucity of data related to durability of CNTs. CNTs used as electrode support have shown better corrosion resistance than Vulcan XC-72 but not much is known about this for extended periods of operations (~1000 hours or more). Biosafety aspects are also not clear. The non-toxicity of CNTs has not been conclusively established and necessitates caution in their handling (Jia *et al.*, 2005).

FABRICATION OF POLYMER SOLAR CELLS

Polymer solar cells based on conjugated polymer and fullerene composites are useful as renewable energy sources as they can be fabricated to cover large areas by means of low-cost printing and coating technologies that can simultaneously pattern the active materials on lightweight flexible substrates (Yu *et al.*, 1995). Unfortunately their low power-conversion efficiencies (about 5%) hinder their commercialization. In this contact, the "tandem cell" architecture (a multilayer structure that is equivalent to two photovoltaic cells in series) offers several advantages. Because the two cells are in series, the open-circuit voltage (V_{oc}) is increased to the sum of the V_{oc}'s of the individual cells. The use of two semiconductors with different band gap is conducive to absorption over a wide range of photon energies within the solar emission spectrum; the two cells typically use a wide band-gap semiconductor for the first cell and a smaller band-gap semiconductor for the second (Wanlass *et al.*, 1989). The power-conversion efficiency of the two cells in series is fundamentally better than that of a single cell made from the smaller band-gap material. Also, because of the low mobility of the charge carriers in the polymer-fullerene composites, an increase in the thickness of the active layer increases the internal resistance of the device, which reduces both the V_{oc} and fill factor (FF). In fact the tandem cell architecture often has a greater optical density over a wider fraction of the solar emission spectrum than that of a single cell without increasing the internal resistance. Such an architecture can significantly improve the light harvesting in polymer-based photovoltaic cells (Kim *et al.*, 2007).

Kim *et al.* fabricated tandem solar cells with each layer processed from solution with the use of bulk heterojunction materials comprising semiconducting polymers and fullerene derivatives. A transparent titanium oxide (TiO_x) layer separated and connected the front cell and the back cell. The TiO_x layer served as an electron transport and collecting layer for the first cell and as a stable foundation that facilitates the fabrication of the second cell to complete the tandem cell architecture. Kim *et al.* used an inverted structure with the low band-gap polymer-fullerene composite as the charge-separating layer in the front cell and the high band-gap polymer composite as that in the back cell. Power-conversion efficiencies greater than 6% could be achieved at illuminations of 200 milliwatts per square centimetre (Kim *et al.*, 2007).

CARBON NANOTUBES AND PLANT GROWTH

Hoping that nanomaterials may possibly produce some beneficial effects to crops, agricultural scientists are keenly interested in exploiting the techniques of nanotechnology for improving crop yields. As CNTs show desirable mechanical, electronic and thermal properties, they have found use in several applications such as sensors, flat panel displays, energy and gas storage, drug delivery and medical diagnostics (see Ohashi and Dai, 2006). Yang and Watts (2005),

Zheng *et al.* (2005) and Lin and Xing, 2007) have studied the effects of nanoparticles on plants. Lin *et al.* (2009), using confocal fluorescence microscopy, reported the ability of SWNTs to penetrate through both the plant cell wall and cell membrane and that SWNTs can act as effective nanotransporters to deliver DNA and small dye molecules into intact plant cells. Unlike plant cell walls and membranes, however, the penetration of nanoparticles into seeds is more difficult in view of thicker seed coats. Nevertheless, Khodakovskaya *et al.* (2009) showed that CNTs penetrate seed coat and influence both seed germination and plant growth, also that treating seeds with CNTs enhances seed germination and growth rate of tomato.

Indeed, tomato plants grown in media containing CNTs not only showed an increase in biomass but also had longer and sturdier stems than those in the control. As the root lengths were the same in both cases, it appears that CNTs do not show toxic effect on development and elongation of roots in the concentration range studied by Khodakovskaya *et al.*

SOME FLAWS OF NANOTUBES

Theoretically, there is no limit to achieving essentially perfect covalent bonding in material structures. But with increasing atom number, a size is eventually reached where the defect-free structure loses some of its high stability. Yet, it may be kinetically stable if there are high barriers to the nucleation of defects. Dumitrica *et al.* (2006) discussed pathways to failure in CNTs caused by tensile load as a function of time and temperature. Since CNTs can show different chiralities, the orientation of the C–C bonds in the different CNTs can critically influence the ultimate strength, the type of defects that nucleate and how they grow or propagate.

The possibility of having entirely defect-free structures seems more likely for small structures than large structures – living organisms typically achieve such perfection. The remarkable mechanics of biological motors (Schmidt and Montemagno, 2004) and viral DNA packaging and ejection (Purohit *et al.*, 2005) have been studied. Possibly evolution may have optimized composite materials found in such biological systems as bone so that they become "flaw-tolerant.". As shown by Pugno and Ruoff (2004), nanostructures having covalent bonding with (relatively) stiff bonds, in contrast, do not tolerate, e.g., point defects (a missing atom in the lattice).

The synthesis of carbon, boron nitride, and metal dichalcogenide nanotubes, and single-crystal inorganic and metal nanowires (and nanorods, ribbons, plates, platelets, sheets, etc.) has allowed study of the influence of small numbers of atomic-scale defects on strength (see Levitt, 1970; Ogata *et al.*, 2004). Nanostructures can be made with a very broad range of compositions and seem to be single-crystal throughout. But it is not known whether they are free of defects. Measurements of strength, such as of CNTs or WS_2 nanotubes (Kaplan-Ashiri *et al.*, 2006) play a role in revealing such defects.

Even with a few atomic-scale defects, the strengths of many nanostructures can be much greater than that of conventional materials. Achieving atomic-scale perfection in bonding could be critical for properties such as electronic or thermal, but we do not know whether this also applies to mechanical strength.

5

CHAPTER

Some Techniques for Nanoscience and Materials Science

INTRODUCTION

It was half a century ago that Richard Feynman in his famous lecture "There's Plenty of Room at the Bottom" advocated the development of atom-resolving electron microscopy as a crucial instrument of what since then has become **nanoscience**.

Recent years have seen transmission electron microscopy (TEM) becoming capable of taking this visionary step forward with the introduction of aberration-corrected electron optics. An entirely new generation of microscopes has made possible critical studies in condensed-matter physics and materials science to be performed at genuine atomic-scale resolution. Spatial instruments can achieve accuracy in the picometer range – dimensions where many physical effects and functions originate. Scientists can now derive values for atomic-position dependent parameters of physical properties of materials by measurements atom-by-atom.

Novel possibilities are now satisfying the growing demand of nanoscience and nanotechnology for the atomic-scale characterization of materials, nanosynthesized products and devices, and the validation of expected functions. Yet, critical understanding of the atomic-scale results is usually not so straightforward and is only possible with extensive quantum-mechanical computer calculations.

NANOLITHOGRAPHY AND SELF-ASSEMBLY FOR NANOFABRICATION DEVICES

The most spectacular successful exploitation of nanotechnology has been in miniaturizing devices. Miniaturized devices excel in terms of high efficiency and sensitivity, less power consumption and better device integration. Photolithography, which uses light, has been a most successful technique that has served the microelectronics industry for long. The resolution (R) of the features achievable by using light of wavelength (l) is expressed by the relation $R = k_1 l/NA$, where k_1 is a constant, and NA means numerical aperture. To reduce the resolution of the features, either l or k_1 has to be reduced or NA needs to be increased. It is this property that prompted the use of X-ray, ions and electrons in X-ray lithography, ion-beam lithography and electron beam lithography respectively, to make smaller devices. These above-mentioned techniques represent a modification of photolithography and have inspired the development of features below 100 nm. But because of optical diffraction, in these techniques, feature size becomes limited by the wavelength of the radiation used for photochemical transformations (Chowdhury, 2009). With accelerating demand for making nanofeatures, the need of the hour is to overcome the fundamental intrinsic resolution problem. Some techniques developed as a viable alternative to photolithography include stamping or moulding methods (soft lithography), nanoimprint lithography (Lucas *et al.*, 2008), nanosphere lithography/colloidal lithography (Ormonde *et al.*, 2004) and scanning probe microscopy-based techniques (Lee *et al.*, 2002).

Patterning technique can help in making templates for assembling materials – the first requirement for device fabrication. Such an assembly of molecules/materials involves directed self-assembly based on the utilization of structure-directing additives, often organic molecules; it differs in principle from spontaneous self-assembly. Template-guided self-assembly sometime involves the use of lithography or patterned surface fabricated by other processes. An ideal situation is selective, periodic assembling of materials. Making an assembly of various smart materials like metal nanoparticles attracts interest in view of many of its applications as photonic materials (Reufer *et al.*, 2007), high-density magnetic data-storage devices (Shaw *et al.*, 2005), microchip reactors (Gau *et al.*, 1999) and biosensors (Nair and Alam, 2008). The property of self-organizing is referred to as self-assembly of organic molecules and can be utilized for this purpose. An organic self-asembled monolayer (SAM) can potentially probe its use as a device. The SAM can be controlled at the molecular level by altering either the head group, or tail group or backbone composition, thereby making it chemically diverse. A densely packed SAM of any functional group can be made. The commonest method of having SAM on silicon is based on siloxane chemistry, where the organic molecule is attached to the substrate through Si-O-Si bonds.

The driving forces involved in molecular self-organization may be ionic, covalent, hydrogen and non-covalent; they generate properties not found in individual components. The material self-assembly can cover all scales, the forces responsible for self-assembly beyond the molecular are capillary, colloidal, elastic, electric and magnetic. Such systems proceed towards a state of lower free energy and greater structural stability. Much recent effort has been spent in developing assemblies of different materials on a variety of surfaces, with a view to realizing the devices and/or circuits with functionalities comparable to conventional integrated circuits. Chowdhury (2009) reviewed some scanning probe lithography (SPL) techniques, especially constructive nanolithography, and how its combination with self-assembly may be gainfully exploited to generate hierarchical structures leading to various 2D and 3D structures.

The merit of various scanning probe-based nanolithography techniques arises from their potential to replace the expensive photolithography process. One such scanning probe technique is constructive nanolithography; it is a nano-electrochemcial process of oxidizing only the terminal $-CH_3$ group to -COOH of the highly ordered monolayer of OTS (n-octadecyltrichlorosilane) self-assembled on silicon with a conducting scanning force microscope tip (Chowdhury, 2009). By using surface self-assembly along with constructive nanolithography, different metals, metal ions, semiconductor nanoparicles and organic compounds can be assembled, over which reactions and reorganization processes can take place for their further use as nanodevices.

Scanning Probe Lithography

Scanning probe lithography (SPL) is well suited for nanofabrication at the nanometer scale. In this technique, writing is done by using sharp probe tips in an atomic force microscope (AFM). Nanolithography can be done under ambient conditions and no sophisticated machinery is needed. By using sharper AFM tips, fabrication with feature sizes even below 10 nm level is possible. Some examples of SPL techniques are dip-pen AFM lithography, local anodic oxidation, mechanical plowing and constructive nanolithography.

In dip-pen lithography technique (DPN), the material is transferred from AFM tip to the surface. Material on the AFM tip serves as the ink and is transferred to the substrate by capillary forces. Various materials can be assembled on different substrates. DPN has been used to deposit nanoarrays of proteins, IgG and lysozome on gold substrate (Lee *et al.*, 2002).

Ding *et al.* (2005) successfully deposited nanoscaled structures of semi-conducting CdS materials by means of DPN. DPN-based generation of nanometer-scale magnetic structures with precise feature size control has also been demonstrated. Magnetic iron-oxide nanoparticles have been prepared by slowly adding an aqueous solution of ferrous chloride to an aqueous solution of ferric chloride at room temperature. Using 16-mercaptohexadecanoic acid (MHDA) templates allowed assembly of the magnetic nanostructures (Liu *et al.*, 2002). DPN has been used for the direct patterning of organic/inorganic composite nanostructures on silicon and oxidized silicon substrates. Al_2O_3, SiO_2 and SnO_2 nanostructures on silicon and silicon oxide surfaces have been fabricated (Su *et al.*, 2002).

Local Anodic Oxidation

AFM anodization lithography produces protruding oxide patterns on a substrate by Faradic current from the cathode tip to the anode substrate. Anodic patterning is realized on a silicon substrate by electrochemical reactions mediated by the scanning probe tip. Electrochemical anodization takes place due to emitted electrons from the tip at the locally scanned area on a silicon substrate in the presence of an atmospheric water column between the tip and the scanned area of the sample (Chowdhury, 2009).

Electron Beam Lithography

Electron beam lithography is commonly employed to pattern the smallest features on semiconducting silicon substrates and it can also modify insulating substrates. However, at typical beam energies, the insulating surface develops a negative charge which deflects the beam so that the desired pattern is distorted. Approaches to addressing this problem all require either additional sample processing steps or complex gas-handling and vacuum equipment. At lower energies, electron beams can instead induce positive charging of insulating surfaces so that a critical energy exists for which the surface will remain neutral (see *Nano Lett.* 6: 10.1021/nl061211q, 2006). For 65-nm-thick poly(methyl methacrylate) films on glass, a critical energy value of 1.3 keV was measured. Successful tuning of the incident beam to this energy enabled creation of features finer than 100 nm on this substrate. In contrast a 5-keV beam produced distortions that are easily seen in scanning electron micrographs.

Constructive Nanolithography

Hoeppener *et al.* (2002) developed a tip-mediated nanolithography technique that differs somewhat from other SPL techniques that make use of organic monolayers as passive ultrathin resists. They used this method to generate nanopatterns of chemically active groups on a highly ordered monolayer of n-octadecyltrichlorosilane (OTS) self-assembled on silicon, via a local electrochemical oxidation process performed with a conducting scanning force microscope tip for converting the surface exposed $-CH_3$ groups of the base monolayer to -COOH sites to increase their utility (Hoeppener *et al.*, 2002; Wouters *et al.*, 2005). This approach limits the local electrochemical oxidation to the selective oxidation of only the top groups of the organic monolayer coating, preserving the compact structure and molecular organization of the monolayer, hence the name 'constructive nanolithography'.

Integrating Constructive Nanolithography and Self-assembly

The tip-inscribed -COOH patterns made by constructive nanolithography show high chemical, thermal and mechanical stability, so increasing their utility as templates in subsequent template-assisted assembly steps. Planned nanostructures of colloidal gold particles, silver particles or semiconductor nanoparticles on silicon are easily obtained via hierarchical self-assembly approach (see Fig. 5.1).

Diverse patterned surfaces exposing different functional groups are generated by the self-assembly of an electrically insulating top monolayer of nonadecyltrichlorosilane (NTS) on the tip-inscribed -COOH sites. The terminal ethylenic groups of the NTS are chemically modified to $-NH_2$ (amine) by photoreacting with formamide which adds a terminal amide group to the hydrocarbon tail of the NTS. Its continued reduction with BH_3-THF leads to terminal $-NH_2$ functionality and promotes template-guided surface assembly over the terminal $-NH_2$ monolayer and site-specific anchoring of gold nanoparticles from a colloidal solution (Hoeppener *et al.*, 2002). Another possibility is to functionalize terminal ethylenic groups of the NTS to -S-H (thiol) top functions, accomplished by photochemical radical addition of H_2S to the reactive ethylenic functions of NTS and further reduction with BH_3-THF of the fraction of

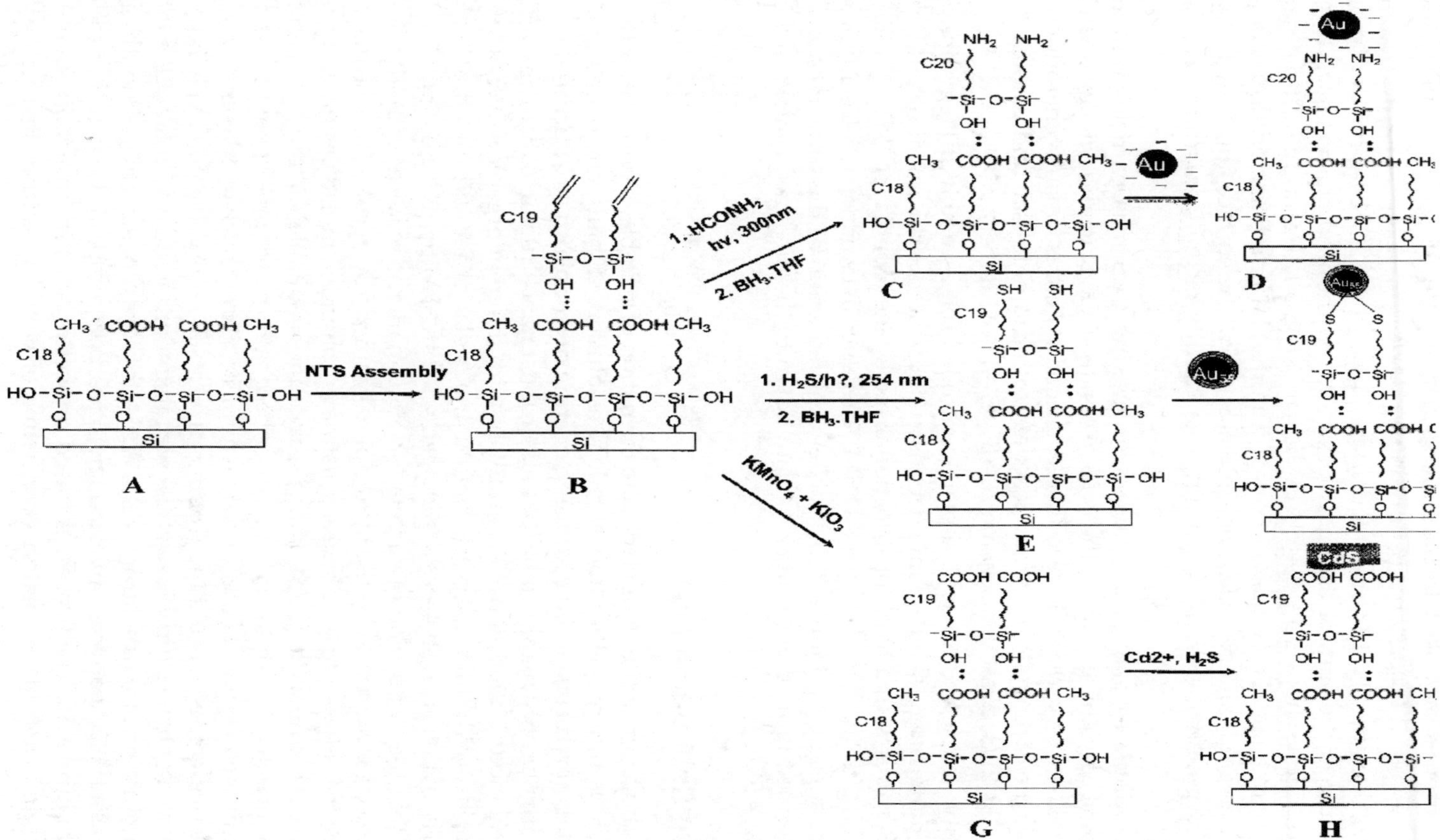

Fig. 5.1 Schematic sketch (not drawn to scale) showing the hierarchical self-assembly of a planned gold nanostructure via template-guided self-assembly of gold nanoparticles over *in situ*-generated top amino functionality (A→B→C→D), gold clusters [Au55' (A→B→E→F) over thiol functionality on a preassembled organosilane bilayer template pattern, and cadmium sulphide nanostructure generated over carboxylic functionalized bilayer pattern (A→B→G→H) (after Chowdhury, 2009, with permission from *Current Science*).

disulphide (-S-S-) groups produced in the process. This creates elevated templates that are replicas of the initial inscribed patterns (Chowdhury, 2009).

Subsequently, like the $-NH_2$ terminated monolayer, these –SH terminated monlayer template patterns direct the surface assembly of *ex situ* synthesized clusters such as [Au55] according to the reaction scheme (A→B→E→F) shown in Fig. 5.1. These are well-defined gold 'supramolecules' with a metal core made of only 55 atoms (1.4 nm diameter) (Liu *et al.*, 2002). The anchoring of the gold nanoparticles and gold clusters to the surface faithfully follows the initial tip-inscribed patterns, thereby confirming the validity of the hierarchical step-by-step assembly strategy.

In fact, such templates may also be used for making CdS nanoparticles by chemical oxidation of terminal ethylenic groups of the NTS region to –COOH (carboxylic) by $KMnO_4$ and KIO_3. Here, the bilayer of carboxylic terminated template guides *in situ* generation of CdS nanoparticles over the carboxylic region by reacting with cadmium salt and passing H_2S gas (Hoeppener *et al.*, 2002) (see reaction scheme A→B→G→H in Fig. 5.1).

It is also possible to use and extend template-controlled self-assembly strategy tip-induced electrochemical patterning to the planned construction of hybrid metal-organic surface nanostructures. Starting with a thiol-top-functionalized silane monolayer (TFSM) with silver ions chemisorbed on its outer surface (Ag+-TFSM), metallic silver nanoparticles are produced at selected surface sites by tip-induced electrochemical reduction of the surface-bound metal ions. Thus, site-defined reduction of silver thiolate down to the nm-sized range is achievable with a conducting AFM tip.

Wetting-driven Self-assembly

Bottom-up nanofabrication strategy can facilitate formation of suitable, highly precise template patterns with various functionalities and excellent resolution, stability and/or compatibility with other crucial requirements posed by the engineering of complex nanostructured materials. These templates are not only highly stable but also robust nanostructures that can participate in various chemical modification steps like treatment with different organic solvents, reagents, or concentrated acids, with full preservation of the molecular order and structure. They also retain their integrity when such surfaces are imaged with an SFM tip in the contact mode. The only difficulty is that the generation of such tailor-made patterns requires specially designed molecules which need to be made by using complex synthetic protocols. This means that the materials used for the self-assembly process must be amphiphilic molecules having the right structure with a specific anchoring group, that make them suitable for producing a closely packed SAM. As repeated use of the initially inscribed patterns for other new post-assembly processes is quite limited in view of the high structural stability of such overlayer surfaces, this limitation may be overcome by using the wetting-driven self-assembly (WDSA) (Chowdhury *et al.*, 2007). This strategy makes direct use of the wettability properties of a material to fabricate planned metal-organic surface nanostructures. WDSA involves constructive nanolithography to fabricate melt-assembled patterns of various hydrocarbon compounds such as octadecane, eicosene, ODT or lauric acid. The process is based on wetting-driven immobilization of such compounds on organic monolayer patterns consisting of wettable features in a non-wettable surface background. WDSA is highly versatile – as any material (not necessary amphiphilic) with any shape, functionality or structure may in principle be used to produce melt-assembled patterns, requiring only the conditions of low melting point, being solid at room temperature, non-volatile, and a fairly high surface tension. The WDSA approach potentially allows constructing 3D structures with various functionalities without utilizing complex (non-commercial) synthetic precursors such as NTS. It attracts broad application as it uses commercially available materials that are already chemically characterized. Besides relying on simple, inexpensive and widely applicable chemical methods, the immobilized material with one or more reactive groups serve as space-confined reactors within which various chemical reactions take place; this offers unprecedented new options in surface nanofabrication of planned metal-organic nanostructures due to their structural and compositional diversity. Fig. 5.2 illustrates the WDSA approach.

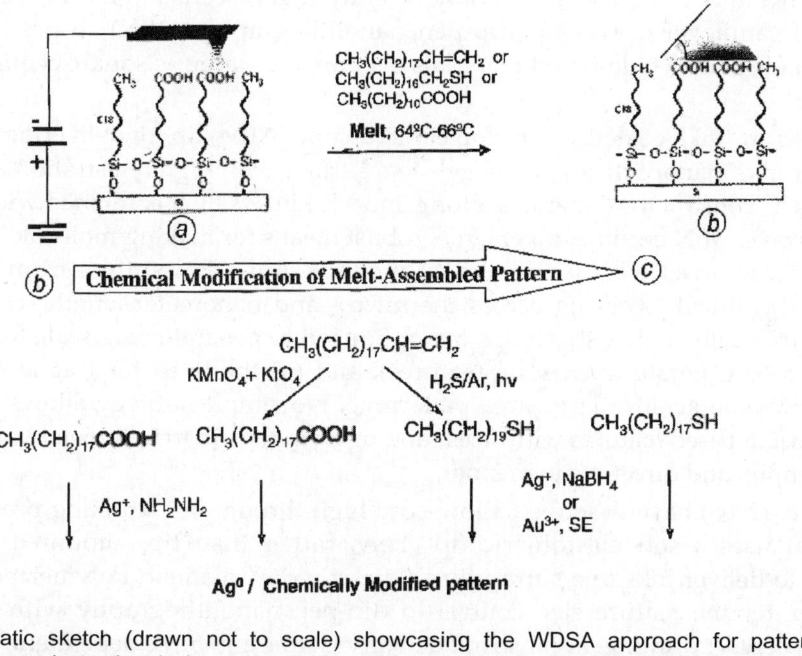

Fig. 5.2 Schematic sketch (drawn not to scale) showcasing the WDSA approach for pattern generation. (**a**) Tip-inscribed nanoelectrochemical, non-destructive patterning of self-assembled OTS monolayer on Si, thereby generating wettable nanofeatures on a non-wettable monolayer background. This acts as template for fabricating wetting-driven nanopattern of melt of functionalized material (**b**) and its subsequent chemical modification (**c**). The various reaction routes carried out to generate Ag^o assembled on these chemically modified patterns are also listed (after Chowdhury *et al.*, 2007, 2008, with permission from *Current Science*).

Essentially, nanopatterning techniques such as constructive nanolithography can effect electrochemical conversion of the top functional entity without affecting the base monolayer and so could be used as template for chemical self-assembly as well as WDSA. The chemical nanofabrication strategy combines the nano-electrochemical pattern inscription on a self-assembled base monolayer with a variety of post-patterning chemical processes performed at different levels of complexity.

Polymer Pen Lithography

Lithography has been used in the production of integrated circuits, information storage devices, video screens, MEMS, miniaturized sensors, microfluidic devices, biochips, photonic bandgap structures, and diffractive optical elements (Qi *et al.*, 2004; Mirkin, 2007). Lithography is divided into two classes on the basis of patterning strategy: parallel replication and serial writing. Parallel replication methods such as photolithography, contact printing and nanoimprint lithography are useful for high-throughput, large-area patterning, but most such methods can only duplicate those patterns which are predefined by serial writing approaches and so cannot be used to arbitrarily generate different patterns (see Chou *et al.*, 1996; Zheng *et al.*, 2006). Serial writing methods – including electron-beam lithography (EBL), ion beam lithography, and many scanning probe microscopy (SPM) based methods (Piner *et al.*, 1999; Gates *et al.*, 2005) produce patterns with high resolution and registration but are limited in throughput (Kramer *et al.*, 2003). It is only recently that researchers found ways to use two-dimensional cantilever arrays for dip-pen nanolithography (DPN) to produce patterned structures made of molecule-based materials over areas as large as square centimetres (Salaita *et al.*, 2006).

DPN uses an "ink"-coated atomic force microscope (AFM) tip for delivering such materials as molecular inks, nanoparticles, or sol gels to a surface with high registration and sub-50-nm resolution in a "constructive" manner (Hong and Mirkin, 2000). If combined with high-density cantilever arrays, DPN becomes a versatile, robust means for making molecule-based patterns over fairly large areas with moderate throughput. But DPN suffers from the following limitations: (i) difficult working across the micro- and nanometer-length scales in a single experiment (typically, either sharp tips are optimized to generate nanoscale features or blunt tips are used to generate microscale features), and (ii) the need for fragile and costly 2-D cantilever arrays to achieve large-area patterning. No simple strategy allows one to rapidly pattern molecule-based features with sizes ranging from the nm to millimeter scale in a parallel, high-throughput, and direct-write manner.

Huo *et al.* (2008) have described a low-cost, high-throughput scanning probe lithography method that uses a soft elastomeric tip array, rather than tips mounted on individual cantilevers, to deliver inks to a surface in a "direct write" manner. Polymer pen lithography (PPL) integrates the feature size control to dip-pen nanolithography with the large-area capability of contact printing. Because ink delivery is time and force dependent, features on the nanometre, micrometre, and macroscopic length scales can be formed with the same tip array. Arrays with over 10 million pyramid-shaped pens can be brought into contact with substrates and readily levelled optically to ensure uniform pattern development.

Whereas DPN and other SPM-based lithographies typically use hard silicon based cantilevers, PPL uses elastomeric tips without cantilevers [typically polydimethylsiloxane (PDMS)] to deliver ink. A typical polymer pen array (Fig. 5.3) can have thousands of pyramid-shaped tips made with a master prepared by conventional photolithography followed by subsequent wet chemical etching. The pyramids are connected by a thin PDMS backing layer (50 to 100 μm thick) which is placed on a glass support before curing. The glass support and thin backing layer markedly improve the uniformity of the polymer pen array over large areas of water surface. Bringing the sharp tips of the polymer pens in contact with a substrate delivers ink at the points of contact (Fig. 5.3). Ensuring uniform ink delivery over large areas greatly increases the amount of light reflected from the tips, when these make contact with the substrate. Indeed the transparent elastomer polymer pen array makes it possible to determine visually when all of the pens are in contact with an underlying substrate. Huo *et al.* successfully demonstrated capabilities with an array that contains either 15,000 or 28,000 pyramid-shaped pens; they have also made and demonstrated patterning capabilities with arrays with as many as ~11,000,000 pens. In a typical experiment, a polymer pen array (1 cm^2 in size) could be inked by immersing it in a saturated solution of 16-mercaptohexadecanoic acid (MHA) in ethanol for 5 min followed by rinsing with ethanol. The inked pen array generated 1-μm diameter MHA dot patterns on a thermally evaporated polycrystalline gold substrate (25 nm Au with a 5 nm Ti adhesion layer coated on Si) when it was brought in contact with the gold surface for 0.1 s. Repeating this process 35 times generated a 6 by 6 array of MHA dots (less than 10% deviation in feature diameter). The exposed gold on this MHA patterned substrate was thin etched (20 mM thiourea, 30 mM iron nitrate, 20 mM hydrochloric acid, and 2 mM octanol in water) to yield raised structures ~25 nm in height, which can be imaged by optical microscopy.

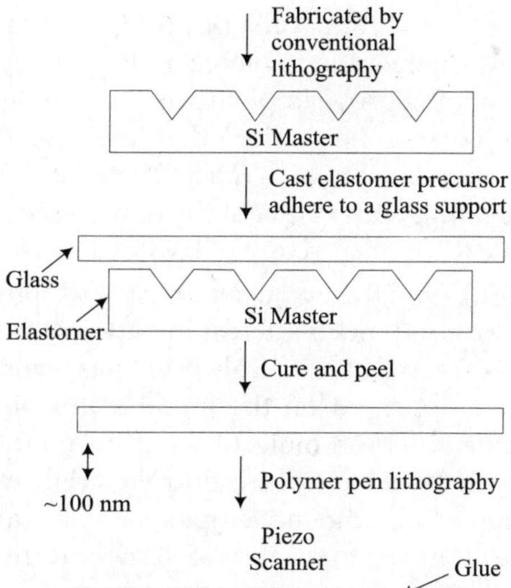

Fig. 5.3 Schematic sketch of the polymer pen lithography setup ((From Huo, F. *et al. Science* 1658-1660, 2008. Reprinted with permission from AAAS).

An important property of PPL, in contrast with DPN and most force-independent contact printing strategies is that it shows both time- and force-dependent ink transport. As with DPN, features made by PPL show that a size depends nearly on the square root of the tip-substrate contact time. This property of PPL tips allowed the researchers to pattern submicrometer features with remarkably high precision and reproducibility. One merit of the force dependency of PPL is that one does not have to depend on the time-consuming, meniscus-mediated ink diffusion process to generate large features. Indeed, either nanometer- or micrometer-sized features can be generated in only one printing cycle by adjusting the degree of tip deformation.

Unlike conventional contact printing, PPL allows for the combinatorial patterning of molecule-based and solid-state features with dynamic control over feature size, spacing, and shape (Huo *et al.*, 2008). As in conventional lithography of images, the polymer tips can be used to create a dot pattern of the intended pattern or structure.

The elegance of the force dependence and maskless nature of PPL make it possible to create many structural variants without the hurdle of designing a new master by a throughput-impeded serial process. Furthermore, PPL can be used with sub-100-nm resolution with the registration capabilities of a closed-loop scanner.

The time- and force-dependent ink transport characteristics of the polymer pen pyramid arrays provide a tunability to PPL that may also facilitate development of other patterning capabilities. PPL is also well suited for fabricating arrays of structures made of soft matter such as proteins. This points to its potential utility for the life sciences as well.

ATOMIC MANIPULATION

Understanding the forces needed to lift up, pull, or push objects and building blocks is crucial in diverse fields ranging from engineering to biotechnology (Galburt *et al.*, 2007). Ternes *et al.* (2008) used a sophisticated scanning probe implementation to determine how much lateral and vertical force is needed to pull one of the smallest building blocks: a single atom on a metallic surface and how these forces are related to each other. They did this by combining the two most powerful techniques for interacting with matter at the nanometre scale i.e., scanning tunnelling microscopy (STM) and atomic force microscopy (AFM) in a single instrument.

Eigler and Schweizer (1990) were the first to assemble a nanometre-scale structure, atom by atom, on a surface. This feat aroused much interest in nanometre-scale devices based on a few atoms and materials engineered at the atomic scale. STM has enabled notable advances toward these concepts - reports have appeared on the initial states of "wiring" a single molecule (Nazin *et al.*, 2003), a logic gate based on molecules has been built (Heinrich *et al.*, 2002), and chemical reactions at the atomic level have been controlled (Hla *et al.*, 2000). AFM has recently been used to manipulate atoms at cryogenic temperatures (Oyabu *et al.*, 2005) and to make complex nanostructures, one atom at a time, at room temperature (Sugimoto *et al.*, 2005).

Ternes *et al.* used an improved version of the STM based on a small quartz tuning fork in which one of the prongs is fixed and the other bears a sharp tip at its free end. The free prong acts as a cantilever. Variations in its longitudinal bending can be detected by measuring tiny piezoelectric currents produced when the quartz deforms (Custance and Morita, 2008).

Ternes *et al.* observed that the lateral force to pull a Co atom on a Pt(111) surface is independent of the vertical force (at least in the tip-surface distance range), with a value that remains almost constant. In fact, it was this behaviour that enabled Ternes *et al.* to compare the lateral forces required to move the same atomic species on other metallic substrates; they could thus quantify friction at a single-atom scale. By applying this approach to full images instead of line profiles, the authors generated maps of the tip-surface interaction potential - maps that provide information about the adsorption stability of the atom (or molecule) on the surface.

The work of Ternes *et al.* points to new avenues for studies on nanostructuring and of friction at the single-atom scale. Combining the powerful tools provided by STM and AFM in a single instrument, and of correlating tip-surface short-range chemical forces with simultaneous tunnelling currents at the atomic and molecular scale could well open a new dimension in nanoscience that will lead to exciting discoveries and technological achievements.

ATOMIC FORCE MICROSCOPY

Atomic force microscopy (AFM) is widely used for imaging surfaces ranging from micro- to nanometre scales. The objective is to visualize and characterize surface textures and shapes. It provides subnanometre resolution under physiological conditions, needed for imaging biomolecules as well as living cells. Measurements of molecular recognition forces can reveal the function and structure of biomolecular assemblies. In combination with fluorescence microscopy, AFM can help to identify biomolecules based on fluorescence labelling and high-resolution topography imaging (Kada *et al.*, 2008).

AFM has been extensively used in several disciplines such as life science, solid-state physics, and materials science. It has evolved into an imaging method which provides structural details of biological materials such as proteins, nucleic acids, membranes, and cells in their native environment (Hörber and Miles, 2003; Kienberger *et al.*, 2006). AFM is particularly valuable in providing subnanometre resolution at a reasonable signal-to-noise ratio under physiological conditions. It has already become a companion technique to X-ray crystallography and electron microscopy (EM) for the determination of protein structures for instance. It complements EM well by enabling visualization of biological samples in buffers that preserve their native structure over fairly long time periods. Unlike EM, AFM yields 3-D maps with an exceptionally good vertical resolution of less than a nanometre.

Besides high-resolution imaging of proteins, nucleotides, membranes, and living cells, the measurement of mechanical forces at the molecular level reveals useful insights into the function and structure of biomolecular systems. Inter- and intramolecular interactions can be studied directly at the molecular level, as exemplified by the analysis of polysaccharide elasticity (Marszalek *et al.*, 1998), DNA mechanics and the function of molecular motors (Tanaka *et al.*, 2002).

The AFM works by scanning a pointed probe over a surface while adjusting the force applied to the sample to minimal values which prevent displacement or destruction of weakly adhering biomolecular complexes. Certain difficulties caused by unfavourable probe-surface interactions in contact mode AFM, especially lateral forces, have been removed by the development of dynamic force microscopy (DFM) methods, chiefly used for the investigation of soft biological membranes and supramolecular complexes (Han *et al.*, 1996). In DFM, the cantilever is oscillated close to its resonance frequency at an amplitude of a few nm as it raster-scans the surface and touches the sample only at the end of its downward movement (Kada *et al.*, 2008). Exploiting direct cantilever excitation in magnetic ac mode AFM (where tip oscillation is achieved by an external magnetic field), lateral forces can be minimized and the tip oscillation properly controlled. Möller *et al.* (1999) even showed that DFM enables the imaging of hard protein surfaces with similar resolution to contact mode AFM.

For obtaining high-quality AFM images of biological samples, tight immobilization on a flat support is essential. The simplest and commonest immobilization method is physical adsorption from solution. Muscovite mica is the most frequently used support for biological applications. It is a layered mineral which is easily cleaved, leading to clean, atomically flat surfaces. Since the mica surface is negatively charged at neutral pH, simple adsorption of positively charged samples suffices for stable immobilization. Even negatively charged molecules (e.g. DNA) can be physically adsorbed onto mica surfaces when multivalent cations (like Ni^{2+}, Mg^{2+}, Zn^{2+}) are present in buffer solution (Kienberger *et al.*, 2007).

The high lateral resolution and sensitive force detection capabilities of AFM have facilitated measurement of inter- and intramolecular forces of biomolecules on the single-molecule level. Methods have been developed for exploring kinetic and structural details of interactions and molecular recognition processes. Data obtained from force spectroscopy include certain physical parameters which cannot be measured by other methods, and new perspectives have opened up for exploring the regulation of biological process dynamics. New instrument developments like recognition imaging enable study of the chemical composition of samples, while the combination of AFM and fluorescence microscopy enables a detailed characterization of cellular processes. Improvements in the cantilever sensitivity and acquisition speed can pave the way to further applications in nanobioscience and nanobiotechnology.

CANTILEVER DYNAMICS IN AFM

Dynamic atomic force microscopy (dAFM) is one of the most important tools in nanotechnology with its unmatched capabilities of: (i) measuring topography and physico-chemical properties of organic and inorganic materials at nm length scales in various ambient media; and (ii) the manipulation and fabrication of diverse functional nanostructures. In dAFM, materials are probed with gentle forces by means of an oscillating nanoscale tip that intermittently interacts with the sample. These capabilities have catapulted dAFM to the status of a leading tool that allows researchers to directly 'see' and 'touch' at the nanoscale.

DAFM essentially involves a vibrating microcantilever with a nanoscale tip. The tip interacts with a sample surface via intermolecular forces. Microcantilevers possess distinct eigenmodes and the tip-sample interaction forces are highly nonlinear. So, cantilevers vibrate in interesting but unanticipated ways; some produce negative effects on imaging stability whereas others can be gainfully used to enhance performance (Raman *et al.*, 2008).

Two main classes of dAFM are frequency modulation (FM) AFM (Giessibl, 2003) or noncontact AFM, and amplitude modulation (AM) AFM, also known as tapping-mode or intermittent contact AFM. In FM-AFM, the phase of oscillation and tip amplitude are kept constant by means of feedback circuits, while in AM-AFM the drive frequency is held constant and the tip amplitude is maintained constant by means of active feedback.

Currently, two of the most active areas of dAFM are (i) the manipulation and imaging of soft matter, especially in biology; and (ii) the accurate quantification of sample properties in materials science. A deeper understanding of cantilever dynamics has become essential for both the active areas. Progress in these areas requires the development of a new generation of

AFM with greatly reduced imaging forces and better material contrast. Many researchers and companies are actively engaged and interested in controlling, tuning or exploiting cantilever dynamics to reduce imaging forces and develop new modes to improve material contrast and sensitivity.

Eigenmodes of AFM Cantilevers

AFM microcantilevers are made from single-crystal Si or Si_3N_4, and possess a sharp conical or pyramidal tip near their free end with which to probe the sample surface. Just as a taut string shows many eigenmodes of vibration, each with its own natural frequency, so too do AFM microcantilevers. The aim of any dAFM system is to drive the microcantilever externally into a resonance of a specific eigenmode.

Three typical examples of AFM microcantilevers are (a) a stiff rectangular cantilever (~50 μm long), (b) a soft rectangular cantilever (~290 μm long), and (c) a soft triangular cantilever (~125 μm long) for biological applications in liquids. The vibration spectrum contains distinct maxima corresponding to the mechanical resonances of the cantilever or of the dither piezo; under ambient conditions, the quality (Q) factors of the microcantilever resonances are usually much larger than those of the piezo resonances, thus allowing one to distinguish between those peaks (Raman *et al.*, 2008).

The cantilever dynamics in AM-AFM are highly nonlinear because typical tip oscillation amplitudes (>5 nm) are longer than the decay lengths associated with short-range interaction forces (<1 nm). So, cantilever dynamics cannot be predicted by linearizing interaction forces about an equilibrium position.

In the coming years, continuing advances in the understanding of cantilever dynamics are expected to facilitate development of the next generation of AFMs. We may assume that the greatest impact of cantilever dynamics in dAFM will lie in (i) applications to significantly improve the quantitative mechanical/electrical/magnetic property sensing using dAFM; (ii) applications for the development of new hydrodynamically streamlined AFM probes for applications in liquids, and (iii) the continued development of new modes for improved contrast with piconewton imaging forces for biological applications in liquids (see Nayfeh and Balachandran, 1995; Raman *et al.*, 2008).

Cantilever Sensors

Cantilever-based devices exemplify how a 'simple' sensor can be tailored by microfabrication techniques and can achieve unprecedented performance. Recent experiments have used different mechanical transduction principles for detecting and analyzing small quantities of materials. The principles of these experiments allow biologists to study surface biochemistry on a nanoscale and offer new, exciting opportunities in developing microscopic biomedical analysis systems with unique characteristics (Raiteri *et al.*, 2002).

Six possible cantilever transducer principles are: (a) force sensor with integrated tip for AFM; (b) 'bimetallic' temperature and heat sensor; (c) mass loading sensor; (d) medium viscoelasticity sensor; (e) thermogravimetric sensor; and (f) stress sensor (Raiteri *et al.*, 2002).

SCANNING TUNNELLING MICROSCOPE (STM)

STM has a sharp needle that traces the roughness of surfaces with such high resolution that individual atoms can be seen. The landmark event of designing STM has opened up a view of the world at nanometer scale and has led to the emergence of a highly diversified field of research - known variously as nanoscience or nanotechnology. In this field, two major research areas are: (a) designing new materials and devices from the bottom up by using nanometre-scale structures as building-blocks; and (b) making new tools for investigating matter - mainly biological structure - at the nanometer scale.

Using an STM, Venema (2006) showed that a two-dimensional array of nanoparticles (gold) with insulating spikes becomes electrically linked up after immersion in a solution of conducting organic molecules. In this system, each pair of neighbouring particles seems to be connected by approximately one molecule. The linking process is fully reversible and repeatable.

USE OF MAGNETIC FIELDS TO ASSEMBLE PARTICLES INTO MEMBRANES

Molecular self-assembly underpins the highly complex form and functions of living systems. An elementary goal of materials science is to make synthetic materials based on particle assembly - its practitioners exploit suitable methods for manipulating and quantifying particle interactions for achieving precise control not only over the morphology of self-assembled particles but also their stability in response to mechanical disturbances and changes in thermodynamic conditions. Osterman *et al.* (2009) reported achieving such assembly by subjecting μm-sized colloidal particles to dynamically modulated magnetic fields. They studied self-assembled particle membranes having several characteristics of biological self-assemblies, including structural self-healing when mechanically disturbed.

Martin *et al.* (2003, 2004) had earlier studied membrane organization of particles that is driven by such external fields as time-varying orthogonally oriented magnetic fields. The value of the work of Osterman *et al.* comes from experimental quantification of the interactions governing particle self-assembly in triaxial magnetic fields -three pairs of Helmholtz wire coils were used to apply a triaxial magnetic field to a central cell containing μm-sized colloidal iron particles. Whereas the vertical pair produces a uniform magnetic field in the vertical direction, the two horizontal pairs generate an oscillating biaxial field in the horizontal direction. A laser beam was made to act as a tweezer for manipulating the particles. Osterman *et al.* confirmed the prediction that these interactions not only result in non-trivial topological changes in particle aggregation, but also in the emergence of characteristic-structural scales as the strength and character of the driving magnetic field are varied (Douglas, 2010).

In the experiment performed by Osterman *et al.*, flat membranes of particles form in the horizontal direction for large biaxial-field strengths (or, equivalently, large negative dipole interactions) - rather than particle chains in the vertical direction, which would form in a uniform field applied vertically. The two fields together provide a precessing triaxial magnetic field having a tunable angle. Under the specific experimental conditions, the dipole interaction induced by the uniform magnetic field on an isolated particle was exactly compensated by a negative dipole interaction, contributed by the biaxial field.

The quantitative measurements made by Osterman *et al.* confirm the predicted van der Waals-like inter-particle interaction and show that the self-assembly of free-standing particle membranes from iron particles exposed to a magic-angle magnetic field is caused by many-body interactions which take place during the particle assembly process (Martin *et al.*, 2003, 2009).

Particle self-assembly into membranes can also occur in the absence of an applied magnetic field and under conditions of thermal equilibrium; e.g., if particles show static charge configurations of a multipolar character, or if they have highly directional interactions on their surface (Martin and Gulley, 2009) (such as patchy local hydrogen bonding or hydrophobic interactions). Directionally dependent inter-particle interactions without applied fields can also appear from many-body inter-particle interactions, e.g., in nanoparticles grafted with polymer-chain layers, in which the emergent directional interaction arises from the distortion of the grafted layer as the particles approach one another (Akcora *et al.*, 2009). Self assembly both from intrinsic directional inter-particle interactions, and from emergent or 'autosteric' interactions (Caspar, 1980) arising from the conformational polarizability of the assembling species, is quite common in biological systems.

Magnetic-field induced particle assembly has been exploited for creating sensor materials and materials capable of efficient, directed heat transfer (Martin and Gulley, 2009). It also leads to strong particle mixing when the strengths of the field's three orthogonal components are dynamically changed so as to generate a collective vortex-like particle motion reminiscent of the fluid motion seen in a washing machine (Martin, 2009).

MICROFLUIDICS FOR MATERIALS SCIENCE AND BIOMEDICINE

Extensive research has been done on microfluidic technologies for the development of sample-to-answer platforms focusing on protein and nucleic acid diagnostics. However, till today, no such systems have reached the market with as much success as, for example, the hand-held blood glucose metre. Several hurdles that still need to be crossed in nucleic acid diagnostics relate to preparation of the biological sample.

The major focus in microfluidics has been on molecular diagnostics. Microscale phenomena can be exploited to efficiently prepare biological samples for the detection of nucleic acids (NA) and proteins (Manz *et al.*, 1990). What this field urgently needs is the development of a single system that can accept a wide variety of biological samples, e.g., blood or saliva; it should perform the required sample preparation and analysis; and it should quickly deliver an answer without requiring significant user input (except for the initial sample introduction). In contrast, reasonably satisfactory sample-to-answer possibilities exist for measuring simpler analytes in biological samples, such as finger-stick blood-glucometers, Abbott's blood gas and electrolytes panels, and urine-based pregnancy tests (St. Louis, 2000).

Modern microfluidic devices are serving as welcome tools in the hands of geneticists or microbiologists engaged in characterizing clonal cell lines and changes in their progeny, particularly phenotypic variations in successive generations. For any critical study of the phenotype of single cells in the context of pedigree we need a method to collect data over many cells and over many generations. The yeast, *Saccharomyces cerevisiae* is a suitable model eukaryotic system: yeast cells divide rapidly, so that multiple generations of cells can be studied. Proteome-wide studies of *S. cerevisiae* have characterized stationary distributions of protein levels across a population by microscopy and flow cytometry, and revealed that expression of stress-related genes is usually more variable ("noisy"), whereas housekeeping genes show less cell-to-cell variation (Huh *et al*, 2003; Newman *et al*., 2006; Bar-Even *et al*., 2006). Unfortunately, these measurements detect neither changes in expression over time nor correlations in protein levels resulting from age or pedigree relationships among individuals. Characterization of cells and their progeny requires tracking single cells and their offspring during growth. This is done either by individually separating cells by micromanipulation or by imaging cells as they grow sandwiched between an agar pad and a cover glass (Kaufmann *et al*., 2007). Manual manipulation of cells is very laborious, whereas accurately determining pedigree and protein expression by microscopy is very difficult because cells grow out of the focal plane after only a few divisions. Modern microfluidic devices maintain cells in a single focal plane as they grow (Groisman *et al*., 2005; Lee *et al*., 2008); however, many of these devices require sophisticated fabrication techniques such as multilayer fabrication with valves, channel height differences, or membranes (Balaban *et al*., 2004; Charvin *et al*., 2008). For optimizing the statistical power of these techniques, the initial placement of cells needs to be controlled. While several other microfluidic devices have successfully achieved single-cell trapping (Di Carlo *et al*., 2006; Luo *et al*., 2008), these trapping mechanisms are not suitable for careful lineage analysis. What we need is the ability to efficiently and repeatedly trap, spatially organize, and track the growth of single cells over many generations in a device which can be easily fabricated, is simple to use, and would enable the data collection over many cell lineages in a single experiment (Rowat *et al*., 2009).

Prompted by this need and with a view to correlating phenotypes with the age and genealogy of single cells over time, Rowat *et al*. developed a microfluidic device (Fig. 5.4) that makes it possible to track multiple lineages in parallel by trapping single cells and constraining them to grow in lines for as many as 8 divisions. To illustrate the utility of this method, lineages of cells expressing one of 3 naturally regulated proteins were studied, each with a different representative expression behaviour. Within lineages deriving from single cells, genealogically related clusters of cells with similar phenotype were observed. Cluster sizes varied markedly among the 3 proteins, suggesting that the time scale of phenotypic persistence is protein-specific. Growing lines of cells also enables one to dynamically track temporal fluctuations in protein levels at the same time as pedigree relationships among the cells as they divide in the chambers. Rowat *et al*. observed bursts in expression levels of the heat shock protein Hsp12-GFP that occur simultaneously in mother and daughter cells. In contrast, the ribosomal protein Rps8B-GFP showed fairly constant levels of expression over time. This method seems to be useful for understanding the time scales of phenotypic variation and correlations in phenotype among single cells within a population (Rowat *et al*., 2009).

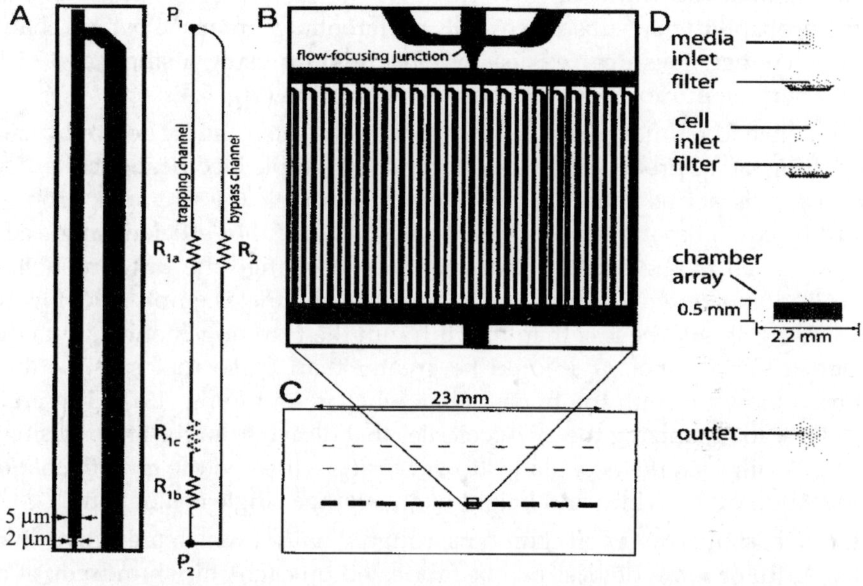

Fig. 5.4 Lineage chamber device layout. (A) A single chamber consists of a long channel with a 5 µm width and a 2 µm width and a 2 µm constriction, as shown in this inset of B. The flow patterns can be understood in terms of an electrical circuit, with R_{1a}, the resistance of the trapping chamber; R_{1b}, the resistance of the constriction; R_{1c}, the resistance of a single cell; and R_2, the resistance of the bypass channel. The pressure drop (P_1-P_2) is equivalent for both fluid paths. (B) The chamber array consists of 50 chambers arranged in parallel, approximately half of which are active and shown in this inset of C. A flow-focusing geometry centres the flow of the cells into the chamber array. (C) Ten individual devices fit onto a 50 x 24-mm coverslip enabling multiple conditions or cell types to be tested in a single experiment. (D) Overview of the entire lineage chamber device. Two inlets allow for separate flow paths of the cell suspension and media. Filters eliminate aggregates of cell (Rowat *et al.*, 2009 with permission from *PNAS*, USA).

A simple microfluidic device was used for following lineages deriving from single yeast cells. Single parental cells were needed into channels fabricated at a high density to maximize the number of lineages tracked in each experiment. To simplify tracking both pedigree and levels of protein expression, the cells were geometrically constrained to divide in a line within a single focal plane. The device was so designed that fluid can constantly perfuse through it – this allowed researchers to replenish media, change environmental conditions, and perform other analyses. It was also possible to fix and stain the cells *in situ*. By studying protein expression in the context of pedigree, Rowat *et al.* could see patterns of expression where phenotype is correlated over multiple generations; such information is not revealed when one studies the population as an ensemble.

For analyzing single cells and their progeny, Rowat *et al.* designed a microfluidic device in which lineages deriving from single cells can be spatially organized in lines. However, when introducing cells in chambers of a fabricated device, the cells distribute randomly and the number of cells per chamber is dictated by Poisson statistics. To achieve a high proportion of single cells seeded in the linear chambers, an array of chambers was fabricated that have a constriction at one end, so that cells are trapped when they flow into the chambers. After one cell enters a chamber, the ratio of flow through the chamber to bypass channels shifts, increasing the probability that subsequent cells preferentially enter the bypass channel instead of the growth chamber. This device is easily fabricated by using a single cast of PDMS and requires only a syringe pump and microscope for operation (Fig. 5.4).

It is also possible to estimate the flow rate through the microfluidic device by using lumped element modelling, an approach often used to analyze simple electrical circuits. The trapping and bypass channels act as 2 lumped resistors in parallel; the pressure drop across both channels must be equal because the end points are the same. For efficient single cell trapping, the presence of a cell in the trapping channel should change the flow in such a way that subsequent cells do not enter. Thus, when the trapping channel is empty, the flow through the bypass channel, Q_2, should be less than that through the trapping channel, Q_1; when a single cell is present in the channel, Q_2 should be greater than Q_1 so that most of the flow, and subsequent cells, flow through the bypass channel (Yen and Fung, 1978; Tan and Takeuchi, 2007). With a view to organizing the microcolonies that derive from the array of single cells and force them to grow in a single focal plane, Rowat *et al*. engineered the growth chambers with a square cross-section of the width and height of an average single cell of 5 μm.

Their device has an array of 50 chambers; roughly half of which are active chambers that fill with cells. As 10 or more devices can be fabricated on each chip, hundreds of cells can be trapped, enabling, simultaneous testing of different flow conditions or cell types in a single experiment. Cells were loaded into the device by activating 2 syringes that contain the cell suspension and growth media. Their device provides adequate control between the 2 fluid streams; the cells flow down the centre while the media flows in from the sides. This elegant geometry prevents cells from flowing into the media line, and so maintains a cell-free source of media for perfusion during the experiment. When single cells are loaded, the cell-loading inlet is deactivated by disconnecting the tubing from the syringe. To allow for metabolite exchange during cell growth, media is continually flown through the device during the experiment; the cells being round and the channels square, media perfuses well through the chambers as cells grow. Such continued media flow also ensures that there is constant flow backward through the cell inlet, preventing cells trapped upstream from entering the chamber array (Fig. 5.4).

The device designed by Rowat *et al*. can fix and stain single suspension cells and their lineages, allowing morphological phenotypes on subcellular scales to be resolved, for example, by immunofluorescence or FISH. With slight modifications in size, the lineage chambers may also possibly be used for culture of other suspension cells, including mammalian blood cells or stem cells. More broadly, this device could facilitate studies revealing correlations among single cells and their progeny that remain masked in the ensemble average (Rowat *et al*., 2009).

Intense research done on NA and protein-based diagnostic tests using microfluidics has materialized largely by the commendable progress recorded in molecular biology and molecular diagnostics (Bissonnette, 2006). In general, development of microfluidic systems for molecular diagnostics has involved developing separate pieces of equipment which address specific steps within the total sample-to-answer process. The emphasis is placed either on the amplification of the NA analyte or on a more sensitive/selective detection of the NA or protein. Little, if any, emphasis is placed on preparing the sample itself, or on how and from where the sample is obtained (Yager, 2008).

In fact, it is the neglect of sample preparation that has largely prevented greater success of a portable, integrated, microfluidic system for molecular diagnostics, especially NA diagnostics. To address this problem, Siegrist *et al.* (2009) described the steps typically involved in an NA diagnostic system, with due emphasis on sample preparation. They described centrifugal microfluidic platforms based on the compact-disc format and stated the merits of tackling each of the steps involved in molecular diagnostics. These authors carefully defined the general steps required in such a sample-to-answer NA system, with special reference to sample preparation, technology description and clarification. They also argued for centrifugal microfluidic devices.

Some microfluidic terminologies being used to describe an integrated system that can accept a sample and perform a complete analysis are: **micro total analysis system (μTAS), lab-on-a-chip (LOC)**, and, when referring to centrifugal platforms, **lab-on-a-CD**. Other terms include **point-of-care (POC)** and **sample-to-answer**. Sample-to-answer and ìTAS are near synonyms. LOC and lab-on-a-CD refer to particular steps or processes adapted onto microscale devices and centrifugal microscale devices, respectively; they do not always indicate a complete system. Point-of-care refers to diagnostic tests performed at the patient's bedside, as opposed to sending a sample to a lab. In practice, all these terms are commonly used synonymously. The terms μTAS, POC, and sample-to-answer can be used interchangeably when describing complete, microfluidic, molecular diagnostic systems.

A common and general sample-to-answer process flow for NA diagnostics is illustrated in Fig. 5.5. Variations in these steps can be introduced by the type of sample collected. For example, a highly concentrated NA sample of a very sensitive detection scheme may preclude the need for amplification.

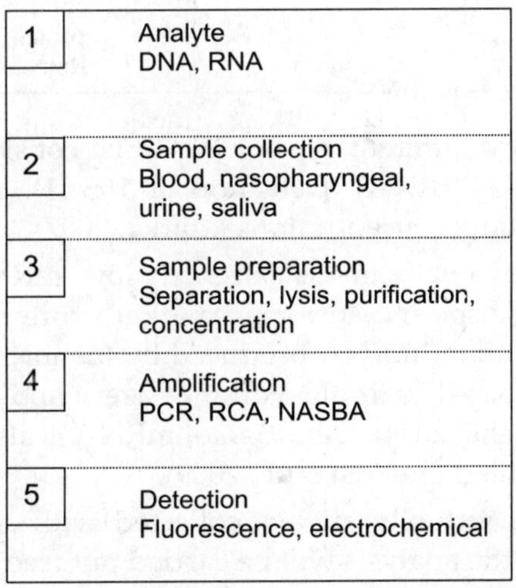

Fig. 5.5 Flow diagram of a typical nucleic acid analysis, listing common examples for each step (after Siegrist *et al.*, 2009, with permission from *Curr. Trends in Sci.*, IASc, Bangalore).

NA diagnostic tests target either DNA or RNA as the analyte (Table 5.1). In the case of DNA, the presence of foreign DNA (e.g., the DNA of a bacterial species causing infection) can be detected, or changes in genomic DNA (e.g., mutations that may be the source of cancer) can be examined.

Table 5.1 Analysis-relevant contrasts between DNA and RNA, listing common characteristics of each (after Siegrist *et al.*, 2009, with permission from *Curr. Trends in Sci.*, Plat. Jubilee vol., IASc, Bangalore).

Property	DNA	RNA
Primary structure	Double stranded (including linear and circular); denaturation or digestion produces single strands needed for amplification or detection	Single stranded; secondary structure is often present; denaturation required for amplification or detection
Type	Genomic plasmidic	mRNA, rRNA, as well as genomic (viral)
Concentration	Always present but concentrations may be low	Concentrations usually higher but there is the risk of having none present if it is not currently expressed
Location	Inside cells and/or viruses	Inside cells and/or viruses
Amplification	Can be directly amplified	In some cases, has to be reverse-transcribed into DNA before amplification (e.g., PCR); can also be directly amplified in some cases
Diagnosis	Can detect the presence of an analyte	Besides detecting an analyte, the condition of the analyte (e.g., protein expression) can also be detected

In the case of RNA, many different types have to be considered: e.g., messenger RNA (mRNA), and ribosomal RNA (rRNA). Detection of 16S rRNA helps in identification of pathogenic strains present at low concentrations (Starke, 2007).

When the concentration of a pathogen is extremely low, mRNA is often a good choice for detection. It enables study of a specific host gene-marker expression response, elicited by the presence of a microbe, by profiling mRNA produced by the infected organism. This makes it possible to test with quite small sample volume sizes, and is highly compatible with microfluidic disposables (Prucha, 2004). Detection of mRNA is also a wise option when tumor tissue is to be genetically profiled (Siegrist *et al.*, 2009).

The biological sample is virtually always collected with a means independent of the microfluidic device in which the analysis will be carried out (see Hosek, 2006).

Commonly collected clinical samples include blood, urine, nasopharyngeal swabs, tissue, fecal, saliva, and interstitial fluid. For efficient downstream analysis, the minimum expected concentration of the analyte must be known. Some critical sample parameters are fluidic properties, such as viscosity, surface tension, and density. These, as well as their variations among different samples have a substantial effect on the microfluidic function of a given disposable. Biological inhibitors in the sample also need to be considered. For example, the haemoglobin in red blood cells inhibits the commonest DNA amplification method, the polymerase chain reaction (PCR) (Radström *et al.*, 2004). Interfering compounds need to be neutralized or removed. The physical and biological homogeneity of the sample largely determines what kinds of sample preparation steps are required.

After collection, the sample is prepared for analysis. The most critical aspect of sample preparation is lysis. The NA needs to be released from the cells or viruses before it can be amplified and/or detected.

Various types of lysis can be enzymatic (digestion via enzymes), chemical (breakdown via detergents), plasma lysis (disruption via electrical charge pulses), and mechanical (breakdown via physical means). The two commonest types are chemical and mechanical. Chemical lysis methods leave behind some residual substances that need to be removed or filtered out. These methods require little or no power. They can be activated just by mixing or reconstitution of reagents. Mechanical disruption requires more energy, leaves behind no residuals, and often requires the sample to come in contact with some energy transfer medium (e.g., beads). Nucleic acids can adsorb upon to this medium and decrease the concentration and efficiency of the degrading amplification and detection steps.

Mechanical lysis seems to be the most effective method for breaking down cells having thick cell membranes, such as Gram-positive microbes, also for extracting intact DNA (Taskova *et al.*, 2006). The bead heating method is most effective – it functions by crushing cells in an agitated mixture of milling beads (Zoetendal *et al.*, 2001). This method can be properly controlled by merely changing bead size and the extent of agitation.

The mechanical lysis methods successfully disrupt all cells and viruses. However, when preparing blood samples, it is often necessary to perform a separation step before lysis. Separation steps usually occur on the cellular or viral level. While processing blood, the red blood cells, white blood cells, and plasma can be separated by using a centrifugal density gradient. However, the host NA present in white blood cells can adversely affect amplification or detection, and so has to be removed or avoided.

Separation steps effectively reduce the amount of interfering compounds but it is necessary to perform a purification step on the NA, in which the NA of interest is isolated. This is done after cell lysis. It usually accompanies a concentration step in which beads functionalized with NA probes bind the NA of interest as the lysed sample flows through. Then, the bound NA can be eluted by heating or pH change into a defined buffer volume. This solid-phase capture technique both purifies and concentrates.

Most or all of the above steps help in ensuring that a very concentrated and clean sample is delivered to the amplification and/or detection steps. This occurs through a very complex, lengthy, and costly system. A good microfluidic system for blood-based microbe detection should accomplish the following tasks: accept a blood sample (say with a volume on the order of several mLs), perform a centrifuge-based separation to isolate the plasma (or perhaps the white blood cells and plasma), mechanically lyse cells on the plasma to release the microbial DNA, flow the resulting lysate over a column of microbeads functionalized with appropriate probes that hybridize only with the microbial DNA of interest, and finally wash an elution solution (with a volume on the order of μL) over the same capture beads to release only the microbial DNA. The end result is a very small volume of solution containing a clean, pure, and concentrated DNA sample.

Amplification

After the sample has been prepared, the target NA is amplified by PCR, in which DNA can be amplified by using DNA polymerase, or reverse transcriptase-PCR (RT-PCR). In the latter, RNA is first reverse-transcribed into DNA which is then amplified by PCR.

PCR requires the sample to be thermocycled between at least 2 different temperatures. Some other NA amplification methods enable amplification at a single temperature (isothermal amplification). Most isothermal amplification methods are enzymatic. One such alternative is Nucleic Acid Sequence Based Amplification (NASBA) (Compton, 1991). Some other isothermal enzymatic amplification methods do not amplify the targets themselves, but amplify the primers. Good examples are rolling circle amplification (RCA) (Fire, 1995); and Expar, which produces a 10^6-fold signal amplification in just five minutes as detected by nanobead agglutination (Tan *et al.*, 2005). A nonenzymatic method of isothermal amplification is branched DNA amplification (Collins, 1997). It depends only on hybridization. Nanosphere's technology makes use of gold nanoparticles to catalyze silver deposition when a specific NA is present (Park *et al.*, 2002). Fluorescent chain reaction (FCR) likewise uses polythiophene polymers to detect the NA target in very small quantities ((Boudreau *et al.*, 2006).

The highly specific PCR achieves intense amplification but consumes much time and power because many temperature transition cycles have to be completed. For many analytes, a non-PCR method is better as it saves time and power.

Detection

The last step in NA analysis is detection of the analyte. This step can be incorporated during amplification, and the commonest method of this is **real-time PCR.** This should not be confused with RT-PCR, which refers to **reverse-transcriptase PCR.** Real-time PCR is performed in several different ways, all involving the monitoring of an increased fluorescence signal as amplification progresses (Fig. 5.6A-C). When accumulated fluorescence signal exceeds a certain background fluorescence (threshold), it points to occurrence of amplification. The cycle number at which such verification occurs is called C_t (threshold cycle). When references are included in a multiplexed fashion (e.g., 10 copies of reference A and 1000 of ref. B), a starting-concentration calibration-curve can be drawn based on C_t values. This is then used to back-calculate the starting concentration of the analyte(s) of interest. In this way the real-time PCR becomes quantitative, and is described as **qPCR.**

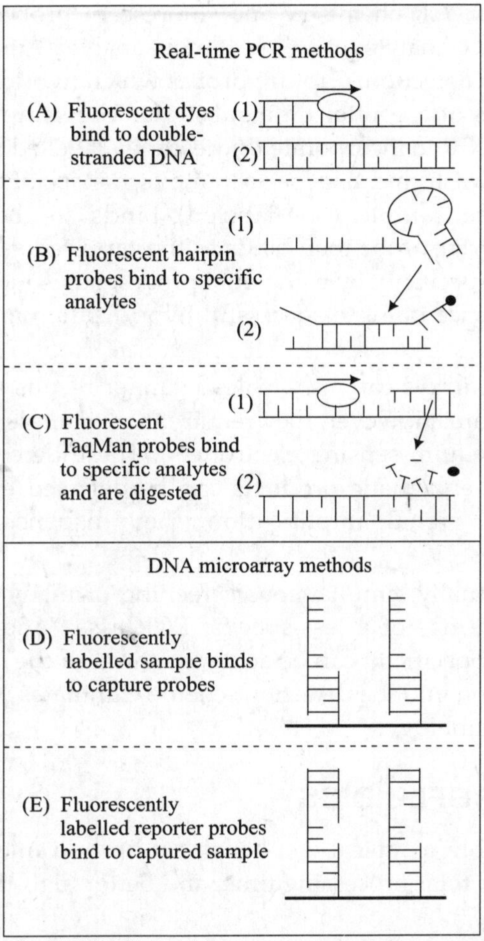

Fig. 5.6 Sketches of the commonest forms of optical nucleic acid detection. (A) real-time PCR using intercalating dyes, (B) real-time PCR using hairpin FRET probes that emit fluorescence upon analyte binding. (C) real-time PCR using TaqMan FRET probes that release the FRET pairing upon digestion by a polymerase, (D) DNA microarray using a labelled sample, and (E) DNA microarray using a capture probe-analyte-reporter probe sandwich structure (after Siegrist *et al*, 2009, with permission from *Curr. Trends in Sci.*, IASc, Bangalore).

Dyes that bind to double-stranded DNA and produce fluorescence serve as fluorescence labels for real-time PCR. The commonly used SYBR green dye (SYBR Green I) is not specific – it will bind to all double-stranded DNAs.

An alternative is to use fluorescently-labelled DNA probes which are not only highly specific but also allow multiplexing so that different fluorophores can be used for each analyte. Förster resonance energy transfer (FRET) probes (i.e., a reporter and a quencher) are commonly employed in this method to minimize fluorescence until a binding or polymerase-extension event occurs. In the former case, a hairpin DNA probe is used with FRET quenching near the probe's end until bound with the analyte. In the latter case, the probe is subject to digestion by the polymerase, which separates the FRET probes. This is now the most widely used real-time PCR method, and is known by the trade name of **TaqMan** (viz., TaqMan probes).

The chief limitation of PCR chemistry and fluorescent probes is that it does not allow robust multiplexing beyond 6 analytes at a time. If more analytes are to be studied, **optical DNA microarray** can be used for detection. Capture probes which hybridize with the NA of interest are immobilized by means of a spotter on a substrate. The sample is fluorescently labelled (usually by application of PCR to incorporate fluorescently labelled primers) and applied to the substrate, followed by a wash and a rinse to remove nonspecifically-bound and adsorbed analyte. Alternatively, the sample (non-labeled) binds to the capture probes, and a fluorescently-labelled reporter probe then binds to the captured sample, forming a sandwich structure. This gives more specificity by using dual probes, but is more complex. Capture probe spots that exhibit fluorescence point to successful hybridization and detection (Siegrist *et al.*, 2009)

Electrochemical microarrays can also achieve similar results as optical microarrays and with less expensive hardware. However, they require more complex electrical multiplexing, as each detection spot will require sensing electrode(s). These detection schemes often utilize enzymatic labels, and the enzymatic products can be detected electrochemically. As these methods involve built-in signal amplification, they dispense with the need for NA amplification in some cases.

Microarrays allow virtually simultaneous screening of many analytes. Unlike real-time PCR, they may necessitate use of a less-specific PCR assay and one that amplifies many different analytes at once. Specificity can be achieved through the microarray capture probes. Building signal amplification into the detection scheme eliminates NA amplification, e.g., with FCR and Nanosphere's technology.

CENTRIFUGAL MICROFLUIDICS

Some merits of shifting from a typical wet-bench set-up to a microfluidic device are lesser reagent use (saving), lesser total processing time, and better high-throughput and/or parallel processing (Beebe *et al.*, 2002).

However, microfluidics also faces several challenges, notably the issue of laminar flow typical for the conditions in many microfluidic conduits. Good mixing is a must for fast reactions but is not easily achieved in some conditions. Other problems are pumping and the ever-present surfactants which often lead to failure of many valves, besides channel-clogging air bubbles. Centrifugal microfluidic platforms could possibly help in overcoming some of the limitations inherent in microfluidic disposables.

In fluidic devices of any kind, the primary concern is how to get liquids to and from the areas of interest in a controlled manner. This problem necessitates the use of two technologies: pumps and valves. The centrifugal compact-disc (CD) platform provides elegant, simple, and effective modes both for pumping and valving.

Fluid propulsion on the CD occurs through centrifugally induced pressure on the fluid as the CD spins (Madou and Kellog, 1998; Madou, 2002; Duffy *et al.*, 1999). The volumetric flow rate depends on the speed at which the disc spins, the distance from the centre of the disc, the geometry of the fluidic channels, and the fluidic properties such as density, viscosity, and surface tension. By using suitable combinations of channel geometries and spin speeds, flow rates ranging from 5 nL/s to 0.1 ml/s have been achieved with high accuracy and precision. Fluid-pumping rotation speeds typically range from 300 to 2000 RPM (Siegrist *et al.*, 2009).

In fact, centrifugal pumping forces on the CD offer distinct advantages over such other pumping methods as syringe, peristaltic and electroosmotic pumping. Although pressure-driven syringe and peristaltic pumps do provide effective control over large flow rates, they become unwieldy when one tries to miniaturize and/or process in parallel (Madou, 2002). Some currently practised methods depend strongly on pH and ionic strength of the fluid being pumped. The high-voltage power supplies (>>1 kV) required for these systems further make them very expensive and unpractical. Centrifugal forces in contrast need only a low-power motor, are not dependent on pH or ionic strength of the fluidic, and do not need fluidic interconnects or tubing for force application. Another merit is that they provide forces across the entire length of a fluid element, conducive to smooth and controlled flow – bubbles have less chance of disrupting downstream fluidic processes. In fact, bubbles do not pose significant problem on CD platforms. As many individual systems can be placed on a single CD parallel processing is easy. When combined with the simple valving solutions available on the CD, powerful platforms are readily developed.

Valving operations can be carried out on the CD by using either hydrophobic valves or capillary valves. The former are again of two types: one utilizing changes in channel geometries and the other utilizes surface modification. In both cases, the fluid is forced past the hydrophobic valve by increasing the spin frequency beyond a critical value. The capillary valve is more commonly used; it can achieve a balance between centrifugal and surface tension forces in a hydrophilic material. Microfluidic structures with channels of varying capillary valve sizes can be designed to control when a valve 'opens' simply by increasing the rotational speed of the CD. This even enables sequential control of valve openings (see Madou *et al.*, 2006).

CDs for microfluidic applications usually are made of multi-layer structures of inexpensive polycarbonate plastic and pressure-sensitive adhesives (PSA). Channel widhs down to 1 mm can be machined into stock polycarbonate plastic. A cutter-plotter cuts channel widths as narrow as 200 μm in thin plastic films. The pieces can then be aligned centrally and radially and laminated together using the PSA layers.

The simplest, microfluidic CD used by leading researchers has at least 5 layers: (1) top polycarbonate CD with CNC-machined sample loading, sample removal, and air venting holes (sealed using a thin adhesive film during operation); (2) pressure-sensitive adhesive with microchannel features cut using a plotter; (3) middle polycarbonate CD with CNC channel features; (4) pressure-sensitive adhesive with microchannel features cut using a plotter; and (5) solid bottom polycarbonate CD to seal off the channels. Many CD platforms, however, have more than 5 layers to accommodate more complex fluidics. Different devices and substances such as beads, lyophilized reagents, or filter can be placed inside the CD during fabrication.

Microfabrication techniques, by providing precise control over fluid handling and temperature, enable new lab-on-a-chip devices to be built. These devices improve the speed and accuracy of biochemical assays. However, monitoring reactions in complex biological environments such as cell membranes is very difficult. Conceivably, biochemical reactions may be performed on the surface of a porous alumina membrane which is then integrated into a polymer microfluidic device. Such a set-up could possibly deliver an enzyme to one side of the membrane and reagents to the other. Work is underway to eventually produce labs-on-a-chip that might contain up to a million test chambers, each capable of screening an individual drug.

NANOELECTRONICS: TECHNOLOGIES FOR ENABLING SYSTEMS ON CHIPS

The past half a century witnessed unparalleled growth in electronics industry since the invention of the first Integrated Circuit (IC) chip in 1958. Its cumulative aggregate growth rate (CAGR) has been about 15% in the last 5 decades (see Bhat, 2007). This became possible because of the exponential increase in the component density integrated on a single chip, as predicted by Moore's law (Moore, 1965). The Moore's law assumes that the feature size of the transistor – the basic building block of an IC – may be shrunk at an exponential rate. (Moore's law is an empirical extrapolation made by Gordon Moore in 1965 – the component density on an integrated circuit doubles every year. Although there has been an exponential growth in the component density in the last four decades, as predicted by Moore's law, the growth rate has now fallen with the component density doubling every two to three years). In 1999, the transistor gate length crossed the 100 nm barrier, catalyzing the large scale production of the nanoelectronics chips.

The past advances in silicon technology development highlight many innovations which made it possible to overcome formidable barriers. This is exemplified by dual poly-silicon gates with n^+ gates for *n-channel* (NMOS) and p^+ gates for *p-channel* (PMOS) transistors, channel engineering with super steep retrograde and pocket halo implants, source/drain engineering through shallow extensions, shallow trench isolation to shrink the active device pitch, silicidation to overcome parasitic resistance, and sub-wavelength optical phase shift lithography to pattern sub-100 nm features by using 248 nm/193 nm deep UV source (see Taur *et al.*, 1997; Thompson *et al.*, 1998). But the silicon technology has mostly depended on the traditional scaling of classical CMOS device architecture. (CMOS = Complementary Metal Oxide Semiconductor). Non-classical CMOS device technologies such as high-k gate dielectrics, strained silicon channel, Silicon On Insulator, multi-gate transistors, and metal gate electrodes scale the CMOS devices to an ultimate limit of 5 nm physical gate length. In the nanoelectronics era, the technology scaling has become nontrivial due to fundamental limits imposed by device physics and materials technology. Non-classical CMOS transistors seem to be essential in the short term and non-CMOS transistors together with new information processing architectures will most probably be required in the long term.

The success of electronics depends on two critical aspects. The transistors have been shrinking in size, and the number of transistors successfully assembled on a chip has grown exponentially. Nomenclatures such as *VLSI – Very Large Scale Integration, ULSI – Ultra Large Scale Integration* and *GSI – Giga Scale Integration* are used to describe this. Over one billion transistors of sub-100 nm size can now be integrated with very high yield on a silicon chip of only 1 cm^2 area. The unique aspect of the modern electronics systems is this capability of giga-scale integration by using nano-scale building blocks. The different aspects involved in realizing the chip are shown in Fig. 5.7. The expertise needed comes from several different domains, broadly divided into application and technology domains. The scaling of the device in nano regime strongly depends on a thorough understanding of the physics of the device to engineer a high performance transistor. The process technology which can fabricate the device requires expertise in materials science, process chemistry and physics. The circuit design and system architecture requires good expertise in electronics design, computer science, communication and certain other application domains. Such a high complexity can only be managed through the tools of Computer Aided Design (CAD) which the experts in computational science and programming can provide.

One interesting development has involved building Systems on Chips (SoC) by combining non-electronic energy domain components with electronics subsystems of computing and storage elements on a single chip (see Fig. 5.7). The Silicon process technology defines non-electronics structures, such as micro/nano machined sensors and actuators, also referred to as Micro Electro Mechanical Systems (MEMS) and Nano Electro Mechanical Systems (NEMS). In fact, the SoCs include electronics and mechanical components as well as optical, chemical and biological building blocks on the same chip. In the nanoelectronics era we are going to see an increasing confluence of multiple energy domain components on a single chip, leading to such SoCs into newer application domains.

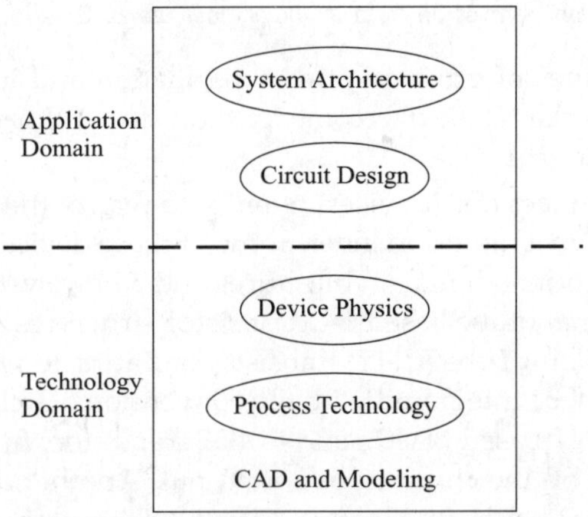

Fig. 5.7 Hierarchical view of building an electronics Integrated Circuit (IC) chip (after Bhat, 2007, with permission from *J. Indian Institute of Science*).

Device Technology

A cross-section of a classical NMOS transistor is shown in Fig. 5.8. The PMOS transistor complements this structure with N type substrate doping and p^+ source and drain regions. The CMOS technology utilizes both NMOS and PMOS transistors to build the circuits and systems. Both NMOS and PMOS transistors are created on bulk silicon substrate with SiO_2 as gate insulator and poly-crystalline silicon as gate electrode. In the MOS transistor, essentially a gate controlled switch, the vertical electrical field controls the channel conductivity (field effect transistor FET). The gate length, L_g, of the transistor determines the switching speed of the transistor.

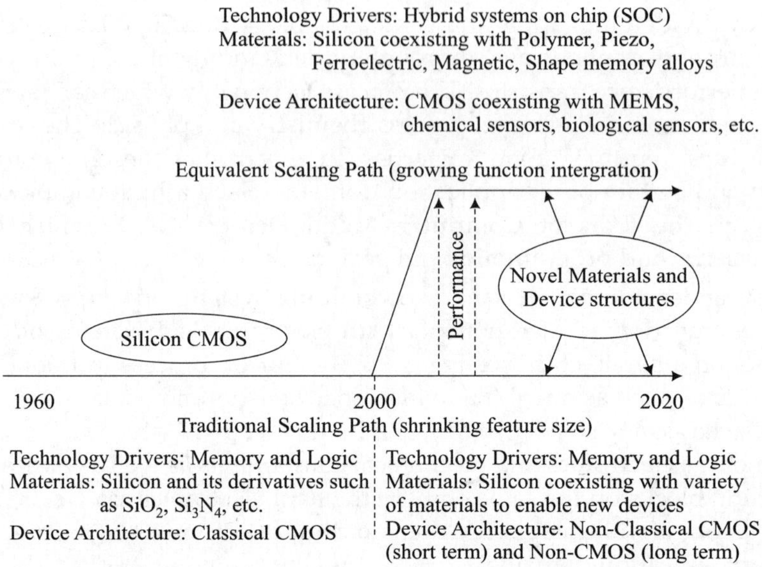

Fig. 5.8 Traditional and equivalent scaling paths in the Nanoelectronics Era (after Bhat, 2007, with permission from *J. Indian Institute of Science*).

Some system level issues of growing power dissipation and increasing device-to-device variability in the chips still need to be overcome for the successful handling of complex systems in nanoelectronics technologies.

Non-CMOS Logic architectures (devices) attempt to realize the fundamental element of a Logic block, the transistor, in a radical manner. Some of these devices are not suitable for being integrated into the conventional digital architectures and a new system architecture paradigm is needed. Nanowire and nanotube based 1-D transistor structures are potential candidates in this context. The Carbon Nano Tube (CNT) transistor has attracted interest in this context (Lin *et al.*, 2004). It may possibly be integrated with the conventional architecture, if the problem of selectively defining the CNT based NMOs and PMOS transistors in giga-scale by growing the CNTs at precise locations on the chip can be sorted out. Another alternative is the Resonant Tunnelling Transistor (RTT) having a negative differential resistance and very high switching speed (Matsuo *et al.*, 2003). Single Electron Transistors (SET) show structural resemblance to the conventional MOSET except that the source-to-substrate and drain-to-substrate junctions are tunnel junctions and the channel is ideally formed by a quantum dot (Park *et al.*, 2005). A SET might possibly be used to represent a binary bit by a single electron, but alternate architectures such as Cellular Nonlinear Networks (CNN) will be needed for realizing such systems. Another class of transistors exploits the spin of the electrons, leading to what is popularly referred to as "Spintronics". Several ideas have been floated to make such devices. The spin MOSFET is a hybrid device that relies on gate control (electron static channel control) and drain control (spin dependent scattering at the drain). However, it requires the efficient injection of spin polarized electrons into the channel, necessitating half-metallic-ferromagnetic contacts at the source and drain (Sugahara and Tanaka, 2005).

Hybrid Systems on a Chip

Traditionally, the electronics system performance could be improved in the context of Moore's law by increasing the computing performance and storage capacity of the chips. But in recent years, the silicon technology infrastructure has been exploited for other application domains also and the functional integration of nonelectronic domain components onto the chip is being driven by what is popularly called, *'the More than Moore'* philosophy. These chips are also referred to as Hybrid Systems on Chips.

Figure 5.9 brings out some emerging possibilities in terms of functional integration. In most cases the same set of semiconductor processes that are traditionally used to build transistors, are utilized to create MEMS/NEMS structures on silicon. These structures can serve as sensors and actuators; they interface with the on-chip electronics. Silicon shows excellent mechanical properties conducive to making robust mechanical building blocks such as beams, membranes and rigid masses. The success of such hybrid SOCs is shown by the fact that some of these chips have already attracted consumer applications. Most of the airbag release systems installed in automobiles actually utilize a MEMS accelerometer chip which consists of a suspended poly-silicon proof mass created on top of Silicon substrate through surface micro-machining technique. The mass is configured as one of the plates of the capacitor, and so the capacitance changes in response to acceleration. The integration provides high reliability, low cost and high performance solution which can be exploited for several applications. Other opportunities in other domains are illustrated in Figs. 5.8 and 5.9). These systems may be expected to be autonomous through harvesting the energy from the environment. In fact, the convergence of electronics and biology for lab-on-a-chip applications in diagnostics and therapy may turn out to be the biggest technological revolution in the 21st century (Bhat, 2007).

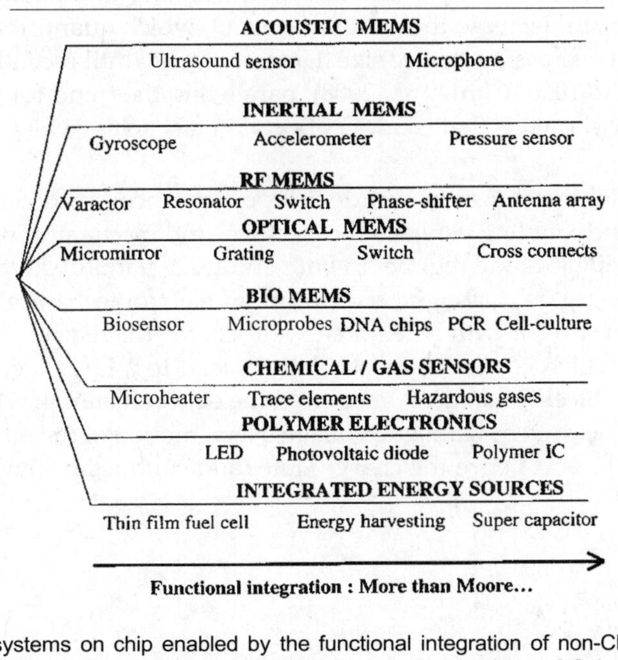

Fig. 5.9 Hybrid systems on chip enabled by the functional integration of non-CMOS components (after Bhat, 2007, with permission from *J. Indian Institute of Science*).

ELECTRONICS AND MECHANICS

The twin fields of electronics and mechanics have enabled the creation of true quantum mechanical devices. Impressive improvements in device performance and measurement techniques have not only been instrumental in facilitating NEMS leading to possible high-sensitivity detection of charge, mass, and spin but have also approached the quantum limit of mechanical motion (Schwab and Roukes, 2005). Likewise, successful manipulation of individual electrons in quantum dots has catalyzed developments in solid-state quantum computing (Hanson *et al*., 2007). Lassagne *et al*. (2008, 2009) and Steele *et al*. (2009, 2009a) integrated both these fields with a view to studying the effect of charge transport on nanomechanical motion in high-performance carbon nanotube mechanical resonators that simultaneously act as quantum dots (Hone and Deshpande, 2009) The resonant frequency and dissipation in the nanotubes both turned out to be highly sensitive to the charge state at the level of single electrons.

CNTs serve as a good model system for nanoelectronics. Adding even a single electron to this small system, however, means a large energetic cost; thus, at low enough temperatures, electrical transport in CNTs can occur through the tunnelling of electrons one at a time. This expresses itself in peaks in the current as a function of the voltage on a nearby gate that modulates the chemical potential of the nanotube - a phenomenon known as **Coulomb blockade**. Recent progress in the growth and fabrication of nanotubes (Cao *et al*., 2005) has facilitated the development of clean, freely suspended devices for observing electronic phenomena such as correlated electron states and spin-orbit coupling (see Deshpande *et al*., 2009).

Being ultrasmall, highly stiff, also having low density, CNTs prove excellent materials for NEMS; another advantage is that they circumvent the surface dissipation mechanism known, in bulk-etched NEMS, to decrease the quality factor Q (which quantifies the sharpness of the resonance peak) with decreasing device size. Unfortunately, until recently, CNTs have shown Q values of only 100 to 1000 (Chiu *et al*., 2008), paralleling the trend for etched devices. What Steele *et al*. have now shown is that clean CNTs can in fact achieve a Q of 100,000 at ultralow temperatures.

The above parallel improvements in device performance prompted Lassagne *et al*. and Steele *et al*. to study the coupling between the electrical and mechanical properties of nanotube NEMS in the Coulomb blockade regime. In both studies, a nanotube was suspended above a substrate that acts as a gate. Lassagne *et al*. used an electromechanical mixing scheme that exploits the change in current with gate voltage - a scheme that is well suited to the Coulomb blockade regime, in which conductance oscillations lead to a large signal, providing single-atom mass sensitivity. Steele *et al*. (2009) measured the dc conductance, which is sensitive to the second derivative of the current with gate voltage. Here, motion of the nanotube modulates the capacitance to the gate, and hence the charge state (and conductance) of the nanotube (Hone and Deshpande, 2009).

When voltage is applied to the gate, electrostatic force produces tension in the nanotube, increasing the resonant frequency. The frequency does not change smoothly; it shows discrete jumps correlated with the charge state as determined by the Coulomb blockade measurement. This happens because the charge on the nanotube (and hence the electrostatic tension) changes in discrete amounts; Steele *et al.* termed this effect "**single-electron tuning**".

The works of Lassagne *et al.* and Steele *et al.* reveal the rich physics involved in the coupling of NEMS and electron transport in quantum dots.

NANOSCALE CONTACTS FOR CONTROLLING FRICTION IN NANOELECTROMECHANICAL DEVICES

It is well known that gears, bearings, and liquid lubricants reduce friction in the macroscopic world, but the origins of friction for small devices such as micro- or nanoelectromechanical systems require other solutions. As the surface-to-volume ratios characteristic of these devices are much greater, they cause considerable adhesion and wear problems, and traditional liquid lubricants become too viscous when confined in layers of molecular thickness (Hu and Granick, 1998). This has prompted several suggestions to reduce friction (Urbakh *et al.*, 2004). Sliding with negligible friction is associated with superlubricity, a term coined by Hirano *et al.* (1990). Dry friction decreases when the sliding speed is reduced below a critical velocity that increases with temperature, the effect being related to thermally activated jumps occurring in the contact area and is termed thermolubricity (Krylov *et al.*, 2005).

A third way to achieve ultralow friction utilizes a sharp tip sliding over an atomically flat surface. When the normal load acting on the tip decreases below a critical threshold, the characteristic stick-slip motion is suppressed and sliding takes place smoothly without abrupt jumps and dissipation while contact is maintained. Socoliuc *et al.* (2004) observed this transition on ionic crystals in ultrahigh vacuum (UHV) (see also Nadai, 1950).

The chief limitation of all these techniques to achieve superlubricity is that they cannot be easily applied in practical situations. Friction can be switched on and off quite efficiently and superlubricity can be achieved at the atomic scale by exciting the mechanical resonances of the sliding system perpendicular to the contact plane (Socoliuc *et al.*, 2006). The resulting variations of the interaction energy reduce friction below 10 piconewtons in a finite range of excitation and load, without significant wear. Without actuation, atomic stick-slip motion, which leads to dissipation, occurs in the same range. Even if the normal oscillations require energy to actuate, the new technique minimizes energy dissipation in nanocontacts.

SINGLE-ELECTRON TUNNELLING AND NANOMECHANICAL MOTION – NANOSCALE RESONATORS

Nanomechanical systems have attracted applications for ultrasensitive mass detection (Craighead, 2000; Lassagne *et al.*, 2008; Chiu *et al.*, 2008). Nanoscale resonators which oscillate at high frequencies have proven useful in many measurement applications. Nanoelectromechanical resonators have potential applications in sensing, cooling, and mechanical signal processing. An important parameter in these systems is the strength of coupling of the resonator motion to charge transport through the device. The combination of a high resonance frequency and a small mass is what makes nanomechanical resonators so attractive for a critical study of mechanical motion in the quantum limit (Naik *et al.*, 2006). Successful observation of quantum motion of a macroscopic object depends on a high-frequency nanoscale resonator having low dissipation (which implies a high quality factor *Q*) and a sensitive detector with minimum back-action (i.e., quantum limited) (Regal *et al.*, 2008).

Steele *et al.* (2009, 2009a) studied a high-quality mechanical resonator made from a suspended carbon nanotube driven into motion by applying a periodic radio frequency potential using a nearby antenna. Single-electron charge fluctuations created periodic modulations of the mechanical resonance frequency. A quality factor exceeding 10^5 enables the detection of a shift in resonance frequency caused by the addition of a single-electron charge on the nanotube. Additional evidence for the strong coupling of mechanical motion and electron tunnelling is provided by an energy transfer to the electrons causing mechanical damping and unusual nonlinear behaviour. Steele *et al.* also reported that a direct current through the nanotube spontaneously drives the mechanical resonator, exerting a force that is coherent with the high-frequency resonant mechanical motion. They demonstrated a dramatic back-action that strongly couples a quantum-dot detector to the resonator dynamics of a carbon nanotube, and which, in the limit of strong feedback, can spontaneously excite large-amplitude resonant mechanical motion.

Whereas nanomechanical resonators have been produced by etching down larger structures, in small devices, surface effects impose an upper limit on *Q*. An alternative is to use suspended carbon nanotubes to avoid surface damage from the (etching) fabrication process. Hüttel *et al.* (2009) developed a mechanical resonator based on an ultraclean carbon nanotube with high resonance frequencies of several hundred megahertz and a *Q* exceeding 10^5. Steele *et al.* used this resonator to explore a strong coupling regime between single-electron tunnelling and nanomechanical motion. Their research was inspired by the use of aluminium single-electron transistors as position detectors (Naik *et al.*, 2008) and atomic force microscopy cantilevers as resonators (Zhu *et al.*, 2005). However, the experiment described by Steele *et al.* is in the limit of much stronger electromechanical coupling, achieved by embedding a quantum-dot detector in the nanomechanical resonator itself.

The device described by Steele *et al.* consists of a nanotube suspended across a trench that makes electrical contact to two metal electrodes. Electrons become confined in the nanotube by Schottky barriers at the Pt metal contacts, forming a quantum dot in the suspended segment. The nanotube growth constitutes the final step in the fabrication process, yielding ultraclean devices (Steele *et al.*, 2009a). Steele *et al.* made all measurements at 20 mK with an electron temperature of ~80 mK.

The resonator can be actuated with a nearby antenna and the resonator motion is detected by its influence on the dc current through the nanotube. *Q* typically exceeds 10^5, representing an increase of more than two orders of magnitude compared with previous nanotube studies (Lassagne *et al.*, 2008). The resonance frequency can be tuned by more than a factor of 2 with the gate voltage. The electric field from the gate pulls the nanotube toward it, and the subsequent lengthening of the nanotube creates more tension, analogous to the tuning of a guitar string.

Witkamp *et al.* (2006), Eriksson *et al.* (2008) and Jensen *et al.* (2008) used CNTs to fabricate mechanical resonators that can operate at ultrahigh frequencies, have widely tunable resonance frequencies, and serve as ultrasensitive inertial mass sensors. Besides this, CNTs have exceptional electron transport properties, including ballistic conduction over long distances or multiple Coulomb blockade-related phenomena that can be observed even up to room temperature. This makes such coupling of the mechanical motion of nanotube resonators to electron transport so highly appealing. Researchers are interested in using such a coupling as a means to control the mechanical motion at the nanoscale.

Lassagne *et al.* (2009) studied the coupling between the mechanics and the electron transport of a SWNT resonator at cryogenic temperature in which conducting electrons enter the Coulomb-blockade regime. Single electrons tunnelling into and out of the nanotube greatly affect the nanotube motion. The coupling could be strengthened significantly over that achievable with NEMS resonators, including those fabricated in silicon by using top-down approaches. Such a coupling is strong enough to drive the oscillation in the nonlinear regime.

MEASUREMENT PROBES FOR THE NANOSCALE

Advanced measurement probes have been designed to determine nanomechanical properties. These probes are not only revealing novel phenomena but also enabling commercialization of devices based on MEMS and NEMS. The resistance of various materials and structures to reversible, irreversible, and time-dependent deformation modes can be measured accurately with high precision and spatial resolution.

Most devices perform mechanically in an elastic (reversible) manner. This makes the elastic modulus extremely important. The greatest spatial resolution is provided by **contact resonance atomic force microscopy (CR-AFM).** In this technique changes in the resonance frequencies of an AFM cantilever probe are measured by bringing the probe tip into contact with a surface. Such a contact stiffens the cantilever response and increases the cantilever resonance frequencies (Cook, 2010). The modulus of the contacted material can be determined. Contact radii of 2 to 3 nm are created by using probe tips of 20- to 40-nm radius and controlled contact loads in the 100 nN range; this enables mapping of the stiffness (and hence the modulus) with better than 10-nm spatial resolution. Stan *et al.* (2008) used this system to map the elastic modulus of a nanocrystalline gold film and confirmed that, as they are less dense, grain boundary regions are more compliant than grains. Such measurements allow one to predict the response of nanomaterials, especially **nanograined polycrystals**, to stress, because they contain a much greater proportion of grain boundary material than their macrograined counterparts and so are more compliant (Cook, 2010).

CR-AFM has also been used to measure the elastic moduli of zinc oxide and tellurium nanowires (NWs) (Stan *et al.*, 2008a). The technique revealed new mechanical phenomena intrinsic to the nanoscale. NWs with radii greater than about 100 nm showed moduli comparable to bulk material whereas those with radii less than 100 nm showed greatly enhanced moduli. This change in the NW properties is opposite to that for grain boundaries – surface tension effects create rearrangement of atomic positions and generate greater atomic densities at surfaces, so leading to a stiffer shell of surface material surrounding a NW core of bulk-like material. As the NW radius decreases, the ratio of shell/core material increases, producing a stiffening effect. For NWs smaller than a critical radius, the bulk core disappears, the NW is all "shell" and the modulus becomes invariant.

Judicious application of the technique which provides the greatest resolution for measurement of material resistance to (irreversible) plastic deformation is **instrumented indentation testing,** or **nanoindentation.** In this, the load and displacement of a hard, stiff probe, usually a pyramidal diamond with a 50-nm tip radius, can be measured as the probe contacts a surface. Nanoindentation measurements, combined with AFM of the real shape of the indenting probe and finite element analyses of the indentation stress field can measure the single-crystal shear yield stress of tungsten (Ma *et al.*, 2009).

Nanoindentation techniques can also measure viscoelastic (time-dependent) properties. It is possible to measure the response of the indenter during fixed-load or fixed-displacement as a function of time. Indeed, researchers can determine the viscoelastic properties of polyurethane by using a spherical indenter and also predict the time-dependent responses over different time scales and indenter geometries. Conical indentation models can describe and predict the time-dependent and irreversible viscous-elastic-plastic properties of certain glassy polymers (Oyen *et al.*, 2003).

Despite great progress in nanomechanical probing, two challenges need to be tackled. One relates to brittle materials and fracture. Nanoindentation with acute probes can generate cracks at all length scales, and the mechanics of acute indentation have been identified and analyzed; we still require postindentation observation of cracks to determine toughness.

The second challenge is posed by biological and other gel-like materials which contain an appreciable fluid phase. In these systems, it is the flow through a porous elastic or viscoelastic matrix that governs the time-dependent deformation properties. Galli *et al.* (2009) used probe-based techniques to characterize the behaviour of hydrogels. Such measurements can also estimate the permeability of the matrix to the fluid and indicate reduced permeability at the nanoscale in bone (Oyen *et al.*, 2008). Development of suitable nanomechanical probes for poroelastic systems may enable rapid development of artificial tissues and biomedical devices (Cook, 2010).

BIOELECTRONICS

Ions and membranes that are common biological signal transducers can be combined with electronic circuitry.

Noy *et al.* (2009) insulated 20-40-nanometre-wide silicon nanowire transistors by coating them with lipid membranes. This was followed by adding proteins that formed channels through the membranes. A flow of ions through the membrane pores triggered an electrical response in the nanowire. Conversely, by applying an electric field to the wire, the membrane's protein channels could be opened or closed.

FLEXIBLE BATTERIES

Recent years have seen several soft portable electronic equipment such as rollup displays and wearable devices which require flexible batteries. Active radio-frequency identification tags and integrated circuit smart cards are used to achieve durability in daily use. This has prompted exploration of diverse options for the development of flexible batteries. In some, batteries are made mostly or entirely from plastic; here the added advantage is avoiding ignitable and toxic substances such as lithium and lead. Huvila (2007) showed how an inorganic primary battery may be bent like a piece of paper for disposable-card applications. However, because primary batteries generate current by a one-way chemical reaction and are not rechargeable, their utility in portable electronic equipment is not much. Instead rechargeable secondary batteries are preferred for powering portable equipment. Dudney *et al.* (2007) described recent efforts to make secondary lithium-ion batteries into thin films while maintaining their high energy capacity.

Making a bendable lithium-ion battery requires suitable soft electrode-active materials, e.g., metal oxide nanoparticles or nanocoatings for cathodes and lithium foil or nanocarbon materials for anodes. Virus-templated Co_3O_4 nanowires can improve the capacity of thin, bendable lithium-ion batteries (Nam *et al.*, 2006; Scrosati, 2007). The charging/discharging process of batteries is determined by the electron and counterion transport at the surface of the electrodes. Using nanostructured inorganic materials increases the rates of the electron and counterion transport (see Nishide and Oyaizu, 2008). As these materials are soft and bendable, bendable lithium-ion batteries can be developed. The problem is that transport rates at inorganic electrodes can be limited by the slow kinetics of ion intercalation and migration in these materials. The transport properties may be improved by using nanoarchitectured electrodes with large surface areas (Taberna *et al.*, 2006).

Plastic batteries contacting organic electrodes show marked advantages over lithium-ion batteries, because the organic materials are not only flexible but their properties can also be adjusted through chemical synthesis.

A flexible plastic battery (Fig. 5.10) uses an electrolyte layer sandwiched between thin layers of polymers that have low conductivity but incorporate redox-active groups, so as to enhance the overall redox capacity of the battery. The R groups in the cathode and in the anode have different redox potentials. During the charging process, charge is stored by oxidizing R groups at the cathode and reducing R groups at the anode. The output voltage of the battery corresponds to the gap between the redox potentials. Polymer chains connect the R groups and give flexibility. Several R groups are bound to the polymer chain, enabling electrons to move between neighbouring R groups to produce the output current (see Nishide and Oyaizu, 2008).

The notable point is that the polymer backbones act as a matrix (rather than a conducting path) to interconnect many redox sites for the hopping of electrons by a self-exchange mechanism, thereby resulting in the storage and transport of charge in a homogeneous solid. Unfortunately, one shortcoming of this design is that some concurrent chemical reactions can result in an electrochemically irreversible reaction characterized by slow kinetics (Berridge *et al.*, 2006). Organic electrode reactions need to be reversible for developing organic batteries that can be rapidly charged and discharged with a large current.

Another option for the development of organic electrode-active materials exploits the high redox capacity of aliphatic redox polymers, which are densely populated with pendant redox-isolated sites (Miller *et al.*, 2004). Purely organic polymers based on open-shell molecules (radicals) such as nitroxides and galvinoxyls facilitate reversible one-electron redox reactions

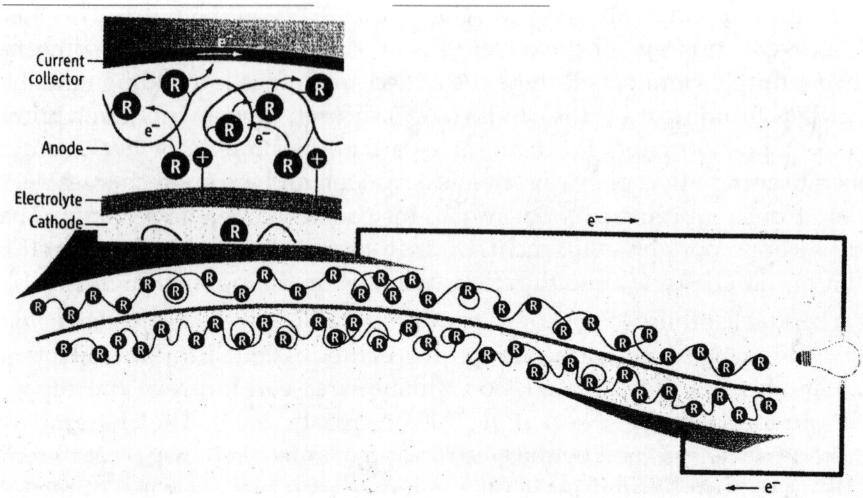

Fig. 5.10 Sketch of a flexible plastic battery (for explanation, see text) (From Nishide, H., Oyaizu, K. *Science* 319: 737-738, 2008. Reprinted with permission from AAAS).

featuring fast electrode kinetics, reactant recyclability, and high redox capacity. These polymers act as both cathode- and anode-active materials – their redox potentials can be adjusted by varying the organic substituents. "Radical polymer batteries" can be charged in less than 30 seconds. They generate burst power at rated voltages, and can be transformed into fully flexible, foldable, and semi-transparent batteries (Suga *et al.*, 2007).

Although the technologies of lithium-ion batteries have matured, they have some limitations because of safety concerns (prompted by accidents involving ignition and explosion) and because it is very difficult to make highly flexible lithium-ion batteries. As compared to their inorganic counterparts, plastic batteries are safer, adaptable to both roll-to-roll and inkjet printing processes, and easier to dispose (they can be burned away without toxic gas and ash formation). They can be fabricated from less-limited resources. Plastic batteries are, however, more bulky than lithium-ion batteries. But if the batteries are sufficiently light, flexible, and environmentally friendly, then bulkiness does not pose a significant problem.

MICROBATTERY ELECTRODES BASED ON VIRUSES

Emerging needs for small scale power sources, including those embedded in flexible substrates and textiles have made the fabrication and spatial positioning of electrodes the most central issue in battery technology. Novel electrode positioning methods could enable the use of nanostructured electrodes and multidimensional architectures in new battery designs showing improved electrochemical performance. Nam *et al.* (2008) demonstrated the synergistic use of biological and nonbiological assembly methods for fabricating and positioning small battery components which enable high performance microbatteries with complex architectures. A self-assembled layer of virus-templated cobalt oxide nanowires serving as the active anode material in the battery was formed on top of microscale islands of polyelectrolyte multilayers serving as the battery electrolyte, and this assembly was stamped onto platinum microband current collectors. The resulting electrode arrays show full electrochemical functionality. This versatile approach to fabricating and positioning electrodes enhances flexibility for implementing advanced battery designs such as those with interdigitated microelectrodes or 3D architectures.

Biological systems are uniquely able to grow and assemble nanostructures with considerable control, best exemplified by biomineralization (Mann, 1996). Knowledge about the interface between biological molecules and inorganic materials prompted novel approaches in the design and fabrication of man-made devices (Dujardin *et al.*, 2003; Nam *et al.*, 2006; Carette *et al.*, 2007). Various kinds of biomolecules such as DNA, peptides, and proteins have been functionalized, assembled into nanostructures, and inserted in electronic devices such as transistors, biosensors, memory devices, and battery electrodes (Zhang, 2003; Seeman and Lukeman, 2005). Whaley *et al.* (2002) showed that the genetically engineered M13 virus is a good model to understand the interaction between peptides and inorganic materials and can be used as a programmable molecular building block to template inorganic materials growth and achieve self-assembly (Nam *et al.*, 2004; Mao *et al.*, 2004). The filamentous M13 virus is about 6.5 nm in diameter and 880 nm long, with a genetically modifiable outer protein coat composed of 2,700 identical protein subunits (Nam *et al.*, 2008). This makes the virus a versatile device for designing materials "from DNA sequences to functional devices"; M13 libraries have already been used extensively for biomolecules selection in biotechnology. The M13 viruses have well established genetic modification protocols, and their fine structure and biophysics have been thoroughly studied (Dogic and Fraden, 2006).

Nam *et al.* (2006) reported that using M13 viruses in electrode fabrication improves the storage characteristics and rate capability of a lithium ion battery. By adding a few nucleotides into the viral DNA, specific functional groups with affinity for cobalt ions can be displayed on the viral major coat proteins. These engineered viruses were successfully exploited to grow nanowires of cobalt oxide; cobalt oxide represents a new type of battery material with a much greater storage capacity than the carbon-based negative electrodes used in lithium-ion batteries. Alternating electrostatic layer-by-layer assembly (LbL) is used with the appropriate choice of polyelectrolytes to generate ultrathin multilayer films having ion transport properties that act as solid state electrolytes (Lowman *et al.*, 2004; Lowman and Hammond, 2005). The fact that the virus is uniquely able to self-assemble on top of polyelectrolyte multilayers via competitive charge interactions and/or enhanced polyelectrolyte chain mobility pointed to the possibility of using the virus to make a fully self-assembled battery. By exploiting the toolkits of genetically engineered virus templates for biomineralization, multilayer assembly methods, and soft lithography, Nam *et al.* (2008) achieved the assembly and characterization of a stamped microbattery.

Many challenges face the daunting task of improving battery performance. One key limitation of using presently available assembly techniques is that less than half of the mass and volume of the battery is occupied by electrochemically active materials. This challenge becomes strong for smaller batteries – as battery size is reduced, the active fraction diminishes more and more. To increase energy density, and specific energy, the relative amount of inactive supporting material needs to be reduced (Cho *et al.*, 2007). Battery performance may be substantially improved (Arico *et al.*, 2005) if nanostructured battery materials are used. As electronics become increasingly ubiquitous in new form factors ranging from flexible paper-thin cards to fibres to miniaturized implantable medical devices, there is a growing but unmet demand for high-performance batteries in matching form factors. Much research has been done to integrate electronic nanodevices (Kang *et al.*, 2007; Yu *et al.*, 2007), but much less work has been done on how to power these devices. Nam *et al.* (2008) proposed a versatile approach to fabricating and positioning small battery components that may lead to designing high performance microbatteries with complex architectures (e.g., interdigitated or 3D structures), and which could even be placed on flexible substrates (e.g., fibres or flex circuits). Their work combines biological assembly and microcontact printing technology to fabricate a hierarchically nanostructured battery anode, which is then successfully positioned together with polymer-based solid electrolyte and microband current collectors onto a substrate (Nam *et al.*, 2008).

Figure 5.11 shows the process of constructing the microbattery electrode, a polydimethylsiloxane (PDMS) stamp of embossed microcylindrical patterns was fabricated by thermal-curing of the prepolymer over a master pattern of photoresist. Two different cylindrical patterns were used to determine the versatility of this process: a 4-μm diameter post with 5-μm center-to-center spacing (denoted as stamp pattern I) and an 8-μm-diameter post with 20-μm spacing (stamp pattern II). Polyelectrolyte multilayers were deposited on the PDMS stamp pattern by electrostatic LbL assembly techniques (Hammond, 2004; Park and Hammond, 2004) and acted as the solid electrolyte and battery separator. Linear-polyethylenimine (LPEI, 25,000 M_r) was used as the polycationic building block and anionic polyacrylic acid (PAA, 90,000 M_r) as a counter-charged polymer. Twelve bilayers of both LPEI and PAA were alternately deposited at pH 5. Ellipsometry showed that the height of the electrostatically deposited polymeric separator was 150 nm. Following polymer deposition, an M13 virus solution was dropcast onto the $(LPEI/PAA)_{12.5}/PDMS$. Tetraglutamate (EEEE-) was fused to the N terminus of each copy of the major coat p8 protein of the M13 virus by adding the corresponding nucleotides to gene 8. Nam *et al.* dipped the assembled viruses in cobalt oxide precursor solution, so that planar micropatterns of cobalt oxide nanowires corresponding to the stamped patterns could be grown on the polyelectrolyte multilayers which act as a solid electrolyte and separator for lithium ion transport (Lutkenhaus and Hammond, 2007). They then transferred the microbattery electrodes comprising the virus-templated cobalt oxide anodes and solid polymeric electrolyte onto four lines of platinum microband current collectors; this could be done by using an adaptation of the multilayer transfer method. The stamp pattern I was transferred onto microbands of 10-μm pitch and 3-nm length. A small pressure of a few bars was applied during the pattern transfer so as to ensure conformal contact between the PDMS mould and the line-patterned platinum electrode. To boost efficient electron transfer, the platinum electrode was lightly treated with air plasma for 20 sec to increase the surface energy and then rinsed with water. Pattern transfers were achieved successfully regardless of the pattern size or the width of the platinum microbands (Nam *et al.*, 2008).

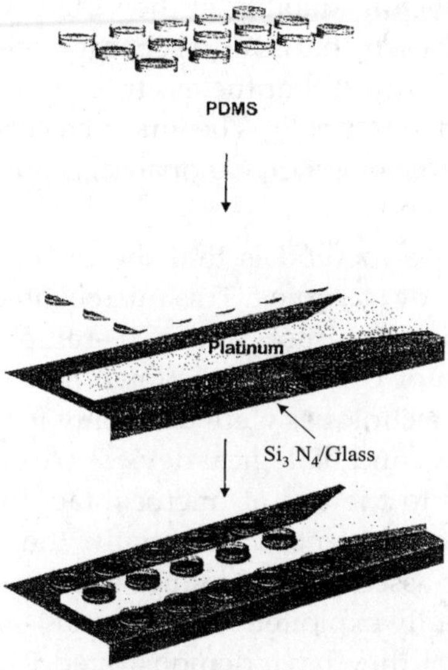

Fig. 5.11 Making virus based microbattery electrodes. The virus-based cobalt oxide nanowires assembled on the polyelectrolyte multilayer are together stamped onto Pt microband current collectors (PDMS = polydimethylsiloxane) (after Nam *et al.*, 2008, with permission from *PNAS*, USA).

In the microbattery anode and solid electrolyte assembled on a PDMS stamp before pattern transfer, the stacking order from top to bottom is cobalt oxide/virus/LPEI/PAA)/PDMS. The increased surface mobility of the polyelectrolytes gained from a polymeric interdiffusion process augments the surface mobility of electrostatically bound M13 viruses on top of the polyelectrolyte multilayers. This leads to the formation of a 2D liquid crystalline assembly of viruses (Yoo *et al.*, 2006). Nanoscale cobalt oxide is a promising class of lithium-ion battery electrodes with very high reversible capacity. Growth of cobalt oxide nanowires on the virus-assembled monolayer generates a characteristic surface morphology. This virus-mediated charge interaction assembly method achieves high-nanostructure packing density of the active materials while also maintaining the structural integrity of the assembled nanostructures (Nam *et al.*, 2006). This is in marked contrast to conventional methods employed to deposit nanoparticles or nanowires on multilayers which result in 3D network structures of low packing density. The 2D structures reported by Nam *et al.* (2008) show good potential for the high-density assembly of nanowires for applications including microbatteries and other microdevices. And, if supplemented by certain adsorption and microfluidics techniques, this method can potentially enable production of oriented, close-packed viral systems for templating structures in three dimensions.

The method designed by Nam *et al.* may also be applicable to the assembly of other devices requiring small-area charge storage or sensing components. The additional stamping of cathodes onto the same substrate as the anodes might aid the designing of batteries with interdigitated electrode architectures. As Nam *et al.* succeeded in arranging as many as 150 microelectrodes on each platinum microband of 20-μm width, it is estimated that each microelectrode may have a storage capacity between 625 and 766 pAh, a feat of no small importance!

Nam *et al.* also tested the cycling stability of the virus-based microbattery electrode. The stable open-circuit voltage after each charge or discharge cycle and the consistent open circuit voltage values between cycles showed that the electrochemical reaction of Li with the Co_3O_4 nanowires is quite stable and reversible. The microbattery electrodes showed a similar charging/discharging curve as that of Co_3O_4 nanoparticles produced by other methods, even at such a high charging rate of 255 nA..

A great merit of Nam *et al*'s method is that the microbattery electrodes can be easily transferred to various substrates by stamping. This microcontact printing technique may enable economical small-scale patterning of viruses and virus-templated nanostructures on different substrates having a vast range of chemical functionalities. It may be recalled that till very recently, microcontact printing techniques were used only for fabricating functional electronic devices, complex optical devices, and biological devices (Ahn *et al.*, 2006; Menard *et al.*, 2007). The work of Nam *et al.* points to the use of microcontact printing to position microbattery electrodes formed by virus-based assembly. Essentially, the approach adopted by Nam *et al.* combines polymer electrostatic assembly techniques and biological self-assembly at multiple length scales, and may be gainfully exploited to make small-scale power sources embedded in various system structures. What they have demonstrated is that traditional soft lithography technology and a knowledge of the interface between polymers, viruses and inorganic materials can be effectively combined for fabricating and positioning microbattery electrodes – electrodes that can be easily stamped on various different rigid or flexible substrates to make microbatteries that perform stably and reliably. In the coming few years we may well see exciting advances in biotemplating, self-assembly, and polymeric nanopattern designs that will provide even greater flexibility in the design of novel battery architectures.

ELECTRONIC EXCITATIONS ON METAL SURFACES AND NANOSTRUCTURES

Single particle and collective electron excitations on metal surfaces and nanostructures can be studied by using the highly versatile technique of **photoemission spectroscopy**. Plasmons are collective oscillations of electrons in a solid. Plasmon loss in photoemission spectra of metals has been extensively studied since long (see Hedin, 1974; Langreth, 1974). Bulk plasmons are longitudinal oscillation modes of the electron gas in the solid. The physics of plasmon excitations in photoemission has aroused interest in view of the fact that different processes contribute to its intensity (Barman, 2005). As reported by Langreth (1971), the sudden change in potential due to creation of a core-hole attracts conduction electrons to screen the core-hole, manifesting in intrinsic plasmon excitation. In contrast, extrinsic plasmon excitation results from Coulomb interaction of the conduction electrons where the photoelectron traverses through the solid from the photoemission site to the surface. Some interference occurs, caused by quantum interference between the intrinsic and extrinsic plasmons. Such an interference can be visualized as the interaction between the localized photo-hole (intrinsic) and the outgoing photoelectron (extrinsic) in which the virtual plasmon created by one is absorbed by the other. Such collective excitations are critical for interpretation of all surface spectroscopies which use electromagnetic fields or charged particles. In **angle resolved photoemission (ARPES)**, surface screening effects can change the line shape of the direct transition from Ag(111)'s band (Hansen *et al.*, 1997), Claessen *et al.* (1999) reported a strong modification of the photoemission line shape in Li near its bulk plasmon frequency. Indeed, surface screening effects can markedly influence the spectral line shape in ARPES (Barman, 2005).

Barman (2005) reviewed photoemission studies on quantization of electron states due to confinement in nanostructures such as epitaxial Na films on Al(111) and the electronic structure of Ar nano-bubbles embedded in the sub-surface region of Al. Ar bubbles in Al constitute a very suitable system for studying the response of conduction electrons in a nearly free electron metal to a core-hole generated by photoemission in an inert solid bubble of nanometer size implanted in the metal. An estimate of the bubble size is crucial in fields such as sputter growth of thin films, electromigration failure in conduction lines of integrated circuits, and life time of reactor walls (Arzt, 1994). For Al, an asymmetric line-shape is seen for both monopole surface and bulk plasmon in agreement with theory. The relative contributions of the intrinsic, extrinsic and the interference processes to the surface plasmon intensity can be determined from theoretical plasmon line-shape calculations and angle-dependent photoemission. Using angle resolved photoemission, Na thin films on Al(111) have been studied for different thicknesses. Some features in the valence band spectra behave like quantum well resonances in a narrow photon energy range where the overlayer collective excitations are noticed. These resonances result from quantum confinement due to a potential step at the Na/Al interface and the dynamical enhancement of the electric field in the overlayer.

Metallic Thin Film Nanostructures – Quantum Size Effects on Work Function

The minimum energy needed to move an electron from a solid into the vacuum, called the work function, plays a crucial role, for instance, in the photo-electric effect which has been the foundation stone on which quantum mechanics rests. The work function (Φ) is the outcome of a complex interplay between quantum mechanics and atomic scale forces. There is an exciting possibility of controlling the work function through quantum engineering of electronic structures at the nanoscale (Paggel *et al.*, 2002; Qi *et al.*, 2007; Miller *et al.*, 2009). Kim *et al.* (2010) provided direct evidence for **quantum size effects (QSEs)** on the work function in ultrathin Pb films, and established the direct correlation of this quantity with the behaviour of **quantum well states (QWSs).** The ability to control the growth of metallic thin films with atomic layer precision enables one to anticipate the possibility of tuning the work function of metallic thin films through the QSEs, thereby influencing chemical processes on surfaces. By using scanning tunnelling microscopy and spectroscopy, they showed that the very existence of QWSs in these ultrathin films greatly affects the measured tunnelling decay constant *K*, resulting in a very rich phenomenon of "quantum oscillations" in *K* as a function of thickness, *L*, and bias voltage, *V*. More specifically, the phase of the quantum oscillations in *K* vs. *L* depends sensitively upon the bias voltage, which often results in a total phase reversal at different biases. On the other hand, at very low sample bias ($|V_s| < 0.03$ V) the measurement of *K* vs. *L* accurately reflects the QSE on the work function (Ö). In particular, the minima in the quantum oscillations of *K* vs. *L* occur at the locations where QWS cross the Fermi energy, thus directly unravelling the QSE on the work function in ultrathin films, which was predicted over three decades ago. The work reported by Kim *et al.* further clarifies certain contradictions regarding the relationship between the QWS locations and the work function.

6

CHAPTER

Silica-based Materials, Electronics, Electrets and Transistors

INTRODUCTION

Silica (silicon dioxide, SiO_2), the basic component of sand and quartz, is the most abundant compound in Earth's crust. It is possible to synthesize it as an amorphous solid with nanoscale pores and an accessible internal surface of up to 1000 square metres per gram (Kresge *et al.*, 1992). If such a carefully designed porous silica framework is combined with certain organic molecules, hybrid 'organosilica' materials result that can perform some impressive chemical feats. For instance, Alauzun *et al.* (2006) fabricated a mesoporous organosilica (MPO) that enables two mutually antagonistic molecular groups - one acidic, one basic - to coexist peacefully. The term 'mesoporous' is applied to nanostructures with pore sizes in the range of 2-50 nm.

Indeed, as acidic and basic molecules are antagonistic, it is not easy to keep them in their place unless they are united under the control of ordered, nanoporous organosilicas (Jaroniec, 2006). Mesoporous silica structures can be fabricated in channel-like and cage-like conformations. If we examine a cross-section through purely siliceous pore walls, a structure consisting of interconnected silicon, oxygen and hydrogen atoms is revealed. Mesoporous silica structures can accommodate organic groups in various ways: e.g., pores decorated with organic surface groups; an outer framework with organic 'bridging' groups; and bifunctionality with both surface and bridging groups.

Silica nanostructures are easily prepared through self-assembly of a silicate precursor material on an organic template gel. Removing or burning the template leaves channel-like or cage-like pores and it is possible to attach or immobilize other molecular groups here (Zhao *et al.*, 1998). In both cases, the pore walls of silica nanostructures are made of randomly connected

SiO_4 tetrahedra in each of which a silicon atom forms four bonds, one to each of four oxygen atoms. Since an oxygen atom has two available bonds, it can link to two silicon atoms. So in most of the framework, each silicon atom is bound to four neighbours through oxygen atoms. The exceptional bonding, however, is on the surface of the framework where the oxygen atoms use one of their two bonds to link up with a hydrogen atom. The characteristic features of silica materials are the presence of siloxane groups (denoted ≡Si – O – Si ≡) in the bulk, and both silanol (≡Si – OH) and siloxane groups on the surface (Jaroniec, 2006).

These reactive silanol groups permit modification of the silica surface if hydrogen is replaced by an organic group. But this surface modification, termed post-synthesis grafting, is by no means the only way to decorate the pore walls of silica nanostructures with surface groups. Stein *et al.* (2000) used the so-called 'one-pot synthesis', which comprises stages of self-assembly and co-condensation, in the presence of an organic template, of the silicon-containing precursors $Si(OEt)_4$ and $RSi(OEt)_3$. The ethoxy group (EtO) in these precursors leaves the precursor during hydrolysis, thereby allowing the condensation and formation of the framework.

The chief merit of such a self-assembly approach is that use of slightly more complex precursors known as silsesquioxanes, for example $(EtO)_3Si\text{-}R\text{-}Si(OEt)_3$, makes it possible to introduce organic groups into the silica framework during templating synthesis, rather than being cobbled on to its surface after the event. So-called periodic mesoporous organosilicas are the result (Asefa *et al.*, 1999), in which organic bridging groups (≡Si – R - Si≡) replace some siloxane bonds. In this way, self-assembly of simple or mixed precursors offers many opportunities for incorporating both surface and bridging groups into silica nanostructures. However, although the self-assembly approach is quite general, it is not suitable for the simultaneous incorporation of two or more groups with antagonistic properties, such as acidic and basic groups, because the precursors of these groups would simply react during organosilica synthesis. It is in this context that Alzauzun *et al.* (2006) designed an elegant method of transforming the framework and surface groups of a specially designed bifunctional PMO into acidic and basic groups, respectively. What these authors have shown is that the self-assembly synthesis of mesoporous organosilicas with framework and surface groups, followed by appropriate chemical transformations, is a promising route towards obtaining nanostructures showing two opposing properties. According to Setoyama (2006), the materials designed by Alauzun *et al.* (2006) may be ideally suited for the acid-base bifunctional catalysis relevant to many industrial chemical reactions.

SINGLE-CRYSTALLINE SILICON

Silicon, in crystalline and amorphous forms, has been used in >90% of the photovoltaic (PV) production (Brendel, 2004; Green *et al.*, 2004; Smestad and Lampert, 2007). However, its widespread adoption is limited by fabrication costs – wafer production accounts for ≈40% of the cost of a module. Such costs may be reduced by using ultrathin silicon films as the basic material for solar cells (Yoon *et al.*, 2008). But thin film solar cells usually have lower efficiencies due to limited absorption thickness – for a single-crystalline Si film, a thickness of ≈125 μm is required to absorb ≈ 90% of the transmitted light incident on it. To obtain good optical absorption, thin-film solar cells need effective strategies not only to prevent reflective 3losses but also appropriate light-trapping (LT) mechanisms to improve the effective optical path length of a cell. This prerequisite is particularly important for silicon due to its high reflectivity, in which >35% of incident photons are reflected from the surface and thereby lost, as well as its fairly low absorption cross-section, as an indirect band gap semiconductor (see Nakajima *et al.*, 2005; Tsakalakos, 2008).

Some ideas concerning modification of both the front and back surfaces of a Si-based device that have been explored include the use of textured surface structures, antireflection (AR) coatings, and LT schemes (Lipinski *et al.*, 2002; Zhou *et al.*, 2007; Obermeyer *et al.*, 2008). Some more complex designs also have been examined for improving LT efficiency by using several 3D architectures, including V-folded thin film organic solar cells, 3D thin film cadmium telluride (CdTe) solar cells deposited over the top of carbon nanotube tower arrays, spherical devices, although from solid Si spheres, and cylindrical shaped solar cells that are made of a thin film of copper indium gallium selenium (CIGS) deposited on a glass tube (Minemoto *et al.*, 2006; Rim *et al.*, 2007; Flicker and Ready, 2008). By using these, it is possible to increase the optical path length through multiple internal and external reflections to enhance light absorption within an optically thin semiconductor absorber. Moreover, unlike traditional flat solar cells, these 3D optical structures absorb light quite efficiently from several different incident directions (Guo *et al.*, 2009).

The fabrication of 3D electronic structures in the micrometre-to-millimetre range is extremely difficult due to the inherently 2D (planar) nature of most conventional wafer-based fabrication methods. But this problem can be tackled by self-assembly, and the related method of self-folding of planar patterned membranes. The most attractive methods are based on planar microfabrication, followed by suitable processes that transform planar objects into the desired 3D structures (Gates *et al.*, 2004). Whereas microelectronics depend on stacking and interconnecting individual layers, folding and self-assembly routes are better for 3D microfabrication (Whitesides and Boncheva, 2002). Folded 3D structures have already been made by using magnetic forces, residual stresses, and electrical activation (Boncheva *et al.*, 2005; Li *et al.*, 2008). Capillary forces can also be exploited to guide the assembly of 3D, submillimetre scale objects (Py *et al.*, 2007).

Guo *et al.* (2009) gainfully utilized the mechanical bendability of thin, single-crystalline Si foils of a few μm thickness, coupled with capillary-driven self-assembly to create folded 3D PV devices. The self-assembly processes studied by them are driven by wetting interactions to shape the contour of a functional, nonplanar PV device. The flexible constructs easily adopt nonplanar shapes, such as triangular prisms, cylinders, and spheres, with less defects than would result from the deformation of thicker Si wafers (Nakajima *et al.*, 2005). The 3D functional architectures so generated function an a LT element that improves efficiency relative to its planar analogues.

Guo *et al.* (2009) also identified the critical conditions for self-folding of different 2D geometrical shapes. This strategy was successfully demonstrated for specifically designed millimetre-scale silicon objects, which could be self-assembled into spherical, and other 3D shapes and integrated into fully functional light-trapping PV devices. Spherical solar cells could be self-assembled from flower-shaped flat Si leaflets with thicknesses of 2 μm. The 3D devices produced by Guo *et al.* potentially provide a promising method of efficiently harvesting solar energy in thin cells by means of concentrator microarrays which function without active light tracking systems.

Unfortunately, this device although functional, does not perform very well from the perspective of Si-based devices. The rather low performance is a result of the non-optimized fabrication processes. The performance is adversely affected by the elimination of the high temperature process needed to provide a back-surface doping field. In view of its absence, the high surface area to volume ratio strongly increases losses due to recombination. To enhance the short-circuit current density, an antireflective coating is needed to minimize the surface reflection loss. Further, the metal contacts used for the foil are not good choices for PV devices. Appropriate process modification unrelated to the folding steps can also enhance performance; and it seems feasible to achieve an efficiency of about 2%, which compares well with that of other transverse and vertical junction thin film Si solar cells of comparable dimensions (Kasai *et al.*, 2001; Liang *et al.*, 2003).

Spherical PV devices made from thin solar cells are quite attractive and their spherical light-receiving surface also intrinsically serves as a passive tracking optic, accepting light from all directions. Yet, this merit of the 3D architecture is offset by the complexity of the fabrication processes required to construct it. The "flower-like" form required for guided self-assembly does not "tie" the planar wafer – its Si content is not effectively used.

Some of the above limitations and problems could be addressed by adopting a strategy for fabricating 3D single crystalline Si structures through a combination of photolithography and self-folding driven by capillary interactions. This approach suggested a route to 3D LT architecture for use in PV applications. It was this approach that enabled Guo *et al.* to make spherical and cylindrical-shaped Si solar cells and to evaluate their improved electrical performances and competencies for LT and mitigation of reflective losses. According to Guo *et al.*, 3D structures of this type can be combined easily with other LT/guiding principles to generate device form factors that are low-cost and simple, yet perform very efficiently (Goetzberger *et al.*, 2008).

NANOPOROUS SILICON NITRIDE MEMBRANES

Membrane technologies play a crucial role in many industrial and other applications (Baker, 2004; Schafer *et al.*, 2005). High porosity membranes with application-dependent pore sizes have been used in desalination, separation of molecules or particles, and several other industrial processes. Current quest for new membrane systems lays emphasis on narrowing the pore size distribution, improving the chemical and mechanical robustness of the membranes, and achieving high fluxes with low energy input (Shannon *et al.*, 2008). Membranes with single nanopores have aroused interest as a biosensing platform for detecting and characterization of single biomolecules such as DNA (Dekker, 2007; Healy, 2007). In fact not only single but also multiple-nanopore membranes have proven valuable for biosensing and separation processes, and also are able to mimic biological membranes. The density of pores, their shape, and their surface chemistry are the critical factors that determine membrane transport and separation capabilities.

Several researchers have prepared membranes with well-defined pore geometry in polymer and inorganic materials (e.g., Han *et al.*, 2008; Savariar *et al.*, 2008; Fornasiero *et al.*, 2008). Some work has also been done on thin membrane platforms containing free standing polymer membranes (Fujikawa *et al.*, 2009) and block copolymer inverse opals (Yan and Goedel, 2004), protein, and block copolymer membranes (Uehara *et al.*, 2009; Nuxoll *et al.*, 2009). It has emerged that silicon-based platforms are well-suited to provide great membrane robustness, allow miniaturization and the possibility to integrate with other ionic and electronic devices (Vlassiouk *et al.*, 2009).

Current methods for making silicon membranes with well-defined pores enable preparation of samples with extreme porosities, either highly porous or a single pore, but nothing in between. It is possible to create highly porous ultrathin silicon membranes by thermally annealing thin silicon films, which results in spontaneous pore formation (Kim *et al.*, 2008). We can precisely control the pore diameter by choosing suitable annealing conditions. But this method does not allow control over the number of pores. It is also possible to fabricate pores one by one by means of ion beam sculpting, focused ion beam drilling, e-beam lithography, or e-beam drilling in a transmission electron microscope (TEM) (Deshmukh *et al.*, 1999; Storm *et al.*, 2003; Tong *et al.*, 2004). Unfortunately, although these serial approaches are applicable to silicon and even polymer materials, they are appropriate only for preparation of membranes with low numbers of pores; also they are not able to tune the geometry of the pores. To address these challenges, Vlassiouk *et al.* (2009) reported a versatile fabrication method for ultrathin, free standing silicon nitride (SiN) membranes containing conical and double-conical pores prepared by the ion track-etching (ITE) technique of Fleicher *et al.* (1975). These SiN membranes had fully controlled porosity, pore geometry, and pore surface chemistry. The ultrathin freestanding SiN platform has conical or double-conical nanopores of diameters as small as several nanometers, prepared by the track-etching technique. This technique allows the membrane porosity to be carefully tuned from one to billions of pores per square centimetre. In the ITE technique, a film is irradiated with accelerated heavy ions, followed by subsequent chemical etching of the irradiated sample. The number of heavy ions corresponds to the number of fabricated pores. Till hitherto, most track-etched membranes have been made in polymer films with thicknesses of at least several micrometres. In contrast, the ITE SiN membranes reported by Vlassiouk *et al.* have thicknesses of ~100 nm, and the pore diameter as small as 1 nm. These membranes withstand markedly harsh chemical conditions not tolerated by polymer membranes. The ultrathin nanoporous SiN membranes described by Vlassiouk *et al.* provide excellent separation capability based on electric charge and size of the permeate species. Indeed, the SiN surface chemistry provides a good starting point for chemical functionalization of the pore walls to tailor these separation capabilities. Furthermore, the membranes are a good platform for biosensing: controlling the number of swift heavy ions that are used in the irradiation directly controls the number of pores. Single-ion irradiation can enable the preparation of single pore membranes. Also, the conical and double conical shapes of the pores with tunable opening angles, together with the ultrathin character of these membranes (free standing 20-nm thick SiN membranes can be routinely fabricated by common microfabrication techniques).

ULTRATHIN SILICON MEMBRANES

Nanoporous membranes are being widely used in medicine for the filtration and separation of blood proteins in an artificial kidney (haemodialysis) and their market has been growing rapidly. Striemer *et al.* (2007) prepared ultrathin nanoporous membranes made from silicon. These silicon membranes can separate molecules far more effectively than traditional polymer membranes and are potentially promising for diverse applications. Conventional nanoporous membranes also function as a mechanical support for desalination membranes being used to purify sea water for irrigation and human consumption.

The customary method of preparing all technologically relevant nanoporous membranes involves initiating the precipitation of a polymer from solution. This is done by adding either a nonsolvent (e.g. water), or by rapid cooling. The solution precipitates into micrometre- and nm-sized entities rich in polymer which form a filter structure. Between these polymer domains, polymer-free areas exist as pores. A wide spectrum of morphologies and geometries can be created from various kinds of starting material (Vogelaar *et al.*, 2005).

These nanoporous membranes have a skin less than 500 nm thick, composed of small bumps (nodules) with a radius of up to 50 nm. The pores are up to 50 nm across. The pore size is determined by the voids between the nodules. The porosity of the membrane as a whole is low. A much thicker layer, having larger pores and greater porosity, gives mechanical support to the nodular skin. While the pore size of the membrane skin can be adjusted by choosing appropriate starting material and processing route, its thickness, porosity and pore-size distribution cannot be so adjusted (Mehta and Zidney, 2005).

In the nanoporous membranes prepared according to these methods the flux through them can be enhanced by increasing the pore diameter, but only at the cost of less effective molecular discrimination. What Striemer *et al.* (2007) have done is to achieve a welcome optimization between flux and selectivity simultaneously. The porous nanocrystalline silicon (pncSi) membranes designed by Striemer *et al.* combine small membrane thickness and pore sizes; they are quite robust; their pore size can be well controlled; and they are fairly simple to produce. Previous attempts to create ultrathin nanoporous membranes exploited sophisticated and expensive nanolithography or were based on colloidal templates (Tong *et al.*, 2004; Yan and Goedel, 2004); also very small, controlled pore sizes could not be achieved easily.

The newly developed nanopores self-form from a deposited layer of amorphous silicon through rapid thermal annealing (Grom *et al.*, 2000). The pore sizes can be controlled between 5 and 25 nm (the range relevant to protein separation) by the choice of appropriate annealing temperature.

Two important proteins, immunoglobulin-γ and bovine serum albumin (BSA), with hydrodynamic diameters of 14 and 6.8 nm, and molecular weights that differ by a factor of less than three, can be separated by using the pncSi membrane. By comparison, for efficient separation using conventional ultrafiltration membranes, a molecular-weight ratio of over ten is essential (van den Berg and Wessling, 2007). The flux through the pncSi membranes is at least ten times faster than the flux through conventional membranes with similar selectivity properties. Moreover, changing the surface charge of the new membrane through chemical modification facilitates separation of proteins that are similar in size, but bear a different charge (Eijkel and van den Berg, 2006).

The greatest merit of the method designed by Striemer *et al.* may be its easy integration into 'labs-on-a-chip', i.e., microfluidics systems that attract growing attention in view of their high potential for medical diagnostics, drug discovery and chemical synthesis (Whitesides, 2006). Such integration with other nanofluidic separation and analysis techniques for biochemical and biomedical applications together with the inherent merits of the silicon-based system highlight the importance of this work.

Most commercial ultrafiltration and dialysis membranes have broad pore size distributions and can be over 1000 times thicker than the molecules they are designed to separate, leading to pore size cut-off properties, filtrate loss within the membranes, and low transport rates (Kuiper *et al.*, 1998; Tong *et al.*, 2004). In contrast, nanofabricated membranes are superior as they offer more precise structural control (Yamaguchi *et al.*, 2004) but transport is also limited by micrometre-scale thicknesses (Martin *et al.*, 2005). This limitation can be abolished by a new class of ultrathin nanostructured membranes where the membrane is roughly as thick (~10 nm) as the molecules being separated, but membrane fragility and complex fabrication prevent the use of ultrathin membranes for molecular separations. Striemer *et al.* (2007) developed an ultrathin pnc-Si membrane by employing silicon fabrication techniques that provide effective control over average pore sizes from about 5 nm to 25 nm. These new pnc-Si membranes retain proteins while permitting the transport of small molecules at rates an order of magnitude faster than existing materials, separate differently sized proteins under physiological conditions, and can separate similarly sized molecules carrying different charges. Even though they are only 15 nm thick, pnc-Si membranes that are free-standing over 40,000 μm^2 can support a full atmosphere of differential pressure without plastic deformation or fracture. As they enable efficient, low-loss macromolecule separations, pnc-Si membranes may be expected to enable several new devices, including membrane-based chromatography systems and both analytical and preparative microfluidic systems which require highly efficient separations.

Striemer *et al.*'s research on pnc-Si membranes may well represent the first use of ultrathin nanomembranes for size-based molecular separations, and attract their use in several near-term applications. The successful separation of BSA and IgG suggests that pnc-Si can be used for membrane-based protein fractionation. BSA and IgG are too close in size (2.2 times MW difference) to be efficiently separated by using conventional membrane process (Li *et al.*, 1997). Being molecularly so thin and as they have a minimal filter surface area, pnc-Si membranes may prove to be quite efficient for separation processes. Indeed, the diffusion measurements recording a transport rate of 156 nmol cm^{-2} h^{-1} for Alexa 546 dye (see Striemer *et al.*, 2007) show this to be more than one order of magnitude faster than those for thick nanofabricated membranes, and >9 times faster than measurements made by Striemer *et al.*, through dialysis membranes. Further, pnc-Si membranes with fixed charges can also be gainfully employed to separate similarly sized molecules with different charges, adding another dimension of control for highly efficient molecular separations (Ku *et al.*, 2004). Finally, the silicon-based platform opens such other routes for future developments, as scalable production of membranes, straightforward integration into microfluidic devices, and surface functionalization by using well-established chemistries (Mace *et al.*, 2006) to modify surface charge, reduce protein adsorption and protect the silicon from chemical attack in harsh environments. The mechanical strength of these ultrathin membranes might facilitate the construction of large-scale dialysis systems, and encourage their use in pressurized filtration devices at the macro- and microscale.

FOLDABLE SILICON INTEGRATED CIRCUITS

Producing high performance electronics that use rigid semiconductor wafers, but in lightweight foldable and stretchable formats, can attract numerous new applications. Some examples are wearable systems for personal health monitoring and therapeutics, "smart" surgical gloves with integrated electronics, and electronic eye type imagers that incorporate focal plane arrays on hemispherical substrates (Lu and Xia, 2006). Circuits that use organic or certain types of inorganic electronic materials on plastic or steel foil substrates can give some mechanical flexibility, but cannot be folded or stretched (see Dodabalapur, 2006; Sun and Rogers, 2007; Yuan *et al.*, 2006). Typically these systems offer only modest electrical performance. In contrast, stretchable metal interconnects with rigid or stretchable (Lacour *et al.*, 2005; Sun and Rogers, 2007a) inorganic device components exemplify alternative strategies that can also, in some cases, perform very well. But in their existing forms, no approach allows scaling to circuit systems with practically useful levels of functionality (Kim, D.H. *et al.*, 2008).

Kim, D.H. *et al.* (2008, 2008a, 2010) reported a simple method of producing high-performance, stretchable, and foldable integrated circuits. The systems integrate inorganic electronic materials, including aligned arrays of nanoribbons of single crystalline silicon, with ultrathin plastic and elastomeric substrates. The designs combine multilayer neutral mechanical plane layouts and "wavy" structural configurations in silicon complementary logic gates, ring oscillators, and differential amplifiers. Kim *et al.* performed three-dimensional analytical and computational modelling of the mechanics and the electronic behaviours of these integrated circuits. Their results open new routes to devices, such as personal health monitors and other biomedical devices, that require extreme mechanical deformations during installation/use and electronic properties approaching those of conventional systems built on brittle semiconductor wafers.

Essentially, Kim *et al.* have developed high-performance, single-crystalline silicon complementary metal-oxide semiconductor (Si-CMOS) integrated circuits (ICs) that are reversibly foldable and stretchable. These systems integrate high-quality electronic materials, such as aligned arrays of silicon nanoribbons, with ultrathin and elastomeric substrates, in multilayer neutral mechanical plane designs and with "wavy" structural layouts. In these systems, the use of commodity bulk silicon wafers (Mack *et al.*, 2006) for cost-sensitive applications provide the source of the required ultrathin pieces of Si. It is conceivable that vacuum-evaporated materials such as nanocrystalline Si (Lee *et al.*, 2006), which also perform well, could possibly offer further advantages in cost. The approaches to stretchable and foldable integrated circuits reported by Kim *et al.* may also be used with these and other related classes of materials.

D.H. Kim *et al.* illustrated the achievement of high stretchability and foldability can be achieved in fully formed, high-performance integrated circuits by using optimized structural configurations and multilayer layouts, even with intrinsically brittle but high-performance inorganic electronic materials. Such designs even offer the possibility of direct integration of electronics with biological systems, medical prosthetics and monitoring devices, complex machine parts, or with mechanically rugged, lightweight packages for other devices.

CLOAK OF LIQUIDITY

Droplets of a liquid alloy on a silicon surface can rearrange the surface in such a way that they mimic the short-range ordering of atoms in the alloy. Remarkably, this effect inhibits freezing of the droplets.

Freezing can be promoted in a liquid by merely adding suitable templates. The templates can be either 'seeds' of the crystalline phase that would form from the liquid, or small crystals of another material whose atomic-level surface structure somehow matches that of the seeds (Greer, 2010).

It is far more difficult to imagine a solid surface that inhibits freezing by acting as a template for the liquid. However, Schülli *et al.* (2010) showed that such a template may be possible.

The cooling of a liquid to below the ideal freezing temperature, termed supercooling, is of considerable interest in several areas ranging from the control of microstructure in metallic welds and casting to the inhibition (or promotion) of ice formation necessary for the survival of living systems (see Spittle, 2006; Zachariassen and Kristiansen, 2000).

Schülli *et al.* studied the freezing of a liquid gold-silicon alloy near its eutectic (lowest freezing point) composition. This alloy is a useful model system and also has practical significance, being used in the vapour-liquid-solid process for growing silicon nanowires. Schülli *et al.* made microscopic islands of gold on a single-crystal silicon substrate and melted them by heating. Some of the silicon reacted with the gold and formed liquid-alloy droplets. On cooling, these droplets froze at a temperature that was reproducible over repeated heating-cooling cycles, but which depended on the upper temperature limit of the heating. By means of a critical analysis of extensive X-ray data acquired at a wide range of angles to the substrate, Schülli *et al.* showed that this freezing point corresponds to a slight, gold-induced reconstruction of the atomic arrangement of the silicon surface.

The researchers observed that silicon can undergo a more radical reconstruction to yield what is known as a 6×6 superstructure. While the freezing point of the alloy in contact with the 6×6 silicon surface is about 120 K below that of eutectic gold-silicon, Schülli *et al.* noted that the eutectic freezing temperature is not the most appropriate reference point from which super-cooling of the system should be measured.

Schülli *et al.* successfully characterized in-plane order in the liquid gold-silicon alloy adjacent to the 6×6 silicon surface. According to them, the liquid is anisotropic, with atomic positions strongly correlated with the structure on the underlying surface. Interestingly, this correlation with a periodic surface pattern impedes the crystallization of the liquid, rather than inducing it.

NANOWIRES

It has long been a cherished desire of electronics experts to find a material that is small-scale, bendy and fairly transparent. This desire is about to be fulfilled. Semiconductor nanowires are diminutive semiconductor crystals, usually a few μm long and 20-80 nm in diameter. They could find wide use in future integrated circuits such as ultra-scaled microprocessors or terabit memory chips. The nanowires' tiny dimensions could possibly be exploited to make such devices at reasonable cost by packing up to 200 billion nanowire transistors onto each cm^2 of silicon substrate! This will of course demand a departure from the traditional planar transistor layout (Rustagi *et al.*, 2007).

However, nanowire transistors are already creating a niche for themselves in applications requiring much smaller integration densities and lower switching speeds (Klauk, 2008). Ju *et al.* (2008) showed that the unique properties of nanowire transistors make them a very good choice for active-matrix displays made from organic light-emitting diodes (OLEDs). In these displays, each pixel must contain several transistors along with the actual OLED. The transistors ensure that all pixels emit the desired colour at the desired brightness, even when the image changes rapidly during full-motion video replay. The number of transistors per unit area in a display is rather small (around $10{,}000/cm^2$) so the area occupied by each transistor becomes less important than in a memory or microprocessor chip.

The practical implication is that transistors in an OLED display may use semiconductors (e.g. amorphous silicon) (Long *et al.*, 2006), or conjugated oligomers (Zhou *et al.*, 2006), in which the mobility of charge carriers is quite low so that the transistors have to be large (wide) to transport enough carriers and drive the necessary currents (about 10^{-5} amps per pixel). These semiconductor materials can be processed at temperatures low enough to allow the use of polymeric substrates that can be both flexible and transparent.

One goal of display development is to increase the aperture ratio – the fraction of the total pixel area available for light emission. With amorphous silicon or organic transistors and a pixel size of 300-500 μm, aperture ratios of 40-50% are achieved. To boost this ratio, the first mass-produced active-matrix OLED displays are using transistors based on polycrystalline silicon. The higher charge-carrier mobilities in this semiconductor, as compared with those in amorphous silicon or organic semiconductors, permit narrower transistors and a larger aperture ratio. However, polycrystalline silicon transistors also demand higher processing temperatures, which makes it difficult to use polycrystalline silicon with polymeric substrates.

This is where semiconducting nanowires come into the picture. As they consist of just a single crystal, carrier mobilities are high: Ju *et al.* reported a value of 250 $cm^2\ V^{-1}\ s^{-1}$ for their nanowires made of indium oxide. A small assembly of properly aligned nanowires therefore delivers the same drive current as a polycrystalline silicon transistor (Klauk, 2008).

Ju *et al.* made their nanowires by pulsed laser ablation that involves vaporizing material from a bulk solid target by using a laser, and transferred them to the display substrate from a liquid suspension. While on the display substrate, the nanowires connected into transistors and pixel circuits by photolithography. Excess nanowires could be shaken off in a bath by using ultrasound.

Employing an optically transparent conductor for the source and drain contacts and for the gate electrodes led to the production of transistors very useful for see-through displays – on car windscreens, for example. The gate electrode has to be isolated from the nanowires by an insulating layer; for this Ju *et al.* used a very thin molecular dielectric with very few defects and a large capacitance. This latter capacitance means that the gate voltage required to induce a sufficient number of carriers in the nanowire is, at about 4 volts, fairly small and minimizes the power consumption of the display.

As the nanowires are not produced on the display but are transferred to the substrate after preparation, the thermal budget for preparing the semiconductor becomes decoupled from that for transistor fabrication. So, even if high energies or temperatures are needed to make nanowires with high enough carrier mobility, the display can be made on low-temperature polymeric substrates. According to Ju *et al.*, nanowires work very well with organic gate dielectrics that can be prepared at room temperature. Amorphous and polycrystalline silicon, in contrast, usually require inorganic dielectrics deposited at temperatures of about 200°C.

The displays reported by Ju *et al.* have a very favourable aperture ratio of 46%. The demonstration of an active-matrix OLED display with nanowire transistors has valuable implications: all that is required now for high-quality flexible displays to finally become a reality is an efficient nanowire placement process and perhaps a few improvements in OLED efficiency and lifetime.

SILICON NANOWIRES – USED AS SOLAR CELLS AND NANOELECTRONIC POWER SOURCES

Solar cells already provide cost-effective clean and renewable power source. If miniaturized, they could also possibly serve as integrated power sources for nanoelectronic systems. Nanostructured materials are generally used to reduce both cost and size and to improve efficiency in photovoltaics (Lewis and Crabtree, 2005; Baxter and Aydil, 2005; Luque *et al.*, 2007; Zhang *et al.*, 2007). Nanoparticles, nanorods and nanowires are used for improved charge collection efficiency in polymer-blend and dye-sensitized solar cells, to achieve carrier multiplication and to facilitate low-temperature processing of photovoltaic devices (Tian *et al.*, 2007). Coaxial nanowire structures could potentially improve carrier collection and overall efficiency in regard to single-crystal bulk semiconductors of the same materials. Unfortunately, the main problem with solar cells based on hybrid nanoarchitectures is that they suffer from low efficiencies and poor stabilities; another drawback is that their use as photovoltaic power elements in nanoelectronics has not attracted critical study. Tian *et al.* (2007) report the creation of p-type/intrinsic/n-type (p-i-n) coaxial silicon nanowire solar cells. Under one solar equivalent (1-sun) illumination, the p-i-n silicon nanowire elements yield a maximum power output of up to 200 pW per nanowire device and an apparent energy conversion efficiency of up to 3.4 per cent, and stable and better efficiencies can be achieved at high-flux illuminations. They also showed that individual and interconnected silicon nanowire photovoltaic elements can function as strong power sources to drive functional nanoelectronic sensors and logic gates. These coaxial silicon nanowire photovoltaic elements provide a new nanoscale test method for research on photo-induced energy/charge transport and artificial photosynthesis (Gust *et al.*, 2001). This points to their potential application as elements for powering ultralow-power electronics and diverse nanosystems (Browne and Feringa, 2006; Avouris and Chen, 2006).

ELECTRONIC PROPERTIES OF SELF-ASSEMBLED MOLECULAR NANOWIRES

In a study of self-assembled molecular nanowires, level switching has been achieved in a multiwire transistor, and their suitability for the production of multi-logic devices has been demonstrated (see Yuyaka Wakayama *et al.*, 10.1021/nl801649v).

These researchers made a thorough study of self-assembled molecular nanowires by examining molecular design and one-dimensional crystal growth. They also made resistivity measurement of individual wire and explored its application to field-effect transistors. Using p-stacked molecules in a single crystalline structure, self-assembled molecular wires were grown with precise tuning of molecular packing and of dimensional control. Its face-to-face packing characteristics allowed synthesis of 6,13-bis(methylothio)pentacene (BMTP), and provided efficient p-orbital overlap and one-dimensional wire growth. Treating the SiO_2 substrate surface with octadecyltrichlorosilane (OTS) prevented angular distribution and enhanced the linearity of the wires relative to the surface. As the wires were made of a single crystal, good electrical transport with low resistivity could be achieved.

Field-effect transistors with single- and doublewire channels could be fabricated and point to the potential application of the molecular wires. Appropriate molecular design and control of interfacial interactions led to single crystalline wire growth with an extensive p-stacking motif. The single- and double-wire transistors could be fabricated by using needle probes to apply gate voltage. With the latter, a double topgate configuration enabled independent control of the drain current through each wire and enabled modulation to four different values by turning gate voltage on and off. Resistivity measurements of an individual molecular wire point to the utility of these structural features for electrical transport. According to Y. Wakayama, four-level switching can be realized by their application to field-effect transistors, with emphasis on a multi-value operation of the multi-wire transistor, where it is possible to modulate electrical current through respective molecular wires independently by topgate configuration.

NANOWIRES FOR NANOELECTRONICS

Today, most leading-edge chips have transistors whose critical dimension is no more than about 100 atoms long, but, this miniaturization cannot go on much longer in view of the compulsion of the Moore's law. This limitation has prompted researchers to seek an alternative to the well known silicon transistor – an alternative that could revolutionize the industry that has fueled the information revolution since World War II (Forester, 1982). Nanowires and transistors made from them are some promising alternatives (Cui and Lieber, 2001). These nanowires are grown as carbon nanotubes or made from silicon as well as a variety of other semiconductors. However, while these nanowires may not replace the silicon transistor, they could potentially provide the paradigm shift that will extend Moore's law. Properly connecting circuit layers with nanowires and nanotransistors can potentially cause a paradigm shift in microchip design.

According to Intel, the latest 45-nm microprocessor (with a gate length – the critical dimension in the direction of current flow – of ~22 nm) contains over 400 million transistors in 107 mm^2; i.e., each transistor occupies a square of silicon real estate that is roughly 500 nm on a side (see Ferry, 2008).

Originally, Moore's Law was driven by three factors: (i) decreasing the transistor size (and hence the square of silicon upon which it sits), (ii) increasing the size of the microchip itself, and (iii) circuit cleverness (by which the number of transistors required to perform a function could be brought down with consequent savings in silicon real estate). In the past decade, thc size of the microchip has remained roughly constant, so that factors (i) and (iii) have assumed greater importance. We are now approaching the atomic limit on critical size; and it is factor (iii) that will have to provide the continuity to Moore's law. It is here that nanowires may play a critical role.

Although transistors have very short gate lengths, they have much larger widths to provide the current necessary to make the circuits work. Should we replace the current planar transistor with nanowires, we will need to use many more such devices in parallel to provide this current. But, we must also keep in mind the economic driving forces. In this context, one possibility is that nanowires in the plane will not effectively compete with novel transistors such as the "fin" field effect transistor, or "finFET", which can more effectively use silicon. However, nanowires can be grown vertically, and this growth can be initiated on a wide variety of substrates (see Bowler, 2004).

Today the transistors are placed in a planar array but are overlaid with many metal lines, with nine or more levels of metal. These lines provide power, clock signals to synchronize the switching of the transistors, and various interconnections between the different functional blocks of the chip. Connections from these metal lines to the transistors are made by downwardly reaching metal fingers termed "vias". When it becomes impossible to reduce the transistor size, the use of Si can be enhanced by moving vertically (Ferry, 2008).

We can enhance "cleverness" of Moore's law by replacing some of the vias with vertical nanowire transistors (Duesberg *et al.*, 2004). These vertical transistors can extend from the silicon to a metal wire or even between different levels of metal wire. It is possible to consider creating reconfigurable architectures in which the connections between different functional blocks are altered by switching just a few of these vertical transistors. In this way we might make real 3-D architectures in a different way from the traditional approach of stacking chips (Koyanagi *et al.*, 2006).

For more effective architectues based on the use of these vertical transistors, there is need to improve microchip design. This chip design is currently achieved with automated transistor layout programs which optimize the planar design placing of the various functional blocks and minimize the necessary interconnections (in the metal layers). For reconfigurable architectures, we will need switchable interconnections, based on vertical transistors, and device specialists will need to collaborate with circuit designers to achieve this.

THREE-DIMENSIONAL NANOWIRE METAL-OXIDE-SEMICONDUCTOR CIRCUITS

Integration of transistors into a single layer has so far constituted the dominant approach to integrated circuits in electronics. Three-dimensional (3D), multi-transistor-layer, integrated circuits represent an important technological innovation that promises welcome advantages in integration density, operation speed, and power consumption as compared with 2D circuits. These merits have prompted much research to realize 3D integrated circuits (Topol *et al.*, 2006). Chemically synthesized semiconductor nanowires (NWs) and carbon nanotubes (CNTs) (Thelander *et al.*, 2006; Lieber and Wang, 2007; Avouris and Chen, 2006) have become the choice approaches to construct 3D integrated circuits. It is possible to assemble these nanoscale building blocks layer-by-layer into 3D structures, where sequential layers can readily be formed with either the same or distinct nanomaterials (Javey *et al.*, 2007). Much capability cannot be easily achieved by conventional top-down approaches. For NWs, suitable combination of distinct nanoscale materials with widely tunable nanoscale compositions and structures might one day enable 3D integrated circuits having unique capabilities (Gudiksen *et al.*, 2002; Fan *et al.*, 2008).

Ahn *et al.* (2006) and Javey *et al.* (2007) worked on 3D integration of nanoscale materials and showed that sequential printing of NWs, nanoribbons, and/or CNTs is a good approach to making 3D integrated circuits on conventional and plastic substrates. Further, complementary metal-oxide-semiconductor (CMOS) configuration could also be attained. These reports contribute toward 3D integrated NW and CNT circuits, yet fall short of showing features central to CMOS integrated circuits, including (i) control of switching thresholds; (ii) realization of full-swing operation for CMOS devices; and (iii) AC operation at high speed. These central issues have been addressed by Nam *et al.* (2009) through the fabrication and demonstration of fully functional 3D integrated NW CMOS inverter logic gates and high-frequency ring oscillators. They used layer-by-layer assembly of indium arsenide (InAs) NW n-type field-effect transistors (n-FETs) and germanium silicon (Ge/Si) core/shell NW p-type FETS (p-FETs) (Xiang *et al.*, 2006) to fabricate vertically interconnected, 3D CMOS inverters and ring oscillators. Their choice of a group-III/V material and group-IV heterostructure was prompted by three points, viz., the high-performance shown previously for the corresponding NW FETs (Jiang *et al.*, 2007); the desire to find out if such distinct materials systems could yield functional circuits through a bottom-up approach, and the considerable interest in hetero-integration of these two systems by conventional top-down fabrication to extend scaling in microelectronics (Heyns and Tsai, 2009).

Based on work done on their fully functional, 3D integrated complementary CMOS circuits made of separate interconnected layers of high-mobility n-type indium arsenide (n-InAs) and p-type germanium/silicon core/shell (p-Ge/Si) NW field-effect transistors, Nam *et al.* (2009) reported that the DC voltage output (V_{out}) versus input (V_{in}) response of vertically interconnected CMOS inverters showed sharp switching at close to the ideal value of one-half the supply voltage and, also showed substantial DC gain of ~45. The gain and the rail-to-rail output switching seem consistent with the large noise margin and minimal static power consumption of CMOS. Vertically interconnected, three-stage CMOS ring oscillators were also fabricated by using layer-1 InAs NW n-FETs and layer-2 Ge/Si NW p-FETs. Measurements of these circuits revealed stable, self-sustained oscillations with a maximum frequency of 108 MHz, which represents the highest-frequency integrated circuit based on chemically synthesized nanoscale materials. These results highlight the flexibility of bottom-up assembly of distinct nanoscale materials and point to high potential for 3D integrated circuits. Indeed, the 3D ring oscillator represents the highest-frequency integrated circuit reported for nanoscale materials, and it is possible that this frequency could be further increased by reducing channel length and parasitic resistance and capacitance of the circuit. Also, their bottom-up approach using optimized n- and p-type NW materials points to considerable promise, including (i) extension to multiple interconnected layers both for increasing device packing density and to open up design and testing of novel circuit architectures; (ii) implementation of 3D interconnected NW circuits on flexible (Friedman *et al.*, 2005) and highly curved substrates (Yu *et al.*, 2007) for high-performance electronics on unconventional substrates; and (iii) exploring general concept of heterogeneous 3D integration of two or more NW and carbon nanotube building blocks for 3D multi-functional nanosystems (Lu and Lieber, 2007).

TOUGH CERAMICS (SIALONS)

Ceramics based on silicon nitride have proved very useful in terms of engineering applications. The core of materials science rests on microstructural control especially when some specific combination of mechanical properties is required. In nitrogen ceramics the final microstructure is produced by 'liquid-phase sintering' and it is difficult to control the morphology of the final product; whatever little can be done is achieved by controlling the starting composition and the processing parameters—mainly time and sintering temperature. Shen *et al.* (2002) devised a technique that makes possible significantly greater microstructural control.

Silicon-nitride-based ceramics seem to be better than their competitors in engineering applications, thanks to their low density and toughness. When silicon nitride, aluminium oxide and aluminium nitride are reacted together, they form ceramics called 'sialons'. Two common forms of sialons are a-sialons and b-sialons: in their atomic arrangements they are like a- and b-silicon nitride. b-sialons readily form with a needle-like (acicular) morphology. Some sialon grains are elongated (with aspect ratios as high as ten), resulting in a tough final product (Thompson, 2002).

Acicular morphologies have also been produced for a-sialon ceramics by using unusual starting materials (Chen and Rosenflanz, 1997) (an excess of liquid phase or an oxygen-rich starting composition). But only limited control of the aspect ratio for either a- or b-sialons prepared by conventional sintering procedures has been achieved.

During liquid-phase sintering of sialons, a chemical force drives the formation of the final phases from the mixture of oxides and nitrides that make up the starting material. The very small grains present in the starting powders usually dissolve rapidly in the early stages of sintering and, except for the large acicular b-sialon grains that grow from the silicon nitride needles present in the initial mixture, most of the remaining grains are of similar size. The result is a slow ripening.

Further, most conventional sintering furnaces reach the final sintering temperature only slowly, and the composition of the liquid phase changes continuously with increasing temperature to remain in equilibrium with the main sialon phase.

Shen *et al.* used a very rapid sintering technique—'spark plasma sintering' (SPS) – that allows samples to reach the sintering temperature very quickly. For both a- and b-sialon phases, this results in elongated, needle-shaped grains; the increase in length of the needles after such short sintering times is indeed remarkable (Thompson, 2002).

This technique has the potential to develop tough, needle-shaped microstructures with greater precision, and very soon a new range of high-toughness sialon ceramics may be created.

MATERIALS FOR ELECTRONICS

The electronics industry has depended strongly on silicon-based devices for many years. Devices have been shrinking in size and have become faster, more efficient, more powerful, and cheaper. But the size of transistors, which are the building blocks of electronics, is now approaching the limits of what can be done on a large-scale industrial basis. It is now no longer possible to avoid or postpone replacement for silicon.

The simple constant-electric-field scaling rules that guided the shrinking of field-effect transistors (FETs) for decades (see Dennard *et al.*, 1974) broke down a few years ago, adversely affecting the computer industry - microprocessor clock frequencies plateaued around 2005, after a long run of ever-compounding improvements in computing speed. As transistors shrank in size, they switched faster and used less power to switch. But constant-field scaling requires that operating voltages should decrease in tandem with transistor dimensions. This has now become very difficult to follow because a minimum gate voltage swing is needed to switch the device from an "off" (low-current) state to an "on" (high-current) one. If the swing is too small, the device designer faces two bad choices: excessive leakage current in the nominally "off" state or low current (slow circuits) in the nominally "on" state. The choice is between computers that run hot while doing nothing or computers that run slower than previous models (Theis and Solomon, 2010). In the past decade, the industry has moved slowly from constant-field scaling toward constant-voltage scaling (Haensch *et al.*, 2006). Companies still shrink the transistor, emphasizing the increasing number of parallel processors (cores) that can be placed on a single silicon chip. But as power supply voltages are stuck at about 1 V, increasing clock frequencies would now result in unsupportable increases in power dissipation and heat generation. Indeed, the transistor is fast approaching its physical limits.

The only way to address such a power dissipation bottleneck is to modify the physics of transistor operation in ways that allow further reduction of operating voltage. This means increasing the nonlinearity of the switching behaviour so that a much smaller swing in gate voltage can switch the device from off to on. There are several ways to gate the flow of carriers through the channel. For instance, the interband tunnel FET (Banerjee *et al.*, 1987) uses band-to-band tunnelling to filter the energy distribution of charge carriers in the source. In those cases where electrons are the charge carriers, the injected electrons become limited in energy from below by the conduction band in the channel and from above by the valence band in the source. This can produce a much sharper turn-on. Appenzeller *et al.* (2004) reported highly promising behaviour in a carbon nanotube device. In contrast, limited success was obtained with silicon-channel devices (Kim *et al.*, 2009). This has shifted interest to direct-gap semiconductors in which it might be possible to engineer larger on-state tunnelling currents (Mookerjea *et al.*, 2009). The best approach may be to incorporate a direct-gap material (e.g. a III-V compound semiconductor nanowire or a carbon nanotube) in a wirelike device geometry, which allows more abrupt modulation of energy bands for increased band-to-band tunnelling and hence increased current in the on state (Theis and Solomon, 2010).

Theis and Solomon (2010) mentioned some possible approaches to developing a new transistor, still based on silicon but involving clever redesign and engineering of the basic structure. This seems to be largely a materials issue - whether it involves pushing the capabilities of silicon farther or developing new materials altogether.

Takagi and Hwang (2010) moved away from silicon and considered the richness of the electronic phases and enhanced functionality afforded by transition metal oxides. The properties of these strongly correlated electron systems can be greatly influenced by even small changes in applied voltage or magnetic field, and can be used as memory storage or active devices.

Rogers *et al.* (2010) reviewed developments in making the electronics wearable or more transportable through a combination of flexibility, bendability, or stretchiness achieved by integrating inorganic and organic materials with suitable substrates. Flexible electronics have been used to measure the patterns of a beating heart (see Osborne *et al.*, 2010).

Mannhart and Schlom (2010) reviewed the progress made with oxide interfaces. The junction of two insulating materials exhibits a vast array of different properties such as metallic, superconducting, and magnetic; here the effects are tunable and devices scribable with an electric field. While it is difficult to exercise control over the interface properties, if successful, it will offer welcome opportunities for technology and device function.

Another possibility of increasing the sharpness of the FET turn-on may be to incorporate an internal gain mechanism so that a small voltage swing on the gate electrode causes a larger swing of the internal potential that gates the flow of current. Salahuddin and Datta (2008) suggested a device with a gate insulator stack consisting of a layer of ferroelectric material sandwiched between conventional dielectric layers. The energy barrier provided by this gate structure to charge injection relies on the polarization of the ferroelectric layer which, because it arises from a collective effect, switches abruptly between two polarization states. Conceivably, layer thickness and material properties might be engineered to strongly suppress the undesirable hysteretic behaviour that would typically be expected from such a gate structure while still amplifying the effect of the gate potential on the channel potential.

According to Theis and Solomon, both the tunnel-FET and the ferroelectric-gate FET promise low voltage and hence ultra-low-power switching, but these devices may not be so fast. Perhaps some new direct-gap materials and novel device structures may reduce the internal device resistance presented by the band-to-band tunnelling barrier.

The fact that materials lie at the heart of all electronic devices cannot be overemphasized. Service (2010) voiced concern at an imminent crisis in supplies of rare-earth elements (REEs) and other materials that are vital to novel electronics and energy technologies. Service (2010a) also mentioned advances in **nitride semiconductors**. These are materials that hold great promise for fast-switching, heat-tolerant transistors, compact chemical sensors, and high-performance solar cells.

The REEs extend from lanthanum (element 57) through lutetium (element 71) to scandium and yttrium. These REEs have become vital to many novel electronics and green-energy technologies. The problem is that, while researchers are steadily finding new applications, the supply of rare earths is dwindling. Today, China supplies about 97% of all REEs, and increasingly also the products from which they are made. Those products range from phosphors in electronic displays, to magnets in disk drives, cell phones, and MRI machines, and motors in missile guidance systems.

China's production of REEs has grown steadily over the past decade. As its domestic demand has grown even faster, the country's REE exports have dropped from 75% of the total produced to 25%. For neodymium dysprosium, terbium, and yttrium – China in the coming few years is expected to use all it can produce, leaving nothing for export.

REEs aren't the only concern. Supplies of other metals vital to electronics and green technologies (such as indium and lithium) are also falling as industrial uses for these materials skyrocket (see also Chapter 9).

In fact, rare earths are not really rare: they are ubiquitous in soils around the globe. But unlike gold and silver, which are heavily concentrated in particular ore deposits, rare earths are sparse. Even the richest veins of REEs contain no more than 9% of the elements. Most deposits have 1% or less of the elements – too little to make their processing economical.

New mining efforts may help to some extent but numerous rare earth oxides are invariably mixed together in ores. They need to be separated and purified, reduced to metallic form, and then alloyed, cast, and shaped.

OXIDE INTERFACES FOR ELECTRONICS

Transistors, lasers, and solar cells all exploit interfacial phenomena. It is interfaces that enable data processing, memory, and electronic communication. Interfaces in semiconductor structures have been the birthplace of many fascinating discoveries in fundamental science. It is remarkable that away from interfaces, in the bulk of the material, the behaviour of electrons in silicon and some other semiconductors is less exciting (Mannhart and Schlom, 2010). The electrons zip through the crystal lattice mostly as independent, free particles that hardly interact with one another. In contrast, there are many oxides in which electron interactions in the bulk of the material give rise to very interesting phenomena, such as colossal magnetoresistance and high-temperature superconductivity. These phenomena operate in oxides from regularly spaced ions interacting with the electrons, from the unique electronic character of oxygen ions, and from those interactions among the electrons which make them deviate from free-particle behaviour.

It is possible to generate extraordinary electron systems at well-defined interfaces between complex oxides. The fundamental properties of such interfaces have been explored and exploited. It has emerged that these electron systems carry high potential for possible future devices.

As the introduction of interfaces into semiconductor structures enabled many highly useful semiconductor devices, it is conceivable that well-defined interfaces might also be incorporated into oxides to activate novel phenomena. Oxide multilayers being much more difficult to grow than semiconductor heterostructures, there was very little progress in this direction for many years, but sustained recent efforts to grow oxide superconductors led to significant progress in the growth of oxide multilayers. Some key steps were (1) the ability to terminate oxide substrates at well-defined ionic planes; (2) the application of pulsed-laser deposition (PLD); and molecular-beam epitaxy (MBE) to the growth of multicomponent oxides containing difficult-to-oxidize constituents; and (3) the development of high-pressure reflection high-energy electron diffraction to monitor the deposition of individual atomic layers (Schlom *et al.*, 1988; Kawasaki *et al.*, 1994; Rijnders *et al.*, 1997). Consequently, it is now possible to grow epitaxial heterostructures of oxides with atomic-layer precision. The chemical abruptness and crystalline perfection of oxide multilayers compare well with semiconductor multilayers; it is in fact possible to change from one material to another over a distance of just a single unit cell (Logvenov *et al.*, 2009). Such oxide heterostructures can also be patterned laterally with nanometer resolution (Cen *et al.*, 2009). The ability to precisely produce interfaces connecting different oxide materials has opened up numerous possibilities to generate novel electronic phases. The phenomena that control interfaces in standard semiconductors, such as the formation of space charge layers, are also applicable at oxide interfaces.

The physical phenomena that operate at oxide interfaces go well beyond those shown by conventional semiconductor interfaces because new physics enables novel electronic phases. Some key factors such as ionicity and oxygen control the electronic behaviour that can arise at precise tailored interfaces between transition metal oxides. These oxides are an important class of materials for emerging electronics (Mannhart and Schlom, 2010). The unique character of oxygen ions is a critical factor in generating novel electronic properties at oxide interfaces. O^{2-} ions are exquisitely polarizable, typically causing a large, non-linear, and nonuniform polarizability of the lattice. The carrier density at oxide interfaces being very high, this polarizability usually leads to electrostatic screening lengths in complex oxides of 1 to 100 nm, which are smaller than typical screening lengths in semiconductors, again giving rise to more local and more confined novel interfacial properties.

Electronic Correlations

The most exciting phenomena at customized oxide interfaces result from correlated electron effects. In correlated electron systems, the interaction of an electron with other electrons cannot be described by the Coulomb interaction of a virtually free, independent particle with an average background charge that represents the other electrons. In a very complex nonlinear manner, electronic correlations determine the phases an electron system possesses, and in turn its characteristic properties (Mannhart and Schlom, 2010).

Reyren *et al.* (2007) and Gozar *et al.* (2008) exploited the so-called "electronic reconstruction" to generate interfacial superconductivity and also for the design and fabrication of ultrathin magnetic layers. They grew heterostructures of antiferromagnets that electronically reconstruct to induce a ferromagnetic state precisely at the interface (Koida *et al.*,

2002). Pushing ultrathin magnetic layers to the limit, strategies to create two-dimensional fully spin-polarized electron systems have been conceived and based, for example, on $LaAlO_3$-EuO interfaces (Wang *et al.*, 2009).

Interfaces involving manganites have attracted intensive research to exploit their strong magnetoresistance. Manganites already show high magnetoresistivity in the bulk but the magnetoresistivity is enhanced by orders of magnitude at grain boundary interfaces (Hwang *et al.*, 1996). Space charge suppresses the local Curie temperature thereby enhancing the magnetoresistance. These experiments bring out the extraordinary sensitivity with which interfacial electronic systems can react to external parameters – the basis of a sensor (Mannhart and Schlom, 2010).

Unlike the simpler ferromagnetic interfaces in manganites used to enhance their magnetoresistivity, the interface between $La_{2/3}Ca_{1/3}MnO_3O_{7\text{-ä}}$ and the high-T_c superconductor $YBa_2Cu_3O_{7\text{-ä}}$ has remarkably surprising properties (Chakhalian *et al.*, 2006). It is across this interface that covalent bonds form, and modify the electronic structure of the CuO_2 layers of the $YBa_2Cu_3O_{7\text{-ä}}$. The Cu ions in the $YBa_2Cu_3O_{7\text{-ä}}$ at the interface are ferromagnetically magnetized. This is impossible for bulk YBa_2-$Cu_3O_{7\text{-ä}}$. The possibility of using interfaces to modify materials is a novel means to design novel superconductors (Chaloupka *et al.*, 2008).

MATERIALS FOR STRETCHABLE ELECTRONICS

Whereas biology is soft, elastic, and curved, silicon wafers are not. Any electronics technology which can overcome this basic mismatch in mechanics and form is sure to attract applications which just cannot be achieved with currently available hard and planar integrated circuits. Examples range from these surgical and diagnostic implements that naturally integrate with the human body and prove active therapeutically, to cameras which use biologically inspired designs to achieve superior performance. It should also be possible to create sensory skins for robotics, structural health monitors, wearable communication devices, and other systems requiring lightweight, rugged construction in thin and conformal formats (Rogers *et al.*, 2010). This future research direction in electronics is markedly different from the one governed by the current trend toward smaller and faster devices that are still confined to the planar surfaces of silicon wafers.

In fact, work toward mechanically unconventional forms of electronics had begun at least 15 years ago with polymer transistors formed on bendable sheets of plastic (Garnier *et al.*, 1994). In these, paperlike displays (Gelinck *et al.*, 2004) represented the chief target application. Advances in printing and related patterning techniques and in organic semiconductors catalyzed the early progress in this field (Forrest and Thompson, 2007). In the past few years, research on ultrathin inorganics, nanotubes, and nanowires has registered considerable improvements (Cao and Rogers, 2009; Fan *et al.*, 2009). Many functional demonstration devices with several hundreds of active components have been made, among which several may be commercialized soon.

The scope of research also has now expanded dramatically to cover more technically challenging opportunities in soft, biointegrated devices and in curved, bioinspired designs. Here, the electronics must not only bend but also stretch, compress, twist, and deform into complex, curvilinear shapes while still maintaining high levels of performance, reliability, and integration that match those of well-developed wafer-based systems. Advances in mechanics and materials have opened new routes to integrated circuits that offer the electrical properties of conventional, rigid wafer-based technologies plus the ability to be stretched, compressed, twisted, bent, and deformed into arbitrary shapes. Inorganic and organic electronic materials in microstructured and nanostructured forms, intimately integrated with elastomeric substrates, offer particularly attractive characteristics with realistic pathways to sophisticated embodiments. Stretchable electronics are achieved in two conceptually different, but complementary, ways. One uses new structural layouts in conventional materials (Kim *et al.*, 2010), the other uses new materials in conventional layouts (Sekitani and Someya, 2010). Rogers *et al.* reviewed these strategies and described their applications in systems ranging from electronic eyeball cameras to deformable light-emitting displays.

Structures that Stretch

Two simple ideas have inspired strategy based on **structure**. The first exploits an elementary result in mechanics: Any material in sufficiently thin form becomes flexible by virtue of bending strains that decrease linearly with thickness. Whereas a silicon wafer is brittle and rigid, nanoscale ribbons, wires, or membranes of silicon are flexible. For instance, ribbons with thickness of 100 nm suffer peak strains of only 0.0005% upon bending to radii of curvature of 1 cm. Even after mounting on 20 μm thick plastic sheets, the strains (~0.1%) at similar bend radii remain well below the fracture limits (~1%) (Park *et al.*, 2008). More improvements are made by moving the silicon away from the surface of the plastic, where the bending strains are strongest to the point in its depth where these strains are nil. Ultrathin circuits using silicon nanoribbons in this type of neutral mechanical plane design can be bent to radii of ~150 μm, with strains in the silicon that are less than ~0.1% (Kim, E. *et al.*, 2008).

Creating such structures in "wavy" shapes and bonding them to elastomeric substrates yield systems that not only flex but also stretch and compress. A stretchable silicon membrane (~100-nm thick) can be made wavy and bonded to a piece of rubber, and an extremely stretchable silicon membrane (~100-nm thick can be patterned into a mesh geometry and bonded to a rubber substrate only at square pads located between arc-shaped bridge structures.

Materials that Stretch

Some new materials provide an alternative route to stretchable electronics. Elastic conductors are used as electrical interconnects between active devices that are rigid or only bendable. Conductive rubbers based on elastomers loaded with carbon black have been known since long but the resistances and their dependence on strain are too large to be useful. In a highly promising recent approach, long, single-walled carbon nanotubes (SWNTs) serve as conductive dopants in a rubber matrix (Sekitani *et al.*, 2008, 2009). The SWNTs are ground in an ionic liquid and mixed with a fluorinated copolymer to yield a black, pastelike conductive substance, termed **bucky gel**. Individual SWNTs form tangled mats in these gels, and can reconfigure in response to applied strain in a way that preserves highly conductive pathways for charge transport. This material may be printed onto sheets of poly(dimethylsiloxane (PDMS) to produce elastic conducting traces with full stretchability.

STRETCHABLE ELECTRONICS, OPTOELECTRONICS, AND INTEGRATED SYSTEMS

The above concepts have been gainfully exploited to produce stretchable, integrated systems. A stretchable silicon circuit has been developed in a wavy geometry. It is compressed in its centre by a glass capillary tube (main) and wavy logic gate built with two transistors. This approach was applied to an ultrathin, neutral mechanical circuit sheet that supports an array of silicon transistors, logic gates, and ring oscillators (Kim *et al.*, 2008). Mechanical responses enable stretchability. The wavy shapes change to accommodate applied strains, and the underlying PDMS substrate provides an elastic restoring force. In another approach, microscale arc-shaped "ribbon cables" of metal and plastic in neutral mechanical layouts are used to interconnect silicon devices located at the nodes of the mesh (Kim *et al.*, 2008a). The circuit is then conformally integrated onto a model of a fingertip, as an example of a surface whose nonzero Gaussian curvature would be impossible to wrap with a system that is only flexible (see Rogers *et al.*, 2010).

STRETCHABLE ELECTRODES

Work in stretchable electrodes, as opposed to electronics, enjoys an enviably long and glorious history as well as broad range of materials and design options. The origins of the field of stretchable electronics can be traced back to observations that films of gold formed by physical vapour deposition directly onto PDMS spontaneously adopt microstructured or nanostructured forms (Bowden *et al.*, 1998) and that these structures act as electrodes which accommodate large applied strains without breaking (Lacour *et al.*, 2003). One-dimensional wavy shapes are easily achieved in conducting and semiconducting nanomaterials such as SWNTs (Khang *et al.*, 2008), formed in a wavy shape on PDMS by using appropriate techniques. With nanowires, wavy deformations form in the plane of the substrate, and lithography can create similar structures in microscale metal wires (Huyghe *et al.*, 2008).

Besides structured wires, similar levels of stretchability can be created directly in electrodes of suitable materials, like some SWNT-based conductors. In a low-melting-point metal solder housed in sealed microchannels in PDMS, the metal can plastically deform and flow in response to large strain deformations, while the PDMS provides an elastic restoring force. Kim *et al.* (2009, 2009a) further showed that graphene on PDMS exhibits reversible responses to large applied strain, probably involving a mechanics similar to that of some gold electrodes. In this case, a stretchable metal wire consisting of a low-melting-temperature solder is sealed in channels in PDMS.

MOLECULAR ELECTRONICS

The possible development of individual molecules as the key components of electronic devices will undoubtedly constitute one of the most outstanding advances in electronics since the invention of the integrated circuit (see Aviram, 1992). Using typical molecular dimensions, such a molecular electronics could enable perhaps as many as 10^{13} logic gates cm^{-2} – an increase of 10^5 over the roughly 10^8 gates cm^{-2} available using current lithographic technology on silicon (Colbert and Smalley, 1999). Increased packing density and shorter distances between elements would correspondingly increase the operating speed—from nanoseconds in present devices to femtoseconds in molecular devices.

Increasing attention has recently been given to the molecular wires needed to interconnect these devices. The largest type of candidate molecules for molecular wires seem to be intrinsically semiconducting conjugated organic molecules (e.g. polyacetylene), which are characterized by highly delocalized electronic states. But stable molecular conductors suitable for molecular electronics have not been made so far and it is here that fullerene nanotubes may help solve these difficulties. Their electrical conductivity is far superior to the undoped conjugated molecules and, more importantly, good enough to serve as device interconnects. Besides, the quantum wire aspect of electron transport in fullerene nanotubes may turn out to be a key feature in two ways: (i) heat dissipation will be a critical issue in molecular-scale components owing to the packing densities sought, so minimizing the energy dissipation intrinsic to diffusive transport in ordinary wires will be an advantage with nanotubes; (ii) coherent transport makes it possible to transmit more information, because phase coherence can be maintained over relatively long distances.

According to Colbert and Smalley, of the several challenges in producing a fullerene-nanotube-based molecular electronics, the most fundamental is how to manipulate chemically short lengths of fullerene nanotubes, to assemble and integrate them efficiently and reliably into devices and circuits. The technologies for doing this make up what may be termed the molecular science and technology of fullerene nanotubes, and include methods to purify, cut, separate, sort by length, diameter or helicity (i.e. conductivity), functionalize open and closed ends and sides, and assemble them with chemical precision (Colbert and Smalley, 1999).

Both molecular electronics and optoelectronics depend critically on the molecular spatial organization. The basic mechanisms underlying recent discoveries in these areas emanate from co-mingling of electronic energies in the molecule with external, macroscopic structures.

Molecular electronics (**moletronics**, ME) have been achieved as the set of electronic behaviours in molecule-containing structures that depend upon the characteristic molecular organization of space (Jotner and Ratner, 1997; Kwok and Ellenbogen, 2002; Ratner, 2002). Being electronics, ME depends upon certain transduction processes in electronic energy, charge or radiation. Since molecules (unlike atoms) are not spheres, the characteristic structural motis of ME may differ substantially from traditional semiconductor electronics. ME behaviour is fixed at the nanoscale. While ME structures and devices may be macroscopic, the fundamental behaviours manifest at the molecular level.

Three chief domains of ME are the following: (1) central issues of molecular charge transport, i.e., how can a molecular junction be made. What determines and controls the transport through such a junction? What structure/function relationships exist between transport processes and molecular architectures? (2) Molecular optoelectronics is related to charge transport, but also involves electromagnetic interactions of higher frequency; (3) Molecular magnetics, actuators, ferroelectrics and elastics (see Sheats *et al.*, 1996).

Ratner (2002) defined the following five different regimes for the transport in molecular junctions:

1. **Elastic scattering** In the case of a thin junction at low temperature, considerations based on the time required to tunnel through the junctions suggest no electron localization within the bridge, and the problem becomes just that of elastic scattering (see Datta, 1995).
2. **Inelastic scattering**. In isolated molecular electron transfer, coupling to vibrations dominates the rate process (see Ratner, 2002). In molecular wire junctions, inelastic behaviour is also expected. Inelastic electron tunnelling experiments have helped to characterize such coupling.
3. **Charge injection**. This should occur for higher temperatures, smaller energy gaps and longer junctions (Donhauer *et al.*, 2001; Cui *et al.*, 2001; Ratner, 2002). Moving charge a long distance in a nonresonant molecular structure always involves one exponential decay or another. In the tunnelling limit, the decay is exponential in length, largely because of poor electronics (bridge assisted) mixing between the molecular termini. For longer junctions, elastic behaviour is too slow, and one expects an inelastic process corresponding to injection.
4. **Multiple wires**. The simplest conceivable molecular junction consists of a single molecule acting as a molecular wire. Preparation methodologies, however, for the most part yield adlayers (Cui *et al.*, 2001) the transport junction actually can have up to a few thousand molecules (with some exceptions).
5. **Dynamical stereochemical gating**. Unlike carbon nanotubes or traditional semiconductor structures, organic molecules can have multiple potential energy minima and low frequency torsional modes that significantly change the intramolecular electronic mixings, leading to domination of charge motion by geometric modifications.

Conjugated organic molecules seem suitable for molecular electronic devices in view of their low resistance electron transport, nonlinear I/V characteristics, and potential for easy fabrication and use. Work on these nonlinear electrical characteristics of single molecules revealed that they depend on the presence – or absence – of certain functional groups. Matsunaga and Sohlberg (2001) modelled a molecular junction. When an electron traverses such a molecule, the latter briefly accesses its anionic state, that is, it has an extra electron. This energetically unfavourable process can occur by quantum mechanical tunnelling. Certain functional groups, however (e.g. the $-NO_2$ group) lower the energy of the anionic state, effectively giving the electron a rest stop as it tunnels across the molecule, but only if the applied voltage is just right. There is no doubt that a better understanding of the electrical behaviour of single molecules will help in realizing the potential of electronics.

A research team led by Lars Hultman (Linkoping University) synthesized a novel fullerenelike material (*Phys. Rev. Lett.* 87, 225503, 2001). The solid form of carbon fullerenes is easily degraded because the individual spheres are only held together by weak van der Waals forces. By substituting nitrogen (N) atoms for some of the carbon (C) atoms, Hultman *et al.* developed a fullerenelike material that is not only strong but also elastic. Using ultrahigh vacuum magnetron sputtering, **nano-onions** of C and N were grown atom by atom on a substrate surface. Each nano-onion is ~7 nm in diameter. Each contains 7-10 concentric shells. The new material seems to be a novel $C_{48}N_{12}$ aza-fullerene.

Nanoindentation tests on the resulting thin films of cross-linked nano-onions suggested a high resilience material with hardness of 7 GPa, Young's modulus of 7 GPa, and complete elastic recovery after loading with 0.5 mN to a depth of 75 nm. Conceivably, even better hardness and modulus values might be achieved by increasing film density and cross-linking.

The properties of the new material could allow fullerenes to attract application as **wear-protective coatings of hard disks, field-emission flat panel displays, or even medical implants.**

Fullerene Nanotubes for Biology

Nanotubes may have an impact on biological investigations in various ways. The realization of true molecular electronics should have a strong impact on biotechnology by enabling new bioelectronic devices that exploit or require a smaller scale of electronics. For instance, many important applications of MEMS (microelectromechanical systems) devices are in the biomedical field.

Nanotubes are already exerting an important impact in biosciences as proximate probes. Multiwall nanotubes attached to the end of a pyramidal atomic force microscope (AFM) tip have significant advantages as a probe tip (Dai *et al.*, 1996). These advantages have been exploited in several investigations of biological systems (see Wong *et al.*, 1998; Colbert and Smalley, 1999). Derivatized nanotube tips have provided much chemical sensitivity for chemical and biological probing in addition to their tactile sensitivity (Wong *et al.*, 1998a). Nanotubes' cylindrical, nanometre dimensions may make them good electrodes to probe inside living cells or directly in synapses, for example, to enable measurements of biological phenomena with unprecedented spatial and temporal resolution (Colbert and Smalley, 1999).

Later, it turned out that buckminsterfullerene is not so difficult to make. In fact it is formed naturally in some combustion processes involving carbon (even candle burning), and its traces are found in soot. Simple ways have been devised to make buckyballs and nanotubes; examples include triggering an electrical arc between two graphite electrodes or passing a hydrocarbon gas over a metal catalyst. Carbon nanotubes (CNTs) have drawn much interest; among their many intriguing properties, they have the greatest tensile strength of any material known. They resist 100 times more strain than typical structural steel.

CNTs are now driving substantial advances in electronics, energy, medicine and materials.

ELECTRET SENSORS, FILTERS AND MEMS DEVICES

An **electret** is a dielectric material that has been electrized (polarized). Eguchi (1925) discovered the thermoelectret effect when he solidified a mixture of carnauba wax, resin and beeswax in the presence of a high DC electric field. The energy absorbed during the charging process manifested in space-charge polarization by trapping of positive and negative charges at the interstitial sites. The photoelectret effect was discovered by Nadzakov (1937) when he exposed a photoconducting dielectric to both light radiation and electric field simultaneously.

Thermoelectrets and photoelectrets have proved to be extremely important in modern sensing and optoelectronic devices as they show persistent polarization and a surface charge, which remain stable for a long time. A landmark in electret applications was achieved in 1966, when G.M. Sessler at Bell Laboratories (USA) devised an **electret condenser microphone (ECM)**. ECMs are widely used in tape recorders, stereos, telephones, cordless and cellular phones and hearing aids (Sessler, 1987; Goel, 2003). Goel (2003) reviewed the progress made in electret formation in dielectric materials and their polarization as well as applications in transducers and air filters as low energy-consuming, cost-effective pollution monitoring and control devices, and their ability to remove/filter toxic gases and suspended particles. Miniature electret microphones integrated on silicon chips as **microelectro-mechanical systems (MEMS)** have been designed. Electrets are also attracting interest for possible biomedical applications such as replacement of heart valve, bone implants and artificial muscles in robotics. These developments have created challenging possibilities in materials science research.

An electret-forming material can be a sheet, ribbon or film coated on the electrode or any other form that is conducive to continuity and which has two flat surfaces. As permanently charged material, an electret is generated by polarization in high electric field or when corona or electron beam injection is applied along with heat or light. Three basic processes that can occur when a dielectric material is subjected to electric field are: (i) space charge polarization near the electrons; (ii) bulk charges trapped within the material at the interfaces or due to migration; and (iii) dipolar polarization due to polar molecules. Fig. 6.1 oulines some different mechanisms of charge trapping in an electret. Fig. 6.2 illustrates electret formation, method of activation, measurement techniques, and the parameters that can be measured.

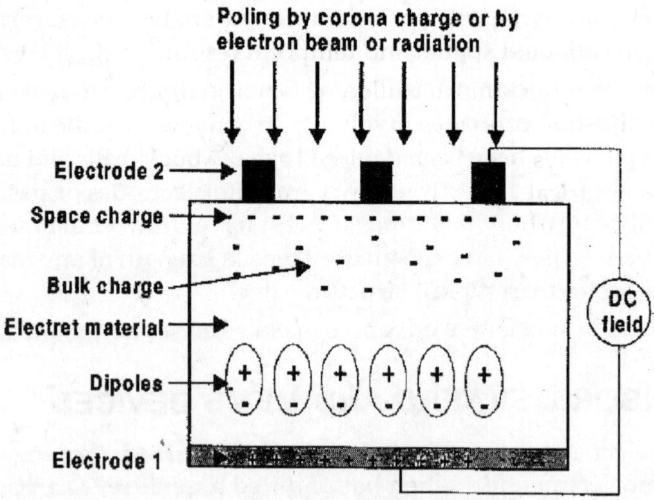

Fig. 6.1 Different mechanisms of charge trapping in an electret material (after Goel, 2003, with permission from *Current Science*).

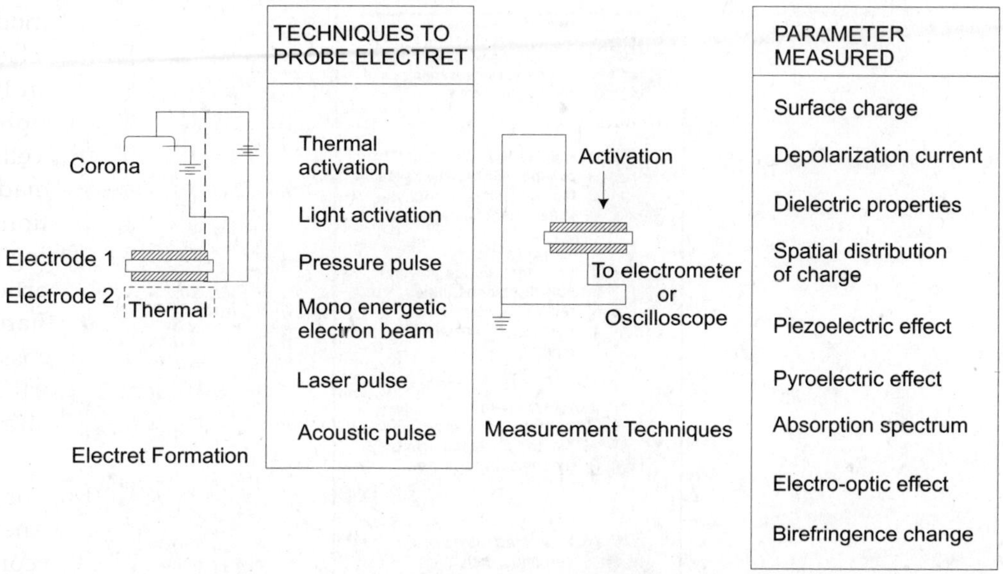

Fig. 6.2 Electret formation, method of activation, measurement techniques and parameters measured (after Goel, 2003, with permission from *Current Science*).

Upon exposure to a pressure wave, a thermal wave or a mechanical distortion, a light wave or a voltage pulse, an electret produces signal in the external circuit. This allows it to be used for sensing temperature, pressure, light, sound and electric signals. Commercialization of electret technology has led to useful electret devices ranging from sensors, transducers and air filters to radiation dosimeters. These devices usually need neither external power supply nor a battery during operation. The critical feature is their long-term charge stability, besides desired material compatibility (Goel, 2003).

The MEMS technology has benefited greatly from developments in microelectronics. It strives to develop products that can perform various different functions including mechanical, optical, biological and electrical, in compact light weight designs, integrated on microchips. **Microgears, micromotors, microactuators, micropumps and microswitches** have been fabricated by exploiting MEMS technology. In MEMS, micromachines are embedded below the surface of a microelectronic device. Microelectronic devices containing over ten million transistors and capacitors on a chip, smaller than a fingernail materialized in the late 1990s. Micromachining techniques can produce miniature mechanical components on a chip. Microfabrication of moving components involves the use of many different materials than those used in microelectronics. Fabrication techniques are quite similar to those used for integrated circuits on chips. Photolithography, chemical vapour deposition, ion implantation and diffusion can be adopted for 3-D shapes of MEMS. MEMS being 3-D structures, complexity in their design and packaging is very high and requires thorough knowledge of mechanical engineering and other disciplines (Goel, 2003). Table 6.1 lists some emerging applications of electrets in MEMS and Fig. 6.3 illustrates a historical summary of the materials focus in electret research.

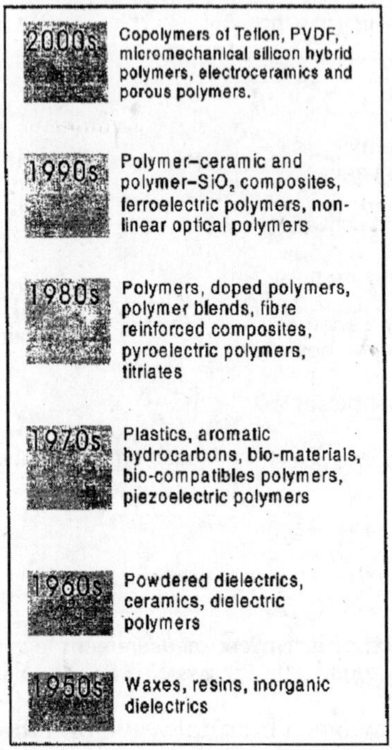

Fig. 6.3 Materials focus of electret research (after Goel, 2003, with permission from *Current Science*).

Table 6.1 Some emerging electret MEMS applications (after Goel, 2003, with permission from *Current Science*).

Biomedical and health sectors	Disposable blood pressure monitor
	Heart pacemaker
	Catheter tip temperature sensor for detection of onset of heart attack and ovulation.
	Pointed drug delivery system
	Artificial muscle and veins
	Detection of polyps
	Kidney dialyser
	Fibre optic sensors for endoscopy
Telecommunications industry	Hearing aids
	Microhydrophones
	Cellular phones
	Sound localization aid for deaf-blind persons
	Optical electret microphone
	Laser scanners
	Holographic storage displays

Contd..

Table 6.1 *Contd..*

Building and transport sectors	Premise information distribution system Automatic room light controller Touch switch Floor sensors and impact sensors Personal verification systems Pyroelectric array detector for night vision Gas flow meter Lubricants

In India, the current focus of electret is on basic materials research, understanding of charge production and its stability in dielectrics, preparation and characterization of novel materials and development of nanocomposites.

FUTURE POSSIBILITIES

Exciting recent advances in materials and mechanics potentially create promising possibilities for stretchable, curvilinear electronic and optoelectronic components as well as complete integrated systems in discrete, distributed, or hybrid forms, but several challenges will need to be addressed before the possibilities actually materialize. The challenges are particularly pronounced for SWNTs and graphene, where larger mismatches in properties occur and we know nothing about even the basic foundational mechanics underlying stretchability. Other notable areas for study in these and other heterogeneous systems are the physics of heat transport, to enable efficient thermal management, and the materials science of interfaces, to ensure mechanical reliability (Rogers *et al.*, 2010).

If and when the scientific and engineering hurdles are crossed, our conception of electronics could well change from hard, rigid, planar chips to soft, stretchable, curvilinear sheets. Here, mechanics design will be no less important than electrical design in the designing of new systems. Of the many potential areas of application, some of the most interesting are in biomedical devices that address important problems in human health. Soft, elastic mechanical properties and curvilinear layouts provide both mechanical modulus and shape matching to biological tissues. These "tissue-like" devices, especially when implemented with biocompatible materials, will provide solutions to long-standing challenges in the establishment of viable, intimate biotic-abiotic interfaces with high levels of functionality. Even today, passive, stretchable electrodes are being used for research applications (Yu *et al.*, 2009) and diagnostic systems with discrete designs (Carta *et al.*, 2009), both with some degree of deformability. In the coming few years this could expand into fully distributed, biointegrated, multifunctional devices for clinical use (Rogers *et al.*, 2010).

Beyond biology, stretchable electronics will prove useful in conformal, active antennas and other components for communications, possibly even in cellular telephones of the future. It is also likely that stretchable sensor tapes will provide thin, conformal monitors of the structural health in the wings of aircraft. The basic ideas may even be exploited in such other semiconductor technologies as photovoltaics and thermoelectrics. In the latter case, devices for scavenging power will be used in the form of sheets that wrap the complex surfaces of engine parts for efficient thermal coupling.

NANOCOMPUTERS

Whereas rapid miniaturization of electronics to the micrometer scale catalyzed both scientific and economic progress over the last three decades of the 20th century, nanometer scale electronics (**nanoelectronics**) moved to the forefront thereafter (Ellenbogen and Love, 2000; Wada, 2001). Bachtold *et al.* (2001) and Huang *et al.* (2001) described dramatic steps toward the realization of electronic nanocomputers. Whereas Bachtold *et al.* demonstrated logic circuits made from individual carbon nanotube molecules, Huang *et al.* successfully assembled logic circuits from semiconductor nanowires.

Various types of molecular-scale wires and switches including molecular-scale transistors based on CNTs and semiconductor nanowires were described by Tans *et al.* (1998), Cui and Lieber (2001); Schön *et al.* (2001); Reed *et al.* (2001) and Collier *et al.* (2000). But the papers of Bachtold *et al.* and Huang *et al.* actually advanced molecular-scale electronics fully from the single-device level to the circuit level (Tseng and Ellenbogen, 2001). Both teams designed new methods to meet two critical device requirements that had previously prevented the realization of transistor circuits. First, the component transistors must produce signal amplification or "power gain" with an output to input ratio much greater than 1. Second, each transistor must be controlled by its own local "gate" contact (Tseng and Ellenbogen, 2001).

Bachtold *et al.*'s work was prompted by their earlier discovery that individual semiconducting nanotubes adsorbed between two metal contacts on a silicon substrate function like the field-effect transistors in today's microcomputers. However, the controlling gate contact in that earlier work consisted of the entire supporting silicon chip. In such a design, multiple nanotube devices placed on a chip all need to be switched simultaneously. Also, the power gain was less than 1 because the silicon oxide insulator between the gate contact and nanotube was quite thick, and prevented high enough capacitive coupling between the gate contact and nanotube.

To create nanotube circuits, Bachtold *et al.* used electron beam lithography to pattern local aluminium gate contacts and exposed them to air to form very thin insulating layers on the aluminium leads. Insulator thickness could be reduced substantially, so that the new nanotube transistors could operate independently with a gain ratio of over 10. By wiring nanotube transistors together with gold interconnects made by lithography, Bachtold *et al.* were able to construct a range of logic circuits.

Huang *et al.* also built on their previous successes in devices to achieve circuits. This team demonstrated diodes and bipolar transistors made from nanowires in a crossed geometry (see Cui and Lieber, 2001). In their recent work, they assembled OR and AND logic circuits with only diodes, but making other circuits needed the development of nanowire field-effect transistors. The new nanowire transistors were created by placing two nanowires in a crossed geometry; heat was then applied to generate an insulating oxide between the nanowires. Like Bachtold *et al.*'s nanotube transistors, these nanowire transistors also feature local gate contacts with thin insulators and so are easily integrated into transistor circuits (Tseng and Ellenbogen, 2001).

Except for the contacts, Huang *et al.*'s nanowire circuits were assembled without "top-down" methods such as lithography. Instead, "bottom-up" parallel assembly tools such as microfluidics were used. This approach enabled them to build and test many devices and demonstrate readily reproducible behaviour in them. Also, Huang *et al.*'s circuits incorporated at least one natural nanometer-scale metric - the constant, small dimension of the crossing points of the nanowires - suggesting that the entire circuits might be shrunk easily to the nanoscale. This capability is particularly important in view of the fact that the circuits in both studies are still micrometer-scale systems.

While the demonstrations of molecular circuits by Bachtold *et al.* and Huang *et al.* (and also those reported by Derycke *et al.*, 2001 and Schön *et al.*, 2001) are very impressive, they build on several critical steps by previous researchers striving to make molecular-scale computers (Ellenbogen and Love, 2000; Wada, 2001). While the concept for molecular electronic devices and circuits was floated by Aviram and Ratner (1974), it is only in the last decade of 20th century that scientists saw crucial experimental demonstrations of molecules which serve as wires and switches; these have been divided into four broad categories, depending on the type of molecule or molecular-scale structure used to make the devices. These categories are semiconductor and metal nanowires, CNTs and fullerenes, small organic molecules, and biomolecules (see Tseng and Ellenbogen, 2001). What we badly need now is to produce a robust, commercially viable computer integrated on the molecular scale. Circuits need to be designed that are molecular scale in their entirety, not just incorporating molecular-scale components. Developing appropriate means for establishing molecular spacing between components and devices would be critical.

In principle, the very small sizes of molecules can allow researchers to fit a trillion molecular devices in a square centimetre. But even if the problem of heat dissipation can be addressed for so many densely spaced electrical devices, how to harness them for useful computation? Another challenge is how to assemble a trillion devices per cm^2 quickly, inexpensively and precisely.

MOLECULAR TRANSISTORS

Transistors make up the fundamental elements of integrated circuits - they control the flow of current between two electodes (the source and drain electrodes) by modifying the voltage applied at a third electrode (the gate electrode). With manufacturers keenly interested in producing ever smaller devices, it has become possible to make transistors from single molecules, where the flow of electrons is controlled by modulating the energy of the molecular orbitals. The construction and characterization of such a device has been a cherished desire of engineers working on nanoelectronics.

Solid-state molecular transistors that have already been made are based on two mechanisms: **Coulomb blockade**, in which the flow of electrons is controlled by the sequential charging of a molecule; and the **Kondo effect**, in which conducting electrons interact with local spin (intrinsic angular momentum) in a molecular junction (Yu and Natelson, 2004). Ghosh *et al.* (2004) floated a third possible approach, based on electrostatic modulation of the molecular orbital energy of a single molecule, and Song *et al.* (2009) actually described the first practical application of this approach by characterizing the resulting devices in great detail.

Although single molecules have been reported as being electrically connected to two electrodes (Xu and Tao, 2005), achieving this feat is extremely difficult. Song *et al.* made their devices by coating gold wires with the molecules of interest, whereafter breaking the wires by electromigration produced nanometer-scale gaps in the wires. Sometimes, the molecules coating the wire somehow become trapped in the gaps. This creates systems of source/drain electrodes – two broken ends of the wire – spanned by the molecules, forming a junction through which electrons can 'tunnel'. A notable benefit of this fabrication technique is that the junction can be formed directly over an oxidized-aluminium gate electrode, thereby providing the necessary three-terminal geometry of a transistor (Kushmerick, 2009).

For critical characterization of the charge-transport properties of their devices, Song *et al.* used a combination of spectroscopy techniques *in situ*. They resorted to inelastic electron tunnelling (IET) spectroscopy for measuring the interactions between the tunnelling electrons and the vibrational modes of the molecules in the devices. This technique proves that the measured currents do in fact pass through the molecules in single-molecule transistors. It also yields valuable information about the pathways taken by electrons as they cross the junction (Troisi *et al.*, 2007).

Another technique used by Song *et al.* was transition-voltage spectroscopy. Electrons cross a molecular junction by using one of two tunnelling mechanisms depending on the magnitude of the source-drain voltage; the transition voltage (V_{trans}) is that voltage at which tunnelling switches from one mechanism to the other. Beebe *et al.* (2006) had shown that V_{trans} is proportional to the difference in energy between the gating orbital of the molecular junction (the orbital that modulates electron tunneling) and the Fermi levels of the source and drain electrodes (the Fermi level is the highest possible energy for a conducting electron in an electrode. Song *et al.* measured V_{trans} by means of transition-voltage spectroscopy at different applied gate voltages to demonstrate that there is a linear relationship between gate voltage and molecular-orbital energy in their devices, as expected for single-molecule transistors.

Song *et al.* reported that both of their transistor types became more conducting when a negative gate voltage was applied. The fact that negative gate voltages lower the energy difference between the highest occupied molecular orbital (HOMO) of the molecular junction and the electrode's Fermi level suggests that in their device, tunnelling electrons couple to the HOMOs of the molecules.

The authors also observed that the current through their device was gated by the HOMO energy of the molecular junction by examining the dependence of the IET spectra on the applied gate voltage. The IET spectra of the transistors that incorporate alkane dithiols remained unaffected by the gate voltage, suggesting that electron tunnelling through the device is always 'non-resonant', that is, there is a large energy difference between the dithiol's HOMO and the electrode's Fermi level.

Conversely, Song *et al.* found that the applied gate voltage greatly modulates the IET spectra of transistors which contain an aromatic dithiol. Applying a negative gate voltage brings the energy of the molecular junction's HOMO closer to that of the electrode's Fermi level;

this increases the signal intensities of the spectra and changes the shapes of the vibrational peaks. The change in peak shape points to increased coupling between the tunnelling charge carrier and the molecular vibrations, owing to a resonance between the HOMO and the Fermi level (Galperin *et al.*, 2004). The work of Song *et al.* clearly shows that resonant and non-resonant vibrational coupling can be tuned in single-molecule transistors.

Thus it now stands proved that a solid-state molecular transistor can indeed function through the relative alignment of its orbital energies with the electrode's Fermi level, and also that this alignment can be tuned by the applied gate voltage.

TINY CELL TRANSISTOR

A nanometre-sized transistor disguised as part of a biological membrane can infiltrate a living cell and measure its electrical activity.

Lieber *et al.* (2010) designed their hairpin-shaped device out of a silicon nanowire with a tiny transistor on the elbow of the bent pin and an electrical contact on each of the pin's two arms. The elbow's tip was coated in phospholipids (the main constituent of cell membranes) so as to trick the membrane into accepting the tip and pulling it inwards. This enabled Lieber *et al.* to make a device less than 50 nm wide – smaller than many virus particles!

The researchers poked the probe into a single cultured embryonic chicken heart cell and used it to record a series of voltage peaks corresponding to the beating of the cell.

SEVEN-ATOM TRANSISTOR

Very recently, the world's smallest electronic switch has been invented in Australia. This switch is based on the world's smallest transistor built with only seven atoms, and only four nanometers in length.

The seven-atom transistor measures four-billionths of a metre. It is embedded in a single silicone crystal. It represents the first step in a "quantum computer" which can make calculations millions of times faster than existing devices.

This revolutionary 7-atom transistor was developed by Michelle Simmons and her associates. According to Simmons, the technology has strong implications for code-breaking, financial transactions and weather forecasting, which involve testing enormous numbers of possible scenarios.

The development of the transistor and the miracle switch involved effective collaboration between the University of New South Wales' Centre for Quantum Computer Technology (CQCT) and the University of Wisconsin-Madison. The two teams jointly designed the transistor by manipulating atoms by using a special microscope.

This major breakthrough can potentially reduce the size of microchips, which contain billions of transistors, by about 100 times, simultaneously accelerating processing speeds "exponentially".

According to Simmons, commercial applications for the technology are only about five years away. Her team is now working towards the first ultra-fast quantum computer, predicted to be the size of a current silicon chip.

By using Simmons' system, the atoms of the quantum dot can be embedded in one silicon crystal. Till hitherto researchers have been attempting how to make such computers with more interesting materials, but Simmons chose silicon in view of its longer lifespan and also because it would be compatible with existing technology.

DESIGNING HIGH-PERFORMANCE TRANSISTORS

To satisfy consumer demands and needs ranging from entertainment to personalized health information in one device, manufacturers are keenly interested in making improved integrated electronic circuits and to bring down their price. They are also interested in building electronics on flexible, lightweight, and cheap substrates such as glass and plastics (Reuss *et al.*, 2005). Typical semiconductor films for making transistors on such materials need to be deposited at low temperatures so as to preserve substrate integrity. This strongly constrains the quality of the organic or inorganic semiconductor films used for large-area electronics (Sun and Rogers, 2007). Disorder and impurities in the films adversely affect analog and digital performance when formed into transistors. Dry transfer printing and solution casting might help in addressing the temperature problem but they have had no impact on mass production of high-performance devices. An alternative approach involves integration at the device or circuit level through substrate transfer techniques (Yuan and Ma, 2006; Stauth and Parviz, 2006). However, this approach is more costly and its assembly efficiency is low.

Ishii *et al.* (2004) showed that for transistors of disordered silicon films, better switching performance is achievable by making the conduction channel in the transistor very thin. A layer of nanocrystalline silicon film forms the channel and ranges from 8 to 2 nm in thickness; the precise film thickness can be controlled during the deposition process (Guo and Silva, 2008). These devices have excellent intrachip uniformity of electrical characteristics, even when the channel thickness is as small as 2 nm. For disordered silicon transistors, the performance is lowered by high leakage of current due to defect states present in the bandgap. In cases where the transistor channel becomes short, the effect of the drain on the channel grows and results in an increased subthreshold leakage current when the gate voltage drops below the threshold voltage. When the silicon film thickness is reduced to a few nanometers, quantum confinement modulates the band structure, generating band-edge shifts in both the conduction and valence bands relative to bulk silicon.

Directly attributable to this widened bandgap, the OFF-state leakage current (I_{OFF}) falls significantly in the ultrathin channel transistors whereas bandgap widening has a lesser effect on the ON-states current (I_{ON}). The gate voltage required to switch the transistor between ON and OFF states also decreases in view of a stronger gate control of the channel surface potential in thinner channel transistors (Walker *et al.*, 2004). These result in a higher I_{ON}/I_{OFF} ratio of over 10^{11} for devices with a 2.0-nm-thick channel. The high I_{ON}/I_{OFF} ratio is critical for designing integrated electronics as it enables using arbitrary flexible substrates for those applications where low power and fast success times are crucial.

Another interesting concept, the source-gated transistor (SGT) was first proposed for amorphous silicon by Shannon and Gerstner (2003). Here a potential barrier is formed at the source contact and governs current transport in an SGT; in a regular field-effect transistor

(FET), an ohmic contact is made. The gate remains under the source barrier and extends across to the drain contact. Upon switching the transistor on, the current in the SGT is determined by the carrier emission over the source-barrier contact, and the channel creates a parasitic resistance. A marked difference between the SGT and a traditional FET is that a reverse-biased source barrier controls the current in the ON-state and current saturation is determined by the electrostatics at the source barrier rather than by pinch-off of the conduction channel at the drain (Guo and Silva, 2008). As compared with the FET, the SGT can be operated with much less excess carrier concentration, combined with much higher internal fields over small dimensions. The low carrier concentration improves stability, whereas high internal fields enhance the carrier velocity, provided that the latter is proportional to the electric field (Shannon and Balon, 2007). In organic semiconductors, the mobility is usually field dependent. In material systems with disorder, to enhance transistor speed very short channels are needed to generate high internal electric fields (Bhattacharyya *et al.*, 2006). In the traditional FET, achieving thin gate insulator layers of high quality to suppress the short-channel effects is very difficult, but the SGT can work with very short source-drain separations even with a thick gate insulator layer, so providing a great advantage in terms of the fabrication process. The current through an SGT is quite insensitive to source-drain separation, thereby enabling current uniformity with simple and imprecise pattern techniques (Guo and Silva, 2008).

Effective engineering of the transistor structure itself rather than the channel material can enhance device performance and should enable the design of high-performance large-area circuits and systems based on low-cost reliable material processes while keeping the cost low and standard mass-production routes available to industry (Guo and Silva, 2008).

MICROSYSTEMS-ON-SILICON – THE MICROELECTRONIC REVOLUTION IN INDIA

The manner in which developing countries will address silicon revolution is a topic of global relevance. India has a huge domestic market and is blessed with talented and skilled human resources. It has enormous potential for silicon revolution but lacks the required will power to seize the opportunity. Microelectronics illustrate a technological revolution on silicon - in layman's language, on grains of sand. Despite much speculation about its future in the face of new nano materials, it may be decades before an alternative can be found for silicon and **microsystems-on-silicon (MOS)** and we should not ignore or undermine its continuing importance.

Striking developments have taken place in the silicon world and are impacting our daily life (Bhattacharya, 2007). Grains of sand form the basic raw material of silicon that has dominated the human civilization for half a century.

For every person on the Earth, the microelectronics industry currently produces, on an average, about 60 million transistors. By 2008, the figure could be one billion transistors. For the past four decades, the average compound growth rate has, hovered at about 15% and the microelectronics business is around 200 billion US dollars. It literally controls a market of about one trillion US dollars. Silicon may well overtake steel in the coming years. The Indian domestic semiconductor market is worth about Rs. 1700 crore and our semiconductor production is around Rs. 40 crore.

The recorded progress and future projections of silicon revolution are shown in Table 6.2 The scaling predicts around 15% growth of global electronics market.

Table 6.2 Semiconductor International Association (SIA) Projected Roadmap: 1999-2014 (after SIA, 2001).

Year	1999	2002	2014
Feature size (nm)	180	130	35
Minimum mask count	22-24	24	30
Wafer diameter (mm)	300	300	450
Memory (Bits)	1G	4G	1T
Transistor/sq.cm (μP)	6.2M	18M	390M
Interconnect level	6-7	7	10
Colck speed (MHz)	1250	2100	10000
μP Chip size (mm^2)	340	430	901
DRAM Chip size (mm^2)	400	560	2240
Power supply	1.5-1.8	1.2-1.5	0.37
Maximum power	90	130	183
No. of pins	700	957	3350

Moore's Law

In 1965, Gordon Moore, the founder of INTEL, used four data points to predict empirically that the number of transistors per chip would double once every two years, but in fact it doubled once every eighteen months. The prediction has survived through 4 decades and the roadmaps closely follow Moore's law. A layman's conjecture is that sixteen transistors in a chip in the mid-60s blow up to 30 crore transistors in a chip in the beginning of the 21st century! Moore's law predicts the half life of any process technology as very small, akin to a technological soap bubble!

The scaling characterized by the roadmap has the following implications:

- FAB cost escalates exponentially.
- Chips become cheaper.
- Design cost, effort and time increase progressively.
- Number of model parameters describing the electrical characteristics of the transistor increase exponentially, from tens to thousands.
- System-on-chip (SOC) could well become the ultimate product frontier.

Moore's (1975) law has survived despite many limiting factors because it has fuelled ever increasing expectation and hunger for cheaper, faster and smaller electronic systems triggering innovation, competition and surprising domain applications. It is conceivable that Moore's Law could survive and sustain the human desire of scaling lofty heights. The transition from microelectronics to nanoelectronics is a natural evolutionary process; in the nanoscale dimension, it is quantum phenomena that dominate the device physics (Bhattacharya, 2007).

Manufacturing activities need to be continuously scaled by new technology. The problem is not only that of developing one technology. Each technology node requires development of multiple technologies. This means that only a selected few are able to undertake the printing work on silicon on a mass scale. Table 6.3 illustrates escalation of Fab costs.

Table 6.3 Fab costs for various technology nodes (after Bhattacharya, 2007).

Technology node (in micron)	Year	Cost (MUS $)
1.2	1988	350
0.8	1990	500
0.5	1992	750
0.35	1996	1500
0.18	2000	1700
0.13	2002	2000
0.7	2008	2500

The table highlights the following points:

- High capital expenditure for entry in technology
- High entry barrier
- Expensive wafer technology upgradation.
- Very few can sustain the manufacturing business.

Indian Scenario in Manufacturing

The public sector enterprise Semiconductor Complex Limited (SCL) was set up with 5 μm technology in 1983. Its present capability is 0.8 μm with inhouse development efforts. SCL has been taken over by Department of Space to meet India's strategic needs and a total investment of around Rs. 400 crore has been made. Its wafer (6 inch) turnover capability is 3000 wafers per month.

Some more state-of-art silicon fabrication facilities are expected to be created in India in the coming years.

Chinese Impact

The Republic of China also inherited similar technological legacy as India. When India set up SCL, it was ahead of China both in the domains of technology and design. But now China has become a dominant player in manufacturing of silicon integrated chips. It is expected to produce and meet over 50% of the global chip requirements in the next few years. Indeed, China does not view silicon technology in isolation but perceives it as a convergence technology of information, communication, entertainment, processing, control and computation.

Impact of Mechanics and Electronics on Silicon

Some driving factor has always sustained the silicon revolution (Table 6.4). Every second decade seems to begin with a new technology driver. Emerging new products were duly standardized.

Table 6.4 Technology drivers (after Bhattacharya, 2007).

1960-1980	1980-2000	2000-2020
Calculator, equipment, mainframe	Personal computer (PC), communication	Digital consumer, Sensor, Embeded systems, Robotics, SOC, NOC

One notable technology relates to sensors, an indispensable element of information technology. Silicon has an important potential role in this.

Silicon has already conquered the electronics world. It is now all poised to integrate with the mechanical world. In the mechanical context, the positive attributes of silicon are outlined in Table 6.5.

Table 6.5 Attributes of silicon (after Bhattacharya, 2007).

Characteristic	Remark
Single crystalline	High quality factor
Elasticity	Comparable to aluminium
Hardness	Comparable to quartz

The microfabrication technique employed for silicon chips is directly applicable to micromachining for fabricating micromechanical sensors, and actuators. Further expansion and diversification of the silicon-based sensor world could include aspects of: acoustics, force, inertia, vision, touch, smell, acceleration, position, and traction. Sensors will play crucial role in robotics which is projected as the technology driver in the coming decade (see Table 6.6).

Table 6.6 Robot intelligence roadmap (after Bhattacharya, 2007).

Year	Intelligence level (of)
2010	Lizard
2020	Mouse
2030	Monkey
2040	Human

For the evolution of intelligence in the future robots the information processing capability will need to be increased from 104 MIPS[3] to 108 MIPS[3] (Bhattacharya, 2007).

Embedded Systems

Embedded systems play a crucial role in the current information age by integrating hardware, software, programmability sensors and actuators, and by performing computations, unaided by the user. These attributes are: economy; flexibility (hardware/software option); individualization; and function integration.

Some of their applications include the following:

Professional areas: Medical system, traffic control, automobile security, healthcare, agriculture, environment, industrial plants.

Consumer products: Home and kitchen appliances, audio and video products, toys.

Communication: Mobile phones.

Computation: Mobile desktop for storage, communication.

Civil engineering: Intelligent highways, intelligent buildings, intelligent bridges, energy conservation.

Design Status in India

Some salient facts related to Indian design status are: virtually all major multinational companies have VLSI (very large scale integration) design operations in India. About 100 design enterprises are functioning. Around 7000 VLSI designers are employed in India. Outsourcing to India mostly relates to labour intensive functions and low end operations. One manyear of design should generate 1M$ in the life cycle of the design. Implementing roughly about ten designs for a product could possibly generate one billion US$ during the lifetime of the designed product in India.

By comparison, the performance of Peoples Republic of China in terms of revenue earned is estimated to be one billion US$ in 2004 and has grown significantly since then.

Prospects

India seems to have almost missed the silicon revolution that is shaping the destiny of the 21st century industrial revolution. India, however, now realizes its importance as symbolized by the setting up of Semiconductor Complex Ltd. as a public sector enterprise in the year 1983. This was motivated by self reliant technology scaled to 0.8 micron, the limit of infrastructural capability as suggested through its internal R&D effort. The country has a range of niche processes with significant intellectual property assets that can be transformed to productive business. India earlier tended to consider silicon technology merely as a strategic need rather than a commercial platform to build the electronics industry. India can now select a pragmatic approach from the following options.

1. Integrated device model (vertically integrated with in-house product, process, tool, equipment development and marketing. Examples: IBM, Motorola, Siemens and Phillips).
2. Foundry operation (manufacturing, serving as a printing house of silicon design). South Korea, Taiwan, Singapore, Peoples Republic of China are anticipated to produce 50% of global chips on silicon.
3. Fabless enterprise: Outsourcing the manufacturing process with strong focus on design, low cost product, and system development, all a knowledge-intensive input to the silicon empire.

Figure 6.4 shows a suggestive roadmap for India that projects different levels of opportunity conducive to rapid response to market opportunity and fast exit from risky ventures. The degree of success will be determined by the ambience in the country for entrepreneurship, hassle-free and corruption-free business practices, global outlook and high quality technical education. A vibrant fabless silicon activity is destined to be the natural nucleus for investment on foundry.

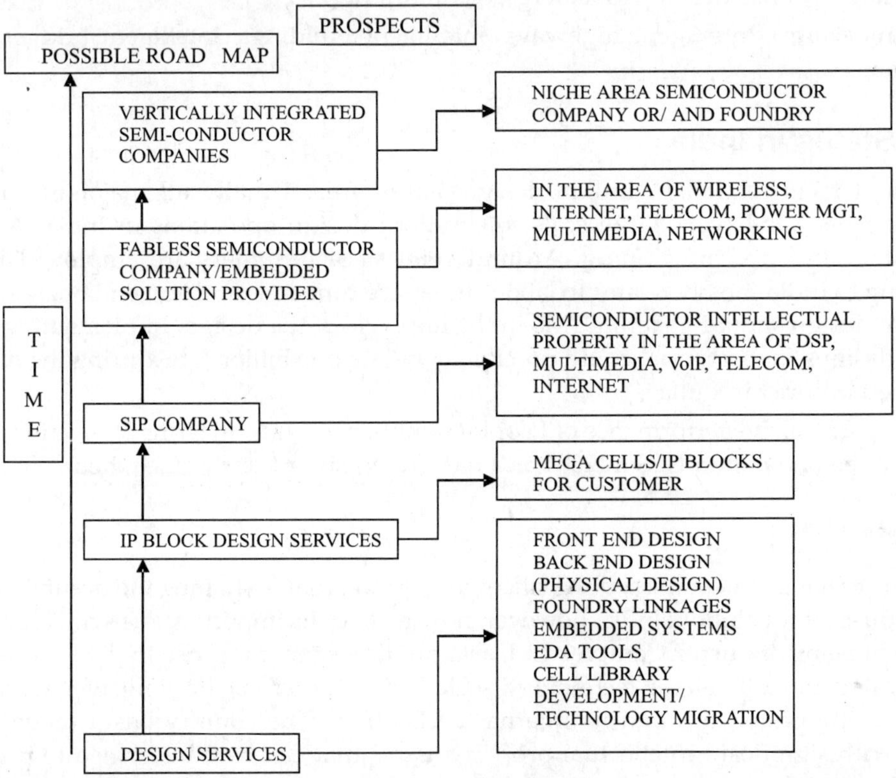

Fig. 6.4 Possible roadmap for silicon enterprise in India (after Bhattacharya, 2007).

Microsystems-on-silicon represent a knowledge-intensive industry in which technical education has decisive role in respect of competitiveness and leadership. Some issues warranting special attention in the field of VLSI education are:

- Problem solving, project oriented approach.
- Learning, rather than teaching orientation.
- Awareness of intellectual property generation.
- Systems approach and complexity management.
- Multidisciplinary focus, and
- Team work (Bhattacharya, 2007).

According to Makimoto (2004), the special attributes of the MOS industry can be encapsulated in 5-dimensional T-space as follows:

Technology – Most subtle and most robust engine of social change.

Time – Secret weapon of business.

Tools – The highest value for the lowest cost in the least possible time.

Task – Civilization advances with tasks performed without thinking of them.

Talent – More from less.

India's greatest assets are the many Indian professionals working abroad who have gained expertise in all sectors of silicon technologies. The stage is ripe for reverse knowledge drain in view of globalization and because India offers welcome opportunities for cost effective operations. The Quit-India trend can be replaced by "destination-India" phenomenon. The so called integrated circuit (IC) technology has the potential of serving as an India-China technology if manufacturing capability in China can be carefully and gainfully integrated with the design potential of India. Creation of an India-China business partnership around silicon can even offset the missed opportunity for both these countries.

7

CHAPTER

Super-functional Materials and Metamaterials

INTRODUCTION

The world around us is made up of material objects with a fascinating diversity of forms and functionalities. The huge variety of physicochemical properties exhibited by various bulk materials and their functional products are not only the result of their chemical constitution but also of different equilibrium or non-equilibrium phases in which matter exists; also the nano- to macroscale physico-mechanical structures of matter produce unimaginable 'metamaterials' showing properties not found in natural materials. A good example is the striking effects such as negative refraction of electromagnetic radiation, which manifest when purposefully created assemblies of subwavelength scale designer components are used – components that simultaneously show negative dielectric permissivity as well as negative magnetic permeability (Pendry, 2006; Veselago and Nasimanov, 2006). The concept of **designer metamolecules** and bulk-structured metamaterials has come in the limelight for diverse application areas such as acoustics.

Naik *et al.* (2009) discussed the potential of structuring surfaces rather than the bulk matter bounded by them, for generating super-functional material properties. There is an enormous range of intriguing and often hierarchically architectured naturally occurring surfaces, for inspirational ideas to create astonishing but desired super functionalities. To a typical onlooker, the sole correlation between the attractive hues of a butterfly or a lizard climbing up the wall, and a water droplet rolling down a lotus leaf, may be attributed to the fact that both occur in nature! In reality, however, each of these phenomena emanate from the presence of micron and sub-micron scale structures on the textured functional surfaces covering the bodies of the animals and plants (Ingram and Parker, 2008; Bhushan, 2009). The current research is focused on two aspects: (a) to understand the underlying science of structure-function relationships, which generate functionalities like superhydrophobicity of a lotus leaf, gravity defying super-adhesion of a gecko on to a vertical wall that enables acrobatic movements, the ability to reuse the adhesive surface, or colourful iridescence of butterfly wing without the deployment of a dye or a pigment; and (b) exploration and development of new fabrication methods for artificially creating surfaces that mimic the naturally occurring ones in terms of texture, morphology and functionality (Naik *et al.*, 2009; Li *et al.*, 2007; Fan *et al.*, 2009). One striking aspect of most biological surfaces is their inherent softness which explains the increasing use of 'soft' materials like polymers and gels and soft lithographies as good fabrication tools for creating surfaces that replicate biological functionalities. Three most remarkable functionalities are: superhydrophobicity, super adhesion with reusability, and structural optical effects.

SUPER WETTING AND SUPER HYDROPHOBICITY

Water is the single largest constituent of all living organisms and has played a crucial role in the evolutionary selection and diversification of living organisms. A range of different designer structured surfaces have evolved that enable plants to perform photosynthesis by which they split water using sunlight, convert CO_2 into carbohydrates, and release breathable oxygen into the atmosphere. Structured surfaces have evolved for dealing with water and help to produce superhydrophilicity or superhydrophobicity effects. For instance, the surface of a superhydrophilic thorn of a cactus *Pelecyphora asseliformis* can condense and capture water vapour from dry air in arid lands. The back of a desert bettle (*Stenocara* sp.) has superhydrophilic bumps for condensing and capturing water from atmospheric vapour, and superhydrophobic channels which facilitate frictionless rolling down of the collected dew drops to the mouth of the creature (Parker and Lawrence, 2001). The silvery lining around fishing spiders (e.g., *Ancylometes bogotensis*) is an air envelope called **plastron**. . Many aquatic insects use a plastron for different purposes (Vogel, 2006). When they dive into water, the plastron is held in position by superhydrophobic hairy structures found on the body surface of these insects. The insects continue to breathe well inside the plastron. When concentration in the plastron depletes, oxygen dissolved in surrounding water diffuses into the plastron to maintain the concentration at a breathable level. The plastron also provides buoyancy whenever required. It also reduces viscous drag during movement.

A water strider (*Gerris remigis*) can perform a gravity defying walk on the surface of water, thanks to its superhydrophobic legs showing multiscaled design features. The self-cleaning leaf of *Colocasia esculenta* can carry a water drop without wetting – a phenomenon termed the **lotus leaf effect**. Similarly, in the **rose petal effect**, the surface texture allows both superhydrophobicity as well as adhesion. A water drop rests on the petal in the shape of a pearl, but does not fall down even when the petal is inverted.

We now have some understanding of the wettability phenomenon underpinning the above wonders of nature. Wettability of a chemically homogeneous, molecularly smooth, flat rigid surface by a pure liquid depends on such parameters as the surface tension of the liquid, surface tension of the solid substrate material, and the interfacial tension between the solid and the liquid. In principle, there can be two distinct wettability regimes: partial wetting (drop formation takes place on the substrate) and complete wetting (the liquid spreads fully on the solid surface).

Various different hues, iridescence and luster effects are produced by physical structures, besides those produced by different pigments. An amazing variety of structures can be seen in the animal kingdom.

The brilliantly blue butterfly *Morpho rhetenor* bears scales on its wings that show cuticle ridges. Each ridge is made up of 10 to 12 fins made of chitin with refractive index $n \approx 1.56$. The resultant surface structure comprises alternating layers of chitin and air ($n = 1.0$), running parallel to the top surface, with a pitch of approximately 170 nm. It selectively reflects blue wavelengths in the visible light incident on the butterfly. The changes in hue of the reflection depend on the viewing angle, which creates an elegant iridescence effect. The brilliant blue colour can be seen by the potential mates of the butterfly from a long distance of over half a kilometre, and enables species recognition besides enhancing mating chances (Naik *et al.*, 2009).

A *Cyphochilus* sp. bettle is commonly found in south-east Asia. Its 5 μm thick scales accommodate a high density of scattering centres formed by a network of cuticular filaments with diameter of 250 nm with volume occupancy of nearly 70%, yet ensures that there is no optical crowding and there is no periodicity among the scattering centres. Light incident on such scales is almost completely back scattered in all directions at all wavelengths of visible light. The super whiteness so created allows the beetles to camouflage against their white background.

Many different physical mechanisms contribute to structural colours. Models based on mutual interference among fractions of rays of light reflected from parallel boundaries formed by a multilayer structure of alternatively stacked transparent materials of two different refractive indices provide a good starting point for understanding the phenomenon and exploiting it (Land, 1972). The parallel plate treatment is applicable to even layered particle arrays by determining and employing effective average refractive index for different layers, so long as the particle diameters are significantly smaller than the wavelength.

Antireflection

Antireflective coatings have several practical applications such as reduction of glare from spectacles or camera lenses, enhancing light output from LEDs, and better harvesting of solar energy. The earliest successful technologies involved coating layers at the interface between two dielectric media with thickness equal to $\lambda/4$. If the refractive index of the coating is a geometric mean of the refractive indices of the two adjacent media, then reflection loss at the interface is minimal. This principle can be extended for creating highly effective multilayer coatings (Rabinovitch and Pagis, 1975; Saitoh *et al.*, 1985). Unfortunately, the applications have been limited because of constraints in material, selection/processing, and high costs. A better option is creating a 'graded index' surface like the one found on the moth's eye, to obtain antireflection effect on any material surface and of any reasonable form (Boden and Bagnall, 2008; Xu *et al.*, 2008; Sun *et al.*, 2008).

SUPER-ADHESION

Over millennia of evolutionary experimentation, nature has learnt that the adhesive property of a surface can be modified by surface structures and patterns. Ranging from ants to geckos, the adhesive pads of many climbing animals can be highly patterned. These patterns and structures are hierarchical. Their length scale varies from a millimetre to a few nanometres. It is these complex structures that enable the animal to climb over vertical surfaces of varying degrees of roughness and surface properties. Not only is the adhesion so strong, it is also easy to detach, reusable, and self-cleaning and leaves no marks. All these qualities are indispensable for rapid, effortless locomotion of the animal. These remarkable qualities of the insects and geckos have fascinated and inspired researchers for long. None of man-made adhesives combines all of these qualities, be it our daily-use adhesive tape or acrylic based reactive glues. Synthetic adhesives can be quite strong but are rarely reusable and are prone to particulate contaminations. The reason is that some overall adhesive strength of man-made adhesives comes from viscoelastic dissipation and plastic deformations (Chung and Chaudhury, 2005). Such dissipation of energy strengthens the adhesive but sacrifices some clean separation and reusability. Once deformed, they never revert to their original shape. On the other hand, natural adhesives rely on van der Waal's force of attraction which is non-dissipative and typically depends on the geometry of the contacting objects and their separation distance. This is why, by conforming to the adhering substrate closely, the natural adhesive pads achieve very high yet reversible adhesion. The close conformity to the adhering surface regardless of its roughness is realized by hierarchical surface patterns of varying length scales (Naik *et al.*, 2009). As these adhesive pads do not typically use any gummy secretion, they are termed 'dry adhesive'. Such a dry adhesive is particularly useful in microelectromechanical systems (MEMS) and microelectronics for integrating micro and nano devices into complete device architecture. Automotive, biomedical and robotics are a few other fields where a biomimetic dry adhesive may find gainful use.

Inspired by the structure and mechanism of 'dry adhesives', several different designs of bio-inspired adhesive have been described. Some tested structures vary from a simple incision/cut to a very complex MWCNT (multiwalled carbon nanotube) based hairy one. Autumn *et al.* (2002) pioneered fabrication of synthetic gecko-like adhesive hair. They prepared conical shaped **nano-hairs** with tip diameter ~230 to 440 nm by micromoulding PDMS and polyester. With a flat AFM probe, the perpendicular adhesive force measured for PDMS and polyester, compared well with that of a biological hair. The first bio-inspired adhesive patch was made by Geim *et al.* (2003) by micropatterning a thin polyimide film (area 1 cm^2) and termed 'gecko tape'. An array of micropillars of height 2 μm, diam. ~500 nm, and a periodicity of 1.6 μm was made and proved that patterning a surface enhances its adhesive strength. While the tape with a soft backing showed a virtual threefold adhesive strength, a similar but smooth polyimide film showed only a small adhesion. Since then, other gecko-inspired adhesives of different shapes and sizes have been manufactured. Some different surface patterns studied include multiple incisions and chocolate bar pattern (Chung and Chaudhury, 2005), micro-fibrils and micro-columns (Hui *et al.*, 2004), fibrils with extended ends or mushroom-shaped columns (Gorb *et al.*, 2007), hierarchical columns (Greiner *et al.*, 2009), fibrils terminating into a thin film

(Glassmaker *et al.*, 2007) and a complex CNT (carbon nanotube) based fibrillar structure (Ge *et al.*, 2007). All these vary greatly in their structures, but converge to a few basic design features of a hairy adhesive: high aspect scale diameter, maximum possible hair density, maximum stiffness, low surface energy and high tensile strength (Naik *et al.*, 2009). High aspect ratio and high hair density enhance the adhesion whereas high material stiffness and low surface energy prevent the slender hairs from bunching together.

Besides the fine and complex structures such as nano-hairs and nano-fibres, even a simple surface cut/incision can enhance adhesion greatly by arresting crack propagation. Whenever the propagating crack comes across a cut or discontinuity in a material, it stops there and further energy has to be supplied to re-initiate the crack. If the pattern dimension is sufficiently small, i.e. the cuts are close-packed, the fracture toughness increases.

Besides polymer fibres, MWCNTs are also being used as adhesive fibres. Yurdumakan *et al.* (2005) first described application of vertically aligned MWCNTs constructed on polymer base as synthetic gecko tape. They prepared MWCNTs of diam. 10-20 nm and length ~65 μm by chemical vapour deposition. The adhesion behaviour was tested with a scanning probe microscope. The MWCNT patch showed weak repulsive forces as the probe approached and high adhesion as the probe was retracted. The calculated minimum force per unit area ($1.6 \pm 0.5 \times 10^{-2}$ nN/nm^{-2}) was 200 times greater than the estimated adhesive force of a gecko's setae (12^{-4} nN/ nm^{-2}). For MWCNT, entanglement improved adhesion unlike polymer-based fibrillar adhesives. The adhesive performs equally well on a variety of adhering surfaces like glass, mica, and Teflon irrespective of their hydrophobicity. According to Ge *et al.* (2007), a multiscale structure of MWCNTs, where nm-scale CNTs are arranged to form ìm scale pattern, generates much greater adhesive force. A 1 cm^{-2} area of CNT pattern can support a shear force of 36 N, nearly 4 times higher than a gecko foot, 10 times higher than polymer pillars and 4 times higher than unpatterned CNT patches on silicon (Naik *et al.*, 2009). CNT based fibrillar adhesive is also self-cleaning and easy to detach in normal load – crucial qualities for a reusable adhesive.

INTELLIGENT GLUE

Artificial glues are mono-functional. Their material properties are so designed that the glue sticks one thing to another. In contrast, spiders' webs are coated with microscopic droplets of glue, but the properties of this adhesive were unclear. It has now been found that the glue's stretchiness underpins its role in catching flies. The 'glue' droplets which coat spiders' webs are multi-functional (Sahni *et al.*, 2010). Depending on the rate at which the webs are extended, the glue droplets act either as a viscous adhesive or as a rubber-like elastic solid (Lee, 2010).

Common examples of adhesive coatings include paint used to cover walls, cars or ships' hulls; gold, silver and other metals plated on jewellery; and dyes used for colouring fabrics and hair. These coatings have various different functions – paint prevents barnacles and mussels from attaching to ships' hulls, and gold- and silver-plated surfaces are anti-corrosive.

But natural adhesive materials do much more. Mussels secrete a material that forms into 'byssus threads'; it is by these that mussels can attach themselves to rocks, shells or fish skin. These threads must be highly elastic as otherwise they could snap when tides lash. Of course they also need some adhesiveness for anchoring the mussels in place. This additional adhesiveness is provided by a coating of a glue-like material. Further, the threads need to maintain their grip under water where most adhesives function poorly.

The work of Sahni *et al.* suggests that the coupling of adhesion with extension is a common design principle of natural adhesives. The droplets are made of a complex mixture of glycoproteins plus various viscous small molecules and salts.

Sahni *et al.* immobilized silk threads on a glass surface, then touched glue droplets to the threads with a tiny glass probe. The probe was then pulled away at a constant speed, and the force exerted on the probe's tip was measured as a function of distance from the droplet, until contact between the droplet and the probe broke. The force-distance response depended on the speed with which the probe was pulled away: rapid extensions of the glue made it highly viscous, whereas slower ones turned the glue into an elastic-solid-like material (Lee, 2010).

This behaviour matches with the dual functions the droplets perform in nature, viz., capturing prey and then retaining it. When a flying insect is captured in a web, the silk rapidly extends on impact. The glue droplets tend to become very viscous under these conditions and provide maximum adhesion to effectively capture the spider's prey.

After the prey has been captured, however, the movement generated by its attempts to escape slowly extend the web. In this situation, the glue droplets turn into a rubber-band-like material which prevents the insect from escaping. All this implies that the glue droplets secreted by spiders behave as an 'intelligent' adhesive whose properties change depending on the extension rate of the underlying silk. Importantly, no human ingenuity or technology has yet succeeded in producing such a smart adhesive material!

In the adhesive used by mussels, the above-mentioned properties arise from the presence in the glue molecules of a repeating pair of amino acids: (a) lysine, and (b) 3, 4-dihydroxy-L-phenylalanine (L-DOPA) (Waite and Tanzer, 1981). This DOPA is highly unusual. Molecules which contain extensively repeating L-DOPA-lysine motifs characteristically stick to several different surfaces by making various types of covalent and non-covalent bonds (Lee *et al.*, 2007).

The glue droplets also probably contain small molecules such as neurotransmitters, amino acids and peptides, as well as macromolecules such as glycoproteins (Vollrath *et al.*, 1990; Choresh *et al.*, 2009). Tsang *et al.* (2006) suggested that the sugars found in glycoproteins could be responsible for the stickiness of glues secreted by other organisms. For instance, N-acetyl-D-glucosamine seems to be important for the adhesion of the holdfast of the bacterium *Caulobacter crescentus*. We still do not know anything about molecular components which cause spider-glue droplets to form a rubbery solid.

FABRICATION OF DESIGNER STRUCTURED SURFACES

A rough surface seems to be essential for superhydrophobicity (Nishino *et al.*, 1999). The critical issue of enhancing the roughness must be tackled by creating patterns or textures on the surface. Practical strategies for achieving this are broadly classified into two categories: 'top down' and 'bottom up' techniques, as well as a combination of the two for some specific cases.

Moreover, multilength scale and hierarchical patterns like hairy fibres grown on top of lithographically created patterns also show extreme superhydrophobicity (Bhushan *et al.*, 2009, 2009a).

Top-down Approaches

Top-down approach refers to certain fabrication techniques involving micromachining, moulding and templating or etching for patterning surfaces. For fabrication of superhydrophobic surfaces, different top down methods like micromachining, various types of lithographies, moulding, plasma and other types of surface etching have been practised. The top-down approach is further subdivided into two categories: (a) direct replication of biological features on to a material and (b) production of patterns or textures on surfaces that replicate the desired functionality of superhydrophobicity.

The simplest fabrication technique involves exact replication of the surface structures present on various types of water repellent plant leaves by moulding and templating. These replications are chiefly achieved in mouldable polymers, the most popular of which are cross linkable PDMS (polydimethylsiloxane) elastomer (e.g., Sylgard 184 from Dow Chemicals), materials that find application in many soft lithographic techniques. Two techniques used to replicate the leaf structures are nanocasting and replica moulding. A mouldable, freely flowing polymer in liquid state is cast on to the surface to be replicated and allowed to crosslink by thermal annealing, UV exposure or other means. The crosslinked and solidified (often flexible) polymer contains the negative replica and is then used as the template for making positive or identical replicas of the original surface by subsequent moulding or casting on it (Naik *et al.*, 2009).

Some techniques for fabricating superhydrophobic surfaces are not based on direct replication of a naturally occurring superhydrophobic surface (like a plant leaf); they are instead based on the natural design concept of roughness induced hydrophobicity. Various surface patterning techniques such as micromachining, lithography and plasma etching can be used for the fabrication of such surfaces. A good example of micromachining techniques is dicing on silicon surfaces followed by hydrophobization and silanization (Li *et al.*, 2007). In lithographic techniques patterns generated by both photolithographic and soft lithographic methods have shown superhydrophobicity. Photolithography based methods employ irradiation of UV light on a photo-curable polymer layer (photoresist) through a mask containing the desired features followed by certain etching steps to produce the surface patterns. Soft lithography is a patterning technique that depends on capillary driven flow of a softened or swollen liquid polymer layer under a confining mould or a stamp to produce a negative replica of the stamp pattern, with or without the application of external pressure (Naik *et al.*, 2009). Major soft lithography techniques used for fabricating superhydrophobic surfaces include nanoimprint lithography, replica moulding, capillary force lithography, micromoulding in capillaries (MIMIC) nanocasting and microtransfer moulding on a variety of polymeric materials like polysiloxanenes, poly-ethylene glycol (PEG), polystyrene (PS), poly-methyl methacrylate (PMMA), and polycarbonate (PC).

Bottom-up Fabrication

These techniques build large structures by integration of smaller building blocks (molecules, particles) or components, either spontaneously by self-assembly or mediated by an externally applied field-guided self-assembly (Martines *et al.*, 2005). Various intermediate, metastable structures can be created by suitably controlling the processing conditions. Some bottom-up techniques adopted for the fabrication of superhydrophobic surfaces include chemical deposition methods like chemical bath deposition (CBD), chemical vapour deposition (CVD), electrochemical deposition, layer-by-layer (LBL) deposition via electrostatic assembly, colloidal assembly and sol gel methods. In chemical deposition techniques, the products of a chemical reaction spontaneously assemble and deposit on the substrate surface, typically in a liquid phase. Thin films of such inorganic species as ZnS, CuSe, and CdS can be made and deposited by this technique. The reaction conditions are controlled in such a way that the deposited compound typically forms aggregates with fine and sharp morphological features, resulting in greater roughness and hydrophobicity of the surface. Needle like, hexagonal ZnO nanorods have been deposited perpendicular to the surface on a glass slide. CVD-induced aligned deposition of CNTs or electrochemical deposition of cadmium, gold or silver on a flat surface have also resulted in superhydrophobic surfaces. This method offers control of film thickness down to molecular processions. It is based on electrostatic charge interactions between the different polyanion and polycation layers. Such films usually require hydrophobization as polyelectrolytes are hydrophilic and typically, nanoparticles are embedded into the system to enhance roughness. Flourination of the polyanions has also been resorted to in some cases to enhance hydrophobicity. Another wet chemistry (sol gel) route has also been widely used for making superhydrophobic surfaces. It enables a material to be deposited by hydrolysis of corresponding oxide in the presence of a solvent or a combination of suitable solvents (sol). Network (gel film) formation occurs during the condensation or gelation, which is then thermally annealed to form the desired material with controlled microstructure. Hydrophobicity in a sol-gel film is typically adjusted by imparting surface roughness, which in turn can be tuned by controlling the hydrolysis and condensation reactions at various stages, along with surface modification by introducing a monolayer resulting from surface condensation reaction (Naik *et al.*, 2009). Optically transparent superhydrophobic silica-based films have found commercial applications in self-cleaning windows (Li *et al.*, 2007).

Surfaces containing self-organized arrays of monodispersed colloidal particles also produce strong superhydrophobicity, as the particle assembly makes the surface rough. Such arrays, which generally possess hexagonal closed pack morphology in absence of any templating strategy, are formed depending on the competition of various interaction forces: steric repulsion between the particles, van der Waal's forces, evaporation of the solvent in which the particles were dispersed and any externally applied fields like centripetal force in spin coating and electric fields (Shiu *et al.*, 2004). This cost-effective approach does not require lithographic techniques; sometimes the method itself is referred to as **nanosphere lithography** (Shiu *et al.*, 2004).

DYNAMIC SUPERLUBRICITY

In the coming years, data storage may depend on the 'write-read' action of a multitude of tiny silicon tips. This is likely to be based on the idea of dynamic superlubricity (see Erdemir and Martin, 2007) which offers a means to avoid the wear that would otherwise cripple them. We cannot imagine a world without abrasive wear – a world that researchers are working towards at least at the nanoscale. Some progress has already been made in creating a wear-free state (Lantz *et al.*, 2009). These authors work on data storage, specifically in thermomechanical systems that involve the use of atomic force microscopes (AFMs) and thin polymer films. Ultrahigh densities of data storage have been achieved in the form of holes in the polymer film created by an array of heated AFM probes. In the reverse 'read' procedure, the data are retrieved by the probes. The silicon tips of the probes need to remain sharp for good system performance, but the problem is that they don't. To cross this hurdle, researchers are seeking ways to extend tip lifetime; one such method is to reduce the wear to which tips are exposed (Gnecco, 2009).

Lantz *et al.* have successfully dragged a silicon tip across a polymer surface for about a kilometre without damaging either the tip or the surface. Here the distance travelled and the tip size differ by eleven orders of magnitude, analogous to a hypothetical case of someone climbing a ladder from the Earth to the Mars without wearing out the soles of ones shoes. Lantz *et al.* estimated that the wear rate in their experiment was seven orders of magnitude lower than what is usually observed on oil-lubricated steel (or five orders lower than on diamond-like carbon film coatings).

To understand the mechanism underlying the feat achieved by Lantz *et al.*, we need to refer to Socoliuc *et al.* (2006) which stated that atomic-scale friction is greatly reduced when the contact between a sharp tip and a solid surface is modulated with superimposed vibrations at well-defined frequencies. While sliding on the surface, the tip has to cross a series of energy barriers, involving abrupt jumps alternating with short periods of relaxation into minima of the surface potential (so-called stick-slip) (Schirmeisen *et al.*, 2005). Crossing a barrier imposes some cost, and it can cause some irreversible damage to a tip if the normal load is not controlled. Manipulating the contact region by inducing vibrations to it addresses the problem by lowering the amplitude of the energy barriers periodically so as to allow motion with negligible friction. This elegant concept, 'dynamic superlubricity', has attracted applications in scanning probe microscopy, and several crystal lattices have already been resolved by AFMs having vibrating contact tips (Gnecco *et al.*, 2009a).

The tips that Lantz *et al.* used form parts of the elaborate thermal probes that can locally heat polymer surfaces above the glass-transition temperature (usually around 400°C), producing tiny holes in the polymer. Each hole can be subsequently located with the same tip and considered as a bit of information by using arrays with numerous nanoprobes; data rates of the order of a few megabytes per second can be achieved, with areal densities of 150 gigabites per cm^2 and very low power consumption (Knoll *et al.*, 2006).

IBM has exploited these arrays as efficient data-storage devices by developing a device termed "Millipede", in view of its many 'legs' in the form of very fine silicon tips. Lantz *et al.* estimated that, under the best-imaginable conditions, the wear rate affecting a Millipede device can still be four orders of magnitude higher than the maximum tolerable value for proper functioning, but this is not the case if the Millipede legs are periodically vibrated while 'walking' by applying an a.c. voltage between the surface and a platform close to the probing tip. The consequence is that the vibrating tip moves smoothly across the surface without losing contact and without losing the possibility of heating the polymer to produce a hole.

Exploitation of the principle of dynamic superlubricity could even be extended to other micro- and nanoelectromechanical systems having tiny contacting parts and, more generally, to lithography and manufacturing processes performed on small scale.

COMBINATION OF TOP-DOWN AND BOTTOM-UP APPROACHES

One popular method for fabricating superhydrophobic surfaces is **electrospinning**; it is normally used for making polymer **nanofibers** (Ma *et al.*, 2005). An electrical bias is introduced between the extrusion and the grounded collection plate to draw the fibre. Electrospun fibres, which are continuous and nonwoven, produce a web of fibres on the surface, resulting in a mat-like rough surface superhydrophobic coating. Low surface energy coating on the fibres and CVD along with electrospinning further improve the hydrophobicity. Controlling the rheology of the polymer melt allows one to tune the morphology of the electrospun film from beads to continuous fibres. The variation in morphology also affects the final wetting behaviour as beads are more hydrophobic than the fibres.

Auxetic Materials

Most materials stretched in one dimension tend to contract in the other two. Auxetic materials, in contrast, expand in a lateral direction when stretched, or correspondingly shrink in a lateral direction when compressed. A 2D rubber sheet punctuated with an array of circular holes has been reported to show auxetic behaviour only under compression (see *Adv. Mater*. 21: 10.1002/adma.200901956, 2009). The holes deform into orthogonal ellipses, as there is insufficient space for all of them to collapse in the same direction. A minimum void fraction of 0.34 is needed in order to obtain auxetic behaviour. Below this value, the compression leads to a macroscopic instability whereby an entire row of holes collapses to accommodate the strain, leaving the rest of the sample mostly undeformed. As the void density also influences the critical strain, it may be possible through engineering the specific void pattern to make a highly tunable compressively auxetic material.

METAMATERIALS

Metamaterials are periodically structured composites in which unit cells are smaller than the wavelength used to interrogate them (Engheta and Ziolkowski, 2006). The unit cells can be tailored to create a wide range of responses. This ability has led to technology that enables computing with light, invisibility cloaks, or superlenses (Stebe *et al.*, 2009). Such materials can be easily created for use at radio frequencies, where the subunits are only a few millimetres large. But use at optical and infrared wavelengths necessitates the assembly of three-dimensional micrometer- and nanometer-scale structures – an extremely challenging task.

Recent studies of particle self-assembly have opened the way to metamaterials at relevant length scales.

Potential uses of assembled colloidal particles range from self-assembled electronics to biosensors. Particle self-assemblies may be a simple route to production of metamaterials at infrared and visible length scales, and these metamaterials could possibly be used to create cloaking devices or light-based circuits based on manipulations of local optical electric fields rather than on the flow of electrons (Engheta, 2007).

Metamaterials have certain inclusions embedded in host media. The size, shape, and electromagnetic properties of the inclusions, inclusion density, arrangement, and alignment all determine expression of their properties in any specific host. Valentine *et al.* (2008) described negative index of refraction metamaterials in the optical range; these were made by serial approaches based on lithography. But if metamaterials are designed to incorporate anisotropic nanoparticles through self-assembly, complicated lithographic approaches are avoided (Stebe *et al.*, 2009).

The existence of an extensive library of anisotropic microparticles and nanoparticles allows inclusion size and shape to be readily selected (Glotzer and Solomon, 2007). It is possible to synthesize particles from many different materials selected for their electromagnetic properties. Unfortunately, incorporating the particles as inclusions in self-assembled metamaterials strongly depends on using assembly techniques that can control particle orientation and spatial arrangement in periodic structures.

Convective assembly may be a suitable technique for creating close-packed assemblies of particles, especially for creating fairly large, defect-free periodically ordered structures of spherical particles. Novel assembly methods are sought to achieve oriented assembly of anisotropically shaped particles in close-packed structures and to control deposition of particles in specific spatial locations on substrates, with potential for creating non-close-packed structures.

Convective assembly involves immersing a substrate in a colloidal suspension. Here the solvent creates a three-phase contact line where solvent, substrate, and vapour meet. As the solvent evaporates, colloidal particles collect at the contact line. Particles that protrude through the liquid-vapour interface generate excess surface area and hence excess surface energy. This energy weakens as particles approach so that some capillary attraction that draws particles together to form an ordered structure (the nucleus) is created. As a result of continued evaporation, particles collect near the nucleus, where they assemble into the growing colloidal crystal (Denkov *et al.*, 1992). Colloidal crystals also arise from evaporating drops of suspensions through what is termed "drop casting" (Stebe *et al.*, 2009).

Ming *et al.* (2008) exploited these techniques to produce close-packed assemblies of **anisotropic particles (nanorods, polyhedra, nanocubes, and bipyramids)** into highly ordered 3-D structures. Work is underway to identify the conditions conducive to such oriented assembly. It is already clear that for this purpose two critical issues must be addressed.

First, the particle orientation in colloidal crystal nuclei must be rigorously controlled, possibly by using shape-dependent capillary interactions.

Second, appropriate means to achieve oriented assembly of particles in solution have to be developed. Conceivably, elongated nanoparticles might readily form oriented liquid crystalline phases driven by the trade-off between translational and orientational entropy (see John *et al.*, 2008). In fact, Rycenga *et al.* (2008) and Zhao *et al.* (2007) have described tuning of particle orientation by tailoring surface properties.

It is also now possible to deposit microscale and nanoscale particles in desired locations from evaporating suspensions to form non-close-packed structures. This should prove useful for metamaterials with unit cells larger than the inclusion – Aizenberg *et al.* (2000) used lithographically defined attractive and repulsive regions on the substrate. When a drop of colloidal suspension is allowed to evaporate on these surfaces, particles readily deposit on the attractive regions, and are equally readily drawn into ordered structures by capillary attraction. In another technique (Yin *et al.*, 2001), particles near the contact line can be swept into grooves that are cut into solid substrates, as the contact line recedes. It is within these grooves that the particles assemble into aggregates when solvent evaporates.

Intense work is underway to design metamaterials whose surface can absorb every photon that hits it. Researchers at Boston College in Massachusetts and at Duke University in Durham, North Carolina, have already produced structures that provide a proof of principle of these perfectly absorbing surfaces. Unfortunately, it has not so far been possible to stop all photons from escaping (Landy *et al.*, 2008). These materials are likely to have applications in ultra-sensitive imaging technologies that use terahertz radiation, with wavelengths intermediate between infrared and microwaves. Terahertz imaging is a useful technique for medicine – fully absorbing materials at these wavelengths might help reveal even the dimmest of features (Ball, 2008).

Terahertz frequencies are notoriously difficult to generate and difficult to detect. But if we use this technology, many exciting applications can be enabled. Metamaterials are formed from arrays of 'designer atoms', each of which is itself a device or structure with special designed properties. The building blocks of most metamaterials have been designed to interact with electromagnetic radiation in unusual ways, giving the assembled metamaterial an optical behaviour that ordinary materials can never achieve. For example, a metamaterial might have a negative refractive index, bending refracted light the 'wrong way'. Superlenses made from metamaterials are more powerful than conventional ones; and they have 'invisibility shields' which bend electromagnetic waves smoothly around hidden objects.

The basic 'atoms' of these structures are C-shaped split rings of patterned metal films. Radiation with a wavelength comparable to the size of the rings stirs up resonant electric currents in the metal, causing the radiation to be absorbed. For metamaterials operating at microwave frequencies, the components are fairly easy to make and are also large enough to be visible.

Landy *et al.* developed a new design for their perfect absorber; it consists of two back-to-back split rings. Their array of these elements absorbs about 95% of incident radiation at the resonant microwave frequency of about 11.5 gigahertz, but more careful fabrication of the split-rings to minimize or avoid errors may produce better performance. Indeed an improved design has been made that, in theory, offers 99.99% absorbance. We badly need near-perfect light absorbers for detectors called bolometres, which sense photons by converting their energy into heat. Bolometres made from silicon are used for infrared sensing, but they need to be cryogenically cooled and only work in a small range of wavelengths. The new metamaterial absorbers work at room temperature, and the wavelength response can be adjusted merely by varying the size and shape of the components. Although such an absorber would still be perfect only within a small range of wavelengths, the range could be broadened, to make the equivalent of a 'perfectly black' surface that absorbs fully over the entire visible spectrum by mixing different-sized split-ring components in the array. Another possibility is that a perfect absorber picks out specific 'colours', that are useful for combining imaging and spectroscopy.

Tao *et al.* (2008) have successfully scaled down their design to work at terahertz and far-infrared frequencies. But making a perfect black in the visible range will be hard because the components would need to be extremely small. In this context, Yang *et al.* (2008) have reported a super-black material made from carbon nanotubes in which the high absorbance comes from the roughness of the surface rather than the antenna action of metamaterials.

SELF-ASSEMBLY OF BINARY NANOCRYSTAL SUPERLATTICE MEMBRANES

The spontaneous organization into superlattices of multicomponent micrometer-sized colloids or nanocrystals is crucial for understanding the assembly process on the nanometer scale. It attracts great interest for bottom-up fabrication of functional devices. Co-assembly of two types of nanocrystal into **binary nanocrystal superlattices (BNSLs)** is now a hot area: it provides a low-cost, programmable way to design metamaterials with precisely controlled properties arising from the organization and interactions of the constituent nanocrystal components (Talapin, 2008). The ability to grow and manipulate large-scale BNSLs is very important for critical exploration of this new class of material. Currently available methods of growing BNSLs depend on the evaporation of a binary nanocrystal solution on a tilted solid substrate (Chen *et al.*, 2007). In this method, BNSL formation occurs when high-boiling-point solvents such as tetrachloroethylene are evaporated and the assembly process typically takes several hours to complete. Further, the substrate wetting properties affect BNSL formation. Till hitherto, growth of BNSLs was limited to the use of hydrophobic substrates such as carbon-coated transmission electron microscopy grids. This made it quite difficult to integrate BNSLs on a desired substrate for device applications. Extended BNSL growth also depends on careful regulation of the solvent evaporation kinetics. To obtain BNSLs with relatively large domain sizes, the assembly process is usually performed under reduced pressure at elevated temperatures (45-75 C). Even under optimized conditions, the as-grown BNSLs usually exist as isolated, micrometre-sized islands scattered irregularly on the substrate, leading to a low surface coverage (Overgaag *et al.*, 2008). Dong *et al.* (2010) described a general method of growing centimetre-scale, uniform membranes of BNSLs which can readily be transferred to arbitrary substrates. Their method is based on the liquid-air interfacial assembly of multicomponent nanocrystals. The method exploits drying mediated self-assembly of multicomponent nanocrystals on an immiscible liquid surface, allowing the growth of macroscopic BNSL membranes with no substrate restrictions under ambient conditions. It avoids the limitations associated with current assembly strategies and enables integration of BNSLs on any substrate for the fabrication of nanocrystal-based devices (Urban *et al.*, 2007). The creation of magnetoresistive devices was successfully demonstrated by incorporating large-area (1.5 mm × 2.5 mm) BNSL membranes whose magnetotransport measurements showed that device magnetoresistance depends on the structure of the BNSLs. The ability to transfer BNSLs also allows the construction of free-standing membranes and other complex architectures that were not accessible earlier (Dong *et al.*, 2010).

METAMATERIALS FOR LENSES

The refractive index of water is higher than that of air. All naturally occurring materials have a positive refractive index. An unusual property of metamaterials is that they bend light; metamaterials are artificially engineered structures with a negative refractive index. One exciting feature of metamaterials is that they may be engineered to 'cloak' objects from electromagnetic radiation such as light – that is, they could be made invisible (Pendry *et al.*, 2006). Now, Valentine *et al.* (2008) have reported a three-dimensional 'fishnet' metamaterial made of alternating layers of silver and magnesium fluoride. This material shows a negative refractive index across a broad region of the electromagnetic spectrum. In recent years, negative refractive index has attracted considerable attention following the proposal made by Pendry (2000) that a material with a negative refractive index would make a 'perfect' lens (see Shalaev, 2007). However, optical negative-index metamaterials have effectively been two-dimensional surfaces, not 3-D.

Hoffman *et al.* (2007) and Yao *et al.* (2008) have manufactured semiconductor metamaterials and composite nanowire materials which show negative refraction but do not have a bulk negative refractive index.

Negative refraction can occur in materials displaying anisotropy, such as calcite, and in two- or three-dimensionally structured materials such as photonic crystals (Cubukcu *et al.*, 2003).

In the material discussed by Valentine *et al.*, use of nano-structuring (tightly coupled nano-plasmonic resonators) simultaneously increased the bandwidth and effectively shifted the losses to a frequency range where the optics is not impaired. Further, this metamaterial has a negative refractive index over a remarkably wide window of the spectrum, ranging from around 1800 nm to less than 1500 nm. One of the first uses for a broadband, low-loss, three-dimensional optical material with a negative refractive index would be to be able to construct Pendry's perfect lens, and thus imaging a 3-D object the size of a molecule (Hess, 2008). Second, it would be beneficial for electromagnetic cloaking and its applications, in that a cloaking material that absorbs less radiation should prevent the cloaked object from appearing as a dark silhouette. Third, the ability to slow down and store broadband light using the 'trapped rainbow' principle has also been virtually realized (Tsakmakidis *et al.*, 2007).

PATTERNING METALS FOR PLASMONICS AND METAMATERIALS

Plasmonic devices make use of electromagnetic waves known as surface plasmons that propagate along a metal interface (Raether, 1988). These waves involve a mixture of both light and charge fluctuations on the metal surface and so have a unique hybrid character. They can exist at metal interfaces because of coupling between light and free electron (Genet and Ebbesen, 2007). Whereas photonic component allows them to interact with optical waves, their electronic component makes it possible for these optical waves to be concentrated in volumes much smaller than the diffraction limit. This combination proves to be useful for applications ranging from biosensing to solar cells. In a typical device, surface plasmons can be created by illuminating a film of silver or gold on its exposed surface. Though this excitation process cannot occur when the interface is perfectly flat, surface patterning on a suboptical length scale allows incident light to create surface plasmons in the film. Patterning also enables manipulating these plasmons after they have been created. Surface structures can direct, channel, or focus the surface plasmons toward specific locations. But since these plasmons are present very close to the interface, they are extremely vulnerable to surface inhomogeneities which cause absorption and scattering, and limit propagation. This is why fabrication of plasmonic devices requires that metal films be patterned while avoiding unwanted roughness and impurities that can seriously undermine their performance (Nagpal *et al.*, 2009).

Even though surface plasmon waves can be channelled, concentrated, or otherwise manipulated by surface patterning, surface roughness and other inhomogeneities have so far limited their propagation in real plasmonic devices, necessitating the use of simple high-throughput methods to fabricate high-quality patterned metals. This prompted Nagpal *et al.* (2009) to combine template stripping with precisely patterned silicon substrates to obtain ultrasmooth pure metal films with grooves, bumps, pyramids, ridges, and holes. Measured surface-plasmon-propagation lengths on the resulting surfaces showed that perfectly flat films are produced in this way. Nagpal *et al.* also demonstrated structures that exhibit Raman scattering enhancements above 10^7 for sensing applications and multilayer films for optical metamaterials.

The new field of plasmonics certainly needs simple, high-throughput, and reproducible approaches to obtain pure metallic films that are smooth yet patterned over large areas. Techniques such as nanoimprinting and nanomolding can pattern metals on the proper length scales (Kumar *et al.*, 2009). These techniques have not proved suitable; in most techniques, patterned polymeric molds are filled with metal to form a replica, a process that introduces marked surface roughness because metals do not easily wet polymer interfaces. This problem can be avoided by substituting other mold materials but then molds must be etched away to release the metal structure, reducing throughput and reproducibility because each mold generates only one device.

The above problem can be addressed by exploiting an alternative technique to generate smooth unpatterned metallic films. This method is known as template stripping. It uses the poor adhesion and good wettability of noble metals on solids such as mica, glass, and silicon (Hegner *et al.*, 1993). Freshly cleaved mica is coated with a film of gold, and the exposed surface of gold is attached to another substrate with an epoxy adhesive. Upon separating the mica and substrate, the gold clings to the epoxy, and a metal interface that has formed on the mica becomes exposed. Since mica can be extremely smooth, ultraflat gold surfaces are readily generated. In fact, these are being used widely in scanning probe experiments and studies of self-assembled monolayers.

Unfortunately, the same approach could not produce smooth patterned metal films. Whereas template stripping did form patterns on mica, very complex multi-step processes had to be used (Jung and Frey, 2008). Nagpal *et al.* used inexpensive silicon wafers instead of mica as the template to exploit the well-developed fabrication techniques for microelectronics. The wafer is patterned, e.g., with lithography and is coated with a thin metal film and a layer of epoxy. The epoxy-metal bilayer is then peeled off the substrate, revealing a patterned structure with a surface roughness determined by the wafer template. They thermally evaporated 275 nm of silver on this substrate, added epoxy, and peeled off the bilayer. A silver "bull's eye" structure resulted.

Besides silver bull's eyes, a variety of silver, copper, and gold plasmonic structures have also been fabricated. From the same patterned substrate more than 30 films could be created without damage (Nagpal *et al.*, 2009).

Whereas bumps, ridges, and grooves are certainly useful for some applications, different structural elements are required for others. A nanoscale hole surrounded by surface structures can allow enhanced transmission of electromagnetic waves through a metal film (Genet and Ebbesen, 2007). It is possible to create such holes in thin smooth silver films. A hexagonal array of deep circular pits was successfully formed on a silicon wafer with nanosphere lithography. By evaporating silver on top of such pits and adding epoxy, a thin silver film with a hole array could be made. Although only one side of this film is ultrasmooth, which would affect its transmission properties, an additional absorbing layer such as Cr can be added to the rough side to minimize its effects. A second useful element is the multilayer structure. As thin patterned films with alternating layers of metal and insulator show a negative refractive index, it leads to new optical phenomena (Smith *et al.*, 2004). The approach described by Nagpal *et al.* can also yield metamaterials. After evaporating a thin film of silver on patterned silicon templates, they deposited alternating layers of alumina and silver. As silver adheres to alumina better than silicon, the entire stack can be removed with epoxy. A bull's eye and a bump array could be easily obtained, both of which might be useful as superlenses. If the adhesion requirement is satisfied, materials other than alumina are also possible. The resulting films can lead to a variety of useful optoelectronic devices.

These results show that template stripping can be effectively combined with silicon microfabrication methods to create ultrasmooth patterned metals for plasmonics and metamaterials. Though simple, it creates a way to fabricating integrated metallic multilayer structures with designed grooves, bumps, tips, and holes with controlled spacing and orientations, while at the same time avoiding problems generated by roughness and impurities (Nagpal *et al.*, 2009).

SUBWAVELENGTH PLASMONIC OPTICS

Exciting recent researches (see Barnes *et al.*, 2003; Maier *et al.*, 2006; Ozbay, 2006) have been instrumental in scaling down the plasmonic waveguides from microscale metal stripes to nanoparticles arrays and nanowires (Sanders *et al.*, 2006; Dickson and Lyon, 2000). Subwavelength plasmonic optics have become feasible. Unfortunately, the inherent metal loss makes it difficult to transfer digital data across the entire photonic integrated circuit (IC) chip (~1 mm) solely with plasmonic waveguides. What is really needed is the ability to integrate plasmonic modules with low-loss, dielectric optical interconnects. For achieving such hybrid plasmonic-photonic circuit, there is need not only to develop nanoscale plasmonic waveguides with reduced losses and small mode volume, but also to ensure compatibility with conventional optical circuitry. Chemically-synthesized Ag nanowires are highly promising candidates for subwavelength plasmonic waveguides. Their high crystallinity and atomically smooth surface might bring down the total propagation loss as compared with those of mcirofabricated polycrystalline Ag waveguides. Consequently, Ag nanowires can support surface plasmon propagation with 2D subwavelength mode confinement. But integrating the low-loss Ag nanowire waveguides into a plasmonic-photonic routing network demands an efficient and versatile method to couple the light field in and out of the Ag nanowire so that the digital information can be launched and harvested (Yan *et al.*, 2009).

Metallic nanoscale structures can significantly support surface plasmon polaritons (SPPs) by propagating collective electron oscillations with tight spatial confinement at the metal surface. SPPs are most promising structures to cross the diffraction limit imposed by conventional dielectric optics. Ag nanowires have drawn increasing research attention due to 2D sub-100 nm mode confinement and lower losses as compared with fabricated metal structures.

The major challenge in optically excited SPPs in Ag nanowires comes from the dispersion relation mismatch for SPPs and photons. Some methods have been suggested to bridge this gap and launch propagating SPPs in Ag nanowires. One method is the total internal reflection (TIR) illumination method. It uses a prism to match the momentum of SPPs and the incident photons (Dickson and Lyon, 2000). Direct local excitation can be achieved by focusing a laser on the end facets of the Ag nanowires or on a metallic nanoparticle attaching to the surface of the Ag nanowire that acts as a scattering center (Knight *et al.*, 2007). However, existing coupling schemes do not elicit easy interconnection with conventional optical components for photonic IC fabrication. Although a recently reported polymer-Ag coupling method represents an important advance toward on-chip integration, it suffers from restrictions on the coupling conditions, including Ag nanowire orientation, position, and incident light field polarization (Pyayt *et al.*, 2008).

To address the above challenges, Yan *et al.* (2009) proposed a simple approach to couple SPPs into the Ag nanowires with an inorganic dielectric nano-waveguide, SnO_2 nanoribbon. They showed that SPPs can be excited merely by contacting a silver nanowire with a SnO_2 nanoribbon that serves both as an unpolarized light source and a dielectric waveguide. The efficient coupling enables measurement of the propagation-distance-dependent waveguide spectra and frequency-dependent propagation length on a single Ag nanowire. Single-crystalline SnO_2 nanoribbons are known to be efficient UV-near (N)IR waveguides (Law *et al.*, 2004). Their high aspect ratio, strength, and flexibility facilitate their easy manipulation on surfaces, and can be assembled into optical networks and components. SnO_2 nanoribbons also show photoluminescence with UV excitation and can act as a simultaneous nanoscale light source and waveguide. Yan *et al.* synthesized their nanowires by a modified polyol synthesis. The nanowires have fivefold twinning planes along their axis, and are bound by five atomically smooth single crystalline surfaces. It could be shown that Ag nanowire SPPs can be excited by simply contacting with a SnO_2 nanoribbon with minimum restriction on the coupling conditions. SPPs can be inserted into the Ag nanowire at arbitrary positions in any orientation. The assembly of the metal-dielectric junction being completely reversible, makes it possible to monitor the properties of a single Ag nanowire while varying the interconnection and propagation condition. This versatile coupling strategy could well be an important step toward realizing photonic-plasmonic hybrid circuitry; it has facilitated the systematic far-field study on the SPP propagation in Ag nanowires (Yan *et al.*, 2009).

MEMORY METAMATERIALS

Metamaterials can create electromagnetic responses not found in nature. This has initiated the new research field of transformation optics (Schurig *et al.*, 2007), with potential applications ranging from electromagnetic cloaking (Pendry *et al.*, 2006) to subdiffraction imaging (Liu *et al.*, 2007). Frequency-agile metamaterials that enable researchers to adjust the electromagnetic response in real time have become an important part of this field. Chen *et al.* (2008, 2008a) and Gollub *et al.* (2009) adopted a hybrid-metamaterial approach to integrating natural materials into the metamaterial composite. Their approach enabled developing frequency-agile metamaterials which respond to the application of voltage, external electric field, light, and heat (Chen *et al.*, 2007; Qazilbash *et al.*, 2008; Degiron *et al.*, 2007). Such a tuning ability makes metamaterial devices more versatile, adapts to shifting input or changes target parameters (Discoll *et al.*, 2009). But these methods for enabling frequency-agile metamaterials work only when a continuous application of an external stimulus maintains altered metamaterial properties. If the external stimulus is removed, the metamaterial reverts to its original response. In principle, any functionality emanating from metamaterials would benefit if the metamaterial tuning persists after the triggering stimulus disappears. Mechanically or geometrically these metamaterials should retain their tuned properties (Driscoll *et al.*, 2007, 2008), but these techniques are very difficult to use at higher frequencies or for complex designs. Recently, however, Driscoll *et al.* (2009) achieved electrically controlled memory effect by means of a hybrid device composed of a resonant metamaterial and a transition metal oxide. The principle underlying the persistent tuning of this hybrid device, memory-capacitance (memcapacitance), was suggested by Di Ventra *et al.* (2009).

The memory-oxide hybrid-metamaterial device consists of a gold split ring resonator (SRR) array lithographically fabricated on a 90 nm thin film of VO_2. Attaching electrodes allow in-plane current-voltage relation transport, and the device is mounted to a temperature-control stage. Simultaneous dc-transport and Far-infrared probing of the metamaterial demonstrated that as VO_2 passes through its insulator-to-metal transition, resistance drops and the SRR resonance frequency decreases. This latter effect manifests because of an increase in the per-SRR capacitance.

This work illustrates a kind of memory capacitance that elegantly interfaces metamaterials with a class of devices known collectively as memory devices.

PLASMON NANOLASERS

Scientists have successfully produced increasingly high-powered, faster and smaller coherent light sources (Altug *et al.*, 2006; Hill *et al.*, 2007; Zimmler *et al.*, 2008; Scalari *et al.*, 2009). Some recent examples are microscopic lasers that can reach the diffraction limit, based on photonic crystals, metal-clad cavities and nanowires. Nanolasers can be produced by exploiting the collective wave-like motion of free electrons on a metal surface. They have a tremendous potential impact on, for example the fabrication of ultra-fast photonic nanocircuits with unprecedented performance and the improvement of techniques such as nanolithography and single-molecule biochemical sensing. Nanolasers may even serve as key elements in quantum-information technologies. They would provide a powerful tool for study of strong interactions between light and matter. Noginov *et al.* (2009) and Oulton *et al.* (2009) have discussed recent experimental advances towards the implementation of functional nanolasers.

A laser differs from ordinary torch in emitting an intense, strongly collimated beam of coherent, monochromatic light. These properties emanate from the interaction between two of the laser's components, an optical cavity that sustains a photonic resonance and a gain medium which amplifies it (Garcia-Vidal and Moreno, 2009). Energy externally pumped into this amplifying medium is released as photons which, through a mechanism called stimulated emission and owing to the laser cavity's optical-feedback property, are all emitted coherently: they have the same frequency and are in step with one another.

The external input of energy involves an electric current (**electrical pumping**) or light illumination (**optical pumping**). The long-cherished desire of physicists to produce ultra-compact lasers of nanometre scale has been obstructed by the diffraction limit. By conventional means, light cannot be focused down to a spot smaller than about half its wavelength. For visible light (wavelengths between 300 to 800 nm), the smallest dimension of both the spot and the laser cavity is limited to only a few hundred nanometers. One possible way to address this limitation is by using a metallic cavity, which can confine light within smaller dimensions than the light's wavelength.

This ability arises from the existence of collective, wave-like motions of free electrons on a metal surface, termed **surface plasmons** (Pile *et al.*, 2005). These oscillations produce electromagnetic waves that are localized at the surface. As surface-plasmon oscillations can be coherently coupled to light outside the metallic cavity, they can be exploited to realize ultra-small lasers (Zheludev *et al.*, 2008).

The challenge facing production of a plasmon-based laser has been the (resistive) energy losses in the metal from which the cavity is made. Noginov *et al.* (2008), however, showed that the incorporation of an optical-gain medium into the system compensates for such losses. Hill *et al.* (2007) had also reported lasing from an electrically driven, metal-coated cavity with dimensions of 200-300 nm. Surface plasmons can tightly localize light, but ohmic losses at optical frequencies have prevented the production of truly nanoscale lasers based on such approaches. What Oulton *et al.* (2009) did is to experimentally demonstrate nanoscale plasmonic lasers that generate optical modes a hundred times smaller than the diffraction limit. They made such lasers by using a hybrid plasmonic waveguide consisting of a high-gain cadmium sulphide semiconductor nanowire, separated from a silver surface by a 5-nm-thick insulating gap. As plasmonic modes have no cutoff, Oulton *et al.* could demonstrate downscaling of the lateral dimensions of both the device and the optical mode.

The works of Noginov *et al.* (2008) and Oulton *et al.* (2009) represent a major advance towards the production of plasmon-based nanolasers. Noginov *et al.* encapsulated gold nanoparticles in silica spheres of 44 nm diameter. The optical gain was supplied by organic dye molecules embedded in the silica shells, which can be optically pumped; the surface plasmons were provided by the gold cores. The energy pumped into the shells is transferred to the surface plasmons. This stimulates the coherent emission and amplification of surface-plasmon waves which become converted into visible laser light of wavelength 531 nm and radiates outside the spheres.

In the laser developed by Oulton *et al.*, the emergent, 389-nm-wavelength laser light is emitted from a strongly confined spot within the gap region. The relevant dimension of the cavity, and hence the size of the laser spot, is of the order of 25 nm i.e., 20 times smaller than the laser's operating wavelength of 489 nm. The size of the laser spot is controlled by the width of the insulating gap and the diameter of the nanowire. In contrast to standard nanowire lasers (Duan *et al.*, 2003), there is no cut-off laser size below which the strongly confined surface-plasmon waves cease to exist. Possibly, even smaller plasmon-based lasers might be realized if the width of the gap can be reduced further.

Plasmonic lasers thus have the potential of exploring extreme interactions between light and matter; they could create new possibilities in active photonic circuits, bio-sensing and quantum information technology (see Akimov *et al.*, 2007; Anker *et al.*, 2008).

This successful demonstration of deep subwavelength plasmonic laser action at visible frequencies points to new sources that might produce coherent light much below the diffraction limit. Extremely strong mode confinement is a crucial aspect of the operation of deep subwavelength lasers. The advantage of plasmonic lasers comes from their ability to downscale the physical size of devices, as well as the optical modes they contain, unlike diffraction-limited lasers. Also, the use of metals in plasmonics could possibly be a natural route towards electrical injection schemes that do not interfere with mode confinement (Duan *et al.*, 2003). It is now emerging that the impact of plasmonic lasers on optoelectronics integration can be potentially very significant because the optical fields of these devices rival the smallest commercial transistor gate sizes and hence reconcile the length scales of electronics and optics (Oulton *et al.*, 2009).

STRUCTURE OF NON-CRYSTALLINE MATERIALS

Neutron diffraction and X-ray studies of several crystalline materials have revealed structures ranging from complex minerals to proteins and DNA (Miller *et al.*, 1993). But there exist many **aperiodic** (i.e., non-crystallized) materials, whose structures cannot be unravelled through these techniques (Rehr, 2006). Juhas *et al.* (2006) addressed this problem with a novel algorithm for reconstructing 3-D structures from 'pair distribution function' (PDF) data. As aperiodic materials represent some of the technologically most interesting nanoscale materials, this approach has potential for a wider application.

Several techniques available for determining the local, atomic-scale structure of materials, range from scanning tunneling microscopy (STM) to spectroscopic methods that use X-rays, such as extended X-ray absorption fine structure (EXAFS) analysis. STM does give beautiful images, but not in three dimensions. For structural information to be inferred from spectroscopic techniques such as EXAFS, an accurate theoretical model relating spectra to structure is essential (Rehr and Albers, 2000).

Some of these problems can be avoided by means of PDF analysis because it only requires data on the distribution of distances between atoms in a structure, and this information is easily obtained from X-ray or neutron-scattering experiments. PDF is not the method of choice for structure determination because of poor (low-resolution) data quality. But this limitation is now being removed by using modern neutron and synchrotron X-ray sources (Rehr, 2006).

Another crucial factor is that an algorithm needs to be found which can solve the "inverse problem"; that is, with a set of experimental data at hand, how to extract the 3-D structure that created it. Determining the structure corresponding to PDF data, the issue addressed by Juhas *et al.* is just such an inverse problem. This often needs considerable computational resources, as it requires the 'direct problem' – that is, a theoretical model for the experimental signal resulting from a trial structure – to be solved repeatedly in the process of finding the minimum (Rehr, 2006). A suitable analogy for solving the inverse problem is an optimization strategy for finding the global minimum of a quantity involving many variables from amongst a forest of possible minima. Some advances made in such strategies include the development of 'genetic' algorithms inspired by the rules of evolutionary biology, and 'simulated annealing' techniques that mimic the manner metals freeze into a state of minimum energy (Wales and Scheraga, 1999). In the case of X-ray and neutron crystallography, the iterative 'shake-and-bake' algorithm is based on the random perturbation of the positions of atoms in a crystal until the lowest-energy state is found. It has already shortened the time required for determining crystal structures from a few months to a few hours (Rehr, 2006).

MATERIAL MICROSTRUCTURE

In the 19th century, Henry Sorby had pioneered the use of optical microscopy for geological and metallurgical samples, leading in due course of time to the recognition that microstructure is fundamental to the properties of not only existing material but also of the next generation materials. Recent growth of the manufacturing sector has depended on this recognition.

Mechanical and physical properties of most materials are strongly controlled by events at the atomic scale. This is why a clear knowledge of atomic structure is essential. Till hitherto we could only see a projection of atoms, and there were many problems in interpretation. In the last few years, however, there has been a revolutionary advancement in our ability to probe atomic structure. R. Feynman, Nobel Laureate, predicted in 1959 that "It would be very easy to make an analysis of any complicated substance; all one would have to do would be to look at it and see where the atoms are… I put this out as a challenge: Is there no way to make the electron microscope more powerful?"

This challenge has, at last, been addressed, but one problem still remains – how does the structure at the atomic scale affect tonnages of material? Can we extrapolate? In recent years, W.M. Rainforth (Univ. of Sheffield, UK) and his associates have attempted to answer these questions through the application of novel techniques for elucidating the structure of nanoscale multilayer coatings, the complex structure of modern high strength steels, and of the surface of hot rolled aluminium.

EXTRA STRONG ALUMINIUM ALLOYS

Al-alloys of high specific strength constitute the most widely used material for structural applications in aviation and transportation industries. Amorphous or nanointermetallic microstructure could potentially yield 2-3 times enhanced compressive strength (~1550 MPa) as compared to the conventional precipitation hardened Al-alloys. A number of Al-rich Al-Cu-TM and Al-TM-Si alloys have been synthesized by mechanical alloying with nano-intermetallic dispersed amorphous matrix composite microstructure (Manna, 2006, 2009). Besides critical structural/compositional characterization by various modern spectrometric and other methods especially quantitative studies with positron annihilation spectroscopy (PAS) and nuclear magnetic resonance (NMR) have made possible the determination of the mechanism of partial/complete solid-state amorphization in $Al_{50}Ti_{40}Si_{10}$.

Critical plotting of profiles of the $Al_{50}Ti_{40}Si_{10}$ powder blend as a function of milling time showed that the final microstructure (at 30 h) is predominantly amorphous with nanocrystalline Al_3Ti. PAS measurements substantiated that amorphization is followed by intermetallic nanophase formation during the process of ball milling. Contrary to the expected trend of increase in positron lifetime with reduction of grain size to nanoscale dimensions, the two lifetimes determined by Manna (2006) gradually decreased and this reduction was attributed to the 2-phase amorphous + nanocrystalline composite microstructure instead of a 1-phase amorphous or nanocrystalline aggregate.

After 10 hr of milling, a sample of $Al_{50}Ti_{40}Si_{10}$ showed a spectrum of a broad structure at a lower frequency with two distinct peaks, indicating the occurrence of at least two different alloy products, both with Al as a component. After 30 h of milling, among the many Al containing alloy components that formed at intermediate stages of milling, only two were retained. The annealed (700°C, 1 h) sample had Al_3Ti and Ti_5Si_3 as the chief constituents. It seems that an amorphous extended solid solution is formed initially. Continued milling leads to the strain-induced nucleation of the stable coordination or environments of the constituent elements which are retained as either nano-intermetallic or amorphous phases as products until the final stages of alloying (Manna, 2006).

Thus, development of nanocrystalline intermetallic dispersed Al-rich amorphous matrix composite seems feasible by mechanical alloying of appropriate Al-Ti-Si powder blend. NMR results have identified certain characteristic new atomic coordinations around Al atoms for amorphous and nanocrystalline intermetallic phases.

NANO-DISPERSED SOLID ALLOYS

Some nano-scale materials have been synthesized by mechanical alloying. The size dependent phase transition and structural and functional applications of such nanomaterials have been investigated, revealing novel nanoscale dispersion phenomena in amorphous Al-alloys and thermal fluids.

Development of high specific strength-structural materials is crucial for the transportation and aviation industry. Several Al-based ternary Al-Cu-TM or Al-TM-Si amorphous alloys have been successfully developed by addition of transition metals (TM) by mechanical alloying. Nano-ceramic particles have been dispersed in these alloys with partial/complete amorphous matrix. Interfacial energy plays critical role in determining the final microstructure comprising nano-intermetallic and amorphous phases. After suitable compacting and sintering, warm compaction, microwave or spark plasma sintering of the composites improved their mechanical properties. This points to considerable potential in developing high specific strength Al-based amorphous and nano-intermetallic dispersed amorphous matrix composites by the fairly low-cost and versatile processing route of mechanical alloying (see Manna, 2009).

POLYMERIC TRIPLE-SHAPE MATERIALS

Shape-memory polymers (SMPs) are materials that can change from one shape to another in response to a stimulus such as heat. In dual-shape materials (e.g. a rubber band) that have found many applications (Lendlein and Kelch, 2002; Metcalfe, 2003), there is a temporary fixation of a second shape by deformation of the polymer network and subsequent cooling under stress. A rubber band is a polymer network that can be elastically deformed; it reverts back into its original shape when the external stress is released. Rubbery polymer networks have covalently cross-linked flexible polymer chains oriented from a coiled state during deformation. The original shape is recovered by regaining the entropy that was lost when chains were oriented. The shape of a polymer network is chiefly defined by its chemical cross-links or netpoints. Depending on the type of chain segments, different macroscopic domains are formed having individual transition temperatures (T_{trans}), like glass transition (T_g) and melting (T_m) temperatures (see Bellin *et al.*, 2006). When a polymer network is cooled below a T_{trans} of a specific domain, this domain solidifies and so forms physical cross-links. These physical cross-links dominate the netpoints, and a new shape is fixed. Bellin *et al.* have reported a triple-shape polymer that can change from a first shape (A) to a second shape (B) and from there to a third shape (C). Whereas shapes B and C are recalled by subsequent temperature increases, shapes A and B are fixed by physical cross-links and shape C is defined by covalent cross-links established during network formation. The triple-shape effect is a general concept that requires the application of a two-step programming process to suitable polymers. It can be realized for various polymer networks whose molecular structure allows formation of at least two separated domains providing marked physical cross-links. These domains act as the switches, that are used in the two-step programming process for temporarily fixing shapes A and B. Different combinations of shapes A and B for a polymer network in a given shape C can be obtained by adjusting specific parameters of the programming process (Bellin *et al.*, 2006).

Polymers having triple-shape functionality could potentially enable applications in diverse fields because their properties can be adjusted over a wide range (Bellin *et al.*, 2006).

SHAPE-SWITCHING POLYMER PARTICLES

Particle engineering for biomedical applications has highlighted the crucial importance of size, surface chemistry, and shape for modulating particle interactions with cells. Dynamic manipulation of such key attributes has attracted attention because of the great need of precisely controlling particle interaction with cells. Interactions (e.g. endocytosis and phagocytosis) of polymeric particles with cells and macrophages govern the efficacy of carriers used for drug delivery and medical imaging (Nel *et al.*, 2009). The fate of these interactions depends on optimal selection of particle properties such as surface chemistry, size, and shape (Mitragotri, 2009). Consequently, many reports on methods to synthesize materials with such precisely engineered functional attributes as size, surface chemistry, mechanical properties, and shape have appeared to promote or mitigate interactions with various cell types (Mitragotri and Lahann, 2009).

Researchers have sought dynamic control over critical particle properties with a view to achieving an interactive interface between the particles and the complex biological milieu. Griset *et al.* (2009) described stimulus-responsive type control over the size of particles which have been gainfully used for the triggered release of encapsulated drugs (see Yoo and Mitragotri, 2010). Lee *et al.* (2005) controlled the surface chemistry of polymeric micelles by considering the environmental pH of solid tumours so as to stimulate the cellular uptake and release of anticancer agents. Lahann *et al.* (2003) achieved dynamic control of surface properties through the use of electric fields. Surprisingly little attention has, however, been given to switching particle's shape.

To fill this gap, Yoo and Mitragotri (2010) described poly(lactide-co-glycolide) (PLGA) particles whose shape can be switched in real time from an elliptical disk to spherical in response to a stimulus such as temperature, pH, or a chemical. The time scale of switching could be controlled over a range of minutes to days, by suitable selection of polymer molecular weight, particle size, and the strength of the stimulus. These particles are able to modulate their interactions with macrophages through shape change.

The shape-switching behaviour was driven by a delicate balance between polymer viscosity and interfacial tension. The balance between the two forces could be modulated by application of an external stimulus chosen from temperature, pH, or chemical additives. The dynamics of shape switch could be precisely controlled over minutes to days under physiological conditions. Elliptical disk-shaped particles which are not phagocytosed by macrophages could be successfully internalized through shape switch, revealing the ability of shape-switchable particles in modulating interaction with cells.

SHAPE MEMORY ALLOYS (SMAs)

The superior properties of many materials often come at the expense of inferior performance in other areas. For example, conventional metals and their alloys are strong and resist stress (i.e., an applied load), but can tolerate only a very small amount of strain (i.e., deformation) before they are irreversibly deformed. In contrast, rubber readily returns to its original shape, even after large deformations, but is much weaker than conventional metals. However, some metal alloys exhibit "**shape memory**"; they are quite strong but can recover from being deformed when heated. Few other materials show this combination of strength and flexibility. Clever engineering has exploited these properties, e.g., in implanted medical devices such as stents. Tanaka *et al.* (2010) developed a **superelastic alloy** that almost doubles the useful range of deformation that can be induced in such alloys.

Superelasticity refers to a particular mechanical response of SMAs. When a stress is applied to an SMA – for example, through pulling or bending – a phase transition is induced, usually from a high-symmetry solid called **austenite** to a solid with lower symmetry, called **martensite**. When this applied stress is removed, the material fully reverts to its original shape. Because superelasticity is stress-driven, it has to compete with other deformation mechanisms such as dislocation motion and deformation twinning. The process that is triggered by the least amount of stress will win.

Superelasticity appears only when the temperature is high enough for austenite to become more stable than martensite, but any more increase beyond this temperature will correspondingly raise the stress needed to trigger the transformation. With a high enough increase in deformation temperature, plasticity sets in, full strain recovery cannot occur, and superelasticity is restricted to a temperature range or "superelastic window" (Liu and Galvin, 1997). In reality, large and fully recoverable superelasticity is very rare.

What Tanaka *et al.* discovered is an iron-based SMA showing almost full recovery of shape change around 15% strain. Such a high recoverable strain has not previously been observed in polycrystalline SMAs. The strength of the new alloy exceeds 1 GPa and enables the use of smaller cross-sections, an advantage both in cost and design flexibility. As it is also extremely ductile, a roll mill can reduce its thickness by over 90% at room temperature without cracking, a feature that many conventional SMAs lack (Otsuka and Ren, 2005).

Engineers have exploited shape memory and superelasticity for assembly steps in remote locations, especially in implanted medical devices. Parts are fabricated and then stored at low temperature; the final shape is recovered when the part is warmed by the body. Superelasticity is best exemplified in stents in arteries which can open to a larger size, and medical guide wires that unfold. Conceivably, the iron alloy developed by Tanaka *et al.* (see Fig. 7.1) may allow for smaller devices or greater ranges of motion (Ma and Karaman, 2010).

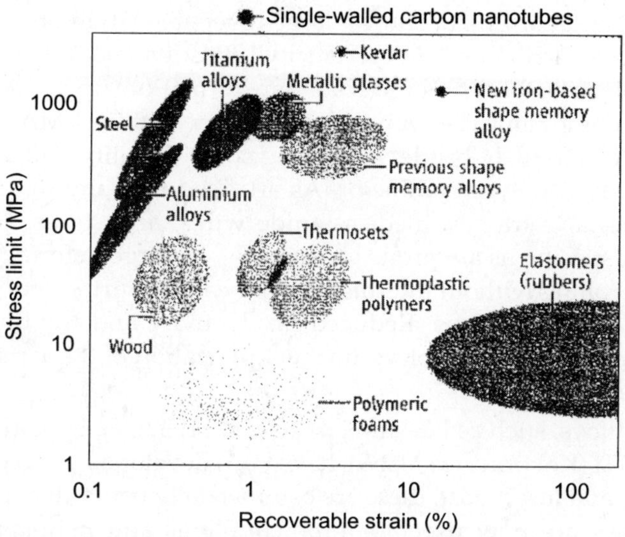

Fig. 7.1 Different materials may be characterized by their strength and their recovery from deformation. Strength is plotted as the stress limit, the maximum pulling force per unit area that a material can withstand without permanently changing shape. Deformation resistance is plotted as the strain – the percentage change in length of the material as it is pulled apart – that a material can take and still recover its original shape. Steels do not easily recover their shape. Most materials that do recover shape readily, e.g., rubbers (elastomers), are not so strong. Shape memory alloys can combine both properties, and the iron alloy described by Tanaka et al. (2010) has pushed their limits even further. The values for metals and ceramics are for either polycrystalline or amorphous structures; some data were taken from Ashby (2005). (From Ma, J, Karaman, I. *Science* 327: 1468-1469, 2010. Reprinted with permission from AAAS).

This truly multifunctional alloy also behaves as a strong ferromagnet in its martensite phase, making it a particularly capable ferromagnetic SMA. Ferromagnetic SMAs (Kainuma *et al.*, 2006) couple mechanical energy to magnetic energy and make possible several applications. They quickly change their shape as a response to a change in the external magnetic field; and are used to sense mechanical deformation or harvest energy from mechanical motion, such as human motion during walking (Sarawate and Dapino, 2008).

Ferrous Shape Memory Alloy

SMAs "remember" the shape they are processed into, and can return to that shape after being deformed by heat. A limitation for most metal-based SMAs is the degree to which they can be deformed elastically. Tanaka *et al.* (2010) demonstrated an iron-based alloy that shows much higher levels of superelastic strain. It surpasses the performance of nickel-titanium alloys. Besides high superelastic strain, this ferrous SMA has much greater strength than NiTi and copper-based SMRs. Consequently, its energy absorption capability is also very high. These properties allow SMAs to be gainfully used as strain sensors or energy dampers.

In metallic materials, elastic strain is usually limited to about 0.2%. Application of stress beyond a critical value causes plastic deformation due to slip or twin boundary motion. Many SMAs, such as Ni-Ti and Cu-Zn-Al, show a large reversible strain of greater than several per cent due to "pseudoelasticity" or "superelasticity" (Otsuka and Wayman, 1998). This strain arises from a combination of stress-induced transformation upon loading and its reverse transformation upon unloading. However, many polycrystalline SMAs having an atomically ordered structure are extremely brittle. Exhibiting some ductility and a superelastic strain of over 7%, the Ni-Ti-based polycrystalline SMAs are used in many products, such as cellular phone antennae, spectacle frames, medical guide wires, and stents. However, Ni-Ti-based polycrystalline SMA specimens are prone to fracturing by large deformations of over 30% that take place after cold rolling without intermediate annealing. Furthermore, their ductility is not high enough for most applications. Reduced productivity and the high cost of fabrication caused by the low cold-workability block further application of these materials to other fields (Tanaka *et al.*, 2010).

Polycrystalline alloys, such as Fe-Mn-Si (Sato *et al.*, 1982), Fe-Ni-C (Kajiwara and Kikuchi, 1990), and Fe-Ni-Co-Ti (Kokorin *et al.*, 1989), have been developed as "**ferrous SMAs**". Because of better workability and lower cost, these are commercially better than the Ni-Ti-based SMAs. Fe-Mn-Si-based alloys are now used for pipe couplings and fishplates. One drawback of current ferrous SMAs is that superelasticity can barely be obtained at room temperature. Maki *et al.* (1984) succeeded in obtaining a thermoelastic transformation in a polycrystalline Fe-Ni-Co-Ti alloy, but the superelastic strain obtained by a bending test at 240 K was only 0.7%. The ferrous polycrystalline SMA developed by Tanaka *et al.* shows a large superelastic strain of over 13% at room temperature. They also found a suitable combination of composition and microstructure to achieve the requisite level of superelasticity.

MATERIALS FOR ENERGY-EFFICIENT SMART BUILDINGS

The current global environmental scenario warrants effective steps to significantly reduce the nonrenewable content and energy footprint of the built environment and to develop "smart buildings" that allow for inexpensive monitoring and self-diagnostic capabilities.

Today's buildings are built from many materials such as reinforced concrete, neoprene, silicone, poly(vinyl chloride) (PVC), ethylene tetrafluoroethylene (ETFE), laminated glass using polyvinylbutyral and fibre-reinforced polymer composites, and many others (Fernandez, 2006, 2007). Many of these materials were being used in industry long before their application in buildings. Pure PVC was synthesized in 1913 and exploited commercially in 1933 but was not used to any significant extent in buildings until the 1970s. ETFE was introduced commercially in 1970 but was not used in buildings until very recently, despite having excellent properties as a glass replacement.

In developed nations, construction and demolition waste from buildings amounts to 60% of the total nonindustrial waste stream, and the world production of concrete alone accounts for about 8% of total CO_2 emissions. Buildings form a component of primary energy use and account for 40% of total energy consumption and 60% of electrical use in developed nations.

Variously reinforced new concretes have been successfully developed, the chief goals being high strength, lower water and cement content, improved toughness, and resource efficiency. These concretes use different additives for strength, workability, controlled and accelerated curing, compaction, shrink minimization, air entrainment, and the reduction of water and cement content. Examples of mineral additives are sands and coarse aggregates, metakaolin, silica fume, industrial slags, and fly ash; these are used in a variety of concrete mixes for volume fillers, compaction, and cement substitution. Chemical additives include superplasticizers of sulfonated melamine-formaldehyde condensates, sulfonated naphthalene-formaldehyde condensates, modified lignosulfonates, and polycarboxylate derivatives. Some additives are helpful in reducing water content and are exemplified by salts and lignins, salts and modifications of hydroxylized carboxylic acids, and certain polymeric materials. These materials allow the production of cement paste at lower water content by deflocculating the cement particles and liberating water. When used along with certain sugars and derivatives, they retard the curing of concrete.

Superior concrete materials are more workable, durable, and reliable. Substantial gains in strength have accompanied the improvement of these properties. A nano-enabled flexible concrete that uses embedded polyvinyl alcohol microfibres coated with a nanoscale film has been developed. It controls the fibre-matrix interfacial bond. As compared to conventional concrete, this material is 500 times more resistant to cracking and 40% lighter per unit weight (Imerito, 2005). Aerated autoclaved foamed concrete is so light and soft that it can be cut with a large handsaw and stacked like masonry (Fernandez, 2007).

Some challenges facing structural materials are how to achieve low cost per unit weight, improved strength per unit weight, and strength per unit embodied energy with reliability and durability. Effective monitoring of structural behaviour is a must. Sensors have been successfully inserted in, e.g., concrete, steel, wood, and fiber-reinforced polymer (FRP) frames (Chong and Garboczi, 2002; Mufti, 2003). Concrete and FRP are specially amenable to devices that can relay information about the internal stress and strain, as well as moisture content and temperature of concrete and FRP matrices. In fact, FRPs, and carbon fibre-reinforced polymers (CFRPs) in particular, are used more frequently as a primary repair system for concrete and masonry structures, and carbon fibres have become a common reinforcing material directly in the concrete matrix (Bencardino *et al.*, 2006). Carbon nanotubes and sheets carry significant potential as monitoring media while also improving strength (Coluci *et al.*, 2006).

As compared to their mineral wool and natural fibre precursors, modern insulation materials can be upto 3 times more resistant to the flow of heat. Typical materials in current use are blankets of loose, nonwoven glass fibres; treated cellulose; and polymer foams including polyisocyanurate, polyicynene, polyurethane and polystyrene. High-performance systems include vacuum- and glass-filled panels encased in aluminized Mylar pillows and aerogel materials of foamed silica that show thermal resistance as low as 0.008 W m^{-1} K^{-1} (Fernandez,

2007). Resistance to water vapour can be conferred by applying a continuous material layer of limited permeability, usually polyethylene sheeting. Air barriers involve various sheet materials, and radiant barriers use aluminized paper and Mylar films (Fernandez, 2007).

Although the so-called smart materials for active glazing have been developed but they are much too expensive for most buildings. These chromogenic materials display electrochromic, photochromic, and thermochromic behaviour and use a wide variety of inorganic and organic materials, e.g. WO_3, NiO, IrO_x, V_2O_5, and others (Lampert, 2005). If the high cost could be brought down, active glazing would lead to effective solutions for low-energy buildings. In fact, materials have been developed that demonstrate a continuously tunable refractive index by varying the use of metal oxide nanoparticles (Krogman *et al.*, 2005). Switchable mirrors which use transition-metal hydride electrochromics are also being developed. Several thermochromic glass laminates may contribute to the production of glazing systems that reject unwanted solar heat gain while maximizing daylighting (Fernandez, 2007).

Novel self-cleaning materials have been produced by using biomimetic micro- and nano-textured surfaces which mimic the behaviour of hydrophobic plant leaves, notably *Nelumbo nucifera* (lotus) and *Colocasia esculenta*. Some artificial biomimetic roughness-induced hydrophobic surfaces have revealed that a combination of surface coating and roughness determines the level of water repellence and thus the self-cleaning capacity of the material (Nosonovsky and Bhushan, 2005).

Materials for Building Services

Recently built buildings deliver carefully modulated amounts of heat, air, artificial light, water (sometimes humidity), telecommunications and data, and natural light to their interior spaces. Further, buildings evacuate wastes, including heat, stale air, and liquid and solid wastes. Building services attempt to create a regulated interior environment tailored to serve the needs of the occupants.

With ever-increasing electricity consumption, harvesting energy from solar radiation becomes important. Silicon-based photovoltaics (PVs) account for more than 99% of production. Single-crystal and amorphous cells are currently marketed for an average efficiency of 12.5% (Green, 2005). Alternative materials, including $CuInSe_2$ and CdTe cells with efficiencies of 20% and organic cells offer new research opportunities. Recently developed PV devices are based on the use of either TiO_2 nanorods together with the conjugated polymer MEH-PPV (poly[2-methoxy-5-(2′-ethyl-hexyloxy)-1,4-phenylene vinylene) or use silicon and silicon dioxide to produce quantum dots that absorb a wide range of light wavelengths (Lin *et al.*, 2006; Lewis and Nocera, 2006).

Materials for Interior Construction

Materials that improve the interior environment contribute to the health effects of our buildings. Both virucidal and bactericidal surface coatings have been developed. A virucidal coating consists of methyl-polyethylenimines (PEIs) and other hydrophobic PEI derivatives (Haldar *et al.*, 2006). A bactericidal coating contains TiO_2 exposed to near-ultraviolet light (Haung *et al.*, 2000). There is need to develop coatings which effectively remove mold and fungus growth from buildings in humid climates.

An interesting material mode of polyamide (nylon) has found applications as a "smart" vapour barrier in exterior envelopes. Its water vapor permeability increases 10-fold in conditions of very high humidity – particularly useful when moisture is trapped inside a wall assembly. The vapour barrier becomes more permeable and allows moisture to escape, reducing the risk of corrosion, rot, and the growth of mold and fungus. Nylon was discovered in 1931 but its properties as a vapour barrier were described only in 1999 since when it has been commercialized (Künzel, 1999).

Latest-generation embedded sensors, self-healing composites, and nanoscale and responsive materials may herald a time when buildings can substantially adjust to changing environmental and functional demands. However, faced with the legal liability resulting from unknown lifetime performance, designers and engineers lack motivation to incorporate new material technologies into building designs. As efficiency issues become more acute, the potential for improvement in performance from new materials, together with partnerships between the materials science community and those entrusted with the design and engineering of the built environment, could potentially offer real breakthroughs for the future (Fernandez, 2007).

HYBRID MICROFIBRE-NANOWIRE FOR ENERGY SCAVENGING

Paradiso and Starner (2005) reported the utility of a self-powering nanosystem that harvests its operating energy from the environment for sensing, personal electronics and defence technologies. This may be quite feasible for nanodevices in view of their extremely low power consumption (Li *et al.*, 2006; Tian *et al.*, 2007). Common sources of energy are solar, thermal and mechanical; these may be scavenged from the environment, but the specific type of energy source is chosen carefully on the basis of particular applications. For military applications such as sensing/surveillance node placement, often involve difficult-to-reach locations, may need to be hidden, and may be used in dusty, rainy, dark environment and/or in dense forest. In a moving vehicle or aeroplane, harvesting energy from a rotating tyre or wind blowing on the body can be chosen to power wireless devices implanted in the surface of the vehicle. Wang *et al.* (2007, 2007a) demonstrated nanowire nanogenerators built on hard substrates for harvesting local mechanical energy produced by high-frequency ultrasonic waves. For harvesting the energy from vibration or low noise from footsteps, or ambient air flow, we need to develop innovative technologies that work at low frequencies (such as <10 Hz) and are based on flexible soft materials. To meet this need, Qin *et al.* (2008) suggested a simple, low-cost method of converting low-frequency vibration/friction energy into electricity by using piezoelectric zinc oxide nanowires grown radially around textile fibres. Entangling two fibres and brushing the nanowires rooted on them with respect to each other converts mechanical energy into electricity through a coupled piezoelectric-semiconductor process (Wang and Song, 2006; Song *et al.*, 2006). Light-wind energy and body-movement energy can be scavenged by using fabrics (Qin *et al.*, 2008).

The fibres used in these experiments were Kevlar 129 fibres. These show high strength, modulus, toughness and thermal stability. ZnO nanowires were grown radially on the fibre surface by using a hydrothermal approach (Hsu *et al.*, 2005). All along the fibre, ZnO nanowires grew radially, showed a uniform coverage and well preserved cylindrical shape. All these ZnO nanowires were single crystalline, had a hexagonal cross-section with a diameter in the range ~50-200 nm and a typical length of ~3.5 μm. Their smooth and clean top and side surfaces made them form the reliable metal-semiconductor junctions required for the nanowire nanogenerators. The space between the nanowires can be a few hundred nanometres – large enough for them to be bent to generate the piezoelectric potential (Gao and Wang, 2007). The nanowires tips remain free from each other because of their small tilting angles (<± 10), but their bottom ends are tightly connected. Consequently, a continuous ZnO film at the nanowires' roots acts as a common electrode for signal output. It was shown that AFM (atomic force microscope) manipulation of a solution-grown ZnO nanowire can generate up to 45 mV output voltage.

As it is important to maintain the high flexibility of the fibre after growing a crystalline film and nanowires, surface coating improves the mechanical performance of the fibre and the binding of nanowires. Two layers of tetraethoxysilane (TEOS) were infiltrated – one above and the other below the ZnO seed layer – as binding agents. The highly reactive Si-O bonds in TEOS have the OH^- groups on the ZnO surface, and its organic chains tightly bind to the body of the aromatic polyamide fibres; the ZnO seed layer and the fibre core were firmly bound with each other by a thin layer of TEOS. Because TEOS could readily form cross-linked chains, the nanowires were bundled and bound together at their roots and fixed on the ZnO seeding layer. This prevented them from scratching/stripping off during mechanical brushing/sliding. Even when the fibre was looped into a circle of ~1 mm diameter, no cracks, loose pieces or peel-offs were observed in the ZnO nanowire coating layer. At the bending area, the radially aligned morphology of the ZnO nanowires was also well preserved revealing its toughness under mechanical deformation and bending (Qin *et al.*, 2008).

The power generation ability of the ZnO-nanowire-covered fibres was demonstrated by means of a double-fibre model system in which two fibres, one coated with a 300-nm-thick gold layer and the other as-grown, were entangled to serve as the core for power generation. The relative brushing between the two fibres could be simulated by pulling/releasing the string by means of an external rotor with a controlled frequency. This was followed by connecting the gold-coated fibre to the external circuit as the nanogenerator's output cathode. The pulling force effectively contacted the two fibres.

In this device, the gold-coated ZnO nanowires behaved like an array of scanning metal tips which deflected the ZnO nanowires rooted at the other fibre; a coupled piezoelectric and semiconducting process created and accumulated charge and released charge (Wang and Song, 2006; Song *et al.*, 2006). The gold coating fully covered the ZnO nanowires as a continuous layer along the entire fibre. When the two fibres became tightly entangled together, some of the gold-coated nanowires penetrated slightly into the interspaces between the uncoated nanowires rooted at the other fibre. So, when a relative sliding/deflection occurred between them, the bending of the uncoated ZnO nanowires generated a piezoelectric potential across their width, and the gold-coated nanowires behaved as the 'zigzag' electrode (as in a DC nanogenerator) for collecting and transporting the charges (Qin *et al.*, 2008).

The output voltage is governed by the characteristic of one nanowire, and the sign of the voltage was found not to change in response to the deflection configuration of the nanowire. The performance of the fibre nanogenerators was judged by measuring the open-circuit voltage. The pulling and releasing of the gold-coated fibre was achieved by a motor at a controlled frequency. The resulting current signals were detected at 80 r.p.m. (revolutions per minute; 60 r.p.m. = 1 Hz). When the current meter was forward-connected to the nanogenerator i.e., the positive and negative probes were connected to the positive and negative electrodes of the nanogenerator, ~5 pA positive current pulses were detected at each pulling-releasing cycle (Qin *et al.*, 2008). But when the current meter was reverse-connected, negative current pulses with the same amplitude were received. (Here reverse-connection means that the positive and negative probes are connected to the negative and positive electrodes of the nanogenerator, respectively.) The inverted output current signals proved that the current indeed originated from the fibre nanogenerator.

The fibre nanogenerator also performed quite well over a range of frequencies. When short-circuit current was measured while the speed of the driving motor was varied from 80 to 240 r.p.m., the current density declined almost linearly with increasing frequency.

Reducing the inner resistance of the fibre and the nanowires turned out to be quite effective in enhancing the output current. By depositing a conductive layer directly onto the fibre before depositing the ZnO seeds, the inner resistance of the nanogenerator was reduced from ~1 G? to ~ 1 k?, so that the output current I_{sc} of a double-fibre nanogenerator was increased from ~4 pA to ~4 nA. The work of Qin *et al.* is an effective approach to increasing the output current. Further, by connecting nanogenerators in series and parallel, the output voltage and current, respectively, can be increased as well (Wang *et al.*, 2007a).

The textile-fibre-based nanogenerator revealed the following advances in comparison to the dc nanogenerators reported previously: (a) using ZnO nanowires grown on fibres, one can fabricate flexible, foldable, wearable and robust power sources in any shape; (b) the output electricity can be sharply enhanced by using a bundle of fibres as a yarn – the basic unit for fabrics; (c) the nanogenerator operates at low frequency, in the range of conventional mechanical vibration, footsteps and heartbeats, thereby significantly expanding the application range of nanogenerators; (d) as the ZnO nanowire arrays were grown using chemical synthesis at 80°C on a curved substrate, this method should be generally applicable to growth on any substrate, so the fields in which the nanogenerators can be applied and integrated can be increased.

HIGH-T_C SUPERCONDUCTING MATERIALS

A conductor is more than just the superconductor. Conductors for power applications are multifilamentary wires or tapes in which many superconducting filaments are embedded in a matrix of a normal metal, such as Cu or Ag, which provides protection against magnetic flux jumps and thermal quenching (Wilson, 1983). These wires must possess enough strength to withstand the fabrication process, device winding, cool-down and electromagnetic stresses, and should lend the machines to being made or cabled to sufficient size to carry operating currents up to thousands of amperes at costs comparable to Cu. Some higher-cost applications include high power density underground power cables in inner cities, environment-friendly, oil-free high **temperature superconductor (HTS)** transformers, or superconducting magnetic energy-storage devices for power networks where low reactance and instantaneous redistribution of power is essential (Hassenzahl, 2001).

Large-scale superconducting devices for electric power industry depend critically on wires with high critical current densities at temperatures where cryogenic losses are tolerable. This restricts choice to two high-temperature cuprate superconductors, $(Bi,Pb)_2Sr_2Ca_2Cu_3O_x$ and $YBa_2Cu_3O_x$, and possibly to MgB_2, which superconducts at 39 K. Crystal structure and material anisotropy place fundamental restrictions on their properties, especially in polycrystalline form. Till hitherto, power applications have followed a largely empirical, twin-track approach of conductor development and construction of prototype devices (Larbalestier *et al.*, 2001). The feasibility of superconducting power cables, magnetic energy-storage devices, transformers, fault current limiters and motors, largely using $(Bi,Pb)_2Sr_2Ca_2Cu_3O_x$ conductor has been well established. Widespread applications now depend significantly on cost-effective resolution of fundamental materials and fabrication issues, which control the production of low-cost, high-performance conductors of these remarkable compounds.

THE SURPRISING SUPERCONDUCTOR

Nagamatsu *et al.* (2001) accidentally discovered that MgB_2 shows superconductivity at $T_c = 39$ K, about twice the temperature of other simple intermetallic compounds.

In fact, new superconductors with higher T_cs have been found. Schon *et al.* (2001) reported 52 K and 117 K superconductivities in the hole-injected surface layers of single crystals of C_{60} and expanded-C_{60} systems, respectively . Cuprates are known to superconduct at temperatures about three times higher than MgB_2 (Yildirim, 2002).

One major reason why MgB_2 attracted so much interest is its simplicity. It has a hexagonal AlB_2-type crystal structure, where the boron atoms form graphite-like sheets separated by hexagonal layers of Mg atoms. Unlike other superconductors, no complications are seen as a result of d- or f-electrons, no spin degrees of freedom, and no structural phase transition. Indeed, MgB_2 is an ideal system to test various theories of superconductivity. The 39 K superconductivity in MgB_2 is also important for practical reasons, even though its T_c is much lower than high temperature cuprates. MgB_2 is made of very light and cheap elements. Unlike cuprates, MgB_2 is a very good metal where there is no high contact resistance between the grain boundaries, eliminating the weak-link problem that hindered widespread commercialization of the cuprates. In the crystal structure of MgB_2 consisting of B and Mg hexagonal layers, neutron measurements indicate that thermal expansion of the lattice parameters show no significant anomalies near T_c. The hexagonal structure remains stable up to 40 GPa and the normalized lattice constant along the *c-axis* decreases with pressure almost linearly and twice as fast as along the a-axis (Yildirim, 2002). Finally, the conduction electron density and normal-state conductivity can be up to two orders of magnitude higher for MgB_2 than for the cuprates used in wires. These features, plus the low cost, light-weight, and easy fabrication of wires and thin films, make MgB_2 attractive for many applications.

For practical applications of a superconductor, such parameters as superconducting temperature, critical current, and magnetic field (above which superconductivity is destroyed) need to meet certain criteria. The T_c of MgB_2 is not high, but is high enough for a closed-cycle refrigerator so that expensive liquid helium is not required. The critical current and field are large for MgB_2, but lower than conventional superconducting wires based on Nb-Ti or Nb_3Sn. This is why much research has focused on improving these parameters and finding ways to form MgB_2 into thin films.

Welcome advances in fashioning MgB_2 into films and wires that can carry high currents greatly increase the commercial prospects of MgB_2. Canfield *et al.* (2001) described how to make wires of MgB_2 by exposing B filaments to Mg vapour. But these wires, being fragile, are not very suitable for commercial use. Jin *et al.* (2001) successfully fabricated Fe-clad MgB_2 wires with current densities of 85000 A/cm^2 at 4.2 K. This fabrication technique overcomes the hard and brittle mechanical properties of MgB_2 that make it so difficult to pull into wires. With current densities of this scale, MgB_2 has entered the realm required for real applications in transmission cables and MRIs (Yildirim, 2002).

MgB_2 surprisingly shows superconductivity at 39 K. It seems that a specific lattice vibration – which pulls two boron atoms so close together that electrons can readily jump the gap between them – leads to the formation of very strong Cooper pairs. According to Yildirim (2001), this coupling is so strong that only creates giant anharmonicity for these phonon modes, but also leads to a large non-linear electron-phonon coupling – a coupling that accounts quantitatively for the high T_c in MgB_2, and its dependence on the boron mass and external pressure. In the crystal structure of MgB_2 consisting of B and Mg hexagonal layers, the in-plane boron phonons are strongly coupled to the boron conduction electrons (Yildirim, 2001, 2002). While the lattice vibrations of MgB_2 are perfect for electron conduction it may be possible to exploit and even improve upon the interaction. Even if it proves impossible to increase T_c any further, the low cost, light mass, and nearly isotropic high conductivity will sure find important technological applications for MgB_2 (Taner Yildirim, see *Materialstoday*, p. 10, Sep. 2001). That no other diboride-based compounds have been reported with high T_cs suggests that MgB_2 is a unique system.

Jin *et al.* (2001) successfully fabricated Fe-clad MgB_2 wires with current densities of 85000 A/cm^2 at 4.2 K. This fabrication technique overcomes the hard and brittle mechanical properties of MgB_2 that make it so difficult to pull into wires. With current densities of this scale, MgB_2 has entered the realm required for real applications in transmission cables and MRIs (Yildirim, 2002).

MATERIALS WITH UNNATURAL OPTICAL PROPERTIES

It has become possible to make materials with optical properties not found in the natural world. Such materials provide unprecedented control over light and enhanced device functionality (Shalaev, 2008).

Recent advances in nano-fabrication methods have generated exciting opportunities to control light in a manner that is not possible with natural materials. Synthetic structures built up from subwavelength elements can be fabricated with a desired spatial distribution of effective electric permittivity and magnetic permeability, generating the potential to guide and control the flow of electromagnetic energy in an engineered optical space. These "metamaterials" have made possible a number of previously impossible applications. We are no long constrained by the electromagnetic response of natural materials and their chemical compounds. Instead, it is possible to tailor the shape and size of the structural unit of the metamaterial and adjust their composition and morphology to create some new functionality (Shalaev, 2008).

The field of transformation optics enabled by metamaterials has necessitated a fresh look at the very foundations of optics. Just as in general relativity, time and space are curved, so also transformation optics show that the space for light can be bent in an almost arbitrary way. Designing and engineering of optical space makes it possible to control the flow of light with nanometer spatial precision. So general relativity could soon attract practical use in certain novel optical devices based on transformation optics, guiding how, using metamaterials, the space for light can be curved in a predesigned and well-controlled way.

MOLECULAR CLOAKING DEVICES

A long-cherished wish of many drug developers is to identify small molecules that strongly and specifically bind to a target enzyme. Binding is usually at the enzyme's active site; the drug nestles into this molecular pocket and so blocks binding of the enzyme's natural substrate. Alternatively, the drug may bind to the enzyme at a site other than the active site, thereby locking the enzyme into a conformation that either severely inhibits or activates substrate processing in the active site itself. The behaviour of most drugs can be explained by these two models of binding.

But there is also a third option: the drug might bind to the enzyme's substrate. Not much is known about this rare third possibility because enzyme substrates are believed to lack drug-friendly binding pockets, and thus are slippery targets for small molecules. Kukar *et al.* (2008) reported that certain small molecules – promising compounds for the development of drugs to treat Alzheimer's disease – operate by just such a substrate-targeted mechanism (see Kodadek, 2008).

Kukar *et al.* studied the enzymatic breakage of a protein substrate by a protease enzyme. Proteases cut other proteins into pieces, but some can be blocked by inhibitors. Proteases attract attention as targets for drug development because their substrate proteins often have little or no biological activity before they are processed by the enzyme; the cleavage reaction represents a convenient point where to block their latent function. Protease inhibitors are extremely valuable as drugs – the development of potent inhibitors of HIV protease that block the conversion of precursor molecules to viral proteins has revolutionized the treatment of AIDS (Flexner, 1998).

Kukar *et al.* studied the cleavage of amyloid precursor protein (APP) by the enzyme γ-secretase. APP is produced by nerve cells. It is cut at several points along its amino-acid chain by γ-secretase. Some of these cleavage products assume a misfolded conformation that causes them to aggregate, forming oligomers and long fibril-like structures that may cause devastating effects on brain function in patients with Alzheimer's disease. Certain small molecules, including some simple anti-inflammatory drugs delay the onset of the symptoms of Alzheimer (Rich *et al.*, 1995) and inhibit the formation of the dangerous cleavage products (Beher *et al.*, 2004). But as the mechanism of inhibition is not known, the clinically useful drugs have not been developed.

Kukar *et al.* tackled this problem by attaching an inhibitor drug covalently to the protein (or proteins) with which it associates, followed by identification of the resulting protein-inhibitor products. They found no evidence for association of the inhibitor with the γ-secretase, which had previously been supposed to be the drug's target. Instead, the authors noted that the binding partner of the drug was the APP substrate. The location of the drug's binding site was narrowed down to a short peptide sequence in APP that includes a γ-secretase cleavage site, and which is also involved in the aggregation of the cleaved products (Kodadek, 2008). This is an extremely valuable finding - there are very few known examples of substrate-targeted enzyme inhibitors and all are peptides rather than drug-like small molecules (Zhang *et al.*, 2001). Should Kukar *et al.*'s findings turn out to be relevant to other proteases and their substrates, it would point to a completely new set of potential drug targets for the treatment of several diseases. It certainly warrants that drug-screening efforts should not only seek compounds that bind to proteases themselves, but also look for those that bind to their substrates.

Substrate-targeted drugs are expected to be far more selective for their proteins than protease-targeted inhibitors. Proteases usually have many different substrates, so that even the most specific protease inhibitors will affect many different reactions. The potential for unwanted side effects *in vivo* may therefore be significant. On the other hand, a substrate-targeted inhibitor could, in theory, specifically block the breakage of the targeted substrate by binding not only to the substrate's cleavage site, but also to surfaces flanking the site (Kodadek, 2002). A welcome feature of such surfaces is that they provide unique 'signatures' to distinguish the protein from the other competing substrates (Fig. 7.2). The cleavage site would thus effectively be cloaked (protected) from the protease by the inhibitor. Proteases break other proteins into fragments. For instance, a protease may have two substrates that share a common cleavage site but have different flanking regions. When an inhibitor is absent, both the substrates can be cleaved by the protease. An inhibitor that targets the protease prevents the cleavage of both the substrates. Inhibitors that target substrates can be so designed as to recognize the unique arrangement of amino-acid residues that flank the cleavage site of a particular substrate allowing them to specifically protect that substrate from cleavage. Kukar *et al.* (2008) successfully identified substrate-targeting inhibitors that prevent the cleavage of amyloid precursor protein (APP) by the g-secretase enzyme, but do not protect similar proteins that have the same cleavage site as APP (Kodadek, 2008).

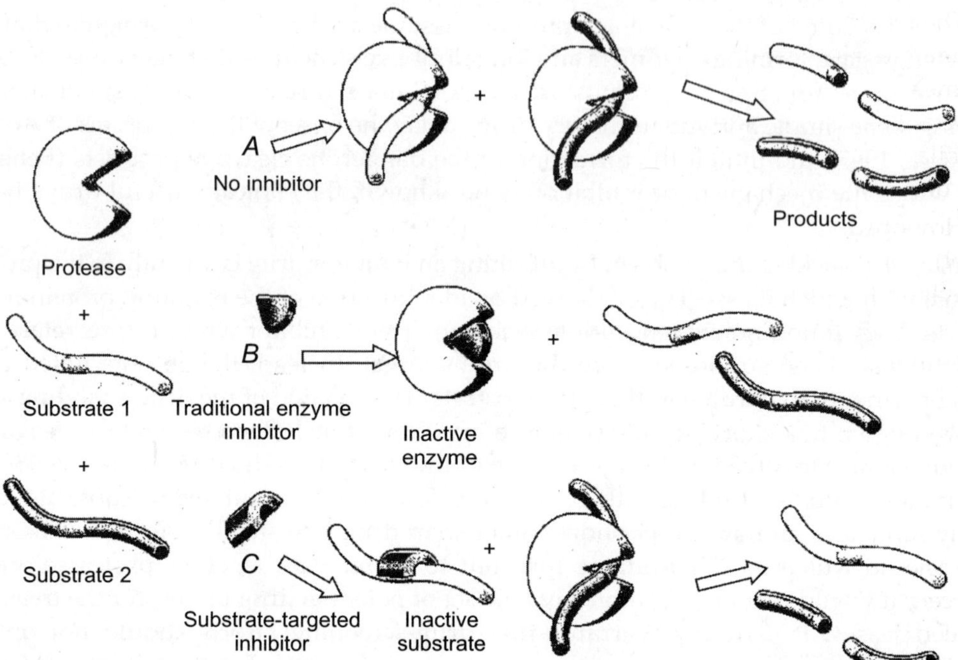

Fig. 7.2 Targets for enzyme inhibition. Proteases break other proteins into fragments. The protease shown here has two substrates that share a common cleavage site but have different flanking regions. A. In the absence of an inhibitor, both substrates are cleaved by the protease. B. An inhibitor that targets the protease prevents the cleavage of both substrates. C. Inhibitors that target substrates can be so designed as to recognize the unique arrangement of amino-acid residues that flank the cleavage site of a particular substrate. This allows them to specifically protect that substrate from cleavage, as shown for substrate 1 (Kukar *et al.*, 2008 identified substrate-targeting inhibitors that prevent the cleavage of amyloid precursor protein (APP) by the g-secretase enzyme, but do not protect similar proteins that have the same cleavage site as APP) (after Kodadek, 2008). (Permission received from author).

POLYMERIC PULSATING VESICLES

Some molecules and molecular assemblies which undergo controllable shape changes in response to an external trigger can be extremely useful for manufacturing nanometer-scale devices. Stimulus-responsive systems are fairly common in nature but it is not often easy to produce their synthetic versions. This is particularly true of vesicles, a class of cell-mimicking structures that have attracted thorough investigation. Yu *et al.* (2009) reported a polymeric vesicle whose structure and properties can be controlled: it 'breathes' in response to pH, becoming twice as big in acid solution as it is in basic solution (van Hest, 2009). This reversible size change goes hand in hand with a permeability change, enabling water molecules to diffuse more easily through the membrane wall of the vesicle when the structure is highly swollen.

Vesicles are analogous to nanometer-scale containers, in which a molecular shell or membrane encloses an aqueous interior. Their biomedicine applications range from drug-delivery devices to nanovessels for chemical reactions. For all such applications, the ability to control the transport of molecules in and out of the container is essential.

The molecular assembly studied by Yu *et al.* exemplifies a special class of vesicle known as **polymersomes** (van Hest, 2009). The membranes of most polymersomes are made up of polymeric molecules which have a hydrophilic block and a hydrophobic block. These surfactant characteristics are conducive to spontaneous aggregation of the molecules in water into bilayers that form spherical shells; the hydrophilic parts of the molecules line the outer faces of the bilayer and shield the hydrophobic parts from the aqueous environment.

However, the polymer used by Yu *et al.* is made of three parts: a hydrophilic poly(ethylene glycol) (PEG) block at one end; a hydrophobic polystyrene block in the middle; and poly(2-diethylaminoethyl methacrylate) (PDEA) at the other end. PDEA has organic bases in its side chains, which show no electrical charges under basic conditions. Consequently, PDEA is hydrophobic in basic solution, and occupies the inner part of the vesicle membrane. On the other hand, when conditions are acidic, the bases become protonated and PDEA is positively charged, so making it hydrophilic. The PDEA block therefore attracts water.

While the polarization of the PDEA block in response to a change from basic to acidic conditions would typically cause the vesicle to break up, elegant design of Yu *et al.*'s of copolymer prevents the vesicle from disintegrating. At high pH, the non-polar polystyrene becomes partly mixed with the similarly non-polar PDEA. At low pH, however, the polystyrene is forced out of the now positively charged PDEA layer because of the difference in polarity between the two polymers. Thereafter the polystyrene forms a specific region of its own between the PDEA and the PEG layers of the vesicle membrane. As polystyrene happens to be a glassy polymer, this region is rigid, and so prevents movement and diffusion of polymer blocks through the membrane. This is why it functions as a kind of molecular restraint which keeps all the loose polymeric components together and in order, thereby preventing reorientation of PDEA towards the water layer and hence stopping the vesicle from irreversibly losing its shape. When the PDEA layer attracts water it swells. This induces the entire vesicle to swell in turn. The water-swollen layer being less densely packed than before, it becomes very easy for water molecules to pass through the polymer membrane.

The unique feature of the capsule reported by Yu *et al.* is that it can 'breathe' repeatedly, i.e., swell and shrink in response to pH changes – without any need for chemical cross-linking of the polymer molecules to stabilize its structure (van Hest, 2009).

To enhance utility of Yu *et al.*'s polymersome in a biological setting, the vesicle's size will need to be controlled under physiological conditions, and its membrane must allow through not only water molecules, but also biologically active compounds. The second issue has already attracted much focus. Kim *et al.* (2009) and Nardin *et al.* (2000) have created polymersomes that irreversibly trap enzymes, but which allow the enzyme's substrates and products to pass through the vesicle membrane. The selective permeability originates from the clever design of the polymer molecules, or from trapping protein channels in the vesicle's membrane.

Soft Polymeric Particles

It is possible for a crystal to grow only if all its atomic or molecular building blocks fit into the periodic lattice. This is so even for colloidal crystals, which form through the ordered self-assembly of μm-sized particles. The need for periodicity places stringent constraints on the variation in the size of particles that can be incorporated into a given colloidal crystalline lattice. But Iyer and Lyon (2009) showed that crystals made of microgel particles are highly tolerant of particle size variations, and that this property could possibly have practical implications for the design of ordered colloidal materials.

Crystalline materials are remarkable: every crystal of table salt consists of as many as 10^{18} sodium and chloride ions, all of which self-assemble into a nearly perfect periodic lattice. Yet atoms or molecules can form crystals only if their size distribution is very narrow (Frenkel, 2009). For crystals made up of simple salts such as sodium chloride, the restrictions are irrelevant, because all particles that occupy equivalent lattice sites have exactly the same size. But strong constraints operate on the relative atomic sizes of metals which, when mixed together, can form substitutionally disordered crystals (in which equivalent lattice sites are not necessarily occupied by the same kinds of atom). When the radius of two otherwise similar metal ions differs by more than 15%, the metals do not mix in the crystalline phase, even though they often do so in a melt.

As for colloidal crystals, hard, uncharged colloidal particles can make substitutionally disordered crystals of a particular cubic arrangement e.g. centred cubic, only if the natural spread in the sizes of the particles is less than 5.7%. If the particle size variation of a colloid exceeds this value, the particles can still form crystals, but each crystal will have particles within a certain size range, such that the variation within every crystal again does not exceed 5.7%. Substitutionally ordered colloidal crystals (in which equivalent lattice sites contain the same kind of particles) can also form from two different kinds of particle in very specific size ratios; but in these cases, constraints are exerted on the size variations of each constituent particle type (Frenkel, 2009).

Considering that most crystals are rather intolerant of the inclusion of particles of significantly different sizes. Iyer and Lyon's findings are surprising: they observed that colloidal crystals of soft 'microgel' particles can incorporate foreign microgel particles of a very different size from those in the host lattice – the volume of the foreign particles may be up to 15 times that of the host particles. But no less surprising is the observation that the oversized particles fit into the host structure without causing any noticeable deformation of the surrounding crystal lattice. Instead, the large particles shrink so much that their size equals that of the host particles. Such a mimicking behaviour is very unusual, even for crystals made up of deformable building blocks.

In colloidal crystals of ultra-soft microgels, microgel particles are not only highly deformable, but also they can swell or shrink dramatically, depending on the temperature, ionic strength or pH. The physical origin of this unusual feature is related to the molecular structure of microgel particles, which consist of weakly crosslinked polymer networks. There is a delicate balance between the osmotic pressures in each particle – pressures caused by polymer mixing, by the presence of confined ions in the particles and by elastic deformation of the polymer network. It is the outcome of the interplay between these pressures which determines whether the particles will swell or collapse.

The importance of the work of Iyer and Lyon is its suggestion that microgel crystals tolerate the presence of small quantities of other microgels. This could make it easier to create microgel-based photonic materials, substitutionally doped with particles having very different optical properties from the host microgel (Frenkel, 2009).

NANOCRYSTALLINE BIOMORPHS

Several examples are known of life being able to create precise, smooth, differentiable shapes made of such polycrystalline minerals as shell, teeth or bones, yet our understanding of the morphogenetical mechanisms leading to the formation of these fascinating architectures is very limited (Garcia-Ruiz *et al.*, 2009). It has not so far been possible to synthesize in the laboratory actual self-assembled structures mimicking the ability of life to make sinuous noncrystallographic morphologies with crystalline materials. Silica/carbonate biomorphs exemplify synthetic self-organized nanocrystalline materials that display a rich wealth of morphologies comparable to that of biominerals (Hyde *et al.*, 2004). Silica/carbonate biomorphs result from the precipitation of barium or strontium carbonates in alkaline silica-rich environments. These crystalline aggregates are named silica/carbonate biomorphs because their morphology resembles that of certain primitive organisms. Like biominerals, biomorphs show nanoscale atomic ordering. Consequently, no characteristic faces or edges are expressed; rather, they are bounded by smoothly curved surfaces. Biomorphs display a variety of curvilinear morphologies, often indistinguishable from the forms of biomaterials found *in vivo.* These aggregates are self-assembled materials of purely inorganic origin, with an amorphous phase of silica intimately intertwined with a carbonate nanocrystalline phase. Garcia-Ruiz *et al.* (2009) proposed a mechanism that explains all the morphologies described for biomosphs. Chemically coupled coprecipitation of carbonate and silica leads to fibrillation of the growing front and to laminar structures that suffer curling at their growing rim. These curls propagate in a surflike manner along the rim of the laminae. All observed morphologies with smoothly varying positive or negative curvatures could be explained by the combined growth of counterpropagating curls and growing laminae.

Silica/barium carbonate biomosphs can be grown routinely by mixing barium chloride solutions with silica solutions and gels within a pH range of 8.5 to 11 at atmospheric pressure and temperature. In alkaline conditions, carbonate from dissolved atmospheric CO_2 reacts with Ba^{2+} to precipitate crystalline barium carbonate (witherite) as pseudohexagonal prismatic crystals tapered by bipyramidal faces. But when barium carbonate crystallizes from silica-rich solutions or from silica gels, it tends to form polycrystalline aggregates showing various micrometer- to millimeter-sized morphologies with noncrystallographic symmetry shaped by "differentiable" surfaces of well-defined and smoothly varying Gaussian curvature (Bittarello and Aquilano, 2007). Its submicrometre structure is made of nanometer-sized rods of witherite, whose mutual arrangement shows a high degree of orientational order. The carbonate nanocrystals are coated by amorphous silica (SiO_2) which cements the structure and forms a composite material, as shown by the ability to selectively dissolve the crystalline or amorphous phase with either dilute HCl or NaOH solutions, respectively (Garcia-Ruiz *et al.*, 2009).

PHOTOVOLTAIC MATERIALS

Most photovoltaic cells respond to only a fairly small part of the solar spectrum – the part that clouds usually block out. This trouble can be tackled by layering different materials in the cell, but that makes the cells quite expensive.

An alternative approach has been taken by Malcolm Chisholm, a chemist at Ohio State University. His group doped a polymer called oligothiophene used for semiconductor applications, with atoms of molybdenum and tungsten. This produced a material that generates power in response to light of wavelengths from 300 (ultraviolet) to 100 nm (the near infrared). From infrared to ultraviolet, this material responds to the full spectrum of sunlight. This contrasts with traditional silicon-based cells which function best in the wavelength range from about 600 nm to 900 nm only. The new polymer can respond at such a wide range because it both fluoresces and phosphoresces (Emspak, 2009).

Most solar cell materials only fluoresce: sunlight impacting them excites electrons into a higher energy state whereafter they fall back to their ground state and emit light. Generally, the fluorescence cannot be seen either because the wavelength of the emitted light is in the infrared, or the light is too weak to see in the sun. (But some solar cell designs reuse the light to boost efficiency). Some of those electrons become so excited that they can break free from the atoms they surround; it is these electrons that can generate the electric current.

Unfortunately, the electrons do not remain free for longer than trillionths of a second. They fall back to the ground state before serving any useful purpose. This is why solar cells cannot operate with 100 percent efficiency.

The polymer developed by Chisholm also phosphoresces like glow-in-the-dark toys. Electrons hold on to their energy longer in phosphorescence than in fluorescence and so remain free longer, on a scale of microseconds. It is the doping that makes the difference. Both tungsten and molybdenum are metal atoms that have more electrons that are available for conduction than those of the polymer alone. The electron configurations of the metals allow for longer-lived free electrons.

The research team laid the polymer down as a thin film, similar to what would be used in a solar cell. But the actual device is still years away. It is hoped that even if these polymer solar cells are less efficient than silicon, they will ultimately be cheaper to produce.

GALLIUM ARSENIDE PHOTOVOLTAICS AND OPTOELECTRONICS

In view of their direct bandgaps and high electron mobility, compound semiconductors such as gallium arsenide (GaAs) are preferred over silicon for certain applications. This is best exemplified by some highly efficient photovoltaic devices (Bosi and Pelosi, 2007), radio-frequency electronics (Chang and Kai, 1994; Schwierz and Liou, 2007) and many optoelectronics (Mokkapati and Jagadish, 2009). However, mass production of large, high quality wafers of these materials, and their effective integration on silicon or amorphous substrates (such as glass or plastic) is expensive. Yoon *et al.* (2010) described materials and fabrication concepts that address some of these limitations through the use of films of GaAs or AlGaAs grown in thick, multi-layer epitaxial assemblies which are then separated from each other and carefully distributed on foreign substrates by printing. This method yields large quantities of superior semiconductor material capable of device integration in large area formats, in a manner that also allows the wafer to be reused for additional growths. Yoon *et al.* successfully demonstrated some advantages of this approach with three different applications: (a) GaAs-based metal semiconductor field effect transistors and logic gates on plates of glass, (b) near-infrared imaging devices on wafers of silicon, and (c) photovoltaic modules on sheets of plastic.

The approaches discussed by Yoon *et al.* are useful in integration possibilities, designs and cost, to generate opportunities which are inaccessible to conventional or established technologies. Besides, further work on exploring other material systems such as gallium nitride and indium phosphide, and other devices, such as light emitting diodes, lasers and sensors exemplify possible potentially valuable directions for future work.

MUSCLE-MIMETIC BIOMATERIALS

The passive elasticity of a muscle is mostly determined by a specific part of the giant muscle protein **titin** which is a complex molecular spring composed of several individually folded immunoglobulin-like domains as well as largely unstructured unique sequences (Labeit and Kolmerer, 1995; Tskhovrebova and Trinick, 2003). These mechanical elements show distinct mechanical properties. Upon combining, they generate the desired passive elastic properties of muscle (Rief *et al.*, 1997; Li *et al.*, 2002). These are a unique combination of strength, extensibility and resilience (Lv *et al.*, 2010). Single-molecule atomic force microscopy (AFM) studies have revealed that the macroscopic behaviour of titin in intact myofibrils can be reconstituted by combining the mechanical properties of these mechanical elements measured at the single-molecule level (Rief *et al.*, 1997, Li *et al.*, 2002). Lv *et al.* (2010) reported artificial elastomeric proteins that mimic the molecular architecture of titin through the combination of well-characterized protein domains GB1 (Cao and Li, 2007) and resilin (Elvin *et al.*, 2005). These artificial elastomeric proteins can be photochemically crosslinked and cast into solid biomaterials which behave as rubber-like materials showing strong resilience at low strain and as shock-absorber-like materials at high strain by effectively dissipating energy. These properties are reminiscent of the passive elastic properties of muscles within the physiological range of sarcomere length, and hence these materials exemplify a novel muscle-mimetic biomaterial. It is possible to fine-tune the mechanical properties of these biomaterials by changing the composition of the elastomeric proteins. This creates the opportunity to develop biomaterials that are mimetic of different types of muscles. Such biomaterials are expected to find applications in tissue engineering as scaffold and matrix for artificial muscles (Langer and Tirrell, 2004).

While protein-based biomaterials, such as those based on **elastin** (Bellingham *et al.*, 2003), **resilin** (Elvin *et al.*, 2005) and **abductin** (Bochicchio *et al.*, 2005) have been engineered from non-globular elastomeric proteins that behave like entropic springs, the GB1-resilin-based biomaterial designed by Lv *et al.* seems to be the first chemically crosslinked biomaterial that incorporates folded, mechanically resistant globular domains in its constituent elastomeric proteins. Critical investigation of this novel biomaterial enabled Lv *et al.* to reveal how the microscopic properties of individual proteins are translated into macroscopic ones in biomaterials.

Indeed, the incorporation of folded, mechanically resistant globular domains into elastomeric proteins is a novel approach for creating biomaterials showing unusual macroscopic mechanical properties. This kind of bottom-up approach enables researchers to tailor the macroscopic properties of biomaterials by fine-tuning the **nanomechanical** properties of their molecular building blocks at the single-molecule level.

To perform their functions, various biological tissues exhibit distinct mechanical properties. For instance, mammalian tendon is highly resilient but relatively inextensible whereas elastin is quite resilient and extensible but lacks toughness (Gosline *et al.*, 2002). Mimicking the biomechanical properties of different tissues has long been a cherished desire in biomaterials research. The muscle-mimetic biomaterial designed by Lv *et al.* is not only highly resilient at low strains but is also extensible and tough at high strain and mimics the passive elastic properties of muscles.

As early as the 1920s, Nobel Laureate Hermann Staudinger had proposed that life is created when sufficiently large molecules are assembled. Later it was experimentally demonstrated that it is indeed the size and structure of these giant molecules that governs their function at both the nano- and macroscopic scales. The work of Lv *et al.* goes even further: they designed protein-based materials whose molecules adopt 3-D dimensional conformations similar to those of one of the biomolecules in muscle. These new materials also mimic the passive mechanical properties of muscle on the macroscopic scale. Lv *et al.* have shown that synthetic biopolymers can be made which emulate and expand on the technologically useful features of their natural counterparts (Chaikof, 2010).

Many structural proteins are elastic. Highly resilient proteins resist structural fatigue and failure when repeatedly subjected to loading forces, and they minimize energy loss. Collagen, elastin and resilin exemplify this category. The proteins in spider silk and mussel byssus fibres are less resilient – a property that allows the deformation energy in the fibres to be dissipated, thereby damping any vibrations produced by loading forces which could lead to structural failure of systems containing the fibres. The elasticity and energy-recovery properties of these structural proteins are well adjusted (fine-tuned) for their biological roles, and these are critical determinants of the normal physiological responses of a wide variety of tissues, including e.g., the cardiovascular and musculoskeletal systems (Chaikof, 2010).

Researchers have been keenly interested in seeking materials that either act like artificial muscle – by converting energy into motion – for biomedical applications, or serve as microvalves or microactuators in non-medical devices. Synthetic (non-protein) elastomers are extremely important but suffer from significant limitations as implantable materials – they not only do not facilitate tissue repair, remodelling or regeneration, but also sometimes provoke maladaptive host responses at tissue-material interfaces. In contrast, protein-based materials that mimic the properties of naturally occurring elastomeric proteins can be an effective alternative to synthetic elastomers. In fact, Urry *et al.* (2002) have already designed various different elastin-mimetic protein polymers at first by using synthetic-chemistry techniques and subsequently by expressing the proteins in genetically engineered organisms. Cappello *et al.*

(1990), Elvin *et al.* (2005) and Rabotyagova *et al.* (2009) designed materials which emulate the features of spider silk and resilin. These naturally occurring proteins contain repetitive oligopeptide domains that also contain the proline-glycine motif – a pair of amino acids that contributes to the flexible turns and/or extended random-coil conformations of proteins – conformations that may be responsible for the elasticity of the natural materials (Kim and Conticello, 2007).

As so presciently speculated by H. Staudinger, life may well be a mortal coil made of a diverse set of biopolymers. By experimentally demonstrating novel methods to wind protein threads into unique conformations, Lv *et al.* have advanced scientific frontiers into a fascinating world of molecular machines, nanoscale devices and molecularly engineered tissues having tailored macroscopic properties (Chaikhov, 2010).

AN INJECTABLE SYNTHETIC BIODEGRADABLE BIOMATERIAL

Polymeric biomaterials have catalyzed notable advances in surgical practice in recent years. Tailoring their macromolecular structure effectively integrates the chemical, physical, and biological properties needed for various medical applications. The rational design of polymeric biomaterials has exerted a welcome impact on drug and gene delivery, orthopaedics, tissue engineering, ophthalmology, and general surgery (Langer and Tirrell, 2004; Putnam, 2006; Wong *et al.*, 2007). Zawaneh *et al.* (2006, 2010) described the design of an injectable synthetic and biodegradable polymeric biomaterial made of polyethylene glycol and a polycarbonate of dihydroxyacetone (MPEG-pDHA). This MPEG-pDHA is a thixotropic physically cross-linked hydrogel that shows rapid chain relaxation and is readily extruded through narrow-gauge needles. It biodegrades into inert products and is well tolerated by soft tissues. Zawaneh *et al.* also showed the clinical utility of MPEG-pDHA in the prevention of seroma, a common postoperative complication following ablative and reconstructive surgeries, in an animal model of radical breast mastectomy. A seroma is defined as an abnormal collection of serous fluid within the body tissues like an internal blister. Surgeries that involve extensive tissue dissection and which create large empty spaces disrupt normal lymphatic flow, so that transduate fluid collects in these poorly drained spaces, leading to the formation of a seroma. Seromas can cause significant patient morbidity, e.g., infection, decreased limb mobility, and reoperation (Agrawal *et al.*, 2006). The common clinical practice involves placing silicone surgical drains in the wound bed by means of separate stab incisions to collect transduate fluid; however, they cause much pain and discomfort to patients and also increase the risk of infection at the surgical site.

Some approaches to reduce seroma formation use surgical techniques, such as collapsing the seroma cavity with sutures. They do not always adequately eliminate seroma formation. Methods such as sclerotherapy (Rice *et al.*, 2000), compression dressing and certain biological adhesives such as fibrin glue (Carless and Henry, 2006) do not significantly decrease the clinical incidence of seroma formation (Zawaneh *et al.*, 2010).

The above problems have prompted exploration of a more promising approach based on the application of synthetic biomaterials to eliminate dead space and reduce seroma formation. What we badly need is to design a surgical biomaterial with adjustable physiochemical properties that is biocompatible, easy to handle and apply, and which can prevent postoperative seromas. A design strategy that bridges the fields of polymer chemistry, polymer physics and medicine has been adopted (Zawaneh *et al.*, 2010). These authors had synthesized and characterized a diblock copolymer, MPEG-pDHA, comprised of an MPEG block, and a polycarbonate (pDHA) block based on the metabolic intermediate, DHA (Zawaneh *et al.*, 2006). Their 2010 paper discusses the further characterization of an MPEG-pDHA-based physically cross-linked thixotropic hydrogel that displays rapid chain relaxation, possesses high porosity and high water content. Its physical and rheological properties can be fine-tuned by adjusting the length of its constituting blocks. This novel biomaterial degrades into biocompatible products, has good *in vivo* biocompatibility and effectively eliminates seroma formation in an animal model of radical mastectomy.

ASSEMBLING BIOHYBRID MATERIALS

Feinberg *et al.* (2007) have reported the assembly of biohybrid materials from engineered tissues and synthetic polymer thin films. They built these constructs by culturing neonatal rat ventricular cardiomyocytes on polydimethylsiloxane thin films which were micropatterned with extracellular matrix proteins to promote spatially ordered, two-dimensional myogenesis. These constructs are termed muscular thin films. These films adopted functional three-dimensional conformations upon release from a thermally sensitive polymer substrate. They were so designed as to perform biomimetic tasks by varying tissue architecture, thin-film shape, and electrical-pacing protocol. These centimetre-scale constructs perform functions as diverse as gripping, pumping, walking, and swimming with fine spatial and temporal control, and can generate specific forces as high as 4 millinewtons per mm^2 (Feinberg *et al.*, 2007).

Muscle cells are microscale linear actuators that are driven by the activation of actin-myosin motors which are coordinated in space and time through excitation-contraction (EC) coupling (Bers, 2001). Structure-function relations are conserved over several orders of spatial magnitude, from the sarcomere to the muscle bundle, by virtue of a hierarchical architecture. Muscle actuation takes place over a wide range of frequencies, spatial dimensions and force regimes (Nishimura *et al.*, 2004). Artificial muscles can sometime match certain temporal, spatial, or force regimes typical of biological muscle but cannot fully replicate all these capabilities, nor can they use the same high-density energy sources. Thus, engineered muscle is important for building actuators and powering devices from the micro to macro scale. Designing such devices with engineered tissues faces the same technical challenges as therapeutic cardiac tissue engineering (Zimmermann *et al.*, 2006). Proper replication of morphogenetic coupling schemes in three dimensions is the most difficult of these. Tissue-engineered myocardium based on cardiomyocytes seeded into gels, rolled up from sheets, or released from surfaces carries the potential to produce actuators, tissue grafts, and power microdevices (see Xi *et al.*, 2005; Tanaka *et al.*, 2007). Recent researches of Whitesides *et al.* (2001) and Tan *et al.* (2004) on soft lithography have enabled replication of cell and tissue microenvironments *in vitro* as an alternative means to achieving the functionality of a three-dimensional device with a 2D tissue (see Feinberg *et al.*, 2007).

8

CHAPTER

Composite Materials

INTRODUCTION

A composite object is made up of two or more distinct parts. In materials science, composite materials are created from two or more components which remain separate within the final product. Some are as simple as a matrix material that envelops a reinforcing material, e.g. when concrete surrounds steel bars which prevent the concrete from failing under tension. Others include layered structures and reinforcing agents capable of functioning in all three dimensions. Composite materials have been typically regarded as materials which can save energy in large structures associated with transport. They are used to produce lightweight structures for fuel-efficient aircraft, lightweight cars, and high-speed trains, speedboats, and racing yachts. Although there are many options for making a composite, actually only a few combinations of materials combine synergistically, and the design criteria are sometimes not clear. A weaker reinforcing material can be added for creating a material that will absorb more energy before breaking. If such a composite fails, it may form many cracks, and it is the additional crack length that further strengthens or toughens the material (Lavine, 2006).

Many contemporary composite materials have attracted practical use in crafting energy-absorbing armour to protect vehicles and people. Here the materials are used to save lives and protect property by absorbing the energy of projectiles, impacts, and crashes (Hogg, 2006). Dealing with a knife attack, e.g., is very different from stopping a bullet and in the case of projectiles, stopping the second bullet or mortar is as important as stopping the first (Lavine, 2006).

Light-weight but super-strong cars can run through obstacles and suffer terrible crashes, yet the driver emerges unscathed. The secret of such amazing ability to protect the drivers is the extensive use of carbon-fibre composites for the cars' chassis and the exploitation of that material's ability to fragment in a controlled way. By fragmentation, a composite absorbs more energy per unit of weight than traditional engineering materials. Many conventional engineering materials such as metals absorb energy in a crash by plastic deformation when the structure crumples. Composites, in contrast, absorb energy by microfragmentation. Most industrial composites are being made from brittle fibres and a brittle polymer matrix, which is not so conducive to developing materials that absorb much energy on fracture. But, in a composite, the existence of innumerable interfaces between the constituents allows for multiple cracking, and the cumulative effect of countless tiny brittle fracture events is to generate a high ability to absorb energy and one that exceeds that of monolithic ductile materials. Composites based on organic fibres, such as Kevlar (polyaramid) and Spectra (ultra-high-molecular-weight polyethylene), also absorb energy by deformation of the fibre itself. Composite structures that absorb energy in crashes by crushing have found extensive use in helicopters and trains as well as light weight cars. They are also being adopted for use in more conventional but inexpensive automobiles.

This remarkable ability to absorb energy is also being exploited in the development of dedicated safety systems being used variously for the protection of structures, property, and people. Fibre-reinforced composites are used to produce the castings of jet engines. Castings produced in this way afford much better protection from broken fan blade debris and so protect the aircraft cabin from catastrophic damage if the blades break in use. Likewise the undershields of some cars can be made from fibre composites to protect delicate parts of the car against stone damage on bad roads. Composite armour is now not only effectively protecting military vehicles but also providing the basis for most forms of personal body armour. The advent of nanomaterials has generated new possibilities for the materials designer to create more and more exotic composites to meet the increasing threats to people and property (Hogg, 2006).

The strong need for a lightweight structure to allow military vehicles to move quickly and to be transported by air has boosted interest in new materials. Composite armour based just on fibre composites has been logically used to give additional protection in light vehicles such as Jeeps and Land Rovers where the panels are typically based on S-2 glass fibres [composition: SiO_2 65 weight % (wt %), Al_2O_3 25 wt%, and MgO 10 wt %]; these have a higher strain to failure and modulus than conventional low-alkali, aluminoborosilicate E-glass fibres (composition nominally SiO_2 54 wt %, Al_2O_3 14 wt %, CaO + MgO 22 wt %, and B_2O_3 10 wt %). Carbon fibres are unsuitable for such applications because the energy absorbed is less for a given weight – the strain to failure of the carbon fibre is lower than that of glass fibre, and the carbon fibre panel does not show as high a level of microcracking before final penetration as a glass fibre composite. Glass fibre composites are quite effective against small-arms fire but are not effective for stronger threat levels posed by large-caliber munitions and armour-piercing rounds. To develop protective armour to meet such threats, different types of composites are being explored, e.g. laminates of different materials which may themselves be fibre-reinforced composites or monolithic systems such as ceramics and metals (Hogg, 2006).

The term **integral armour** describes a laminated material consisting of a front-facing layer for blunting and abrading the incoming round. Meanwhile, a second layer supports the facing material during this initial impact; it then deforms and absorbs energy. Usually, the front-facing material is a ceramic, usually Al_2O_3, but some other materials e.g. TiB_2 have been considered. A hard and dense ceramic is essential because the material becomes pulverized into tiny fragments ahead of the projectile, which is worn down and abraded during its passage through the layer. Originally the backing layers used to be light metal such as aluminium, but today a fibre composite provides the best combination of energy absorption (stopping power) and weight.

A multi-hit capability necessitates the ceramic to be in the form of small tiles; the shattering of the outer tile should affect only a small zone. The use of intermediate layers between the ceramics and the fibre-composite backing layers minimizes the spread of damage. Some novel ideas are using a peg type of ceramic, almost like a short bolt and a hexagonal head which connect together on the surface and slot into holes in a backing structure. The lightest armour is being marketed by IBD in Germany which uses a nanocrystalline ceramic tile to minimize the spread of damage. Superior ceramic properties have been claimed for systems manufactured using a pressureless sintering technique, followed by hot isostatic pressure for consolidation to achieve highly dense >99% ceramic material. Other novel materials for the ceramic tile include functionally graded Ti-TiB_2 systems (Pettersson *et al.*, 2005). If a very thin fibreglass polymer composite is bonded onto the front face of a ceramic tile, it improves the ballistic performance by constraining the channel through which the plume of debris is ejected against the advancing projectile (Sarva *et al.*, 2005).

Some exciting developments in fibres used for composites with a bearing on armour are the extremely tough and stiff rigid-rod organic fibres such as *p*-configured poly (*p*-phenylenebenzobis-oxazole (PBO) (Van de Jagt and Beukers, 1999) and poly-{2,6-diimidazo [4,5-b:4′,5′-e]pyridinylene-1,4(2,5-dihydroxy) phenylene} (PIPD, better known in fibre form as M5) (Larsson and Svensson, 2002). These materials seem better suited to flexible armour as opposed to the rigid armour used in tanks. These fibres need to be able to extend to a larger extent than glass or carbon fibres so as to get the maximum benefit from their inherent toughness – this is restricted in the rigid panels developed for structural armour by the low strain to failure of the polymer matrices used (Hogg, 2006).

Flexible composite armour is used for body protection. In this, the material is mainly made of very stiff organic fibres in a dense fabric stack. The high stiffness ensures that the load is distributed over a large area of the material, and penetration of the fabric requires stretching of the fibres, which absorbs much energy. Flexible armour system must absorb the energy of the projectile, but must not deform so extensively that the wearer of the armour is crushed. Aramids such as Kevlar and Twaron and aliphatic polymers such as Spectra and Dyneema dominate this field. The new module of M5 and PBO fibres are much greater than those of many carbon fibres, and they also have very high strengths. There has, however, been concern regarding the degradation suffered by PBO fibres because of the effects of moisture, which has alarmed potential users. American Army tests suggest that M5 fibres can provide weight savings of upto 60% for protection equivalent to that provided by Kevlar fibres (see Hogg, 2006).

In the so-called "liquid armour" systems, the ability of a liquid-like system to stiffen rapidly under a shear stress is exploited to produce an instant "solid". This interesting phenomenon is being studied by using nanoscale silica particles thickening liquids such as water and ethylene glycol (see Tan *et al.*, 2005). Rapid loading makes the system stiff and strong. Use of these liquids to impregnate ballistic fabrics, during the shock loading by a projectile, surrounds the flexible fabric by an instant stiff matrix, forming a composite providing ballistic protection. The fabrics can be greatly improved by this shear thickening action of the surrounding liquids.

Other novel nanomaterials are sure to be used as parts of more complicated composite armours in the coming years. Whereas carbon nanotubes show extremely high strength and stiffness, inorganic fullerenes effectively absorb energy (Zhu *et al.*, 2005). These materials might well be incorporated into armour panels to reduce weight if the costs are not too high.

It is extremely difficult to simulate diffuse and complex fracture patterns in tough structural composites but some conceptual and computational advances in fracture modelling can help in overcoming the difficulty. Some success has been reported and includes the refinement of cohesive models of fracture and the formulation of hybrid stress-strain and traction-displacement models that combine continuum (spatially averaged) and discrete damage representations in a single calculation (Cox and Yang, 2006). New hierarchical formulations have created the potential of tracing the damage mechanisms down through all scales to the atomic. As the models approach the fidelity required for their use as virtual experiments, some opportunities are generated for reducing the number of costly tests required to certify safety and extending the design space to include material configurations that are too complex to be certified by purely empirical methods.

One long-standing problem of engineering has been how to predict the failure of materials. Predictions have been based mostly on empirical data gathering. For structures upon whose integrity human lives depend, the burden of testing to prove safety is immense. A typical large airframe, for instance may require $\sim 10^4$ tests of materials specimens, plus tests of components and structures up to entire tails, wing boxes, and fuselages, to fulfill the safety criteria for certification. Stress analysis is an excellent method for predicting the distribution of loads throughout the structure when its behaviour is linear-elastic, but if and when damage begins, prediction becomes difficult. The damage in tough engineering materials (excluding brittle materials such as glass) occurs through very complicated nonlinear processes acting from the atomic scale through the microscale (e.g., microcracking, crazing in polymers, plasticity in metals) and right up to the scale of the structure itself (e.g., large cracks and buckling modes).

The difficulties in solving such a problem arise not merely from computational power – the stronger challenge is how to categorize and characterize the various possible mechanisms of damage and incorporate them into a mode.

BOTTOM-UP AND TOP-DOWN MODELS

Current research can be broadly divided into two approaches to failure prediction: (a) the bottom-up and (b) the top-down methods. The **bottom-up** method has attracted many physicists and engineers (see Hirsch, 1975). It aims to simulate failure by building up detailed models of atomic and molecular processes by means of quantum mechanics and classical molecular dynamics. Some processes modelled satisfactorily are crystal plasticity, bond rupture in brittle materials for simple crack geometries, and the deformation of collagen-molecules and fibrils in natural composites (Buehler, 2006, 2006a; Gumbsch, 2001). The problem with this method is that the intervals of time and the size of the material amenable to modelling are many orders of magnitude less than the duration of a test and the size of a structural test coupon, let alone a structure; this means that the mechanisms revealed by the model (rather than written into the model explicitly) may not be exhaustive of those that arise in large-scale, long-term experiments. Although bottom-up simulations yield very good pictures of small-scale failure mechanisms, they do not represent the complex evolution of damage toward failure in engineering materials (Cox and Yang, 2006).

The **top-down** method focuses on engineering necessity to satisfy the constraint imposed at the structural scale that the displacement fields predicted for given boundary loads match those measured in tests. This can be assured for single cases by curve-fitting data, but their problem relates to embellishing the models with representations of the physics of failure that allow prediction for these cases (loads and geometry) which have not been tested. The top-down method begins with a macroscopic engineering model, which is progressively augmented by incorporating just those successive levels of detail that are required to account for features of engineering tests (Cox and Yang, 2006). As the model is calibrated against engineering tests, designers can use the predictions at any stage of model augmentation, subject mainly to some current restrictions. It is for this utility that the top-down method has interested engineers.

A balanced and wise option is to explore top-down and bottom-up methods concurrently and to use the sensitivity analyses of the top-down method to identify those mechanisms for which modelling at finer scales might be pursued, while employing the bottom-up method to yield the optimal parametric forms to be used when representations of mechanisms are to be incorporated in a top-down model. Indeed, research on hierarchical and multiscale models has facilitated formal linking of top-down and bottom-up approaches in single codes (Buehler, 2006, 2006a). These developments have concerned with the mathematical issues of embedding calculations representing fine-scale phenomena in those representing larger-scale phenomena. The mathematical challenges are how to devise hierarchical meshing strategies that are coarse enough at the largest scales to cover perhaps an entire structure, while cascading down through finer and finer meshes down to the atomic scale; and the different boundary-matching problems surrounding the integration of models executed over different time scales and built on meshes of different gauges.

DAMAGE EVALUATION

With carbon-fibre composites becoming the preferred choice for large commercial aircraft, the financial motivation for replacing experimental tests by virtual experiments has increased. Virtual tests make it possible to liberate the design space for composite materials. In the commonest strong composite materials, fibres are used in unidirectional plies (plies within which all fibres have a common orientation), which are stacked in oblique (angled) orientations to achieve high stiffness and strength in different directions. The stiffness and strength of an individual ply being very much higher in the fibre direction than in the transverse direction, tremendous performance improvement is achieved by tailoring the ply orientations to suit the anticipated load distribution. In the most general laminate concept, fibres that follow curved paths within plies are often preferred (Brown, 1998), for instance, to follow load paths around windows or doors. However, the requirements of current certification procedures compel designers to restrict themselves not only to unidirectional plies, but also to some combinations of ply orientations.

In general, a tendency among many champions of top-down methods is to fit engineering data with many different parameters and state that the resulting model is unique and loaded with predictive power. On the other hand, those who believe in bottom-up approaches claim that their favourite mechanism is the one that propagates up among all others to control engineering behaviour.

NANOPARTICLE POLYMER COMPOSITES

The mixing of nanoparticles (nps) with polymers to form composite materials has been practised for long. The clay-reinforced resin called Bakelite was introduced in the early 1900's as one of the first mass-produced polymer-np composite (see Balazs *et al.*, 2006). It radically transformed the nature of practical household materials. Indeed even before Bakelite, nanocomposites were used in the form of nanoparticle-toughened automobile tires prepared by blending carbon black, zinc oxide, and/or magnesium sulfate particles with vulcanized rubber. However, nanocomposites moved into the limelight only in 1993, following reports by Toyota researchers that adding mica to nylon led to a fivefold increase in the yield and tensile strength of the material (Usuki *et al.*, 1993; Kojma *et al.*, 1993). Since then, the interest in polymer-np composites has shot up: the increasing availability of nps of precise size and shape, e.g., fullerenes, carbon nanotubes, inorganic nps, dendrimers, and bionanoparticles, coupled with the development of instrumentation to probe small length scales, such as scanning force, laser scanning fluorescence, and electron microscopes, catalyzed research aimed at probing the influence of particle size and shape on the properties of np-polymer composites.

The mixing of polymers and nps has paved the way to engineering flexible composites endowed with highly desirable electrical, optical, or mechanical properties and researchers have sought design rules that would make it possible to engineer materials which combine the desirable properties of nps and polymers. However, they faced several challenges in producing nanocomposites that show a desired behaviour. The greatest hurdle in the large-scale production and commercialization of nanocomposites is the paucity of cost-effective methods for controlling the dispersion of the nps in polymeric hosts. The tendency of nanoscale particles to aggregate negates the merits associated with the nanoscopic dimensions. What we badly need is to design processing techniques that are effective on the nanoscale yet can also be applied to macroscopic processing. Another problem in the broader use of nanocomposites is the absence of structure-property relationships. There is an urgent need to correlate the morphology of the mixtures with the macroscopic performance of the materials. Such relationships may be elucidated through better understanding of how cooperative interactions between flexible chains and nanoscopic solids can lead to unexpected behaviour, like the improved mechanical behaviour of mica-reinforced nylon (see Balazs *et al.*, 2006).

Recent advances reveal routes to exploit both enthalpic and entropic interactions so as to direct the spatial distribution of nps and thereby control the macroscopic performance of the material. By tailoring the particle coating and size, self-healing materials for improved sustainability and self-corralling rods for photovoltaic applications have been created. A challenge for the future is how to create hierarchically structured composites in which each sublayer contributes a distinct function to yield a mechanically integrated, multifunctional material (Balazs *et al.*, 2006). As the interactions of nps with polymers are mediated by the ligands on the nps, the ligands significantly affect particle behaviour and spatial distribution. Considerable recent work has related to nanoparticle surface modification and the interactions between these particles and polymer matrices, including homopolymers, diblock copolymers, and blends (Balazs *et al.*, 2006).

Surface Functionalization of Nanoparticles

The surface chemistry of np functionalization progressed from studies on functionalized planar surfaces, including self-assembled monolayers (SAMs) and polymer brushes (Husseman *et al.*, 1999) on substrates ranging from gold to metal oxide. Functional small molecules and polymers can be bound to nps by physical adsorption or covalent attachment. Certain synthetic strategies that give polymer-functionalized nps conduct the particle synthesis directly in the polymer matrix, replacing small-molecule ligands inherent to a np synthesis with functional polymers in a "**grafting-to**" process, and growth of polymers from functionalized nps in a "**grafting-from**" process. It is, however, important that the conditions used preserve the specific characteristics of the nps.

Many state-of-the-art np preparations yield high-quality particles by decomposition of organometallic precursors in a solution environment that leads to surface coverage with small-molecule alkane-based ligands (e.g. *n*-alkyl thiols, amines, or phosphine oxides). Simply mixing these alkane-covered nps with most polymers leads to nonuniform particle clustering or aggregation that adversely affects the properties of the composite. Replacement of these conventional ligands with those that favourably interact with the polymer host is crucial for achieving uniform dispersions. In grafting from polymerization, the alkane ligands are replaced with functionality that supports polymerization radially outward from the nanoparticle surface. **Ring-opening metathesis** (ROMP) polymerizations exemplify mild polymerization methods which have proven useful in grafting-from experiments, as they give well-dispersed nps within the composite (Skaff *et al.*, 2002). On the other hand, the grafting-to approach is simpler to perform and is suitable for low-molecular-weight polymer ligands or when the nps are sensitive to polymerization conditions (Korth *et al.*, 2006). Grafting-to chemistry is also suitable for preparing polyethylene glycol-grafted (PEGylated) nps, and for the power which imparts hydrophilicity and biocompatibility to the nps.

Composites based on conductive or semiconductive nanoparticles have proved useful in electronic applications. They augment the use of conjugated organic polymers in flexible electronics, light-emitting displays, and photovoltaics. However, traditional alkane ligands insulate nps from their surroundings, and this undermines their performance. Organic-inorganic hybrid photovoltaics composed of polythiophene-quantum dot (or CdSe nanorod) composites show better performance with tailored ligand coverage, such as amine and phosphonic acid-terminated polythiophenes (Liu *et al.*, 2004). The greater dispersion of particles or nanorods in the polythiophene matrix enhances device performance as seen in much higher power conversion efficiencies than those of conventional blends, yet use only half the nanocrystalline material by volume.

Although uniform nanoparticle dispersions are desired for many applications, particle clumping can be used in some cases. In CdSe nanorods functionalized with alkane ligands in polystyrene (PS), because of the incompatibility of the ligands with the PS, the system phase separates into domains of PS and nanorods. Applying an electric field to a solution of nanorods mixed with PS aligns the nanorods in the direction of the applied field and, as solvent evaporates, the PS and oriented nanorods phase separate. Minimization of the interfacial energy between the domains leads to a close packing of the nanorods into a hexagonal array oriented normal to the film surface (Ryan *et al.*, 2006; Gupta *et al.*, 2006). Such assemblies have some potential use in photovoltaic devices where close packing is used along with surface patterning and interfacial interactions to direct the lateral positioning of the ordered arrays of nanorods onto an underlying circuit (Balazs *et al.*, 2006).

Nanoparticles have also been employed for changing the flow characteristics of homopolymer melts. Hong *et al.* (1999) reported that the addition of some nps to homopolymers eliminates "shark-skin" effects (i.e. instabilities created during the extrusion of polymers). In thin polymer films, large nps are used to nucleate dewetting, whereas smaller ones retard dewetting (Mackay *et al.*, 2002). In all these cases, the segregation of the nps to the interface between the polymer and the solid surface modifies the flow conditions at the interface.

Nanoparticles and Block Copolymers

The importance of diblock copolymer and np mixtures stems from the fact that the microphase separation of the copolymer can guide the spatial distribution of nps and hence tailor the properties of the composite. Bockstaller and Thomas (2004) showed that the optical performance of composites is very sensitive to the specific location of the particles within the matrix, and it is possible to gainfully use the microphase separation of block copolymers. But block copolymers do not just "template" the arrangement of the nps. The nps can act on the block copolymer, and nps can change both the orientation and the morphology of the diblock copolymer microdomains (Lin et al., 2005; Yeh *et al.*, 2005). Two notable features of these composites are: (i) the role that excluded volume effects of large nps play in the structure of the composite; and (ii) the effect of the particle dispersion on the macroscopic properties of the composite.

The nps being solid, the polymer chains need to stretch around these obstacles, thereby suffering a loss in conformational entropy that increases with particle radius (Thompson *et al.*, 2001). In the absence of specific interactions, larger nps are expelled from the bulk of the copolymers, whereas smaller particles are not. This significantly influences the spatial distribution of nps within homopolymers and block copolymers and the global structure of the particle-filled system.

It appears that the spatial distribution of nps in the microphase-separated morphologies can be controlled by tailoring the np ligands and the size of the nps relative to the radius of gyration of the polymer. This allows creation of gradient materials where the largest particles tend to be localized in the centre, bordered by the next larger particles, which in turn are neighboured by smaller particles (Lee *et al.*, 2002). Such gradient materials generate unusual optical or electrical properties (Bockstaller and Thomas, 2004).

If particle-filled diblock films are confined between two surfaces, the influences of chain stretching become conspicuous (see Lee *et al.*, 2003). If the system contains particles differing in size as well as in surface chemistry, it is often possible to carefully place nanoparticles within a diblock matrix (Bockstaller *et al.*, 2003) and so make structures that are clearly decorated with particles. It is therefore possible to achieve precise nanoparticle placement.

What is the role of the spatial distribution of nps within a material on its optical and mechanical properties? Buxton *et al.* (2003, 2003a) simulated the propagation of light through a diblock/np composite to find that the optical properties strongly depend on the spatial distribution of the fillers within the polymer matrix; they also suggested guidelines for tailoring the particles and polymers to yield the desired performance.

Nanoparticles and Mixtures

Lin *et al.* (2003) showed that nps often behave like surfactants and localize at the interfaces in oil/water mixture. In fact, the particles formed a monolayer around the dispersed droplets, and allowed the preparation of novel nanoparticle capsules. This kind of interfactal activity can also be exploited in polymer blends. Si *et al.* (2006) reported a significant reduction in the domain size in phase-separating binary blends by using clays modified with organic ligands as a result of the localization of the clay sheets at the interface between the immiscible components. When the domain size is controlled in this way, it improves the general mechanical integrity of the material.

Nanoparticles affect the phase-separation kinetics in polymer mixtures. They also significantly alter the coarsening dynamics of mixtures. Several workers have noted a substantial slowing of phase separation with the addition of nps to a polymer mixture.

As in the case of nanoparticle-filled diblocks, one unsolved problem concerns how the morphology of the filled blends affects the macroscopic properties of the composite. Buxton and Balazs (2004) integrated different computational approaches with a view to relating the morphology of a particle-filled mixture to mechanical behaviour and electrical conductance. In these rod-filled binary blends, the reinforcement efficiency of the nanorods and the electrical conductivity of the materials were significantly greater than the respective behaviour for homopolymer composites.

BIO-INSPIRED SYNTHETIC POLYMER NANOCOMPOSITES

Nature creates adaptive materials in various ways. When facing some danger, many echinoderms defend themselves. Sea cucumbers, e.g., rapidly, reversibly and strongly change the stiffness of their inner dermis (Thurmond and Trotter, 1996). This inner dermis is made of rigid, high aspect ratio of collagen nanofibres in a viscoelastic matrix of fibrillin microfibres (Wilkie, 2002). Upon encountering some danger, sea cucumbers secrete soluble macromolecules which trigger the formation of transient interactions among the nanofibres, facilitating rapid and reversible change of the stiffness of inner dermis whose tensile modulus changes by ~5 to ~40 MPa (Trotter *et al.*, 2000). Prompted by the striking self-defence mechanism of sea cucumber, Capadona *et al.* (2008) synthesized stimuli-responsive polymer nanocomposites. These synthetic nanocomposites not only have similar architecture but also similar chemoresponsive mechanical adaptability. The control of nanofibre interactions in solid polymer materials by chemical regulators was very similar to the tunable mechanical property of poly (acrylic acid) coated carbon nanotubes (Grunlan *et al.*, 2006) or cellulose nanofibres in aqueous dispersions (Lima and Borsali, 2004). According to Loveless *et al.* (2007), these materials suffer a marked viscosity change with variation of pH or ionic concentration.

The polymer nanocomposite described by Capadona *et al.* is composed of cellulose nanofibres embedded in a matrix made either by an ethylene oxide-epichlorohydrin (EO-EPI) 1:1 copolymer or from poly(vinyl acetate) (PVAc).

EO-EPI matrix has a fairly low tensile modulus and can absorb several chemical stimuli. Cellulose nanofibres isolated from the mantles of certain sessile marine tunicates have been used as the reinforcing component. These highly stiff 'whiskers' have nanoscale dimensions of ~ 26 nm x 2.2 μm. Tunicate whiskers used by Capadona *et al.* tend to aggregate. The whisker-whisker interaction could be moderated by introducing sulphate surface group, which promotes the dispersibility in hydrogen-bond-forming solvents. A proper balance between attractive and repulsive interactions is crucial to the fabrication of cellulose-whisker nanocomposites.

In the synthetic bioinspired polymer nanocomposites, self-interactions among whiskers can be switched off by introducing a chemical regulator that allows competitive hydrogen bonding. Also, the interaction among the whiskers can be switched 'on' by evaporating the solvent. The nanocomposite described by Capadona *et al.* shows a reversible change of tensile modulus by a factor of 40 (from 800 to 20 MPa) through the addition or removal of a chemical regulator which can modify the extent of hydrogen bonding of the whiskers (see Sur, 2010). These properties are potentially useful in applications such as brain implants in which it is crucial for the implanted material to be sufficiently stiff to facilitate insertion through the pial layer of the brain, but, once inside, the material can soften to match the stiffness of brain avoiding and minimizing injury.

The synthetic nanocomposite described by Capadona *et al.* shows excellent tunability by biologically relevant solvents such as water and cerebrospinal fluid. Being biocompatible, it is sure to attract potential applications in biomedical implants, robotic elements, adaptive protective clothings and as orthopaedic devices.

NANOCOMPOSITE PAPER FOR ENERGY STORAGE DEVICES

Ultrathin, flexible, safe energy storage devices are the need of the day to satisfy the design and power needs of modern gadgets. To create fully flexible and robust electrochemical devices, several components with specific electrochemical and interfacial properties need to be integrated into single units. Conway (1999) described flexible safe energy devices, based on supercapacitors and batteries for this purpose. Electrochemical energy has been stored in two different ways. In a battery, the charge is stored by electron transfer that produces a redox reaction in the electroactive materials. In an electric double-layer supercapacitor, the charge-storage process is nonFaradic – ideally no electron transfer occurs across the electrode interface, and the storage of electric charge and energy is electrostatic. Because the charging and discharging of such supercapacitors do not involve any chemical phase and composition changes, these capacitors show much cyclability. However, in certain supercapacitors based on pseudocapacitance, the essential process can be Faradic, like that in a battery, yet a fundamental diference from battery behaviour arises because in such systems, the chemical and associated electrode potentials are a continuous function of degree of charge, unlike the thermodynamic behaviour of single-phase battery reactants. The rising demand for efficient power devices to meet the high-power and energy applications has focused attention on the possibility of an ideal compromise that combines some of the storage capabilities of batteries with some of the power-discharge characteristics of capacitors in devices that store useful quantities of electricity that can be discharged very quickly. Pushparaj *et al.* (2007) showed that the basic components (the electrode, separator, and electrolyte) can all be integrated into single contiguous **nanocomposite** units that function as building blocks for a variety of thin mechanically flexible energy storage devices. **Nanoporous cellulose paper** embedded with aligned **carbon nanotube** electrode and electrolyte makes up the core or basic unit. The units can then be used to build various flexible supercapacitor, battery, hybrid, and dual-storage battery-in-supercapacitor devices. The merit of the thin free-standing nanocomposite paper devices is that they offer complete mechanical flexibility during operation. The supercapacitors operate with electrolytes including aqueous solvents, room temperature ionic liquids, and bioelectrolytes and over record temperature ranges. These easy-to-assemble integrated nanocomposite energy-storage systems could potentially provide unprecedented design ingenuity for various kinds of devices operating over a wide range of temperature and environmental conditions (Pushparaj *et al.*, 2007).

The performance characteristics of energy devices are primarily governed by the structural and electrochemical properties of electrode materials (Arico *et al.*, 2005). Electrolyte choice (aqueous or nonaquous), limiting high-power capability and packaging designs, are other important factors in supercapacitors and batteries (Hammami *et al.*, 2003). Previous designs of flexible energy-storage devices were based on separated thin-electrode and spacer layers. These designs performed not so well because of the existence of multiple interfaces between the layers. What Pushparaj *et al.* demonstrated is the fabrication of electrode-spacer-electrolyte-integrated nanocomposite units to build various thin flexible energy-storage devices. They combined two essential materials (cellulose and carbon nanotubes, CNTs) that suit the characteristics of spacer and electrode and provide not only flexibility but also porosity to the system. Cellulose, the chief constituent of paper, is an inexpensive insulating separator structure with excellent biocompatibility and can be made with adjustable porosity. CNTs, which show excellent flexibility, have been widely used as electrodes in electrochemical devices (Che *et al.*, 1998; Endo *et al.*, 2001; Frackowiak, 2007). The major problem in fabricating CNT-integrated cellulose composites is the insolubility of cellulose in common solvents. This problem has been solved by using a room temperature ionic liquid (RTIL) 1-butyl,3-methylimidazolium chloride ([bmlm][Cl]) described by Welton (1999); it dissolves up to 25% (wt/wt) of unmodified cellulose when microwave irradiation is used. Interestingly, the ionic nature of RTIL is quite conducive to its use as an electrolyte in supercapacitors (Kim *et al.*, 2005), and allows the assembly of all three components via a simple scalable process.

Pushparaj *et al.* grew uniform films of vertically aligned thin-walled multiwalled nanotubes (MWNT) on silicon substrates by using a thermal-chemical vapour-deposition method. Unmodified plant cellulose dissolved in RTIL ([bmIm][Cl]) was infiltrated into the MWNT to form a uniform film of cellulose and ([bmIm][Cl]) embedding the MWNT. After solidification on dry ice, this nanocomposite can be immersed in ethanol to extract excess RTIL and was dried *in vacuo* to remove residual ethanol. The resulting nanocomposite paper that serves as the basic building unit in the devices, can be peeled off from the substrate for use as the supercapacitor. The mechanical flexibility of the nanocomposite paper (CNT cellulose-RTIL) is excellent. It is possible to roll, twist, or bend this paper to any curvature and yet the paper is completely recoverable. The cellulose is porous with randomly distributed pores of 50 ± 5 nm. The nanocomposite paper, typically a few tens of microns thick, contains MWNTs as the working electrode and the cellulose surrounding individual MWNTs, as well as the extra layer as the spacer and the RTIL in cellulose as the self-sustaining electrolyte. Two of the nanocomposite units when bonded back-to-back produce a single supercapacitor device. The thin lightweight ($\approx$15-mg/cm^2) design of the device is made possible by avoiding the use of a separate electrolyte and spacer, typically used in conventional supercapacitors. The use of RTIL electrolyte ($\approx$3 wt/wt % of cellulose in RTIL) also makes the device environment-friendly. In a final package, operating devices can be readily fabricated by laminating multiple stacks of individual nanocomposite layers.

STIFF COMPOSITE MATERIALS

In most elastic systems, a deformed object encounters a force with a component in the same direction as the deformation, and the stiffness is positive. Negative structural stiffness (i.e., force/displacement ratio) can manifest in prestrained objects such as tubes buckled to form kinks, which contain stored energy at equilibrium (Thompson, 1979). Negative structural stiffness has been found in lumped buckled systems made of rubber tubes and models of single foam cells (Lakes *et al.*, 1993; Lakes, 2001).

The elastic modulus, a stress/strain ratio, is used as a measure of material stiffness. A negative modulus may occur in a predeformed object and then indicates a negative incremental modulus. An object with all free surfaces and a negative modulus is unstable. It is possible to stabilize a solid object with a negative bulk modulus (inverse compressibility) by a constraint of the surfaces, in contrast to fluids (see Jaglinski *et al.*, 2007). Negative incremental compressibility has been reported in small-cell foams. It differs from negative thermal expansion - the latter represents the stable contraction of an unconstrained object because of a temperature increase.

Jaglinski *et al.* (2007) showed that composite materials can have a viscoelastic modulus (Young's modulus) that is much greater than that of either constituent. The modulus, but not the strength, of the composite was significantly greater than that of diamond. These composites contain barium-titanate inclusions, which undergo a volume-change phase transformation if they are not constrained. These were prepared with inclusions of barium titanate ($BaTiO_3$) in a tin matrix – $BaTiO_3$ is not only a ferroelastic but also a ferroelectric solid that shows a crystal volume change and a crystal shape change during phase transformations, specifically cubic-to-tetragonal transformations at 120°C. In the composite, the inclusions are partially constrained by the surrounding metal matrix. It is this constraint that stabilizes the negative bulk modulus of the inclusions. The negative modulus arises from stored elastic energy in the inclusions, in contrast to periodic composite metamaterials that show negative refraction by inertial resonant effects. Conventional composites with positive-stiffness constituents have aggregate properties bounded by a weighted average of constituent properties; their modulus cannot exceed that of the stiffest constituent (Jaglinski *et al.*, 2007).

ADAPTIVE COMPOSITES

Exciting work is currently underway in developing materials that can respond dynamically to changes in their environment. Two interesting conceptual possibilities are to develop: (1) a search-and-rescue robot that can change shape to move through crevices with the suppleness of an octopus; or (2) an aircraft skin with a circulatory system conducive to temperature regulation, cooling, and self-healing similar to an animal. Such concepts are inspiring the designing of adaptive composites that mimic biological responsive functionality while working in extreme environments (Vaia and Baur, 2008).

Most high-performance, load-bearing substructures of aircraft, satellites and robots have so far been designed for structural efficiency, i.e. are rigid and passive. Such active functions as sensing, energy harvesting, and propulsion can be added by attaching suitable components to the structure. Such a compartmentalization of functions into attached subsystems not only streamlines manufacturing and maintenance but also facilitates upgrades (Pfeifer *et al.*, 2007). Some advanced passive material technologies, e.g., continuous-fibre organic-matrix composites, have been gainfully applied to sporting equipment, prosthetics, satellites, and aircraft. But this approach is unlike the situation in natural organisms in which flexible, jointed frameworks and complex materials confer functionality at multiple length scales within the materials. Analogous synthetic structures may be conceived by the combination of new materials that deliver active properties and autonomic response, as well as new computational tools that facilitate design, analysis, and optimization of the collective and hierarchical dynamic character.

The information of a rigid substructure to a dynamic, articulated structure is specially relevant to the needs of the aerospace industry. Large-scale actuated bending, extending, and folding structures could enable the use of large-scale antennae in space as well as the development of morphing wings on unmanned aircraft. Energy-efficient locomotion and concealment are some other applications for new adaptive materials. Monolithic piezoelectric ceramics are materials that show reversible property changes (Leo, 2007). Now, novel multicomponent, hierarchical material systems have been inspired by biology. For instance, a cell's ability to change shape and recover from large deformations comes partly from the arrangement of actin filaments in the cellular cytoplasm. Analogous highly deformable networks can be made from cellular materials, bistable composite laminates, and bimorph strips. The elastic buckling of carbon nanotube mats and arrays has been used to store and recover substantial amounts of strain energy (see Cao *et al.*, 2005). Integrating these concepts with materials having a reversible stiffness allows trapping and subsequent release of the mechanical deformation, e.g., the stiffness of shape-memory polymers can be reversibly changed between a thermoplastic glass and an elastomer if triggered (by temperature or other means). In fact, they even "remember" their original shape upon undergoing softening. Hinges made from an elastic memory composite (consisting of predeformed carbon fibres in a shape-memory polymer matrix) can be a viable replacement for current complex mechanical hinges often employed for unfurling structures such as solar arrays (NASA, 2007).

An important point is to trigger reversibility easily and controllably. The trigger for mechanical property changes is usually temperature – it can slow down the transition and complicate control systems. Certain other triggering stimuli might be better: Ebron *et al.* (2007) used a coating of catalytic nanoparticles on a shape-memory alloy wire for converting enough chemical energy to thermal form to trigger a shape recovery of the wire. Better understanding of natural systems has also highlighted chemical triggers. For example, addition of acetylcholine and calcium ionophores shifts the iridescence of the mantle iridophores of the cephalopod *Lolliguncula brevis* from red to blue, thereby providing dynamic coloration. Recombinant derivatives of comparable proteins can be self-assembled *in vitro* to make fibres and diffraction gratings with dynamic coloration (Kramer *et al.*, 2007).

It is also possible to use radiation as a trigger for adaptive response. Optimically triggered reversible colloid surface chemistry could potentially lead to rheological fluids which do not require a bias to maintain an ordered electrical or magnetic state (Nelson and Braun, 2007). Buckley *et al.* (2006) have discussed shape-changing medical devices that can be activated by radio-frequency radiation.

Materials for adaptive composites need not always be solid constituents; fluid networks may also be incorporated. Toohey *et al.* (2007) have developed a self-healing system based on a 3-D microvascular network embedded in the substrate. This network can autonomously deliver a healing agent which repairs cracks in a damaged polymeric surface sheet.

Natural systems also offer exciting possibility for designing and pumping concepts for circulation. The heart of the zebra fish efficiently pumps fluids along a longitudinally graded channel via resonate coupling of an excitation (Vaia and Baur, 2008). The xylem and phloem vessels of vascular plants are good examples of the utility of axially graded channel walls.

Inclusion of fluids and associated mass transport within a composite could potentially generate novel active functions such as thermal management, variable modulus, and dynamic optical transmission. Unfortunately, the use of fluids also has some negative aspects – the resulting materials tend to be more complex and less durable.

POROUS CERAMICS

Highly porous structures are widespread in the natural world (Perkowitz, 2000). Their design enables efficient optimization of characteristics such as the strength-to-density and stiffness-to-density ratios. Synthetic porous ceramics are better than metallic or polymeric components, especially when resistance to high-temperature of corrosive environments or compatibility with biological materials is required. The wide range of morphologies and properties achieved recently for porous ceramics through progress in fabrication methods has resulted in their use in a vast range of applications, including catalyst supports and chemical reactors, biomedical tracking and delivery platforms, electrodes, insulators, and heat exchangers (see Oh *et al.*, 2006; Scheffer and Colombo, 2005). It has now become possible to introduce nanopores in ceramics, opening exciting possibilities for the development of smart devices such as photoactivated sensors and switches and drug delivery capsules (Colombo, 2008).

For these various applications, the porous component should have a specific range of values for different properties (Fig. 8.1). These can only be achieved through careful choice of starting materials and well-controlled processing (Gibson and Ashby, 1999). A wide range of physical properties can be achieved by changing the amount of porosity through processing. Pore size, connectivity, and distribution also affect the properties of the component and can be used to further tailor its characteristics (Colombo, 2008). The actual manufacturing process greatly influences the amount of porosity (from a few percent to more than 95% in volume), the pore size (from nanometers to millimeters), the distribution of the pores within the solid, the shape and interconnectivity of the pores and such other properties as the amount, size, and morphology of defects, and the size and cost of the component. All these considerations make the introduction of porosity a highly versatile and robust tool for greatly extending the range of properties of a ceramic component.

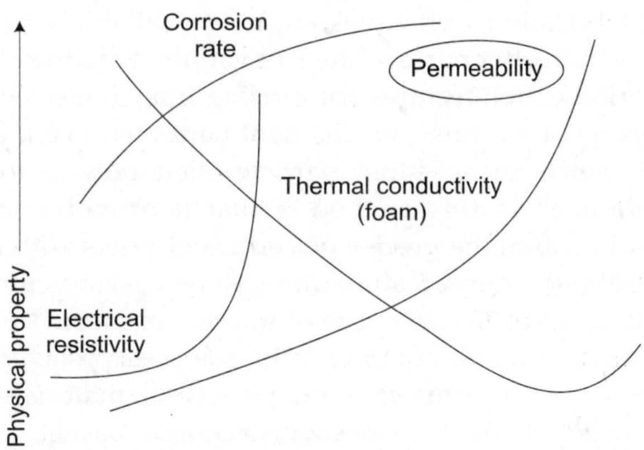

Fig. 8.1 Relations between porosity and physical properties. A wide range of physical properties can be achieved by changing the amount of porosity through processing. Pore size, connectivity, and distribution also affect the properties of the component and can be used to further tailor its characteristics (From Colombo, P. *Science* 322: 381-383, 2008. Reprinted with permission from AAAS).

Macroporous cellular (foam-like) ceramic structures have traditionally been produced by dip-coating a polymeric foam into a ceramic slurry, followed by burnout of the preform and sintering. This method leads to high-porosity, low-cost parts of limited strength suitable for molten metal filters or kiln furniture. Similar porosity is also generated through the elimination of "sacrificial filler" materials by burnout or dissolution, offering tighter control on the average pore size distribution, with a wider range of cell size and amount of porosity. It is also possible to use direct foaming of a ceramic slurry (for example, through mechanical frothing, gas injection, or *in situ* gas generation by decomposing an organic additive and followed by setting (e.g., by gelling or particle stabilization of the liquid-gas interface) and sintering (Studart *et al.*, 2006). This process makes very strong components conducive to the manufacture of smaller parts, such as thinner filters, and the use of porous ceramics in load-bearing applications, such as implants for bone substitutes but precise control of the cell size cannot be achieved.

Diesel particulate filters require unidirectional macropores (as in honeycombs) in which alternate channel openings are blocked. Such a structure forces the gas to pass through the channel walls, depositing the soot on their surface. Such "wall-flow filters" are obtained by extrusion of a ceramic paste and their production has been based on needle-shaped mullite crystals that show improved performance at high porosity volumes (Pyzic and Li, 2006).

Two other modern processing methods that make possible fabrication of mainly unidirectional macropores are freeze casting and pyrolysis of wood or assembled cardboard in an inert atmosphere (Deville, 2008). In the former process, the porous structure is templated by solidifying a solvent, leading to unique morphologies in which the lamellar surfaces show dendritic-like features along the solidification direction. The second process yields complex carbon-based porous structures which may either be used as received, or can be further reacted with gaseous or liquid precursors to form SiC or other advanced ceramic compositions that mimic the morphology of the parent component. These materials have potential use in diesel trap applications, for biomedical applications (bone implants), as sandwich panel cores, in filtering devices, and as lightweight ceramic components (Colombo, 2008, 2008a).

Computer-assisted fabrication techniques are now available for making highly regular macroporous structures which offer narrow range of variability in their morphology and hence their properties. 3-D periodic architectures comprising interconnected cylindrical rods have been made by the extrusion of a continuous filament patterned in a layer-by-layer computer-controlled building sequence, using either particle-filled polymeric inks or concentrated colloidal gels. The advantage of this method is that it provides components with well-controlled properties and enables the precise placement of pores with desired size and shape within the component. Submicrometre feature dimensions, and structures with an area of 1 m^2 can be printed at speeds of up to 10 cm/s. Use of multinozzle arrays can significantly reduce manufacturing times, making this a commercially viable approach for producing ceramic membranes and filters, catalyst supports, and structural materials with well-controlled properties. Infiltrating the polymeric structures with a ceramic or metallic material, followed by sintering, generates inverse wood-pile structures (a solid matrix with a periodic array of tubular holes), with possible applications as photonic crystals, in low-cost micro-electromechanical systems, as microfluidic networks for heat dissipation, as well as in biological devices (Colombo, 2008).

Ceramic powders are used as the main starting materials in the above processes. If components with pores in the microscopic and mesoscopic range are to be built, ceramic precursors have to be used. Two chief approaches are used to fabricate ceramics with pores in these size ranges: sol-gel precursors, which mainly give oxides, and preceramic polymers, which yield more complex high-temperature ceramics.

Sol-gel precursors allow fabricating of unique nanostructured porous morphologies and components. Either via a self-assembly route using surfactants, or via nanocasting, a highly ordered array of uniform pores in the mesoscopic range can be formed, typically based on oxides such as a silica or tiania. Using certain precursors such as silsesquioxane leads to periodic mesoporous organosilicas, allows the precise tailoring of the size, thermal stability, and surface reactivity of the pores through the addition of organic chains and functional groups (Hunks and Ozin, 2005). Powders, fibres, monoliths, and thin and patterned films can be produced, although the stability of the nanometre-scale pores at temperatures above a few hundred degrees Celsius is not so good and also achieving a uniformly well-ordered structure in large components is difficult. Applications include catalysis, gas sensing, optics, and photovoltaics. An alternative is to use tunable mesoscale pores, coupled with very high specific surface area, achieved by etching carbides to selectively remove non-carbon atoms (Gogogtsi *et al.*, 2003). These mesoporous carbons find use as separation media, catalysts, and advanced electronic components for energy conversion and storage (Colombo, 2008).

Preceramic polymers can lead to SiC, SiCN, or BN mesoporous materials that offer better high-temperature tolerance than oxides. These preceramics are inorganic or organometallic polymers that can be converted into advanced ceramics via high-temperature pyrolysis. In view of their polymeric behaviour at low temperatures, these precursors are easily processed by blowing, infiltration or coating of performs, or warm pressing with sacrificial fillers to form macroporous components. In fact, ceramics with a hierarchical pore structure and high specific surface area can be fabricated in a single process (Colombo, 2008a). By means of a suitable fabrication method such as electrohydrodynamic spraying, a variety of miniaturized polymeric porous structures such as bubbles, capsules, nanofibre mats, microtubes, patterned microchannels, all with controlled surface characteristics, can be made and subsequently transformed into a ceramic (Nangrejo *et al.*, 2008). They find use as lightweight fillers for syntactic foams, implantable capsules for drug delivery, components for tissue engineering, and microfluidic devices.

Continuing progress in fabrication methods can increase component sizes, reduce their cost, increase the range of pore sizes, and achieve a tighter control on their morphology. Recent advances in 3-D imaging techniques (e.g., computer-assisted tomography) may enable critical quantification of the pore morphology, providing valuable data for modeling of the pore structure and properties and assisting in the design of reliable porous components. The coupling of porosity with other functionalities (such as electrical conductivity, or ferromagnetic properties) in the same element can lead to highly versatile filters and to components with an even broader range of properties, for use in advanced applications such as solid-oxide fuel cells, absorption of electromagnetic radiation, regenerable adsorption, and such acoustic devices as hydrophones (Colombo, 2008, 2008a).

OSTEOINDUCTIVE CERAMICS

Biomaterials were earlier used as biologically passive medical devices and provided only some structural support. In recent years, their properties have been exploited to facilitate the process of tissue regeneration. This change has been driven by several reports demonstrating that it is possible to modulate cellular behaviour by material properties such as surface texture, elasticity, and chemistry (Engler *et al.*, 2006; Place *et al.*, 2009). In biomaterials for the restoration of bone defects, surface topography influences osteogenesis and proliferation of bone marrow-derived multipotent mesenchymal stromal cells (MSCs) *in vitro* (Dalby *et al.*, 2007). However, though tissue instructive materials carry much potential as off-the-shelf bioactive medical devices, the concept has not progressed beyond the proof-of-concept phase in which *in vitro* assays show an effect on cellular differentiation or proliferation (Kohn, 2004). This is why transplantation of autologous bone still continues to be the reference standard in bone repair strategies. It is autografts which guide the in-growth of osteoblasts, the primary cell type responsible for bone matrix apposition, from the adjacent tissues into the defect – a process termed **osteoconduction** (Yuan *et al.*, 2010). Further, autologous bone stimulates *de novo* bone formation by triggering the differentiation of undifferentiated progenitor cells into the osteogenic lineage, referred to as osteoinduction. This latter process is crucial for the repair of major bone defects. Some drawbacks of autografting are the limited availability of autologous bone and the adverse side effects of bone harvesting; these intensify the search for bone graft substitutes (Damien and Parsons, 1991).

It was reported by Urist (1965) that osteoinduction can be achieved by **devitalized demineralized bone matrix (DBM)** and later **bone morphogenetic protein (BMPs)** were identified as an alternative to bone autografts. Both DBM and BMPs are widely used in the clinic; their biological nature has significant implications for the production process leading to rather high batch variability and high production costs (Wang *et al.*, 2007; Bormann *et al.*, 2010). Furthermore, the *in vivo* delivery of soluble molecules such as BMPs is inefficient. One alternative to the biological approach to bone regeneration may be to develop a synthetic material with intrinsic osteoinductive capacity. Several porous calcium phosphate biomaterials as well as some metals possess significant osteoinductive capacity (LeGeros, 2008). The osteoinductive potential of biomaterials can be controlled by tailoring such material properties as chemical composition, surface topography, and geometry, which affect resorption rate and cell-material interactions (Yuan *et al.*, 2010).

Porous calcium phosphate ceramics can be used to treat small bone defects. They usually do not induce stem cell differentiation which is essential for regenerating large bone defects. Yuan *et al.* have prepared calcium phosphate ceramics with varying physicochemical and structural characteristics. They showed that microporosity is correlated to their tendency to stimulate osteogenic differentiation of stem cells *in vitro* and bone induction *in vivo*. Implantation in a large bone defect in sheep demonstrated that osteoinductive ceramics are as efficient in bone repair as autologous bone grafts. These results provide proof of concept for the clinical application of "smart" biomaterials.

PORE-FREE CERAMICS

Pore-free ceramics having grain sizes in the nanometer range attract considerable interest for their potential and unprecedented optical, mechanical, electrical, and other properties for use in lasers, health care, and electrical devices (Messing and Stevenson, 2008). In most of today's dense ceramics, pores with diameters of 100 to 1000 nm occupy no more than about 5% of the volume of the material. Ongoing progress in nanoparticle processing, multi-step sintering cycles, and novel densification procedures could potentially yield fully dense nanoceramics.

Large pores often impair the properties of ceramic materials and components. A good example is that of the zirconia or alumina balls used in hip prostheses, which must support load without failure for at least 15 years; pores act as fracture initiation sites that decrease mechanical strength and sometimes cause catastrophic failure. The effect of pores is also exemplified by transparent neodymium-doped yttrium-aluminium-garnet (Nd:YAG) ceramics, which show better lasing performance than single crystals (Ikesue *et al.*, 2006) and could even supplant Nd:YAG single crystals in futuristic high-power lasers. Pores scatter light and reduce optical transmission in laser ceramics; so their applications require 100% density. Ceramic components, like barium titanate ($BaTiO_3$) capacitors in electronic devices also become increasingly susceptible to the effects of pores as devices shrink, because pores in the ceramic create weak links that lead to dielectric breakdown and hence to device and system failure (Messing and Stevenson, 2008).

Fully dense ceramics with grain size below 100 nm may be made by starting with 10- to 50-nm particles, which are consolidated into product-specific shapes from colloidal suspensions, dried and then heated (sintered) to remove the porosity between the particles. The mechanism of pore elimination during sintering is the reduction of surface free energy of the ceramic particles, with a magnitude roughly equivalent to 1 megapascal (MPa). Pores are removed by mass transport to the pore surfaces. As mass transport along grain boundaries is much faster than that within grains, pores need to be located along grain boundaries for a ceramic to become fully dense (that is, pore-free). As grains grow, pores are trapped inside grains if their intrinsic mobility is less than the mobility of the grain boundaries – a process called pore-boundary separation (Messing and Stevenson, 2008).

The best approach to prevent pore-boundary separation is to ensure that all pores are smaller than the average particle size after the starting powder is consolidated into the required shape. The best particle packing (and hence the smallest pore size and the narrowest pore size distribution) can usually be achieved by consolidating discrete, narrowly sized, fine particles to random close-packed density of around 64%.

Nanoparticles substantially increase the driving force for sintering and reduce the sintering temperature. They can potentially lower sintering temperature and reduce sintering time. Yet, only in a few cases has it been possible to achieve small-grain-size, fully dense ceramics by pressureless densification of nanoparticles (Krell *et al.*, 2003). The work of Mizuno *et al.* (2007) showed how to retain extremely small grain size while achieving near full density. In this process, particles with diameters of 10 to 30 nm could be dispersed in open organic solvents to avoid particle solubility and to limit interparticle forces. With suitable atmosphere control, multistep sintering cycles, and dopants, fully dense $BaTiO_3$ particles with grain sizes of <100 nm can be commercially produced; experimentally, $BaTiO_3$ with <50-nm grain size has been achieved. These techniques apply equally well to other ceramic systems but have actually been demonstrated only in $BaTiO_3$.

The successful demonstration of transparent, polycrystalline Al_2O_3 exemplifies sintering pore-free, nano-grained ceramics endowed with novel properties. Fully dense polycrystalline Al_2O_3 with micrometer-sized grains is translucent, but by refining the grain size to below 100 nm and sintering to near 100% density, the major light-scattering mechanism is changed, allowing the material to become transparent.

For better quality ceramics such as hip prostheses and optical components, pressure facilitates densification. Mechanical hot pressing and hot isostatic pressing (where gas is used as the pressure-transmitting medium) are being used commercially to create fully dense ceramics. At pressures of ~200 MPa, hot isostatic pressing has become the standard for commercial applications requiring total pore elimination, but cost and contamination limit its general applicability to only a few ceramic systems.

Microwave heating and spark plasma sintering have attracted much interest for sintering ceramics. Microwave sintering enables faster densification than is possible with conventional sintering. Spark plasma sintering applies pressure like a hot press but at the same time heats the ceramic by pulsing an electric current through it. It limits grain growth by reducing densification time from hours to minutes.

POROUS MATERIALS

Since the dielectric constant of gases does not differ much from that of vacuum, the incorporation of free space, or pores, is a good way to decrease the dielectric constant of films. This is done on a submolecular level by using bulky groups such as fluorenyl in the repeating units of polymers to decrease packing density and increase free volume. Calculations of dielectric constants indicate that this should result in lower dielectric constants (Van Krevelen, 1990).

Foamed films with pores in the 10 nm range show quite low dielectric constants. Aerogels or xerogels with SiO_2 as matrix with porosities above 90% and dielectric constants close to 1 have been reported (see Maier, 2001). However, large volumes of solvent have to be removed to obtain such extremely porous structures. Since the pores must not interconnect, the solvent or solvent vapour has to diffuse through the matrix material. Control of the process without shrinkage and formation of macroscopic cracks is difficult.

A new intriguing methodology has been developed by taking advantage of the phase separation of block polymers (Maier, 2001). This results in the formation of spherical domains of the minor component at certain compositions under appropriate conditions of film formation. Polyimides or PPQs [poly(phenylquinoxalines)s] with thermally labile blocks or grafts from poly(propylene oxide) poly(methyl methacrylate), polystyrene, poly(α-methylstyrene), and polyesters have been prepared. Up to 30% (volume) of the labile blocks were incorporated into the films. Heating of the films to temperatures of 250-300°C degrades the thermally labile blocks leaving behind pores which are identical in shape and volume to the space previously occupied by the labile polymer blocks. The exact decomposition process depends on the nature of the labile blocks. Porosities up to 30% have been achieved in this way, but one major problem is that the pores often collapse. In addition, when the temperature used to generate the pores or to fabricate the interconnects is too close to the glass transition temperature of the polymide phase, softening of the matrix collapses the pores by lowering the surface tension.

It appears that dielectric constants below 2.5 will be only reached with porous materials, as long as the absence of fluorine and thermal stability above 400°C can be assured as key requirements; whether the matrix will be organic or inorganic materials is not known.

The situation changes dramatically when processing temperatures are reduced to 350°C or even 300°C when a large number of polymers become applicable.

The further development of dielectric materials is tightly linked to chip manufacturing processes. There appear to be three scenarios for the future. First, the requirement of $E<2.5$ and for future chip generations $E<2$ is maintained. In this case, the use of porous dielectrics is unavoidable, and all back-end-of-line (BEOL) manufacturing steps need to be adjusted accordingly. Second, porous materials are not acceptable, in which case polymers with bulk dielectric constants of approximately 2.6 must be used, and BEOL-manufacturing steps will have to be optimized accordingly. Some optimization of the wiring scheme is required as well. Third, if manufacturing cannot be adjusted to the needs of polymers there will be no materials with $E<3.5$ (fluorine doped SiO_2) and the wiring scheme, geometry (pitch, thickness, length), number of metallization levels and die size will be the only parameters which can be used to overcome the various problems (Maier, 2001).

NANOCERAMICS AND NANOCOMPOSITES

Nanoceramics/nanocomposites are novel bulk materials or coatings having microstructural architecture in which at least one of the ceramic phases is in the nanoscale (1-100 nm). Nanoceramics and their composites have attracted interest because they can be manipulated to show better, even some unusual, material properties. Improved performance and newer applications of the materials, e.g., durable bone replacement materials by using polymer-ceramic nanobiocomposite have become possible.

Research on nanoceramics is focused on production of nanoceramics and nanocomposites, and their properties (e.g. creep and superplasticity), dielectric and ferroelectric properties, nanostructured hard coatings, metallic nanocomposites, biomimetics and tissue engineering, and nanofluids.

Multi-function integrated nanocomposite materials are being developed in several different ceramic systems with a view to improving the mechanical and thermal properties of structural ceramics. Nanocomposite structure carries high potential to develop various kinds of multifunctional materials for the energy and environment sectors. A polymer-based nanocomposite with a new sensing function, like the human finger tip, has been industrialized as a special sensor of the intelligent robot. Recent years have also seen the development of 'intermaterials' which represent a special class of multifunctional materials containing inorganic, organic and metal materials, incorporated in the nano, molecular and atom level. Some other design concepts on molecular/cluster/lattice-level composites are attracting keen research interest.

It is now possible to form high-performance ceramics at low processing temperatures (<1200°C) by using polymer-derived nanostructured ceramics. Several new processing methods have been developed for metal and ceramic nanopowders, such as metal and oxide nanopowders. One such method is called **pulsed wire discharge (PWD)** method which may possibly enable large-scale production in the industrial environment. This highly energy-efficient method facilitates production of passivated metal nanopowders, which find use as conducting paste for electromagnetic radiation shields and in electric circuit boards.

The Korea Institute of Materials Science, Korea has worked on two-step sintering to produce nanocrystalline bulk silicon nitride ceramics which show enhanced densification with minimum grain growth for silicon nitride ceramics with liquid phase.

A new laser-based chemical vapour deposition (LCVD) technique can achieve 100-1000 times greater deposition rates than conventional CVD for wide-area coatings (see Basu, 2008). The LCVD process may prove helpful in obtaining various kinds of oxide coatings by changing the precursor source materials. In fact, this novel technique for thick oxide coatings may be applicable to a wide range of engineering coatings.

Biomimetic nanocomposites are being considered for tissue engineering, following exploitation of molecular cavities present in globular proteins to carry out *in situ* synthesis of magnetic and bioceramic nanoparticles. Conceivably, such nanoparticles might be further transformed into aqueous ferrofluids and microporous hydroxyapatite granules for biomedical applications.

There is keen interest in using nanostructured ceramics for fuel cell and battery applications. The major motivation for the use of such nanostructured ceramics is to enable the operation of **solid oxide fuel cells (SOFC)** in the temperature range 500-650°C. The doping of nanosized silver (Ag) in Li-CoO_2 cathode increases the electrical conductivity by upto 4-3 orders of magnitude with reference to undoped materials. Nanocrystalline $Li_4Ti_5O_{12}$ has been synthesized by using an aqueous combustion process with alanine fuel.

Polymer Nanofibres for Ceramics

Polymer nanofibres can be spun into free-standing, hollow cylinders which look as if they have been shaped on a tiny pottery wheel. An electric field has been used to tease a nanometer-scale jet of polyethylene oxide solution from a capillary tube. When jet was dried in mid-air in less than a second, it coiled up into a spool a few μm in diameter upon hitting a sharp stainless steel tip 2 mm below the capillary tube. Such structures have potential use in nanoscale magnets, bioscaffolds or nanochannels (see Kim *et al.*, 2005, 2010).

Some strong materials, e.g., ceramics are often brittle and prone to snapping – but those that deform gracefully under tension, such as metals, are weaker. Jang and Greer (2010) overcame this trade-off with zirconium-based metallic glasses; these are alloys composed of disordered metal atoms and can be moulded when heated. When the material is reduced in size to pillars measuring 100 nm in diameter, it assumes **both a metal-like ductility and a ceramic-like strength.**

It appears that internal flaws in the glass – which would shear the alloy apart in bulk form – find it much harder to propagate through the nm-scale structure.

STRUCTURAL MATERIALS

Fibre-reinforced composites (FRCs) are being used as structural materials in view of their good performance at low weights. For any given FRC, the mechanical properties depend on such variables as the distribution and orientation of the fibres within the matrix material. However, for most composites, nondestructive visualization of the response of the fibre to stresses is ruled out either by the opacity of the matrix material or by low or weak contrast between the fibres and the matrix.

This problem has been addressed by probing an FRC through diffraction of a high-flux, microfocused x-ray beam from a synchrotron source. The sample consisted of woven poly(*p*-phenylene phthalamide) (PPTA) fibre mats embedded in an epoxy matrix. A hole was drilled into the center of the specimens to model the rivet holes typically used to attach composite panels in aircraft. The diffraction data could resolve the interleaved pattern of the woven fibres. In this method, damage to the fibres can be localized near the hole, and the tilt angle can be increased for fibres parallel to the strain direction as the stresses are transferred to neighboring fibres away from the hole. Since PPTA forms oriented fibres, the researchers could determine changes in orientation of the yarn and hence observe reorientation of the horizontal fibres caused by stress transfer (*Appl. Phys. Lett.* 91: 044102, 2007).

STRUCTURAL NANOCOMPOSITES

Some have predicted that composites produced from such nanoscale reinforcing materials as nanotubes, platelets, and nanofibres could show exceptional mechanical properties. Currently, however, the emphasis is on advanced composites reinforced with high-performance continuous fibres (see Coleman *et al.*, 2006; Dzenis, 2008). Nanocomposites have not so far proved up to the mark because of inadequate dispersion and alignment of the nanoreinforcement, low nanoreinforcement volume fraction, and poor bonding and load transfer at interfaces. But some recent work shows that some applications in reinforcement of small structures may materialize soon and could lead to longer-term work on nanocomposites.

Most of the work on structural nanocomposites has focused on ultrastrong nanoreinforcement such as single-walled carbon nanotubes (SWCNTs). Unfortunately, the high SWCNT strength has not yet manifested as bulk strength: attempts to create strong interfacial bonds could introduce defects into the SWCNTs that weaken their intrinsic strength. Multifunctional applications that do not depend only on the mechanical superproperties can, however, materialize (Kalaitzidou *et al.*, 2007). Controlled anisotropy and tailorability are also useful properties of many nanocomposites. Multiscale modelling may be helpful in achieving the required balance between various functions.

One valuable feature of nanoreinforcement is its small size. A variety of miniature and confined structures stand to gain from the better performance and tailorability of nanocomposites, including ultrathin coatings, membranes, films, and fibres; dental prostheses; structural elements in micromachines and micro- and nanoelectromechanical systems (MEMS/NEMS); thin walls in foams and aerogels; and connections in microcircuits (Dzenis, 2008). Except for nanoreinforced fibres, not much has been done on the use of nanoreinforcement in microstructures but this neglected aspect is expected to come in limelight as and when more robust nanomanufacturing processes are developed.

Using nanocomposites in microscopic structures or critical volumes requires not only critical multiscale structural design but also development of techniques to incorporate the nanocomposite in the desired microcomponents. The resulting applications might involve significantly lower cost as compared to the bulk applications, while improving the overall structural performance by applying the nanoreinforcement where it is most needed and/or where good alternatives are lacking.

Dzenis and Reneker (2001) created an advanced laminate composite with nanoreinforced interfaces. One major problem for laminated composites made of anisotropic fibre-reinforced plies is delamination. Most previous attempts to tackle this problem and to improve delamination resistance failed as they resulted in cost and weight increases.

Nanofibres or nanotubes may be gainfully used to reinforce interfaces in advanced laminated composites. Entangled nanofibres have been shown to improve interlaminar fracture resistance. A minute quantity (less than 1% by weight) of polymer nanofibres could improve the static and fatigue peel and shear interlaminar fracture toughness of aerospace-grade carbon-epoxy composites. In this crack deflection, nanofibre pull-out, plastic deformation, and crack bridging all had a role, and improvements were achieved without significant increase of weight or decrease of other composite properties. Since delamination exerts a negative effect on the overall composite performance (Jones, 1999), the reported improvements could potentially increase composite strength, durability, and impact resistance.

In this and other similar hybrid nano-microcomposite designs, the small-scale reinforcement strengthens the ultrathin matrix-rich layers. These microvolumes are not only weak (nonreinforced) but also experience high stress concentrations. The 3-D nanoreinforcement designed by Dzenis and Reneker with the combination of interlocked in-plane and out-of-plane nanofibres successfully addresses these problems.

ULTRASTRONG POLYMER NANOCOMPOSITES

A major difficulty in nanocomposite fabrication is the ability to realize materials that allow the transfer of the exceptional mechanical properties of the nanoscale materials to the macroscale properties of the bulk materials. Nanoparticle-filled polymer composites based on desired structural elements have mechanical properties that are much below the expected theoretical and experimentally determined values of the individual building blocks, except at low volume fractions of the reinforcement (Treacy *et al.*, 1996; Yu *et al.*, 2000). The deficiency in the properties of the composite is caused by the difficulty of obtaining well-dispersed large volume fractions of the reinforcing nanomaterials as well as a lack of structural control. The difficulty is also associated with achieving an effective load transfer from the polymeric matrix to the nanoscale components and the lack of understanding of the mechanical interactions of the two constituents at the nanoscale. Podsiadlo *et al.* (2007) showed that composites having properties that approach the theoretical maxima can be produced by using spatial and orientational control of clay platelets in a polymer matrix at the nanoscale and retain this order at the macroscale.

Bottom-up assembly of a clay/polymer nanocomposite has enabled the preparation of a homogeneous, optically transparent material with planar orientation of the alumosilicate nanosheets. The stiffness and tensile strength of these multilayer composites are an order of magnitude greater than those of analogous nanocomposites at a processing temperature that is much lower than those of ceramic or polymer materials with similar characteristics (Podsiadlo *et al.*, 2007). A high level of ordering of the nanoscale building blocks, combined with dense covalent and hydrogen bonding and stiffening of the polymer chains causes highly effective load transfer between nanosheets and the polymer.

Decher (1997) prepared the clay nano-composite by using a bottom-up process called layer-by-layer (LBL) assembly. This process is based on sequential adsorption of nanometer-thick monolayers of oppositely charged compounds (such as polyelectrolytes, charged nanoparticles, and biological macromolecules) to create a multilayered structure with nanometer-level control over the architecture.

They characterized the assembly by means of atomic force microscopy (AFM) and scanning electron microscopy and verified dense coverage of the nanoplatelets and their strictly planar orientation. Measurements indicated an average of ~5 nm of thickness per bilayer and the cross-section revealed a well-defined layered architecture.

It was concluded that reinforcement in polymer-nanoplatelet systems such as polyvinyl alcohol (PVA)/Montmorillonite (MTM) clay results from several mechanisms operating at the nanoscale. The extent of structural organization (afforded by the LBL process) of the clay platelets in the composite maximizes the number of polymer/MTM interactions and constrains the polymer-chain motion; this results in a highly efficient load transfer between the polymer phase and the stiff MTM platelets (Podsiadlo *et al.*, 2007).

DEFORMATION OF ULTRA-STRONG MATERIALS

In situ electron microscopy of the extrusion of single nanocrystals from graphitic cages has revealed that these crystals deform at a certain point near their theoretical strength limits. Several nanostructured materials sustain specimen-wide stress up to more than a tenth of their ideal strength for a fairly long time. Lee *et al.* (2008) tested a monolayer of graphene. The test yielded a strength value close to its ideal strength as calculated by quantum mechanics. The precise mechanism underlying the response of these ultra-strong materials to deformation at high temperatures is not known because performing temperature-controlled mechanical tests of nanostructures *in situ* is very difficult (Suresh and Li, 2008).

Sun *et al.* (2008) reported observations of plasticity – the permanent, irreversible deformation of a material, as opposed to elastic deformation in which atomic bonds are stretched but not broken – of nanometre-sized metallic crystals inside a transmission electron microscope (TEM). These observations led them to suggest that the observed deformations are caused by the activity of short-lived, string-like defects in the crystals, known as dislocations.

Fairly large and typical structural dimensions such as micrometre-sized grains in bulk materials are suited to the continuous generation, entanglement and storage of dislocations during plastic deformation. But as the crystal grain size or the smallest dimension of a thin film shrinks below 100 nm, dislocations are 'fatally attracted' to such internal interfaces as crystal grain boundaries and surfaces of the specimen. It then becomes very difficult to sustain a permanent population of mobile dislocations – which are the vehicles of plastic deformation during straining – inside the material (Shan *et al.*, 2008). In these cases, deformation is achieved only if new dislocations are nucleated afresh, commonly from the same internal interfaces and surfaces that also absorb and destroy them (Zhu *et al.*, 2007). Repeated nucleation of new dislocations in these tiny crystals significantly increases the material's strength.

Sun *et al.* at first confined individual, 3-D crystals (as small as 10 nm in diameter) of gold and platinum, in spherical graphitic shells. When this was followed by puncturing and irradiation of the shells by a focused electron beam at different temperatures, the crystals were extruded from the capsules. Based on direct comparison of the lattice spacings in the gold nanocrystals inside and outside the capsules, the authors estimated a prevailing pressure of about 20 gigapascals (about 200,000 times the standard atmospheric pressure) in the capsule when this was irradiated at about 300°C. This certainly represents an extremely high stress for gold where ideal shear strength is only about 1 gigapascal. It shows that these systems are ultra-strong. But what is not so clear is the mechanism of deformation during extrusion. TEM revealed a highly perfect atomic structure with occasional grain boundaries and planar stacking faults but no dislocations. According to Sun *et al.*, deformation arises from individual, transient dislocations that are freshly nucleated and disappear so fast that they cannot be seen with a TEM.

Unfortunately, a woeful lack of understanding of the deformation mechanisms which operate in ultra-strong materials greatly limits the ability to create nanometer-scale systems showing the desired mechanical properties. Computational studies conducted by Zhu *et al.* (2007) have shown that low-temperature deformation of ultra-strong systems, such as the nanocrystals studied by Sun *et al.*, become highly sensitive to strain rate and temperature. The mechanism involves the nucleation, absorption and desorption of dislocations from interfaces and free surfaces, with a consequent reduction in activation volume, typically 2-20 times the volume of a single atom.

Moving Polymer Chains

During deformation of materials, certain changes take place within them all the way down to the atomic and molecular level. The classic Eyring model predicts that when rubbery polymers are pulled apart the molecular chains become more mobile. Lee *et al.* (see *Science* 323: 231, 2009) measured chain mobilities in a rubbery polymer, poly(methyl methacrylate), using an optical photobleaching method. They found that while the Eyring model does work at low stress, once the material starts to flow, mobility values shoot up and the distribution of values narrows. Mobility decreases when application of greater stresses cause strain hardening. These results may prove helpful in developing more general theoretical treatments of plastic flow.

REINFORCED POLYMER MATRIX COMPOSITES

When nanomaterials (NMs) are used as reinforcements in composites, they produce enhancements in mechanical properties of the composites. One unique property of nms is their large surface to volume ratios. Nanomaterials are used for reinforcing polymer matrices as they have ultra-high stiffness and hardness. Small additions (up to ≈ 1 wt%) of certain NMs such as CNTs enhance the mechanical properties very markedly, even up to 100% (Vivekchand *et al.*, 2006; Rao *et al.*, 2007, 2009; Ramanathan *et al.*, 2008; Behler *et al.*, 2009; Das *et al.*, 2009). Molecular level interactions between the nms and polymer matrices seem to play a major role in such enhancement (Prasad *et al.*, 2009) and the large interface area available for such interactions is crucial for the substantial enhancement of mechanical properties.

Various types of NMs characterized in recent years include nanocarbons of different dimensionalities, such as **nanodiamond, nanotubes** and **graphene** with dimensionalities of 0, 1, and 2, respectively (Rao *et al.*, 2009). Much research has been done on the mechanical properties of composites made of polymer matrices with one of the nanocarbons as the reinforcement phase. The nature of interaction of the nanocarbon matrix constituent may be expected to vary with the dimensionality. For instance, CNTs added to a polymer interact over the length of the polymer chain whereas a nanodiamond particle interacts only at a point, e.g., at the ends of the chain. Though each of the nanocarbons enhances the mechanical properties of the polymer matrix, it seems logical to infer that incorporation of 2 nanocarbons might produce some synergistic effects in the mechanical properties, because each of them interacts with the matrix differently. It was this reasoning that prompted Prasad *et al.* (2009) to study the effect of incorporation of different binary combinations of the nanocarbons, nanodiamond (ND), SWNTs, and a few-layered graphene (FG), on the mechanical properties of polymer matrix composites (PMCs) formed with polyvinyl alcohol (PVA). They examined all the 3 possible combinations of reinforcements, viz., ND plus FG, FG plus SWNT, and ND plus SWNT.

The mechanical properties of the resulting composites were studied by the nanoindentation technique and revealed extraordinary synergy – the stiffness and hardness were improved by as much as 400% over those obtained with single nanocarbon reinforcements. These results point a way to designing advanced materials with extraordinary mechanical properties by incorporating small amounts of 2 NMs such as graphene plus nanodiamond or nanodiamond plus CNT (Prasad *et al.*, 2009).

However, the work of Prasad *et al.* was limited to the measurements of hardness and elastic modulus and did not include ductility and toughness. Nevertheless their results do suggest that it should be worthwhile to explore polymer composites with such binary reinforcements.

DISPERSING GOLD NANOPARTICLES

Dispersing gold nanoparticles through plastic could enable several unusual materials such as durable paint, faster computers and cheaper TV sets. Adrian Fuchs (Queensland Univ. of Technology) designed a new method for dispersing metals in nanoparticle form throughout polymers or plastic materials.

As the properties of metals change when they are in nano form, combining the unique property of nanoparticles with plastic leads to a new range of composite materials that could be used for novel catalysts, drug delivery and coatings. Paint is essentially a plastic; when gold nanoparticles are added, it makes colours more intense across the whole visible spectrum. This paint also becomes more durable when exposed to a harsh environment.

If we put gold nanoparticles with titanium dioxide by using a plastic mould, we can make a very efficient catalyst to purify water – the titania absorbs the light and converts it into electricity, which is then passed into the conductive gold. This technique of dispersing nanoparticles through plastic may also be used for encapsulating drugs into plastic casings so that they could be ingested and used for cancer detection and destruction.

STRONGLY-DISORDERED CERAMIC HIGH-TEMPERATURE SUPERCONDUCTORS

Optimally doped ceramic superconductors (cuprates, pnictides, etc.) exhibit transition temperatures T_c much larger than strongly coupled metallic superconductors like Pb (T_c = 7.2 K, E_g/kT_c = 4.5). These ceramic layered high temperature superconductors (HTSC) usually show several phases and must be doped to be not only superconductive but even metallic. In contrast layered undoped MgB_2 with T_c ~40 K is an "ideal" example of a metallic BCS (Bardeen, Cooper and Schreiffer) crystalline superconductor where T_c can be calculated quite accurately. When MgB_2 is substitutionally doped with ~10% Al (on the Mg sites) or C (on the B sites), T_c decreases drastically to ~10 K (see Choi *et al.*, 2009). Optimally doped ceramic superconductors have several universal features that appear to contradict the BCS theory of superconductivity based on attractive electron-phonon pairing interactions (Phillips, 2010). These complex materials are strongly disordered and contain several competing nanophases which cannot be described effectively by parameterized Hamiltonian models, yet their phase diagrams also show some universal features in both the normal and superconductive states. Phillips reviewed the rapidly expanding experimental results that suggest that these anomalously universal features could be the result of marginal stabilities of the ceramic electronic and lattice structures. These dual marginal stabilities favour both electronic percolation of a dopant network and rigidity percolation of the deformed lattice network. This so called double percolation model has earlier accounted for many features of the normal-state transport properties of these materials and has also successfully predicted strict lowest upper bounds for T_c in the cuprate and pnictide families. Phillips has now extended it to include Coulomb correlations and percolative band narrowing, as well as an angular energy gap equation, which rationalizes angularly averaged gap/T_c ratios, and shows that these are similar to those of conventional strongly coupled superconductors.

COMPOSITE MATERIALS BASED ON GRAPHENE

Graphite oxide is a layered material formed by the oxidation of graphite. Unlike natural graphite, the graphene-derived sheets in graphite oxide (graphene oxide sheets) are strongly oxygenated; they bear hydroxyl and epoxide functional groups on their basal planes and carbonyl and carboxyl groups are located at the sheet edges (He *et al.*, 1998). These functional groups render graphene oxide sheets strongly hydrophilic, so allowing graphite oxide to readily swell and disperse in water (Hirata *et al.*, 2004, 2005).

Graphene sheets are one-atom-thick, two-dimensional layers of bonded carbon and have several unusual properties. Their thermal conductivity and mechanical stiffness are quite similar to the remarkable in-plane values for graphite (~3000 m^{-1} K^{-1} and 1060 GPa, respectively) and their fracture strength may be comparable to that of carbon nanotubes (CNTs) for similar types of defects (Dresselhaus and Dresselhaus, 2002). Individual graphene sheets possess extraordinary electronic transport properties (Novoselov *et al.*, 2004). Stankovich *et al.* (2006) had shown that a mild ultrasonic treatment of graphite oxide in water exfoliates it to form stable aqueous dispersions consisting almost entirely of 1-nm-thick sheets. It is thought that these represent fully exfoliated graphene oxide sheets. Today as exfoliation of graphite oxide is the only way to make stable suspensions of quasi-2-D carbon sheets, this is a good starting point for large-scale synthesis of graphene sheets. As such, graphite oxide is now attracting great interest as a filler for polymer nanocomposites (Kovtyukhova *et al.* 1999; Cassagneau *et al.*, 2000).

Nanocomposites are being hailed as a 'radical alternative to conventional filled polymers or polymer blends, particularly in the fields of transportation and electronics (Vaia and Wagner, 2004). The problem is that because of their hydrophilic nature, graphene oxide sheets can only be dispersed in aqueous media which are incompatible with most organic polymers. Further, unlike graphite, graphite oxide is electrically insulating. This limits its usefulness for the synthesis of conductive nanocomposites. Du *et al.* (2004) demonstrated that the electrical conductivity of graphite oxide can be significantly increased by chemical reduction. Unfortunately, reduction of exfoliated graphene oxide nanoplatelets in water irreversibly coagulates them, which renders impossible their dispersion within a polymer matrix at the individual sheet level.

Stankovich *et al.* (2006) reported that the exfoliation behaviour of graphite oxide can be changed by modifying the surface properties of graphene oxide sheets through chemical functionalization. Several chemically modified graphite oxides could be produced by treating graphite oxide with organic isocyanates. Such a treatment weakens the hydrophilic property of graphene oxide sheets by forming amide and carbamate ester bonds to the carboxyl and hydroxyl groups of graphite oxide, respectively. Consequently, the isocyanate-derivatized graphite oxides do not exfoliate in water but do form stable dispersions in polar aprotic solvents (e.g., *N,N*-dimethylformamide, DMF), consisting of completely exfoliated, functionalized individual graphene oxide sheets with thickness ~1 nm, as determined by atomic force microscopy (AFM). Such welcome dispersions of isocyanate-derivatized graphite oxide enable graphene oxide sheets to be well mixed with many organic polymers, facilitating synthesis of graphene-polymer composites. Together with the possibility that graphite oxide can be chemically reduced, the hope was that nanocomposites of polymer and isocyanate-derivatized graphite oxide might be made electrically conductive (Stankovich *et al.*, 2006). And, indeed, electrically conductive graphene-polymer nanocomposites have been prepared by solution-phase mixing of the exfoliated phenyl isocyanate-treated graphite oxide sheets with polystyrene, followed by their chemical reduction. These composites contain individual graphene sheets well dispersed throughout the polymer matrix. Similar dispersions have also been attained with other styrenic polymers, e.g., acrylonitrile-butadiene-styrene and styrene-butadiene rubbers. Besides chemical reduction, which is essential for inducing electrical conductivity (as composite samples with un-reduced phenyl isocyanate-treated graphite oxide sheets were insulating), the presence of the polymer in solution during the reduction step is crucial to preventing the agglomeration of the sheets. As the reduction proceeds, the sheets become coated with the polymer and remain individually dispersed. On the other hand, if the reduction step preceded the introduction of the polymer, considerable agglomeration of graphene sheet took place.

Stankovich *et al.* (2006) adopted a bottom-up approach conducive to molecular-level dispersion of individual graphene sheets as observed by scanning electron microscopy (SEM).

The manufacturing of such composites requires not only that graphene sheets be produced on a sufficient scale but also that they are not only incorporated but also homogeneously distributed, into various matrices. Graphite is inexpensive and available in large quantity, but does not easily exfoliate to yield individual graphene sheets. What Stankovitz *et al.* did was to develop a general approach for the preparation of graphene-polymer composites through complete exfoliation of graphite and molecular-level dispersion of individual, chemically modified graphene sheets within polymer hosts. A polystyrene-graphene composite formed by this method exhibits a percolation threshold of ~0.1 volume per cent for room-temperature electrical conductivity, the lowest reported value for any carbon-based composite except for those involving carbon nanotubes; at only 1 volume per cent, this composite has a conductivity of ~0.1 S m^{-1}, good enough for many electrical applications (see Chung, 2004). The bottom-up chemical approach of tuning the graphene sheet properties shows a very broad new class of graphene-based materials and their possible use in a variety of applications.

The quality of nanofiller dispersion in the polymer matrix is strongly correlated with its efficiency in improving mechanical, electrical, thermal, impermeability and other properties. A composite's properties are also linked to the aspect ratio and surface-to-volume ratio of the filler. The properties of the graphene-sheet-based composites designed by Stankovich *et al.* are potentially valuable because of the very high aspect ratios of the sheets as determined from SEM images, in which the average lateral dimension was around 1 μm, similar to the values determined by AFM. But, in contrast to the typically flat sheets seen by AFM when deposited onto atomically flat substrates, the sheets in the composites are crumpled and wrinkled, and sometimes folded.

Till hitherto, most research in nanocomposite materials using carbon-based nanofillers has been dominated by CNTs. The electrical properties of the composites made by Stankovich *et al.* compare well with the best values reported in the literature for nanotube-polymer composites.

Absolute conductivities of the graphene-polystyrene composites were essentially the same as those reported for SWNT-filled polystyrene composites (Grossiord *et al.*, 2005). But unlike previous reports, which feature thin film composites, the samples made by Stankovich *et al.* by using a quasi-industrial method (hot-pressing, or injection moulding) show similar values of electrical conductivity. Also, graphene sheets have higher surface-to-volume ratios than SWNTs because the inner nanotube surface is quite inaccessible to polymer molecules. This is why graphene sheets seem potentially better for changing all matrix properties - e.g., mechanical, rheological and permeability properties, and degradation stability. Another drawback with SWNTs is their exorbitant cost. By comparison, the much cheaper graphite is a commodity material with about 1,000,000 t being used annually, and which is sold for only a few US dollars per kilogram. Graphite oxide also can be easily prepared from a vast array of graphites, and its industrial production is now generating keen interest (Titelman and Bron, 2005).

9

CHAPTER

Nanotechnology: Some Chemical Dimensions

SUPRAMOLECULAR CHEMISTRY

Supramolecular chemistry, or chemistry beyond the molecules, opens the best route to molecular nanotechnology by allowing the design of versatile functional nanomaterials with the remarkable characteristics of the molecular centres, including their electronic, optical and chemical properties (Lehn, 1995). It is a robust and effective tool for assembling complex molecular systems exhibiting functionality or mimicking biological systems in their organization, cooperative functions, energy conversion and information storage. This is done by connecting molecular entities displaying suitable properties with appropriate groups so as to transmit information and activate effective interaction between them. Some options for this include hydrogen bonding, dipole-dipole and van der Waals interactions, or by forming coordination bonds. Such intermolecular interactions can be quite strong. Supramolecular structures can also be made by self-assembling weakly interacting species such as amphiphilic molecules, as observed in membranes, vesicles and Langmuir-Blodgett (LB) films.

A supramolecular system contains structural and active components. The components perform some action and promote functionality when put together in an organized way, e.g. by forming and breaking chemical bonds, absorbing or emitting light, or transferring electron or energy. The structural components are needed for assembling and connecting the components in an organized way, in order to get molecular recognition, allosteric effects, or to provide energy gradients for promoting directional processes.

Molecular devices designed through a supramolecular approach have their components functionally integrated to perform an action or a process. Synthetic strategies have been developed for the assembly of supramolecular systems encompassing porphyrins, porphyrazines, clusters and polyimine complexes. They attract many applications in nanotechnological devices, such as amperometric sensors, smart windows, photoelectrochemical cells and logic gates (Toma, 2008). A simple design involves a molecular film inserted into a pair of optical or conducting windows. Molecular devices can also be assembled in complex ways in analogy with the electronic devices (Balzani *et al.*, 2003).

Exploiting metal-ligand interactions for supramolecular design has many advantages – it can be guided by the self-assembly character of the metal-ligand interaction, allowing the selectivity control based on simple acid-base principles, while incorporating a variety of structural, electronic and kinetic properties from the molecular systems. Metal complexes can be kinetically labile or inert and their structures often show such different coordination numbers and geometries as trigonal, square, pyramidal, octahedral, capped octahedral and pentagonal bipyramidal, depending on the element and its oxidation states. This allows using the coordination chemistry-based approach in metal ion induced assembly of highly organized structures.

The supramolecular assembly of symmetric coordination compounds shows several strategically distributed binding sites (Fig. 9.1). Various molecular building blocks can be built up in different linear, planar or tridimensional systems which incorporate or combine pertinent physico-chemical properties for use in nanotechnological devices. For instance, the tris-(2, 2′-bipyrazine) ruthenium (II) complex (a) belonging to the class of ruthenium (II) polyimines have been extensively investigated in recent years in view of their notable photochemical, photophysical and redox properties (Balzani and Scandola, 1991; Camera and Toma, 2002). The 6 available coordinating pyrazine N-atoms in an octahedral arrangement may be directly used to build up supramolecular systems in the presence of metal ions. If more bridging ligands (BL) are added, mixed complexes of the type [Ru(bipy) (BL)2] and [Ru(bipy)2(BL)] (bipy = 2.2′-bipyridine), appear; these are used in assembling new polynyclear systems. Indeed, Balzani *et al.* (2006) focused this strategy on 'complex as metals and complex as ligands', for developing several interesting supramolecular systems.

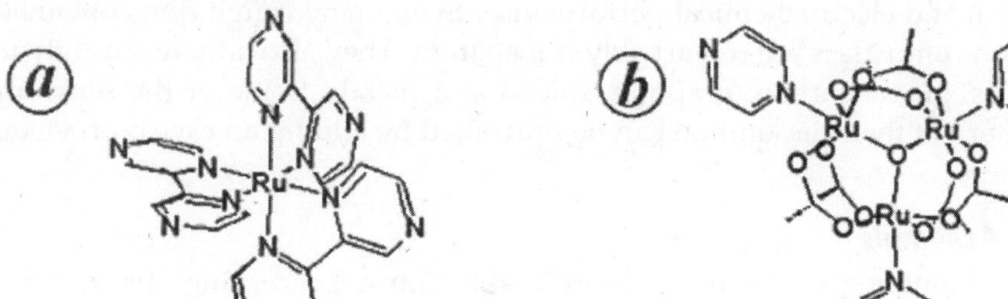

Fig. 9.1 Two metal-organic building-blocks constituted by metal polyimines (***a***), and metal-clusters (***b***); these can be used in the assembly of functional supramolecular nanomaterials, based on the coordination chemistry approach (after Toma, 2008, with permission from *Current Science*).

The important class of complexes shown in Fig. 9.1b, contains triangular μ-oxo-bridged ruthenium-acetate clusters. Their special features are the strong electronic delocalization and the occurrence of five successive monoelectronic redox processes in the range of -2.0 to 2.0 V. The electronic properties of the triangular μ-ORu_3 centre are modulated by the axial ligands, allowing fine tuning of the redox potentials. As the electronic effects and the donor/acceptor characteristics strongly depend on the oxidation state, contrasting redox species are readily generated by changing the applied potentials. Whereas by applying negative potentials the triangular clusters become stronger π-donor species, reversing the potentials make them acquire strong π-acceptor properties. As their structures are robust and symmetric, these triangular clusters provide versatile building blocks for producing structures connected by suitable bridging ligands such as pyrazine (pz) and 4, 4'-bipyridine (4, 4'bipy).

The propensity of certain supramolecular species to form films, even by such simple methods as dip-coating and drop casting provides versatile interfaces for many nanotechnological applications.

SUPRAMOLECULAR FILMS

Direct Assembly

Certain M(TRPyP) and M(TCPyP) (metallated tetraruthenated pyridylporphyrin, TRPyP, or tetracluster pyridylporphyrin, TCPyP) compounds can form homogeneous thin films by direct self-assembly based on dip-coating, spray coating, drop-casting or electrostatic assembly procedure. These films also (Araki *et al.*, 1999, 2000) show a special feature: they preserve most of the characteristics of the supramolecular species and hence provide useful interfaces for applications in molecular devices. This useful aspect of M(TRPyP) and M(TCPyP) extends their utility as functional materials encompassing a broad range of possibilities not found in most supramolecular systems.

Porphyrins generally exhibit a self-association tendency that is also observed in the supramolecular species, contributing to the formation of very stable thin films. The films show excellent optical and electrochemical performance. Even many rough films obtained by drop casting solutions onto glass are remarkably transparent. They also adhere strongly to surfaces of glass, oxides, glassy carbon, graphite, silicon and metals. Some of the films are slightly soluble in water but their dissolution can be prevented by adding an excess of X^- counter ions into the solution.

Electrostatic Assembly

Electrostatic assembly can build up more stable films, preserving the conduction and electrochemical properties. This is based on the strong interaction between the tetracationic species $[M'(TPPS)]^{4+}$ and the tetraanionic $[M'(TPPS)]^{4-}$ molecules, by means of electrostatic and π-interactions. Molecular recognition directs the ion-pairing process because the structural and electric charge of those species is complementary. A face-to-face dimer is initially formed when they are mixed together in dilute methanol solution. In more concentrated solutions, the ion-pairs aggregate and an insoluble material is deposited. It is possible to assemble this material in a controlled manner, layer-by-layer, to make ultrathin films on substrates such as silicon wafers (Azevedo *et al.*, 1998). A similar approach is generating electrostatically self-assembled polymeric films, as electrode modifiers, which display high catalytic activity. The electrochemical behaviour of the electrostatic assembled films is quite similar to that of pure M(TRPyP) modified electrodes.

Electropolymerization achieves better control of the location and amount of the deposited film. It is accomplished by introducing such electropolymerizable groups as 5-chlorophenanthroline (5-C1Phen).

Another option is to perform electropolymerization directly on the films previously formed by dip-coating (Araki and Toma, 2006). The high concentration of the monomer on the electrode surface and the presence of suitably pre-oriented molecules improve the electropolymerization, and only one or two scans are needed in the 0.0 V to -1.3 V range for completion. Also, a more compact and reticulated structure results.

SUPRAMOLECULAR DEVICES

Molecular machines and devices have moved to the frontline of nanotechnology and are in the focus for energy conversion, sensors, displays and as nanosize version of the microscale components of MEMS and microelectronics (Lehn, 1988; Merkle, 2000; Sauvage, 2000; Stoddart, 2001; Pease and Stoddart, 2001; Whitesides and Grzylowski, 2002). The ability of a molecule to efficiently capture light or electrons and respond to electric or magnetic impulses enables it to process information. Specialized molecules can conduct electrons (molecular wires), or pump electrons/photons (photonic pumps). Molecules also serve as effective probes for molecular recognition. By integrating such diverse abilities and properties, molecules can be tailored to communicate with each other; they signal this by means of colour changes, emission of light and/or sound or release or uptake of protons and ions. It is also possible to interface them to a metal wire or fiber optics - this can transduce their response into measurable electrical or optical signals (Toma, 2008).

In general, a functional device is made of electrical and optical elements. The best way to visualize it is by imagining two conducting windows or electrodes (Fig. 9.2), one of which is coated with a thin film of the active molecular material. The electric contact between the electrodes can be made with a suitable liquid or polymeric electrolyte containing a charge-transport mediator. This description applies well to several devices, such as the LCDs (liquid crystal displays), OLEDs (organic light emitting devices), DSC (dye solar cells), smart electrochromic windows, amperometric sensors and logic gates.

In this device, additional layers of semiconducting materials (e.g. TiO_2) can also be inserted along with the molecular materials. The molecular interfaces have a strong role in such devices - they make contact with the active films and provide efficient directional access and transport of electrons and photons. This makes the nature and quality of the molecular films so critical. These films can be prepared by several methods, e.g. thermoevaporation, dip-coating, spin-coating, drop-casting and Languir-Blodgett. Thermoevaporated supramolecular calixarene films used in OLEDs display a tunable electroluminescent emission between 470 and 510 nm

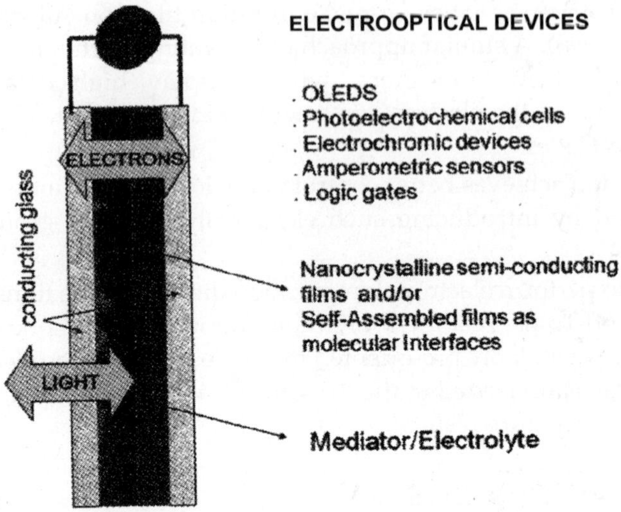

Fig. 9.2 Schematic sketch of a molecular device activated by photons or electrons, showing the conducting windows/electrodes, the semiconducting and/or functional molecular film and the mediator/electrolyte (after Toma, 2008, with permission from *Current Science*).

by changing the applied voltage bias from 4.3 to 5.4 V. Another method of making molecular films exploits electropolymerization of suitable oligomeric species, or the layer-by-layer deposition of opposite electrostatically charged species (electrostatic assembly).

Sensors are special interfaces for communicating with the molecular world. They involve some transduction process for expressing the interaction/reaction of the molecular species with the sensing elements. An example: the impedance changes of conducting polymers such as polypyrrol, polythiophene and polyaniline when in contact with various analytes, are used to sense the quality and origin of wines and coffee. But in this case, the sensors rely on non-specific interactions of multiple compounds with the active materials, and only give an average response.

Thin films that are specially engineered provide selective molecular interfaces for many substrates; their requirements and properties depend on the measurement techniques employed. In the amperometeric technique, the detection is based on the measurement of faradaic currents, produced by the oxidation/reduction of a given analyte at the electrode surface. Suitable materials for amperometric sensors should show (a) electron conducting (redox or electronic) and electrocatalytic sites, (b) size controlled pores through which the cations and/or anions from the electrolyte can diffuse rapidly, (c) good chemical and mechanical stability, (d) selective/specific response to substrates, and (e) resistance to surface poisoning. The sensibility depends on the electron transfer rates and diffusion of electrons/ions through the molecular material, as also on electrocatalysis efficiency (Toma, 2008).

Supramolecular porphyrin films have found use in the determination of drugs such as acetaminophen in pharmaceutical formulations (see Quintino *et al.*, 2002).

Supramolecular Photoelectrochemical Devices

In photoelectrochemical cells, a conducting glass electrode coated with a thin film of a semiconducting material is placed in contact with some suitable redox couple, such as I_3^-/I^-, which carries the electron transfer to a second electrode. The most commonly used semiconductors are based on ZnO, WO_3, SnO_2 and TiO_2, all of which are typically wide band gap materials (Tan *et al.*, 1994).

Light absorbed by the semi-conducting material drives electrons to the conduction band (CB) and creates holes at the valence band (VB). The redox mediator provides the electrons to the valence band, undergoing oxidation, and is regenerated at the cathode, thereby closing the electric circuit.

Conventional semiconductor photoelectrochemical cells are not very efficient, partly because the optical absorption of the wide band gap semiconducting films requires UV light excitation, thus making poor use of the available solar near-infrared radiation. To remove this limitation, dye-sensitizers are used along with wide band gap semiconductors. The ideal dye is one that absorbs a broad spectrum of light from the near UV to the NIR region.

The most extensively investigated dyes were based on ruthenium(II) polypyridine complexes, and the standard dye currently used in experiments involving the use of dye-sensitized solar cells (DSCs) is the $[Ru(dcbipy)_2(NCS)_2]$ complex, also termed N3. The use of dye has made it possible to extend the photoinjection response to the visible region; however, despite this, the efficiency of such DSC in terms of the irradiated area has been fairly weak, but a sharp improvement took place in 1991 when Gratzel and O'Reagan used nanocrystalline TiO_2 films having high surface-to-area ratio, good light transparency, charge separating properties and electronic conductivity (see Kroon *et al.*, 2007).

TiO_2 surface modification is achieved by using very thin tunnelling barrier layers and co-sensitization strategies employing dye multilayers.

SELF-ASSEMBLY

The self-assembly route to nanotechnology is usually defined as creating a material from its constituent components in a spontaneous 'natural' way, by an interaction between the components or by their specific rearrangement in such a manner that it proceeds naturally without any special external impetus. In the living world, this is seen in lipid bilayers which surround each of our cells or the cytoskeleton within them. This raises the hope that in due course using biological molecules is likely to provide a possible route to building novel nanoscale devices.

One particular route being explored is to use amyloid fibrils as nanowires, making them conductive by suitable functionalization. Amyloid fibrils are constituents of plaques found in diseases such as Alzheimer's Disease. They contain misfolded protein molecules in a cross-β-sheet structure (reflecting the packing of neighboring β-sheets). Dobson (2001) suggested that all proteins can, under the appropriate conditions and when usual biological control has been lost, misfold and form such fibrils. So, we should not assume that amyloid fibrils are necessarily always toxic – they can be found in such harmless eatables as yogurt (formed by the milk protein β-lactoglobulin) (Gosal *et al.*, 2002). The fibrils are highly stable and can be made to conduct via various projected routes, including covalently linking colloidal gold (Au) to the fibril (Scheibel *et al.*, 2003).

Although amyloid fibrils seem to be excellent candidates for self-assembly of useful devices, they do have some drawbacks. Because of large surface area, any fibrillar system is subject to nonspecific aggregation. This problem has also plagued attempts to use carbon nanotubes, which have a striking tendency to aggregate in solution and resist subsequent dispersion. Some slight success can however be achieved by treating the nanotube surfaces with surfactant. The aggregates are formed through nonspecific interactions. In the case of amyloid fibrils, the aim is to exploit the biological tendency to self-assemble into fibrils initially, but then to use them as nanowires without any higher-order aggregation (Donald, 2005). Both *in vivo* and *in vitro*, aggregation into larger units does seem to occur. Probably such aggregation is not nonspecific; the factors that control it are not known. In various different protein systems, large, quasi-spherical entities are sometimes seen, including in sections from diseased animal brains and in benign systems such as insulin and milk protein (Krebs *et al.*, 2004). Such so called spherulites are known in many crystallizable systems including metals and are an expression of polycrystalline growth. Granasy *et al.* (2004) proposed that this ubiquitous spherulitic structure results from competition between the local crystallographic symmetries and the randomization of the local orientation because of growth-front nucleation. However, it is not clear how this translates into spherulites in amyloid-fibril-containing systems, where spherulitic growth seems to take place simultaneously with fibril elongation, with a presumed equilibrium between the two types of aggregate (Donald, 2005).

SELF-HEALING MATERIALS: SUPRAMOLECULAR ASSEMBLY

Cordier *et al.* (2008) synthesized a rubber like material that, when severed, can self-heal at room temperature if the severed surfaces are brought together. The underlying supramolecular assembly operates via non-covalent interactions. The material maintains its mechanical properties even when repeatedly broken and repaired.

Researches on self-healing materials examined systems that undergo reversible transformation from individual components to a composite, or those that incorporate self-healing materials within the composite. Some potential applications include tears in clothes that effectively stitch themselves together; durable coatings and paints for houses and vehicles, and self-repairing artificial bone and cartilage (Mynar and Aida, 2008).

White *et al.* (2001) mixed microcapsules containing healing monomers into an epoxy polymer matrix containing a catalyst. A break destroys microcapsules, releasing the monomers into the crack where they come into contact with the catalyst and trigger a polymerization reaction that swells the healing material until the crack is closed. While this system does not need intervention when healing, if a fracture occurs again in the same place, there is too little healing agent to effect a repair. This problem is being tackled by means of 'microvascularization' for resupplying healing agents. However, it is not easy to build the needed channels.

Another strategy was developed by Chen *et al.* (2002) who used polymers modified with moieties that undergo reversible cross-linking; varying decomposition and formation of the composite take place and a self-healing system is created. Reversible systems such as this are able to self-mend repeatedly, even after several fractures have occurred at the same place. The only difficulty is how to synthesize these polymers; also an external stimulus (e.g., heat) is required. According to Chen *et al.*, the system can repair itself under fairly mild conditions.

The elegance of the work done by Cordier *et al.* is the use of supramolecular assembly to produce a material that has the properties of rubber and is also self-healing. Conventional rubbers are usually long-chain, cross-linked polymers that can stretch and then recover to their original size and shape. Cordier *et al.* achieved those properties with a mixture of small ditopic molecules (i.e., which can associate with two other molecules) and multitopic molecules (which can associate with more than two). The resulting supramolecular network shows partial crosslinking, through hydrogen bonds; the material does not crystallize and is elastic. Upon fracturing the active ends of the hydrogen-bond network become exposed because the strength of the self-assembly association is less than that of covalent bonds. The broken ends may be thought of as living; they can be recombined by just bringing them together: microscopic association then produces macroscopic healing (Mynar and Aida, 2008).

Bosman *et al.* (2004) produced materials that exploit the sensitivity of reversible, noncovalent forces to various environmental factors. As these materials are rather easy to make and reprocess, industrial houses are greatly interested. In fact, SupraPolix produces supramolecular polymers that can self-heal, although temperatures of 140°C have to be used. Cordier *et al.* have gone a step further by using supramolecular chemistry to make a material that heals at room temperature. Cordier *et al.* used fatty acids from vegetable oils as starting ingredients; these offer much variety for fine-tuning their material. Even now that material can withstand multiple fractures, requires no catalysts and is quite easy to make. It can also be broken down with heat and easily recycled so that it is environmentally friendly as well.

SUPRAMOLECULAR POLYMERS

Whereas most polymers are composed of long molecular chains made up of numerous units connected by covalent bonds, the highly dynamic supramolecular polymers are quite different: their dynamic properties arise from the reversible bonds that hold their chains together. These useful properties attract many potential new applications (de Greef and Meijer, 2008).

No doubt, even conventional polymers do have excellent properties as materials, but upon melting they become highly viscous as a result of entanglement of their macromolecules. High temperatures and pressures are usually needed to enable a melt of sufficiently low viscosity for processing, but this limits their application. In contrast, supramolecular polymers combine good material properties with low-viscosity melts that are easier to handle. Some supramolecular polymers also show typical characteristics unique to their class, such as the ability to self-heal fractures in their structure.

Supramolecular polymers are nothing but polymeric arrays of monomer units held together by reversible, highly directional secondary interactions – that is, non-covalent bonds, such as hydrogen bonds. The resulting materials maintain their polymeric properties in solution. The directions and strengths of the interactions are so precisely tuned that the array of molecules behaves as a polymer. The non-covalent bonds ensure that supramolecular polymers are always formed under conditions of thermodynamic equilibrium, with the lengths of the chains being related to the strength of the non-covalent bond, the concentration of the monomer and the temperature (Ciferri, 2005).

Most of the material properties of conventional polymers stem from the covalent nature of their spaghetti-like molecules. For example, hydrogen bonding in nylon influences the properties of the polymer's crystals. Supramolecular polymers, in contrast, exploit reversible interactions in the main chain and so all their mechanical properties arise a consequence of secondary interactions, in particular the strength, reversibility and directionality of these interactions (de Greef and Meijer, 2008).

Useful supramolecular polymers must have at least 100 monomers to be connected at any one time. Longer chains can be created by strengthening the interactions between the monomers; i.e., the average lifetime of the bonds has to be increased. But the window of opportunity is small – if the secondary interactions are too strong, the polymers lose their dynamic flexibility, and their bulk properties will resemble those of covalent polymers.

In this context, the question arises whether hydrogen bonds are the best way to hold supramolecular polymers together? Hydrogen bonds are ideal in combining high strength with excellent reversibility. Because they form bonds only in the direction of the polymer chain, unwanted interactions in other directions are minimized. It is possible to precisely adjust the strength and reversibility of hydrogen bonding between monomers by introducing an array of hydrogen-bonding groups in the monomers – this allows several bonds to form. The interactions between such strongly bound units persist even when the polymer is dissolved in water. However, in water, the hydrogen bonding between monomers is weaker than in the solid material. In order to make sufficiently long supramolecular polymers in water, the monomers must form hydrophobic pockets in which hydrogen bonds can be protected from the detrimental influence of the solvent.

Some polymers require several steps of organic synthesis whereas others can be easily made even on a large scale. Fortunately most of the interesting materials fall into the second category.

It is also possible to combine two or more kinds of monomer, just by mixing different monomers together. The properties of the resulting supramolecular co-polymers can be easily adjusted in a modular approach by changing the ratio of the monomers. In the case of co-polymers with an alternating sequence of units, the two different monomers should have complementary binding motifs whereas, for random sequences, self-complementary monomers are better (de Greef and Meijer, 2008). If the monomers strongly repel each other (e.g., by hydrophobic interactions), they might not form mixed chains, so that the resulting material becomes a blend of two different polymers. The modular approach to making supramolecular co-polymers is specially useful for tuning the properties of materials for medical applications, such as tissue engineering and drug delivery. These materials need to be fine-tuned to meet the specific requirements of biological systems.

The melt viscosities of supramolecular polymers strongly depend on temperature – above the material's melting point (or glass transition temperature), small increases in temperature cause a large reduction in viscosity. This allows the materials to be easily processed – a property that will be useful for several applications, such as inks for inkjet printers.

In fact, it is also possible to modify this temperature-dependent behaviour. Second-generation supramolecular polymers are being designed to incorporate additional interactions (such as lateral hydrogen bonds) between groups at the ends of the polymer chains to strengthen the materials. These end-groups are usually attached to oligomers that have molecular masses of only a few thousand Daltons; many polymer chains usually have molecular masses of several thousand Daltons. In this way, supramolecular polymers have been made that melt at higher temperatures, but which remain elastic at temperatures below their melting point. The improved material properties of these polymers come from lateral hydrogen bonding between the end-groups, which causes the material to segregate into long nanofibres (see Binder, 2007).

Supramolecular polymers are expected to find their first commercial use as coatings for delicate, heat-sensitive substrates. It is not often possible to coat wood and fabrics with covalent polymers because they can suffer significant damage from the hot polymer melts. But this is not a problem for supramolecular polymers with their relatively low melting points. Supramolecular polymers also are excellent adhesives. They are also attractive for any applications that require processing from a solution or a gel, such as cosmetics and personal-care products (de Greef and Meijer, 2008).

Examples of naturally occurring supramolecular polymers are also known. Some protein molecules form small networks in order to function, yet do not make long enough chains to be classed as polymers. Others (e.g. G-actin and tubulin, which serve as building-blocks for the scaffolding of cells) do polymerize to form filaments, but the mechanism involves an irreversible reaction with phosphate-containing 'fuel' molecules. One example: tubulin dimers form long, rigid polymers known as microtubules, which can synchronize their growth (or shrinkage). But the strongest resemblance to synthetic supramolecular polymerization is found in the self-assembly of the tobacco mosaic virus – the assembly process here takes place without the use of fuel molecules.

In the coming years we may see supramolecular polymers becoming an integral part of the general field of self-assembling polymers which will find applications in areas ranging from electronics to medicine. The complexity of their molecular systems will increase rapidly and lead to the development of even more unconventional polymeric properties. But the greatest challenge will involve the search for self-organizing (rather than self-assembling) supramolecular polymers (deGreef and Meijer, 2008).

SELF-ASSEMBLY OF COMPLEX MICROSTRUCTURES

In nature, folding of proteins and formation of complex membranes involve self-assembly: in an aqueous medium, tiny building-blocks come close together by virtue of their shape and interactions. Inspired by such natural processes, Onoe *et al.* (2007) designed a method for assembling parts just 10 micrometres across into complex, three-dimensional objects, and built up chains of interlocking rings. Their work marks an advance towards the lofty goal of creating electrical or optical circuits, or even microscopic machines, from components at μm and smaller scales (Pine, 2007).

One elegant example of natural self-assembly is the folding of a protein molecule from a long, linear sequence of linked amino acids. In a watery medium, certain amino acids along the polypeptide chain tend to attract each other in various ways – van der Waals attraction, hydrogen bonding or hydrophobic interactions. Others repel each other through their electrical charges or hydrophilic interactions. Geometrical constraints can limit how amino acids along the chain interact with each other; this means which amino acid occupies a given position in the chain is a critical determinant of the final topology and shape of the protein. It is geometry which conspires with attractive and repulsive forces to fold the amino-acid chain into the complex shape that gives a protein its specific function.

Many simpler examples of self-assembly are also known: one is soap molecules, or 'surfactants' which consist of a hydrophilic head connected to a hydrophobic tail. When placed in water, the hydrophobic tails of different surfactant molecules aggregate together to produce the core of a sphere, with their hydrophilic heads at the surface. This enables both heads and tails to live in their desired surroundings. While arranging themselves, the tails often isolate a particle of dirt, so that it can be easily washed away.

These processes of self-assembly were exploited by Onoe *et al.* (2007) but with the exception that their building-blocks are not organic molecules, but particles of silicon formed into slightly tapered cylinders 10 μm in diameter at the wider end. This wider base was coated with a layer of hydrophobic molecules, whereas a thin layer of silicon dioxide (SiO_2) was created naturally by oxidation of the remaining surfaces in air. When the SiO_2 surfaces were placed in water with a pH of 6.5, they became negatively charged and repelled each other when two particles came close. The hydrophobic bases of two cylinders, in contrast, avoid contact with the water by pairing up flush against each other (Pine, 2007). This is precisely what happens when the water is stirred to bring particles close together: the hydrophobic surfaces pair up to form barrel-shaped 'dimers'.

Upon adding a small drop of acid to the water, pH fell to 2.0, the SiO_2 surfaces lost their charge, and the barrel-shaped dimers did not repel each other. Rather, when close together, they experienced an attractive van der Waals force, which was strongest between the exposed flat tops of the dimers, which provide the largest area of contact. When the acid solution was gently agitated, the barrel-shaped dimers self-assembled into long, thin cylinders.

Earlier, Terfort *et al.* (1997), Breen *et al.* (1999) and Clark *et al.* (2001) had exploited hydrophobic and hydrophilic interactions, as well as capillary action, to build such complex structures on scales of 10 μm and larger as rings, tetrahedral and tiny rugby balls; chiral helices; and stacked plates and three-dimensional crystals. Here also, the geometry of the building-blocks largely determined the objects that were assembled. In fact, particles may even be shaped to fit together in a lock-and-key manner. This mechanism has also been exploited on length scales of a micrometer to build up an 'alphabet soup' of particle shapes (Hernandez and Mason, 2007). These smaller particles can spontaneously self-assemble by thermal Brownian motion.

The elegant contribution of Onoe *et al.* is the sequential control over a self-assembly process involving different interactions – first hydrophobic forces, then electrostatic repulsion and finally van der Waals attraction. Although sequential self-assembly using artificial sequences of DNA has been carried out before, e.g., in experiments by Chen *et al.* (1989) and Taton *et al.* (2000), those experiments used the coding of DNA base pairs rather than different forces to control the sequence of self-assembly. Moreover, they borrowed heavily from biological processes. The approach adopted by Onoe *et al.* (2007) is more general.

To translate self-assembly into a practical technology, researchers will need to go beyond nature and not only exploit all the interactions at our disposal but also create novel pathways.

SOLID STATE MATERIALS CHEMISTRY

Solid state materials chemistry is a distinct discipline of chemical sciences (Rao and Gopalakrishnan, 1997). The field may be deemed to have originated in the 1960s with systematic correlations between the properties and the structures of solid compounds. It now covers a vibrant and welcome interface between chemistry, physics and biology. In its early phases, it dealt with the synthesis, structure and properties of solid compounds, and the solids were classified on the basis of the observed properties (West, 2003). A solid that manifests useful properties is required as a **material** with potential applications. Traditionally the solids have been classified on the basis of their electrical response to the flow of current, into conducting, semi-conducting, insulating and superconducting solids. Bardeen (1947) discovered transistor behaviour; it revolutionized the study of solids. This was followed by the more recent discovery of superconductivity in compounds (based on copper oxides) which has been a landmark in solid state chemistry research (Bednorz and Muller, 1986). Superconductivity refers to a remarkable electron behaviour in solids that arises from the formation of pairs of electrons which allows the electric current to flow without resistance. Subsequent observation of several properties in the same solid warranted a change in the classification. The study of another class of solids having large open spaces, cavities and channels (**porous solids**), which have useful applications in the areas of sorption, separation and catalysis, also catalyzed a change in the classification of solids. Today, solid materials are broadly classified as *dense* and *porous* solids (Natarajan and Gopalakrishnan, 2009).

A solid can be put to gainful use if we tailor some desirable electrical, magnetic or dielectric property. In this context, various types of new materials including thermoelectric, superconducting or ferroelectric materials (multi-ferroic materials), materials for hydrogen storage (energy materials), inorganic-organic hybrid materials and materials for other related properties have been developed (see Zuttel, 2004; Rao *et al.*, 2004; Fiebig, 2005). One structural family of dense solids called perovskites, ABO_3 (Fig. 9.3) have come into limelight for their utility as a suitable platform for actual expression of some of the materials properties. Perovskite structure is formed by the corner sharing of the BO_6 octahedra with the A ions being accommodated at the cubo-octahedral site. This structure shows much flexibility in terms of dimensionality and can accommodate many elements as part of the structure – a justifiable explanation the observation of many important properties in this family of compounds. Table 9.1 lists some important properties exhibited by compounds having perovskite related structures.

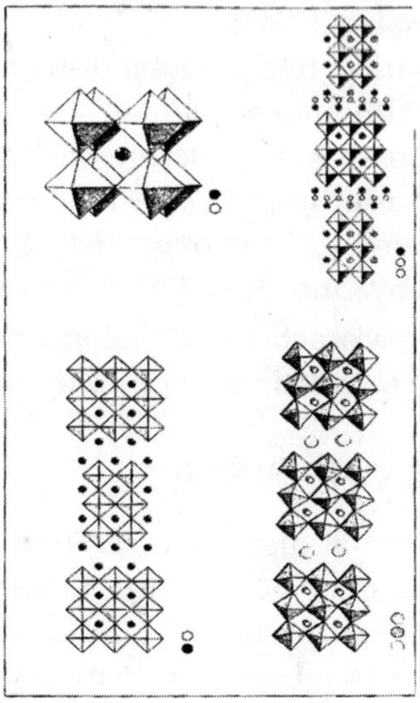

Fig. 9.3 The perovskite structure along with some of its layered variants. (a) the pervoskite, $SrTiO_3$, (b) the Aurivillius phase, $Bi_4Ti_3O_{12}$ {$(Bi_2O_2)[A_{n-1}B_nO_{3n+1}]$} (e) the Ruddlesden-Popper phase, $Sr_4Ti_3O_{10}$ ($A_2[A_{n-1}B_nO_{3n+1}]$) and (d) the Dion-Jaconson phase, $CsCa_2Nb_3O_{10}$ (A'[An-1B_nO_{3n+1}]) (after Natarajan and Gopalakrishnan, 2009, with permission from *Curr. Trends Sci.*, IASc, Bangalore).

Table 9.1 Some notable properties manifested by solid materials based on perovskites (after Natarajan and Gopalakrishnan, 2009, with permission from *Curr. Trends Sci.*, IASc, Bangalore).

Property	Example
Metal (metallic conductor)	$LaNiO_3$
Superconductivity	$(Ba,K)BiO_3$, $La_{2-x}Sr_xCuO_4$
Ferroelectricity	$BaTiO_3$, $PbTiO_3$
Relaxor ferroelectricity	$Pb_3MgNb_2O_9$
Piezoelectric	$PbTi_{1-x}Zr_xO_3$
Ferromagnetism	$SrRuO_3$
Colossal magnetoresistance	$La_{1-x}Sr_xMnO_3$
Spintronics/half-metallicity/ferromagnetism	Sr_2FeMoO_6
Fuel-cell electrode	$Sr_2(Mg,Mn)MoO_6$

The so-called **dense materials** have attracted strong research interest in view of their valuable magnetic, electrical, multiferroic, superconducting and other properties. These materials have found use in communication and data storage. On the other hand, porous solids have served as the work-horse for catalysis and in the oil industry where they are used in the cracking of crude. Porous solids happen to be most suited for gas separation, sorption and hydrogen storage. Many recent developments in materials have been oriented towards energy conservation as the fast depleting fossil energy resources have necessitated development of newer materials. The properties manifested by a solid critically depend on its synthesis and its chemistry.

DENSE SOLIDS

Most traditionally used magnetic materials are dense inorganic solids. The observation of magnetic behaviour in the mineral magnetite, Fe_3O_4 generated much excitement. Structure-property correlations in magnetic materials have been studied by many researchers. Many molecular materials with defined magnetic behaviour can be designed. Such solids are used as hard drives in computers. The combination of charge and spin information in a solid and its response to the applied field have proven important for information storage and related purposes. This new field is known as **spintronics**. In general, semiconducting and half-metallic ferromagnetic solids manifest this kind of behaviour (Wolf *et al.*, 2001). There has been much recent interest in understanding ferromagnetism and ferroelectric behaviour in the same solid, especially in view of the fact that ferroelectricity and ferromagnetism seem to be mutually incompatible.

Materials with conducting properties have attracted intense research. Research on superconducting materials witnessed a sharp surge following the description of cuprate superconductors with superconducting transition temperature above 40K. The discovery of cuprate superconductors was important not only for crossing the crucial barrier in terms of the transition temperature, but also provided a welcome opportunity to experiment solid state chemistry with other related phases, resulting in many different variants of structures based on layered copper oxides. The search for new superconducting materials is an area where solid state chemists can play a leading role. Table 9.2 lists some important superconducting materials discovered in recent years. The manifestation of superconducting behaviour in two-dimensional materials (layered solids) suggests that superconducting properties in solids might be observed even at 450°C. This will, of course, boost experimentation and facilitate the search for newer materials via novel synthetic approaches (see Natarajan and Gopalakrishnan, 2009).

Table 9.2 New superconductors discovered since 2000 (Natarajan and Gopalakrishnan, 2009, with permission from *Curr. Trends Sci.*, IASc, Bangalore).

Material	Year discovered	T_c (max, K)
MgB_2	2001	42
$Na_xCoO_{2-y}H_2O$	2003	5
Y_2C_3	2004	18
CaC_6	2005	12
$LaO_{1-x}F_xFeAs$	2007	36

Materials having high dielectric constant are eagerly sought in view of their application as capacitors in microcircuits, for instance. Further, the desirability of creating ever-smaller electronic devices can increase the demand for highly insulating solids for use as interlevel dielectrics (ILD) with a view to separating the conducting parts from one another. These solids also attract applications as gate dielectrics in integrated circuits (IC) and in memory elements in microcomputers.

POROUS SOLIDS

Porous solids are those whose properties arise from their porous architecture. They are classified on the basis of their pore size. Zeolites (aluminosilicates), aluminophosphates and other related compounds with pore diameters in the range of 2-20 Å are termed **microporous** materials. Compounds with pore sizes in the range 20-100 Å are **mesoporous**, and those with pore sizes >100 Å are described as **macroporous**. Porous solids now cover compounds that include most of the metals in the Periodic Table of elements. Among these, however, aluminosilicates are the most important and thoroughly studied compounds, as they show good thermal stability.

Aluminosilicates are solids possessing regular pore architecture. They show various different properties, e.g., as ion-exchange, separation and catalysis. Many of these emanate from the size of the pore window, the accessible void space and the dimensionality of the channel system. The empirical formula for an aluminosilicate zeolite is written as $A_{x/n}[Si_{1-x}Al_xO_2].mH_2O$, where A is a metal cation of valence *n*, generally located along with the water molecules in the channels and can be easily ion-exchanged (Fig. 9.4). One important property of these compounds is that they avoid Al-O-Al linkages (the Lowenestein's rule). Zeolites have been used as principal catalysts in many industrial catalytic processes (Tosheva and Valtchev, 2005). Some important developments in this area have involved functionalization of the channels within the zeolites, preparation of supported oriented thin films, synthesis of polycrystalline membranes and colloidal zeolites with narrow particle size distribution. Of these, **nanozeolites** offer large external surface areas and activities, shorter diffusion pathways and readily accessible catalytic sites. Their particle surface can be functionalized by organic groups which make them disperse easily in organic medium; so they can be used for building extended structures. Some problems in the use of nanozeolites are their low synthesis yields and lack of reproducibility.

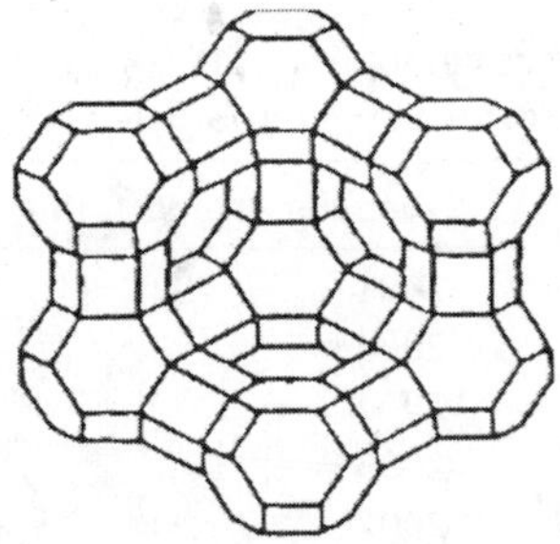

Fig. 9.4 Structure of most widely used aluminosilicate material, Zeolite-Y, showing the large pores (after Natarajan and Gopalakrishnan, 2009, with permission from *Curr. Trends Sci.*, IASc, Bangalore)

Aluminophosphates ($AlPO_4$) were discovered in the 1980s. These show good thermal stability (Wilson *et al.*, 1982). These phosphates offer a possibility of substituting Al^{3+} by bivalent ions such as Mg, Zn or Co (MeAlPO) and at the P^{5+} sites by tetravalent Si (SAPO), which can incorporate acidic protons. The possibility of generating redox centres in MeAlPOs together with the availability of acidic protons may be exploited for both oxidation-reduction and acid catalysis simultaneously (Thomas *et al.*, 2005). AlPOs can be used effectively in industrially important processes. The weaker thermal stability plus reduced recyclability, however, hinder commercial use of AlPOs. Another exciting development took place in the 1990s leading to the successful designing of porous metal-organic frameworks (MOF) (Maspoch *et al.*, 2007; Natarajan and Mahata, 2008). These are crystalline solids with 3-D extended structures in which the metal ions or clusters are connected through molecular bridges (Fig. 9.5). These compounds combine the coordination versatility of the central metal ions and the functionality of the bridging organic molecules (Natarajan and Gopalakrishnan, 2009). These structures not only retain the integrity of the building blocks but also the chemical functionality and rigidity during their synthesis. Pore sizes can be controlled by incorporating non-volatile guest species or through interpenetration or interweaving in the structures. As the metal organic frameworks have weaker bonds, it allows generating solvent driven structural transformations. Besides utilizing the coordinatively unsaturated metal centres, MOF compounds also make it possible to functionalize the bridging ligands. Indeed, MOFs offer the twin advantages of both organometallic and organic chemistry. Being good adsorbents of hydrogen, some MOF structures might be used as hydrogen storage materials (Fig. 9.5B).

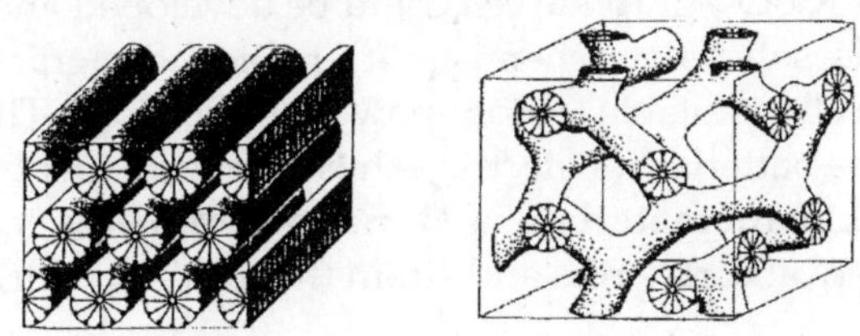

(a) (b)

Fig. 9.5 Structures of the mesoporous silica material. (A) MCM-42 and (B) MCM-48. The extra-large pore sizes have found many applications (after Natarajan and Gopalakrishnan, 2009, with permission from *Curr. Trends Sci.*, IASc, Bangalore).

Some **mesoporous** compounds are proving useful for catalysis, separation, optics, magnetics and sensing applications. The development of periodic mesoporous organic silicas (PMOs) and the production of non-silicate mesoporous compounds has aroused much interest. Carbon replicas of mesoporous silica can be used as electrode materials in secondary lithium batteries, super-capacitors and biosensors (Rao and Govindaraj, 2005).

The **macroporous** materials (pore sizes greater than 100 Å) are conducive to enhanced mass transport of the guest species and offer reasonable space for pore wall modifications such as internal surface modifications by using functional groups, small molecule anchoring and nanoparticle coatings. Such functionalization is useful for catalytic studies. These compounds are also helpful in producing hierarchical structures and multifunctionality. 3-D ordered macroporous solids possess insulating, semiconducting and conducting (metallic) properties, and have been synthesized by the **colloidal crystal templating technique** (Stein and Schroden, 2001). Some macroporous solids have length periodicity in the range of visible radiation and serve as photonic crystals. Other applications are in the areas of bioglasses, catalysts and battery materials.

MATERIALS FOR ENERGY CONVERSION AND STORAGE

The rapidly depleting reserves of fossil fuels are a matter of serious concern today. The need for alternative energy sources and efficient ways of generating power cannot be overemphasized. This can be met at least partly if we exploit the extensive knowledge of the structure-property relationships in the areas of catalysts, batteries, fuel cells, thermoelectric and optical materials, and design newer, efficient materials for energy.

For catalysts, the primary challenge is how to prepare new materials that could facilitate selective oxidation of alkanes and for the conversion of methane to higher hydrocarbons (Thomas, 1994). New electrocatalysts are badly needed for fuel cells. One important pointer to developing suitable materials for emerging needs is to fine-tune the ion mobility and electronic defects in a solid. As oxygen and hydrocarbon adsorption, activation and redox processes critically depend on the structure and composition of the surface of the solid/oxide, high priority should be given to clearer understanding of the surface structures in the context of improved performance of catalysts.

In the area of battery materials, the discovery of redox lithium ion mobility in layered $LiCoO_2$ was a landmark. $LiCoO_2$/graphite cell could be developed and commercialized thanks to the contribution of the solid state chemists; it enabled designing of portable electronic devices. A $LiCoO_2$ based cell is suitable for low-power applications. The requirements for high-power applications such as batteries for electric vehicles may be met by designing nickel-metal hydride batteries. Here, **nanomaterials** and **nanocomposites** can prove very useful. Nanocomposites effectively absorb stress and strain that develop during the charge-discharge cycles (Arico *et al.*, 2005).

Hydrogen Storage

Using hydrogen as a viable fuel in fuel cells and other related applications depends critically on the availability of suitable storage materials. The storage of hydrogen must consider the various constraints such as weight, volume, efficiency, safety, and cost. Solid state chemistry has attracted some role in developing such materials. Hydrogen is generally taken up by adsorption and chemical reaction mechanisms. CNTs having large surface area provide excellent opportunities for the adsorption process. Hydride forming alloys (hydrogen stored in the interstitial sites) facilitate absorption. Metal hydrides use the chemical reaction routes for the storage, and chemical reaction routes for the storage and release of hydrogen (Natarajan and Gopalakrishnan, 2009).

Up to ~5% hydrogen can be adsorbed on CNTs at 77K and about 1% at room temperature (Delacourt *et al.*, 2006). Crystalline microporous metal-organic framework (MOF) compounds comprising metal ions or metal oxoclusters might conceivably provide an alternative. Hydrogen adsorption of up to 2-4 wt% can occur at low temperature and high pressures (Rowsell and Yaghi, 2005).

Intermetallic compounds and alloys for hydrogen storage chemically break H_2 bonds at the surface and intercalate the atomic hydrogen in lattice interstitial sites. The compound, $LaNi_5$, by absorbing one atom of hydrogen per metal atom at room temperature and 1-2 atm pressure, forms $LaNi_5H_6$. Hydrogen is released by heating the sample to temperatures <100°C. The uptake and release of hydrogen can be recycled and about 2 wt% can be adsorbed in these compounds. The hurdle in utilizing intermetallics and alloys as hydrogen storage and transporting agents is their weight and cost. Possible alternatives might be to use the lighter elements (Li, Na, or Mg) as constituents in intermetallics.

Clathrate hydrates have been explored for possible hydrogen storage applications. Clathrate hydrates of methane are presumably abundant on the ocean floor where the pressures are high enough to stabilize the clathrates (Sloan, 2003). In gas hydrates, the water molecules usually form cages in which guest molecules, such as methane, can be trapped. Hydrogen molecules are also entrapped in such clathrates, albeit at very high pressures (about 2 kbar), which precludes their practical application for hydrogen storage.

However, Lee *et al.* (2005) reported that the pressure for storing hydrogen as gas hydrates might be significantly reduced by using THF (tetrahydrofuran) as co-occluding molecule (100 bar). In a class-II clathrate, water molecules form two different sized cavities – the larger one with a free diameter of ~0.67 nm and the smaller one is ~0.5 nm. For a clathrate to be stable, the larger cages must be almost completely filled with suitable guest molecules while the smaller cages may either remain empty or be filled with guest species. The larger cavity can be filled with THF, leaving the smaller cavities available for storing hydrogen. Thus, by using binary clathrate hydrate, up to ~4 wt% of hydrogen can be stored. Clathrate hydrates appear to be superior to cryosystems that use adsorbents with a large surface area.

Fuel Cells

Fuel cell technologies have an important role as an alternative source of energy (Ralph *et al.*, 2001). A H_2/O_2 fuel cell requires good conducting membranes for protons and oxygen (electrolyte) as well as suitable electrodes for the reaction $H_2 \rightarrow 2H^+ + e^-$ or $O_2 + 4e^- \rightarrow 2O_2^-$. A suitable electronic conductor is also needed for the delivery and removal of electrons. Direct methanol fuel cells (DMFCs) are unsuitable in view of the expensive catalyst (Pt) and issues related to crossover of the reactants and product. Intemetallics such as PtBi and Ru doped Pt have been explored as alternate catalysts in DMFC (Casao-Rivera *et al.*, 2004). There is need to find new types of catalyst which would also reduce the overpotential at the cathode. Work is also underway on proton conductors or oxygen ion conductor.

Photovoltaics

Polycrystalline thin film solar cells have attracted keen interest in view of their high efficiencies and ease of preparation (see Natarajan and Gopalakrishnan, 2009). $Cu(InGa)Se_2$, CdTe and other semiconductors have been examined for photovoltaic behaviour and showed reasonable efficiencies. The high cost of the In and Ga and the toxicity of Cd have precluded their use for harvesting solar radiation. New solid state materials having band gaps in the range 1-1.7eV are badly needed.

Transparent Conductors

These show optical transparency and electrical conductivity in the same material, which makes them useful for applications such as automobile and aeroplane window defrosters and transparent contacts for solar cells and displays. The first example of a transparent thin-film transistor (TTFT) was described by Hoffman *et al.* (2003). The challenge is to produce transparent conductors that could be used in high performance complementary devices as in silicon (CMOS) technology (Mitzi *et al.*, 2004).

PRODUCTION OF NOVEL MATERIALS

Solid state materials have been traditionally made by the ceramic technique which employs high temperature solid state reactions. Combinatorial, high throughput and parallel methods of syntheses can be very useful for optimizing experimental conditions. Extended framework compounds can now be created by directed assembly of molecular building blocks, low-temperature synthesis of oxides and chalcogenides in the nanometer sizes. Considerable progress has been recorded but it is difficult or impossible to design and predict new phases.

Synthesis at high temperatures generally occurs through increased diffusion of the reactants. Cuprate superconductors, dielectric semiconductors and half-metallic ferromagnets exemplify the versatility and utility of the high temperature method.

High temperatures routinely used in ceramic synthesis to drive the reaction to completion lead to formation of thermodynamically stable phases as products. This, however, is a limitation because many of the interesting inorganic materials have metastable phases. Further, high temperature synthesis precludes creation of the products in the form of free-standing fine particles, nanomaterials, or thin films.

POLYMERIC CRYOGELS

Cryogels are gel matrices which form in moderately frozen solutions of monomeric or polymeric precursors. Their characteristic feature is a network of interconnected macropores. These macropores not only allow unhindered diffusion of large particles of solutes but also mass transport of nano- and even microparticles. The unique combination of structural properties with their osmotic, chemical and mechanical stability makes them good matrices for chromatography of biological nanoparticles (plasmids, viruses, cell organelles) and even whole cells. Polymeric cryogels serve as efficient substates for the immobilization of biomolecules and cells (Lozinsky and Plieva, 1998; Lozinsky, 2002). They also serve as suitable matrices for electrophoresis and immunodiffusion, and for solid cultural media.

Several polymeric materials form 'gels' whose morphology (homo- or heterophase) is determined by the method of gel preparation, and the nature of the bonds is determined by the chemical structure of the polymers. The physical structure of gels is such that they can withstand considerable reversible deformation without flowing or destruction (Lozinsky *et al.*, 2003; Kaetsu, 1993). Cryogels can be of any chemical type—covalent, ionic or non-covalent. A network of large interconnected pores characterizes cryogels; some cryogels have a spongy morphology.

Cryogels for Bioseparations

Modern process biotechnology greatly relies on the separation and purification of some desired product from a fermentation broth or cell rupture supernatant. Although both analytical and large-scale chromotography is commonly employed in downstream separations, traditional packed-bed chromatography with immobile stationary phase suffers from a serious limitation: incapability of processing particulate-containing fluids, as for instance, cell suspensions or crude cell homogenates. Likewise, expanded-bed chromatography requires special columns and equipment and cannot be fitted in traditional packed-bed chromatographic systems (see Lozinsky *et al.*, 2003).

It is conceivable that continuous supermacroporous beds produced by cryotropic gelation could facilitate chromatography at flow rates similar to those in HPLC or even higher than those used in expanded bed chromatography, while using only minimal pressures typical for low-pressure protein chromatography. Another merit of the large pore size in such cryogels is that crude homogenate can be processed directly without any need for clarification.

As a suitable chromatographic material, cryogels achieve direct separation of proteins from unprocessed crude extracts or even from fermentation broth in the case of extracellulary expressed proteins. Cryogel-based chromatographic materials are attracting focus in bioseparations in which viruses, microbial cells and even mammalian cells, can be isolated and separated.

In many cases, the use of cryogels as matrices of immobilized biopolymers (enzymes, polysaccharides, nucleic acids) is more effective than traditional gel carriers. Kumakura (1997) showed that the biocatalyst produced by the immobilization of enzymes in cryogel macropores allows ready diffusion of macromolecular substrates. Supermacroporous continuous bed cryogel columns are perfectly suited for the separation of bioparticles in plug-flow bioreactors.

An interesting virgin area is the use of cryogels produced from smart polymers. These dynamic gels change their swelling in response to such stimuli as temperature, pH, ionic strength and the presence of certain chemicals (Galaev and Mattiasson, 1999). Enzymes immobilized in smart cryogels change their activity in response to the swelling state of the matrix and to any stimuli that change the swelling.

Various cryogels have proved suitable as carriers for cell immobilization, either by attachment (covalent or adsorption) or entrapment techniques. In the former case, supermacroporous matrices are used. In the case of entrapment there are two possibilities depending on the type of carrier: entrapping the cells within thin pore walls of supermacroporous cryogel, or entrapping them inside the macropores in cryogels.

For entrapment in spongy cryogels, biomass (microbes, spores, organelles) can be suspended in a solution of gel precursors (monomers for polymerization gelation (see Kaetsu, 1993) or reactive polymer and cross-linking agent (Scardi, 1987). When the mixture is cryogenically treated, it gives rise to the supermacroporous matrix containing immobilized cells entrapped in the cross-linked gel phase of pores.

SQUARE ARRAYS OF POLYMERIC NANOPARTICLES

Nanotechnology has created the ability to carry around an entire music collection in our shirt pocket, or our complete genome on a key ring. These amazing feats have materialized from the efforts of nanotechnologists to develop miniature information-storage devices, which in turn are based on tiny integrated circuits. Although current technology cannot scale down circuits much further, self-assembling nanoparticles could be exploited to produce much smaller circuitry elements. The challenge here is that nanoparticles self-assemble into hexagonal arrays that are incompatible with the square arrangements used in industry-standard circuits. This challenge has now been addressed by Tang *et al.* (2008) who described a method in which appropriately designed polymers form nanoparticles that assemble into square arrays.

Till hitherto the tiny structures found in integrated circuits were made by photolithography. This technique uses ultraviolet light to create patterned masks from light-sensitive polymers. The latter are used to control the etching of structures into the surfaces of silicon chips. The resolution of this technique is limited to about 100 nanometres (the wavelength of the light used to make the masks). But round nanoparticles can self-assemble into smaller patterns, packing together much like apples in a box. Should this process prove effective in making smaller features on circuits, it would sharply increase the amount of data that can be stored on a chip: a tenfold decrease in feature size could increase information density 1,000-fold (Ryan, 2008).

Self-assembled structures in nature are best exemplified by molecules such as lipids which form both nanostructures (vesicles and micelles) and thin films (membranes). Block copolymers are nothing but synthetic analogues of lipids. Park *et al.* (1997) had shown that fairly cheap processes can be designed to facilitate spontaneous formation of vast arrays of these synthetic polymeric nanoparticles, creating structures ten times smaller than can be made by expensive photolithography. Although electronics industry was excited by the potential benefits of this, it could not find a cost-effective method of reworking circuit-design software and fabrication protocols to cope with the hexagonal arrays formed by self-assembly. This led Semiconductor Industry Association to challenge researchers to develop a suitable method for making square arrays of self-assembled block-copolymer particles. Tang *et al.* (2008) rose to this challenge by using a judicious combination of theoretical predictions and methods for controlling non-covalent interactions between molecules.

Most blends of two different polymers separate into separate phases because of nonspecific, dispersive interactions between the two types of molecule. By linking pairs of different molecules (A and B) with a covalent bond (to form an A-B diblock copolymer), phase separation into two layers can be prevented, and also some control can be exercised over the nanostructures formed as the molecules aggregate. In fact, the nanostructures formed by such copolymers are the result of competition between molecular chain-stretching and interfacial effects around the covalent bond connecting the two polymer domains.

The phase behaviour of diblock copolymers is fairly well understood (Russell, 1996). In bulk material, the conditions required to make 3-D cubic structures with long-range order have been well described. According to Russell, in thin films deposited on substrates, where the thickness of the film is quite similar to the spacing between nanoparticles in the arrays, the preferred patterns are: hexagonal arrays of spheres, stripes of cylinders of block A embedded in a matrix of block B (with the stripes laid parallel to the film's surface) or single bilayers (or trilayers) of nanoparticles. The actual patterns observed depend not only on the composition of the copolymers but also on the interactions of the polymer molecules with the substrate and with the air above the film.

More complex molecules, e.g., A-B-C triblock copolymers (containing three types of polymer chain linked in series) can assume several different geometries inaccessible to simple A-B diblocks, including chiral and tetragonal arrays formed from cylindrical nanoparticles. For forming patterned masks on circuits, A-B-C copolymers create stable, tetragonal arrays of cylinders in thin films. The cylinders align themselves perpendicular to the film's surface, unlike A-B diblocks, in which they typically lie parallel to the surface.

Unfortunately, A-B-C triblocks are more difficult to synthesize, and they suffer from greater packing 'frustration' than do simple A-B diblocks, because the different blocks in each molecule tend to pack together with blocks of the same type from other molecules. Domains of each block-type thus result. This means that each molecule must pass through two interfaces between different domains. Certain triblock compositions can make square arrays in thin films but long range order is suppressed. Many such problems can be addressed by making polymer blends of different diblock copolymers (A-B/B-C or A-B/C-D), which offer the advantage of having more than two distinct polymer blocks while avoiding the problems associated with connecting all the blocks in one molecule. However, long-range order in such blends has not been achieved in view of their propensity to generate separate phases, just as homopolymers do.

Tang *et al.* could cross this hurdle by using hydrogen bonding between a pair of A-B/B′-C diblocks. In their blend, blocks A and C happen to be chemically different, and both are mutually incompatible with the B and B′ blocks. Tang *et al.* introduced small amounts of complementary hydrogen-bonding groups into the B and B′blocks so that the blocks form hydrogen bonds to each other. The resulting complex of diblocks behaved like a triblock, and phase separation of the diblocks was suppressed because the hydrogen bonding can overcome the nonspecific dispersive interactions.

The success of Tang *et al.* in making their diblock copolymers came from accurately controlling both the molecular weight of each block and the number of incorporated hydrogen-bonding groups. The A-B/B′-C blend could be prepared simply by mixing solutions of the two diblocks, and so thin films of the blend (about 50 nm thick) could be created on silicon wafers. Use of atomic force microscopy to examine the nanoscale structure of the films revealed that cylindrical nanoparticles in the film formed perpendicular, square arrays with long-range order (over an area of about 5×5 micrometres).

Tang *et al.* gainfully used their block-copolymer films as lithographic masks to transfer a template image into a silicon substrate. Scanning electron microscopy revealed that cylindrical pores (with diameters of about 22 nm) in the substrate were spaced 50 nm apart.

Conceivably, a careful combination of Tang *et al.* technique with the modern technologies for making patterned diblock copolymers on patches and in channels might be used to create arrays of magnets which store trillions of bits of information per square centimetre. Further developments could well lead to even greater miniaturization of silicon chips, and to the creation of dielectronic devices.

POLYMER PROCESSING

One discovery in polymer chemistry that stimulated explosive progress in macromolecular engineering is now mature enough to be gainfully utilized by industry – chemists can grow tailored polymers having controllable size and architecture. These polymers are attracting growing importance as major players in the field of nanotechnology.

The discovery was the development of a new method called **atom-transfer radical polymerization**, which made it easy to grow polymers of a desired design (Aida, 2007). The method is based on the use of a catalyst containing a transition metal such as copper. By interacting with the polymer, the catalyst turns it into a reactive radical; this radical then binds another monomer. As each molecule grows one step at a time, it is possible to produce polymers with uniform properties.

The only problem with this method is the large amount of catalyst needed to perform the reaction; also, it leaves residues which are costly to remove. But two groups of researchers have addressed this concern and cut the concentration of catalyst by up to 1000-fold (Jakubowski and Matyjaszewski, 2006; Matyjaszewski *et al.*, 2006). Another advance is based on an unexpectedly active oxidation state of the catalyst's copper that enables production of polymers with an ultra-high molecular weight and a narrow molecular-weight distribution (Percec *et al.*, 2006).

More commonly known as plastics, polymers are long-chain macromolecules of molecular weights greater than 10,000 g/mol. Their macromolecular size is some tens of nanometers. Some very useful properties of polymers include excellent mechanical strength, low density, thermal and electrical insulation, ease of shaping, low energy consumption per tonne of material processed, and reusability. Their service life is quite long. It is also possible to tailor (introduce) such other unique properties as electrical conductivity and biocompatibility. Their remarkable versatility has made polymers the materials of choice for diverse engineering applications. The global consumption of polymers currently exceeds 200 million tonnes (Lele, 2010). They are being widely used in consumer products, packaging, appliances, transportation, construction, medical products and strategic and space applications.

Growing demands for lower prices, higher performance, and sustainability have stimulated new innovations in polymers. Some recent developments in polymerization chemistry have facilitated the development of polymers of different chemical structures, macromolecular architectures, and topologies which can enhance the understanding of how the molecular structure and topology of polymers influence their dynamics and, in turn, how the macromolecular dynamics influence the flow and solid-state properties of polymers. Empirical structure-property relations in polymers have been investigated for long but it is only in the past few decades that some quantitative understanding of macromolecular dynamics has emerged. We now understand how the mobility of long, flexible and entangled polymer chains governs their rheological properties. Our knowledge of how flow controls crystallization of polymers from melt and how the segmental mobility of polymers in their glassy state affects their mechanical response, has also improved (Lele, 2010).

SYNTHETIC POLYMER LUBRICANTS

The basis of all biological lubrication systems is water, together with surfaces containing sugar chains. Yet humans usually lubricate their machines with oils and fats. Our understanding of biological lubrication is not high enough to enable application of these principles to systems of technological importance by using synthetic polymers (Lee and Spencer, 2008, 2008a).

While water lubrication is being used in such special applications as reservoir pumps, oil-water emulsions are commonly used for metal cutting by exploiting the effectiveness of water as a coolant. In the mining industry, hydraulic fluids are typically based on water with a view to excluding flammable materials from underground working areas. The food, textile, and pharmaceutical industries, where product contamination by oil can be a cause of much concern also attracts interest in water lubrication.

Water by itself is generally a weak lubricant. Unlike oil, its viscosity does not rise much with pressure – a property that is essential to the mechanism by which oils can form a lubricating film in high-pressure, nonconformal contacts of hard materials such as gears or ball bearings (Dowson, 1998).

It is possible to overcome the low viscosity of water at high pressure by biological lubricant additives, usually glycoproteins, in which many sugar chains are attached along a protein backbone. Mucins, for example, are found in certain parts of the human body that need lubricating, e.g., eyes and knees. These mucins seem to facilitate lubrication both through their intrinsic viscoelastic properties in solution and via their behaviour when adsorbed on the sliding surfaces. The typical bottlebrush structure of the molecules is important in this context. While the hydrophilic sugars immobilize large amounts of water within the contact region, the backbone either interconnects to other bottlebrushes or to some surface. Proteoglycan aggregate serves as a natural hierarchical bottlebrush with an important structural role in cartilage. Analogous synthetic systems might be used for water-based lubrication of technical applications such as sliding contacts or bearings (Lee and Spencer, 2005, 2008, 2008a). Hierarchically structured sugar-based bottlebrushes also have a crucial role in the mechanical properties of cartilage (Han *et al.*, 2007).

One good biomimetic approach involves decorating the sliding surfaces with a dense mat of brush-forming polymer chains. Klein *et al.* (1994) showed that when two mica surfaces bearing polymer brushes rub against each other under compression in "good solvents", the interfacial friction forces cannot be detected. The remarkable lubricating effect of such hairy polymer layers is due to interchain repulsion, which leads to the incorporation of large quantities of solvent. The resulting fluid-like cushioning layer on the surface can sustain the externally applied pressure, thereby lowering the friction forces. This behaviour results from the interplay between the polymer and solvent; it is not an intrinsic property of either component. It can also be observed for hydrophilic polymers such as polyethylene glycol (EPG) in water (Kobayashi *et al.*, 2007).

What is the maximum pressure that brush-like polymer layers can withstand? Müller *et al.* (2003) attached dense PEG brushes to steel and glass surfaces sliding in water to study their lubricating properties under conditions approaching those of industrial bearings. The formation of a lubricating film was found to be significantly increased by the brushes at contact pressures as high as 0.5 GPa (conditions where water alone cannot form a lubricating film), but the brush layer tended to break free during sliding contact; also it was difficult to fully suppress direct contacts between peaks in the surface roughness.

One remarkable property of natural tribological systems is that they often involve soft surfaces such as eyes, tongues, and cartilage-coated articular joints. In response to external loads, these soft surfaces deform elastically and increase the contact area, which lowers contact pressure. This is the reason why liquids whose viscosity increases only slightly with pressure, such as water, can form lubricating films in soft contacts. This property has prompted much study of the aqueous lubrication of elastic polymers (rubbers or elastomers) (Moore, 1972) and has attracted applications in tires, seals, windshield wipers, and biomedical implants (Lee and Spencer, 2008, 2008a).

The synergistic combination of hairy polymers and soft surfaces for water lubrication has been explored only recently. End-grafted PEG chains greatly increase the water lubrication of silicone-rubber surfaces at low speeds (Lee and Spencer, 2005). In contrast, short-chain surfactants only slightly reduced friction in comparison to brush-like systems. Gong *et al.* (2006) even used hydrogels for aqueous lubrication. Besides being soft, hydrogels allow water to permeate into the network, just as happens in cartilage. Hydrogels with brush-like polymer chains at their surfaces are particularly effective for aqueous lubrication.

The combination of brushes with soft surfaces is a hallmark of biological lubrication. However, we do not know whether carbohydrates possess any specific or unique properties which are lacking in other, synthetic brush-forming hydrophilic polymer chains. Moreover, natural lubricant additives seem to form hierarchical bottlebrush structures more readily than the synthetic water-soluble brushes investigated so far. This points to the need to clarify the role of both the composition and structure of sugar-based, bottlebrush-structured molecules in natural lubrication.

THINNER CATALYSTS

The synthesis of solid, crystalline materials having ordered arrays of pores has long attracted industrial chemists. Such materials have been used as molecular sieves or 'sponges' to store other compounds. By fine-tuning the pore dimensions and the composition of the crystal network, they might also find use in other applications that require specific molecules to be recognized and bound – such as in catalysis, in substrate-selective membranes, or in chemical sensors (Davis, 2002). For some of these applications, thin layers of porous materials are being sought. Choi *et al.* (2009) described one such method for synthesizing exceptionally thin sheets of porous materials known as **zeolites**. These are crystalline, inorganic solids formed from interlocked, tetrahedral TO_4 units (where T can be one of many different ions, such as Si^{4+}, P^{5+}, Al^{3+}). Each oxygen is shared between adjacent tetrahedra and the general molecular formula of the framework is TO_2. Zeolites are microporous crystalline aluminosilicates that are widely used in petrochemistry and fine-chemical synthesis in view of the fact that strong acid sites within their uniform micropores enable size- and shape-selective catalysis. However, the existence of the micropores, with aperture diameters less than 1 nm, introduces some diffusion limitations (Egeblad *et al.*, 2008) which exert negative effect on catalytic activity. This problem can be minimized by reducing the thickness of the zeolite crystals, which reduces diffusion path lengths and hence improves molecular diffusion, e.g., by synthesizing zeolite nanocrystals (Tosheva and Valtchev, 2005), by exfoliating layered zeolites (Corma *et al.*, 2000), and by introducing mesopores in the microporous material through templating strategies (Wang and Pinnavaia, 2006; Xiao *et al.*, 2006), or demetallation processes (Pavel and Schmidt, 2006; Groen *et al.*, 2007). Except for the exfoliation, none of these strategies has yielded 'ultrathin' zeolites with thicknesses less than 5 nm. Zeolites are typically prepared in hot water under pressure, often by using organic cations as structure-directing agents (SDAs) which guide the assembly of the inorganic components. Commonly practised procedures are based on mixing a source of Tatoms, a mineralizing agent (hydroxide and/or fluoride ions) and the SDA in water at temperatures of 100 to 200°C. The resulting materials have cavities filled by SDAs, whose removal leaves behind empty pores. The self-assembly of zeolites by using an organic SDA is a complex process consisting of many weak interactions of the reaction components with the SDA. SDAs seem to directly affect pore topology by closely matching the geometrical arrangement of the inorganic framework.

Till recently, zeolites had pore dimensions of 1 nm or less, which limit the size of molecules that can be stored or reacted within zeolite pores. Researchers have been seeking ways of making materials having larger cavities and have focused on using organic SDAs of different dimensions to direct the synthesis towards making certain secondary building units (Cheetham *et al.*, 2001). A three-dimensional zeolite having extra-large pores of about 1.2 nm diameter has been created by Corma *et al.* (2006) and has generated interesting possibilities for catalysis. In 2009, Sun *et al.* reported the synthesis of the first mesoporous zeolite which contains pores of about 2 nm in diameter.

Recent work on improving the functional properties of zeolites has focused on finding ways to make new zeolite morphologies such as nanocrystals, multicrystal arrays, and amorphous mesoporous materials (Corma, 2009). Among these, single layers of zeolites having high surface areas have great potential for several catalytic applications – for example, when reacting large molecules that cannot readily access the interiors of bulk zeolites; in catalysis that occurs at pore mouths, rather than in pore interiors; and enhancing the efficiency of catalytic processes where the internal pores of 3-D zeolites form undesired side products or deactivate the catalyst (Corma, 2009).

Single layers of zeolites have been made by the post-synthetic treatment of three-dimensional structures, and have served as catalysts.

What Choi *et al.* (2009) reported is that suitably designed bifunctional surfactants can direct the formation of zeolite structures on the mesoporous and microporous lengh scales simultaneously and so yield MFI (ZSM-5, a most important catalyst in the petrochemical industry) zeolite nanosheets which are no more than 2 nm thick; the large number of acid sites on the external surface of these zeolites makes them very active for the catalytic conversion of large organic molecules. Their reduced crystal thickness is conducive to good diffusion and so greatly suppresses catalyst deactivation through coke deposition during methanol-to-gasoline conversion. It is expected that such a synthesis approach could also be applied to other zeolites to improve their performance in many important catalytic applications (Choi *et al.*, 2009).

They used an organic molecule having a long hydrocarbon chain and two cationic groups that are separated by a shorter linkage. The cationic groups serve as SDAs to crystallize the ZSM-5 structure, whereas the long hydrophobic chain limits the growth of the zeolite crystal along that direction (Corma, 2009). By this method, using their SDA, Choi *et al.* obtained a layer of ZSM-5 that was just 2 nm thick, corresponding to the thickness of a single unit cell (the smallest subunit of the crystal structure). When heated, the SDA exits from the pores, leaving a unilamellar zeolite with surface area of roughly 700 m^2 g^{-1} – almost twice that of the conventional form of the zeolite. The material has strong acidity and high thermal and hydrothermal stability; it maintains a sizeable fraction of the aluminium in the zeolite framework even after heating in steam at 700°C – properties that are crucial for catalytic applications. Indeed, the catalytic activity of their nanosheets in several reactions of industrial interest is much higher than for typical ZSM-5 (Corma, 2009). Also, their unilamellar ZSM-5 is a much more durable catalyst than conventional forms of the zeolite in a process that transforms methanol into petrol.

According to Corma, Choi *et al.'s* direct synthesis of zeolite layers will lead to improved catalysts for processes requiring pore-mouth catalysis. The nanosheets are useful not only for processing large molecules in oil refining and petrochemistry, but also for the production of bulk and fine chemicals. Unilamellar and multilamellar ZSM-5 materials may also turn out to be effective catalysts (or co-catalysts) for 'cracking' processes that convert large hydrocarbon molecules into petrol and diesel, or which give large yields of propylene (a valuable feedstock used by the chemical industry). The synthetic strategy for making nanosheets could also be applied to other zeolites that have pores running along only one direction. Such materials would be very suitable for making selective membranes that allow only certain molecules through. Current research aims at reducing the thickness of membranes so as to increase permeability, while maintaining a preferred orientation of pores, for separation applications. Zeolite nanosheets could prove valuable in this context. Finally, catalytic transition-metal complexes may also possibly be grafted to the surfaces of zeolite sheets to generate multifunctional catalysts (Corma *et al.*, 2008).

ELECTROCHEMICAL H_2 EVOLUTION BY USING MoS_2 NANOCATALYSTS

Advances in the field of heterogeneous catalysis have been hampered by the difficulty of identifying the active site on a catalyst surface (Somorjai, 1994). In homogeneous catalysis, the active center is usually very clearly defined and quantified, and spectroscopic studies enable direct insight into reactive intermediates. Solid-state catalysts, in contrast, generally show various different surface sites that cannot easily be identified or quantified. This situation is further complicated when many sites work together in turning over a reaction. Identifying the most active site(s) is crucial for designing and developing better catalytic materials. Many useful *in situ* and *ex situ* experimental techniques, as well as computational methods, are now available (Jaeger, 2001; Campbell, 2001) to tackle this problem, but identifying the active site continues to be a daunting task (Jaramillo *et al.*, 2007).

Jaramillo *et al.* used such methods to identify the active site of nanoparticulate MoS_2 for the hydrogen evolution reaction (HER), $2H^+ + 2e^-$? H_2, which is very important for various electrochemical processes, fuel cells (as the reverse reaction), and solar H_2 production (water splitting), especially where there is a need to replace precious metal catalysts such as Pt (Hinnemann *et al.*, 2005). MoS_2 in bulk form is a poor HER catalyst. Nanoparticulate MoS_2 is a more promising system, as the edges of MoS_2 nanoparticles are quite active for hydrogen evolution.

Jaramillo *et al.* determined the active site for hydrogen evolution on nanoparticulate MoS_2 by atomically resolving the surface of this catalyst before measuring electrochemical activity in solution. By preparing MoS_2 nanoparticles of different sizes, they systematically varied the distribution of surface sites on MoS_2 nanoparticles on Au(111), which was quantified with scanning tunnelling microscopy. Electrocatalytic activity measurements for hydrogen evolution correlated linearly with the number of edge sites on the MoS_2 catalyst.

THE IRREPLACEABLE METALS

Until half a century ago, "engineering materials" mainly meant only metals (see Ashby, 2005). But since then, the share of metals in engineering materials has been declining. So much so that the world's most highly fuel-efficient jetliner (Boeing 787 Dreamliner aircraft) is mostly made from carbon fibre-reinforced polymeric composites (Lu, 2010). Traditional metals are now being substantially replaced by composites having higher strength/weight ratios. Also, aluminium usage has fallen to only about 20%.

The chief limitation of metals as structural materials is their low specific strength (the strength/weight ratio). Most engineering designs demand structural materials possessing high strength, fracture toughness (a measure of the energy required for propagating cracks), and stiffness while minimizing weight. Most metals do have high strength and stiffness. However, as they are very dense (e.g. steels are much denser than ceramics and polymers), their strength/weight and stiffness/weight ratios are low relative to competing materials. This is why metals in aircraft and sporting goods are being replaced - in these goods weight is a primary concern. Aluminium and magnesium are quite light, but are too soft for many applications and also have low toughness and stiffness. Titanium alloys partly overcome these problems. Being about half as dense as steels, these alloys are much stronger and very tough.

While metals can be strengthened by controlled creation of internal defects and boundaries which obstruct dislocation motion, these strategies compromise ductility and toughness, in contrast to the increasing toughness at higher strength seen with polymeric composites. Strengthening can also compromise conductivity and corrosion resistance of metals. Although metals can be strengthened without losing toughness by grain refinement (grain size reduction), when the grain sizes fall below ~1 μm, strengthening is usually achieved by a fall in ductility and toughness (Meyers *et al.*, 2006). Kimura *et al.* (2008) have overcome this problem. In a low-alloy steel containing ultrafine elongated ferrite grains strengthened with nanosized carbides, toughness and strength both increased when temperature was decreased from 60° to -60°C. This sharply contrasts with conventional metals which become strong but brittle at lower temperatures. The observed toughening may be attributed to the unique hierarchical anisotropic nanostructures in their steels (Kimura *et al.* 2008).

Nanotwinned metals exemplify other hierarchical nanostructured metals having extraordinary mechanical properties (Lu *et al.*, 2009). When a high enough density of twin boundaries (highly symmetrical interfaces between two grains of the same lattice structure) is incorporated into polycrystalline copper grains, with nanoscale boundary spacing, the material becomes much stronger than coarse-grained copper. It also becomes highly ductile. The electrical conductivity of ultrastrong nanotwinned copper compares well with that of a high-conductivity copper and also has a much enhanced resistance against electromigration. This is why nanotwinned copper is so attractive for applications in microelectronics (Lu, 2010).

Corrosion is another problem with metals. For protection, metals are usually coated with a layer of some corrosion-resistant material. Metal corrosion is resisted by forming a continuous protective passivation layer on the metal surface. Yamamoto *et al.* (2007) added 2.5% Al to conventional austenitic stainless steels; this formed a protective aluminium oxide layer which resists further oxidation at elevated temperatures. Because of their increased oxidation resistance, these steels might well be used at the high temperatures and strong oxidizing conditions of energy conversion systems to replace the presently used expensive nickel-based superalloys (metals specifically developed for high temperature applications). Another way to boosting the corrosion resistance of metals is to modify the chemical composition of surface layers. Most such processes require high temperatures which deteriorate the metal substrates. The processing temperatures can be brought down greatly if the grains of surface layers are refined into the nanoscale via surface mechanical attribution treatments (Si *et al.*, 2009).

Another problem with metals is that they soften at elevated temperatures. This is why they cannot usually be used above 1000°C. Superalloys have higher operating temperatures (up to ~ 1150°C) for nickel-based superalloys), so can be used in high-temperature applications such as jet turbine engines.

Despite the above drawbacks, however, metals continue to be the chief workhorse of our society in view of their unique properties which make them irreplaceable. Metals have a much greater fracture toughness than other materials. Steels are the toughest known materials. This is why metals are greatly used for critical components with the highest requirements for reliability and durability, such as bridge cables, concrete reinforcement in buildings, and vehicle body frames. The Boeing 787 also has a significant amount of metals (45%), chiefly in those critical parts that require the highest reliability, such as engines, wings, and alighting carriage (see DPFS, 2007).

The properties of metals are uniform in all directions. Their stretch is the same in tension and compression. The strengths of metals are quite predictable. These features are crucial for predicting fracture in engineering structures. In contrast, fracture of composites and ceramics cannot be predicted; their fracture can be catastrophic (with the material failing all at once) and can cause huge economic loss and even loss of life. This is why advanced technologies rely on high-performance metals. Body implants also must withstand severe corrosive and loading conditions that call for advanced tough metals.

Yet another advantage is that most metals are more conductive than ceramics and polymers. Copper and aluminium are the best materials for overhead electricity transmission lines. Conducting lines and thermal spreaders used in information technology are usually made from copper and its alloys. Metals also have unique magnetic properties that are not seen in other materials.

Metals also have the best overall mechanical properties at temperatures up to a few hundred degrees – a window that covers most of the operation temperatures in chemical engineering processes, power stations, and various engines. Finally, being recyclable, most metals are more competitive for quantity applications (Lu, 2010).

CRITICAL MINERALS

Very recently, cries of alarm have been raised about the scarcity of some critical materials/ minerals used in the latest cell phones, electric vehicles, or critical weapons systems (see NRC, 2008). One such critical element is lithium whose supplies are essential for batteries in electric vehicles. Bolivia is virtually the only country that supplies lithium. Rare earth elements are similarly crucial for essential military applications and they are vulnerable to shortages.

One cause for concern is that modern mineral-based materials are becoming increasingly complex (Eggert, 2010). It has been estimated that computer chips had ~11 mineral-derived elements in the 1980s, 15 elements in the 1990s, and could potentially contain over 50 elements in the near future. General Electric is using 70 of the first 83 elements in the periodic table in its products. Emerging technologies and engineered materials could boost the demand for some minerals that were previously used in only small quantities. Such a list has elements such as lithium in automotive batteries for electric vehicles; rare-earth elements in compact-fluorescent light bulbs and in permanent magnets for wind turbines; and cadmium, indium and tellurium in photovoltaic solar cells (Eggert, 2010).

Serious concerns affect the supply side. Some mineral markets are becoming more and more fragile. The United States, for instance, has become strongly dependent on foreign sources for many minerals. China has imposed export restrictions on primary raw materials to encourage domestic processing and fabrication of mineral-based materials into final products. Some mineral markets have production that is concentrated in a small number of countries or companies. Production of platinum-group metals is mostly concentrated in only a few companies in South Africa, creating vulnerability to geopolitical risks and price escalation. Supply chains tend to be more fragmented because mining, processing, and manufacturing operations are increasingly performed in different countries. All factors generate serious supply risks for essential mineral-based elements.

Concerns relate mostly to security of supply. Concerns about security of oil supplies in and since the 1970s have focused mainly on the risks of **higher prices** and the consequent economic costs on the economy as a whole. These risks could be attributed mostly to the Organization of the Petroleum Exporting Countries (OPEC) in a politically unstable part of the world (the Middle East).

Current concerns about the security of mineral supplies mostly center on the **physical availability** of essential inputs for various different products. These concerns relate to supplies of rare-earth elements used in military hardware and compact-fluorescent light bulbs, lithium used in automotive batteries, and platinum-group metals used in pollution-control equipment. Usually, the cost of these elements is only a small part of the overall manufacturing cost of the product; this means that a significant increase in price will probably have a fairly small effect on manufacturing costs. China currently produces almost all of the rare-earth elements used industrially and has enacted export restrictions. Bolivia, which has large and promising undeveloped resources of lithium, does not welcome foreign investment.

Today's concerns are no longer focused mainly on major metals (e.g., aluminium, copper, iron, lead, or zinc) but also on rare or specialty metals which are produced primarily as byproducts. Most indium, for instance, is recovered during the processing of zinc ores, and most tellurium is recovered from the processing of copper ores. The markets for rare or specialty metals (see Tables 9.3 and 9.4) are much smaller and more fragile than those of the major metals.

Table 9.3 Characteristics of selected elements of current concern (source: U.S. Geological Survey, Mineral Commodity Summaries, minerals.usgs.gov). (With permission from author as well as *Issues in Science and Technology*).

	Main application	Production characteristics	Top primary-producing counties (in rank order)
Cadmium	Batteries Solar panels	Byproduct of zinc-concentrate processing	Refined metal: China, Republic of Korea, Kazakhstan, Japan, Mexico
Lithium	Ceramics, glass Batteries	Mostly recovered from subsurface liquid brines, also produced from mining of lithium- carbonate rocks Little recycling at present but may increase (lithium batteries)	Mine production: Chile, Australia, China, Argentina, Portugal Platinum: South Africa, Russia, Zimbabwe, Canada,
Platinum-group metals (PGMs)	Catalysts, especially for pollution control in motor vehicles Jewellery	Most PGMs are produced jointly with one another at the same mineral deposits Significant recycling of spent automotive catalysts	U.S.A. Mine production Palladium: Russia, South Africa, U.S.A., Canada, Zimbabwe
Rare-earth elements (REES)	Permanent magnets Batteries Catalysts Phosphors	Most REEs are produced jointly with one another at the same mineral deposits, but relative concentrations of the elements vary from deposit to deposit Small quantities recovered through recycling of spent permanent magnets	Mine production: China, India, Brazil, Malaysia
Tellurium	Alloying element in steels Cadmium-tellurium based solar cells	Byproduct of copper refining Essentially no recycling of post-consumer scrap	Refined metal: Japan, Peru, Canada, U.S.A.

N.B.: Primary production refers to mining and subsequent processing but does not include recovery of materials from the recycling of post-consumer scrap. Net import dependence as a per cent of consumption = (imports – exports + inventory adjustments) as a per cent of consumption).

Table 9.4 Reserve/production ratios for selected minerals (lifetimes of remaining reserves at then current rate of annual production) 1, 2. (Source: U.S. Geological Survey, Mineral Commodity Summaries, minerals, usgs.gov.; BP Amoco, Statistical Review of World Energy, 2009; Eggert, 2010, with permission from author).

	1978	1995	2009
Copper	65	32	41
Crude oil	29	41	42 (for 2008)
Iron ore	183	150	70
Cadmium	38	29	31
Lithium	Insufficient information	350	550
Rare-earth minerals	221	1390	798

Note: 1. Reserves represent those mineral resources that are known to exist and capable of being extracted with existing technologies under current market conditions. 2. The information in respect of indium, platinum-group metals and tellurium is insufficient.

Geographic location is critical in understanding supply risks. Mineral resources in concentrations high enough to allow commercial extraction are not distributed evenly worldwide. Other things being equal, supply risks tend to be higher the more geographically concentrated production is in the hands of a small number of countries. However, geographic location is not the only determinant of supply risk. It is the concentration of production by a few companies or in a small number of mines that also makes users vulnerable to supply disruptions or high prices.

NANOMINERALS

Minerals are far more complex than believed previously. Their chemical properties vary as a function of particle size when smaller than about 100 nanometers, to perhaps as much as many nanometers. These variations are very likely due, at least partly, to differences in surface and near-surface atomic structure, as well as crystal shape and surface topography as a function of size in this smallest of size regimes. These variations could mean much in important geochemical and biogeochemical reactions and kinetics. This recognition enhances our understanding of how minerals influence the hydrosphere, pedosphere, biosphere, and atmosphere.

Most processes on Earth, whether physical, chemical or biological, are either influenced or fully driven by the properties of minerals. Minerals produce their influence by constituting the bulk of the Earth. They have a wide range of composition and structure as expressed in a substantial diversity of physical and chemical properties. Banfield and Navrotsky (2001), Waychunas *et al.* (2005) and Wigginton *et al.* (2007) have enriched our knowledge of another aspect of mineral diversity that is expressed in the nanoscale size range where atomic and electronic structure of nanoparticles (nps) tends to vary with size even without a phase transformation, and surface-to-volume ratios change markedly (Hochella *et al.*, 2008). These nps are minerals that are as small as about 1 nm and as large as hundreds of nanometers in at least one dimension. When size is limited in one, two, or three dimensions, it results in a nanofilm (or nanosheet), a nanorod, or a nanoparticle, respectively. Minerals are found in all these shapes but nps have attracted the greatest interest. Nanominerals are minerals that only exist in this size range. Their best known examples are certain clays as well as iron and manganese (oxyhydr)oxides (with ferrihydrite, an iron oxyhydroxide, as a typical example). Mineral nps are minerals that can also exist in larger sizes, and these probably include most of all known minerals (Hochella *et al.*, 2008).

The importance of certain kinds of nanominerals and mineral nps, namely clays and the smallest mineral colloids, has been known for long. It has become known recently that nanominerals and mineral nps commonly behave differently as a function of their size within the nanoscale size range. Mineral nps also behave differently than larger micro- and macroscopic crystals of the same mineral. Indeed, this observation violates certain aspects of the widely believed formal definition of a mineral. The general consensus is that minerals are naturally occurring, crystalline susbstances with a characteristic, defined chemical composition. For any particular composition, each mineral expresses a set of specific physical and chemical properties. Nanominerals and mineral nps do satisfy these criteria except that even with a fixed composition, they express a range of physical and chemical properties depending on their size and shape (Hochella *et al.*, 2008).

Nanominerals and mineral nps are widely distributed in the atmosphere, oceans, groundwater and surface waters, soils, in and on most living organisms, and even within proteins such as ferritin. They also exist in crustal or mantle rocks and affect deep Earth processes. The oceans may be the principal reservoir for nanominerals and mineral nps. The large supplies of nps in riverine, glacial, and aeolian contributions to the ocean have significant global biogeochemical consequences (Poulton and Raiswell, 2005).

In the atmosphere, the mass of mineral dust derived annually from windblown arid and semiarid lands, and secondarily from agricultural lands, greatly exceeds the average mass of mineral dust derived from volcanic emissions, as well as biological debris (Seinfeld and Pandis, 1998). The total volume of atmospheric mineral particles is dominated by particles larger than 100 nm, but numerically, most are nps less than 50 nm in size. Mineral dust, including mineral nps is distributed globally from its source regions via atmospheric circulation patterns.

In small mineral nps, even the interior structure can be affected. Gilbert *et al.* (2004) reported that the atomic structure of 3.4-nm-diameter sphalerite (ZnS) nps differed from that of bulk sphalerite. Structural coherence was lost at distances greater than 2 nm, suggesting that disrupted surface environments drive inhomogeneous internal strain, resulting in structural stiffening that greatly exceeds that expected by the mild overall bond-length contraction that is observed (Hochella *et al.*, 2008).

Size Range of Nanominerals and Mineral Nanoparticles

The upper size range of nanominerals varies, depending on the mineral. This limit is the size above which the properties of the particles become indistinguishable from those of the bulk material, or (equivalently) the size where, as the mineral particles become smaller and smaller, properties begin to show variations relative to those of larger particles of the same mineral. This size should depend on the nature of the material - whether metal, semi-conductor, or insulator - the particle shape, and the property studied (e.g., whether chemical, electrical, optical, mechanical, or thermodynamic). We do not know below what size these properties begin to vary for which minerals - so far no such measurements have been made.

The lower size limit of nanominerals and mineral nps is equally challenging to determine. Minerals, by definition, are crystalline, i.e, require a repeating arrangement of atoms. The smallest atomic clusters cannot qualify as nanominerals or mineral nps, but an atomic arrangement greater than a few unit cells in one or more dimensions does qualify (see Luther and Rickard, 2005).

Researchers are exploring the size dependence of the properties of atmospheric nps. Atmospheric mineral nps [typically halite (NaCl) and hydrous sulfate evaporates from sea spray] are hygroscopic and serve as water droplet nucleation and growth centres - a critical step in cloud formation. The size and density of the droplets govern the solar radiation scattering ability and the longevity of clouds which in turn influence average global temperatures. The hygroscopic growth factor for NaCl nps decreases for sizes below 40 nm (Biskos *et al.*, 2006). Therefore, only relatively large mineral nanoparticles (>40 nm) may contribute to cloud condensation nuclei.

Most human irritants and toxic materials in the atmosphere come from anthropogenic sources (most notably particulates emitted from the combustion of fossil fuels (Zheng *et al.*, 2005) and many other emissions from the numerous industries associated with metals (Utsunomiya *et al.*, 2004). Many trace metals in airborne particles, including highly toxic metals (e.g. cadmium, lead, and arsenic) occur in the very fine (<1-µm) to ultrafine (<100-nm) size fraction, and their characteristics influence the bioavailability of the metals in these particles.

Interestingly, the complexities of nanomineral and mineral np composition, structure, stability, and reactivity are equally applicable to the human body, where normal and pathological mineralization (from the nanosize to the micron-size) not only affects the calcification of bones, teeth, arteries, and veins, but also the formation of stones in the kidneys and joints (Sahai and Schoonen, 2006). The important field of nanotoxicology is concerned with the safety evaluation of engineered nanostructures through a knowledge of the mechanisms and biokinetics of nanomaterials causing adverse effects in humans (Oberdörster *et al.*, 2005). The long-known impact of asbestos-human lung interactions exemplifies the scope and importance of this field.

HALOCARBON MINERALIZATION BY METAL NANOPARTICLES

The Earth's surface has an extensive stockpile of halocarbons. As chlorocarbons pollute soil and water even at fairly low concentrations, their detection and removal have attracted considerable research. Many halocarbons are toxic, mutagenic and resistant to microbial degradation. Chlorofluorocarbons are quite inert in the troposphere but when they reach the stratosphere, they cause catalytic destruction of ozone. Although several methodologies for the destruction of halocarbons have been proposed (Burdeniuc and Crabtree, 1996; Nicoll and Francisco, 1999; Chien *et al.*, 2001), reductive dehalogenation and mineralization are particularly important. In mineralization, sodium oxalate can be used to generate sodium halides from halocarbons. Halocarbon can be degraded by activated carbon (Nicoll and Francisco, 1999; Chien *et al.*, 2001) as per the following reaction. In this reaction metal oxide nanoparticles are used.

$$2CuO + CCl_4 \rightarrow 2CuCl_2 + CO_2$$

Nair and Pradeep (2003) described a promising reaction of metal nanoparticles, which mineralizes halocarbons efficiently and economically. These authors studied the reaction with silver and gold nanoparticles to result in the catalytic destruction of halocarbons, forming silver halide (gold chloride) and amorphous carbon. The efficiency of the reaction was greater with silver nanoparticles. The reaction proved efficient for particles in the size range of 2-150 nm, but does not occur for bulk metals. This reaction may be suitable for mineralizing halocarbon stockpile on the earth's surface.

ANNEALING

If we combine a conducting polymer with a small molecular acid, we get high conductivity but poor processability. While a polymeric acid improves processability, the conductivity declines. Yoo *et al.* (2010) showed that it is possible to have the best of both worlds by using a polymeric acid as the dopant and then performing a post-processing solvent **annealing** step which substantially boosts the conductivity. Polyaniline becomes conductive through proton doping with poly(2-acrylamido-2-methyl-1-propanesulfonic) acid. Exposure to dichloroacetic acid (DCA) decreases film roughness because the colloidal particles that form during polymerization can blend, leading to chain conformations with improved charge transport. A different effect is seen in poly(ethylene dioxythiophene) films doped with poly(styrene sulfonic) acid (PSS). Treatment with DCA increases both conductivity and surface roughness. This result is attributed to disruption by DCA of the insulating surface layer of PSS that forms during spin casting. The DCA-treated films have been tested in organic thin-film transistors, solar cells, and light-emitting diodes. In all cases, the treated films exhibited much better characteristics than the untreated ones. Although the properties did not match those of devices made with indium tin oxide (the current benchmark material), the annealed films may well serve as practical alternatives to a range of metals and metal oxides, especially if cost is a concern.

DESIGNING GIANT MICELLES

Exercising precise control over the molecular structure of cylindrical micelles can yield novel supramolecular architectures. Covalently joining two distinct polymer segments at their ends produces hybrid macromolecules called block copolymers (Hadjichristidis *et al.*, 2002). If these "amphiphilic" block copolymers are placed in a solvent in which one segment is soluble and the other is not, they self-organize into supramolecular micelles in which insoluble cores are shielded from the solvent by a corona formed by the soluble segments. The commonest micellar structure consists of a spherical core protected by radiating corona chains. Elongated cylindrical micelles (Zana and Kahler, 2007) or bilayer vesicles analogous to liposomes can also be made. The elegant structural and chemical diversity of block copolymer micelles make them suitable for use in drug delivery, personal care products, and nanolithography.

If these self-assembled soft materials are to be most effective in applications, their size, shape, and chemical structure need to be controlled very precisely. Wang *et al.* (2007) demonstrated how the length and chemical composition of cylindrical block copolymer micelles can be controlled through a new growth mechanism. Cui *et al.* (2007) showed that the assembly of appropriately designed triblock terpolymers (i.e., made from three monomers) can be guided by external agents, providing good control over the resultant cylindrical nanostructures (Hillmyer, 2007).

The cylindrical variants of the micellar structures are specially attractive for several applications, e.g., as artificial tissue scaffolds, in drug delivery, or to toughen cross-linked polymer resins (thermosets) (see Geng *et al.*, 2007). Long worm-like micelles formed by amphiphilic block copolymers have a characteristic distribution of lengths determined partly by the energetics of forming end caps. This makes it difficult to achieve precise control over their mean length (Castelletto *et al.*, 2007).

Massey *et al.* (1998) had reported that block copolymers containing crystalline, organometallic core-forming blocks can form cylindrical micelles and Wang *et al.* have shown how the length of these cylindrical micelles can be controlled. Wang first prepared a set of fairly short cylindrical micelles of a block copolymer in a selective solvent. The same block copolymer was then added, dissolved in a suitable solvent for both blocks, to this micellar dispersion. This protocol is conducive to controlled growth of the micelles; the more block copolymer provided in this second step, the longer the micelles become (Hillmyer, 2007). Such a linear dependence of average cylinder length on block copolymer added brings to mind a living polymerization, a chain polymerization where the average chain length is linearly proportional to the amount of monomer converted to polymer. These crystalline-core micelles appear to grow by nucleation of the free block copolymers at the ends of the cylinders. The elegant control over the micelle length is truly remarkable. Like the "living" chains in a typical anionic polymerization, the cylindrical micelles remain stable over long periods; upon adding a fresh block copolymer supply they can grow again.

Wang *et al.* also explored the effect of adding a different block copolymer (with the same core-forming crystalline block but a distinct corona block) to perform cylindrical micelles. The "block co-micelles" so produced had a cylindrical midsection of one type, flanked on either end by cylindrical assemblies of the other type. These two-component assemblies illustrate the beauty of the growth mechanism. Just as living polymerization made possible a diverse range of molecular architectures, the process designed by Wang *et al.* can potentially pave the way to unprecedented supramolecular assemblies.

In a different approach to controlling micellar structure, Cui *et al.* (2007) showed that it is possible to precisely manipulate additives and by carefully controlling solvent quality. In fact, block copolymer micelles can be coaxed to form supramicellar assemblies with distinct morphologies and chemical compositions,. In the specific triblock system used by Cui *et al.*, the solvent-loving block contained pendant carboxylic acids that form ionic complexes with amine species. Upon treating a mixed aqueous/organic solution of these triblocks with an organic diamine, the spherical micelles were formed. By decreasing the relative amount of water in the solution by the addition of organic solvent, these assemblies can be made to controllably aggregate and generate long multicompartment cylindrical micelles which have an alternating lamellar structure in the cores of the micelles (Li *et al.*, 2006).

Cui *et al*. illustrated the versatility of their specific triblock system by replacing the organic diamine with amine-functionalized gold nanoparticles. Adding the nanoparticles to a dispersion of spherical micelles made from a related ABC triblock terpolymer (where A, B, and C represent chemically distinct segments) caused aggregation of the micelles into segmented cylindrical nanostructures. Alternating organic polymer and gold nanoparticle domains were seen to form along the long axis of the cylindrical domains. Further, by combining ABC and A′BC triblocks (where the A and A′blocks are incompatible with both the solvent and each other) and using similar preparation procedures, undulating cylindrical micelles were produced whose cores contained nanoscopic regions with alternating A-rich and A′-rich domains (Hillmyer, 2007).

The works of Wang *et al.* and Cui *et al.* show how the self-assembly of block copolymer amphiphiles can be controlled very precisely and how the integration of organic, inorganic, and polymer chemistries can facilitate interesting new and exotic nanometer-scale assemblies.

References

Ago, H. *et al. Adv. Mater.* 11: 1281-1285 (1999).

Agrawal, A., Ayantunde, A.A., Cheung, K.L. *ANZ J. Surg.* 76: 1088-1095 (2006).

Ahmad, A. *et al. Langmuir* 19: 3550-2553 (2003).

Ahmad, A. *et al. Colloid Surf.* B 28: 313-318 (2003a).

Ahmad, A. *et al. J. Am. Chem. Soc.* 124: 12108-12109 (2002).

Ahn, J.H. *et al. Science* 314: 1754-1757 (2006).

Aida, T. *Nature* 446: 235 (2007).

Aizenberg, J., Braun, P.V., Witzius, P. *Phys. Rev. Lett.* 84: 2997 (2000).

Ajayan, P.M. *et al. Science* 265: 1212-1214 (1994).

Akcora, P. *et al. Nature Mater.* 8: 354-359 (2009).

Akimov, V. *et al. Nature* 450: 402-406 (2007).

Alauzun, J. *et al. J. Am. Chem. Soc.* 128: 8718-8719 (2006).

Ali, F. *et al. Curr. Sci.* 97: 683-685 (2009).

Alivisatos, A.P. *J. Phys. Chem.* 100: 13226-13239 (1996).

Alivisatos, A.P. *Endeavour* 21: 56-60 (1997).

Alper, A. *Science* 296: 1224-1226 (2002).

Altug, H., Englund, D., Vuckovic, H. *Nature Physics* 2: 484-488 (2006).

Anastasio, C., Martin, S.T. In: *Nanoparticles and the Environment*, pp. 293-349. Banfield, J.F., Navrotsky, A. (eds.). Mineral. Soc. America, Washington, D.C. (2001).

Angell, C.A. *et al. J. Appl. Phys.* 88: 3113-3157 (2000).

Anker, J.N. *et al. Nature Mater.* 7: 442-453 (2008).

Appenzeller, J. *et al. Phys. Rev. Lett.* 93: 19 (2004).

Araki, K., Lima, S.D., Winnischofer, H. *Anals Acad. Brasileira Ciencias* 72: 27-32 (2000).

Araki, K., Toma, H.E. *Electrochim. Acta* 44: 1577-1583 (1999).

Araki, K., Toma, H.E. *N4-Macrocyclic Metal Complexes*. Zagal, J., Bedioui, F., Dodelet, J.P. (eds.), pp. 255-314. Springer, New York (2006).

Arico, A.S. *et al. Nature Mater.* 4: 366-377 (2005).

Arzt, E. *J. Appl. Phys.* 76: 1563 (1994).

Asaro, R.J., Suresh, S. *Acta Mater.* 53: 3369-3382 (2005).

Asefa, T. *et al. Nature* 402: 867-871 (1999).

Ashby, M.F. *Materials Selection in Mechanical Design*. Butterworth-Heinemann, Amsterdam (2005).

Autumn, K. *et al. PNAS* (USA) 99: 12252 (2002).

Aviram, A. (Ed.). *Molecular Electronics: Science and Technology* (Conference Proceedings No. 262). American Institute of Physics, New York (1992).

Aviram, A., Ratner, M.A. *Chem. Phys. Lett.* 29: 277 (1974).

Avouris, P., Chen, J. *Mater. Today* 9: 46-54 (2006).

Azevedo, C.M.N., Araki, K., Angnes, L., Toma, H.E. *Electroanalysis* 10: 467-471 (1998).

Bachtold, A. *et al. Science* 294: 1317 (2001).

Badaire, S. *et al. Adv. Mater.* 17: 1673-1676 (2005).

Bagley, B.G. *Nature* 225: 1040 (1970).

Baglioni, S. *et al. J. Neurosci.* 26: 8160-8167 (2006).

Bahr, J. *et al.* Single wall carbon nanotube based catalysts and electrodes for proton exchange membrane fuel cells. Fuel Cell Seminar, Palm Springs, CA (2005).

Bai, J., Virovets, A.V., Scheer, M. *Science* 300: 781-783 (2003).

Baik, I.-S. *et al. Appl. Phys. Lett.* 87: 263110 (2005).

Baker, R.W. *Membrane Technology and Applications*. Wiley, Hoboken, NJ (2004).

Balaban, N.Q. *et al. Science* 305: 1622-1625 (2004).

Balazs, A.C., Emrick, T., Russell, T.P. *Science* 314: 1107-1110 (2006).

Ball, P. Near-perfect 'black'. *Nature* 453: 834 (2008).

Ball, P. *Chem. Rev.* 108: 74-108 (2008a).

Balzani, V. *et al. Coord. Chem. Rev.* 250: 1254-1266 (2006).

Balzani, V., Scandola, F. *Supramolecular Photochemistry*. Ellis Horwood, Chichester (1991).

Balzani, V., Venturi, M., Credi, A. *Molecular Devices and Machines – A Journey into the Nanoworld*. Wiley-VCH, Weinheim (2003).

Bandaru, P.R. *et al. Nature Mater.* 4: 663-666 (2005).

Banerjee, S. *et al. IEEE Electron Device Lett.* 8: 347 (1987).

Banfield, J.F., Navrotsky, A. (eds.). *Nanoparticles and the Environment*. Mineralogical Society of America, Washington, D.C. (2001).

Banhart, F., Ajayan, P.M. *Nature* 382: 433 (1996).

Banhart, F., Li, J.X., Terrones, M. *Small* 1: 953 (2005).

Banik, S. *et al. Physical Review* B 74: 085110 (2006).

Barber, A.H., Cohen, S.R., Wagner, H.D. *Appl. Phys. Lett.* 82: 4140 (2003).

Bardeen, J. *Phys. Rev.* 71: 717 (1947).

Bar-Even A. *et al. Nature Genet.* 38: 636-643 (2006).

Barman, S.R. *Curr. Sci.* 88: 54-61 (2005).

Barnes, W.L., Dereux, A., Ebbesen, T.W. *Nature* 424: 824-830 (2003).

Barth, J.V., Costantini, G., Kern, K. *Nature* 437: 671-679 (2005).

Basu, B. *Curr. Sci.* 95: 570-571 (2008).
Batchelor, G.K. *An Introduction to Fluid Dynamics*. Cambridge Univ. Press, London (1967).
Baughman, R.H. *et al. Science* 284: 1340-1345 (1999).
Baxter, J.B., Aydil, E.S. *Appl. Phys. Lett.* 86: 053114 (2005).
Becerril, H.A. *et al. ACS Nano* 2: 463 (2008).
Becker, E.W. In: *Metal Clusters*. Trager, F., Putlitz, G. (eds.). Springer, Berlin.
Beckstein, O., Sansom, M.S.P. *PNAS* (USA) 100: 7063 (2003).
Bednorz, J.G., Muller, K.A. Z. Phys. B. *Cond. Matter*. 64: 189 (1986).
Beebe, D.J., Mensing, G.A., Walker, G.M. *Ann. Rev. Biomed. Engg*. 4: 261 (2002).
Beebe, M.J. *et al. Phys. Rev. Lett.* 97: 026801 (2006).
Beena, K.B., Sreekuttan, M.U., Sreekumar, K. *J. Phys. Chem.* C. 113: 17572-17578 (2009).
Beher, D. *et al. J. Biol. Chem.* 279: 43419-43426 (2004).
Behler, K.D. *et al. ACS Nano* 3: 363-369 (2009).
Bein, T. *MRS Bull.* . 30: 712 (2005).
Bekyarova, E. *et al. J. Phys. Chem.* B 107: 4681-4684 (2003).
Bellin, S. *et al. PNAS* (USA) 103: 18043-18047 (2006).
Bellingham, C.M. *et al. Biopolymers* 70: 445-455 (2003).
Bencardino, F. *et al. Cement Concrete Composites* 28: 832 (2006).
Bera, D., Kuiry, S.C., Seal, K. *J. Phys. Chem.* 108: 556-562 (2004).
Bernstein, E.R. (Ed.). *Chemical Reactions in Clusters*. Oxford Univ. Press, New York (1996).
Berridge, R. *et al. J. Phys. Chem.* B110: 3140 (2006).
Bers, D.M. *Excitation-Contraction Coupling and Cardiac Contractile Force*. Kluwer, Boston (2001).
Bethune, D.S. *et al. Nature* 363: 605-607 (1993).
Bharali, D.J. *et al. PNAS* (USA) 102: 11539-11544 (2005).
Bhat, J.S.A. *Curr. Sci.* 85: 147-154 (2003).
Bhat, N. *J. Indian Inst. Science* 87: 61-74 (2007).
Bhattacharya, A.B. *Ann. Ind. Acad. Engg.* IV: 13-18 (2007).
Bhattacharyya, S. *et al. Nature Mat.* 5: 19 (2006).
Bhushan, B. *Phil. Trans. R. Soc.* A 367: 1445 (2009).
Bhushan, B. *et al. Langmuir* 25: 1659 (2009a).
Bi, S., Zhang, J., Cheng, J. *Nature* 461: 593 (2009).
Binder, W. (Ed.). *Hydrogen Bonded Polymers*. Springer, New York (2007).
Biochicchio, B. *et al. Macromol. Biosci.* 5: 502-511 (2005).
Biskos, G. *et al. Geophys. Res. Lett.* 33: L07801 (2006).
Bisoyi, H.K., Kumar, S. *J. Mater. Chem.* 18: 3032-3039 (2008).
Bisoyi, H.K., Kumar, S. *J. Indian Inst. Sci.* 89: 101-110 (2009).
Bissonnette, L. *Expert Rev. Mol. Diagnostics* 6: 433 (2006).
Bittarello, E., Aquilano, D. *Eur. J. Mineral.* 19: 345 (2007).
Bleuse, J., Carayon, S., Reiss, P. *Physica E* 21: 331 (2004).
Blow, N. *Nature* 452: 901-904 (2008).
Bockstaller, M.R. *et al. J. Am. Chem. Soc.* 125: 5276 (2003).
Bockstaller, M.R., Thomas, E.L. *Phys. Rev. Lett.* 93: 166106 (2004).

Boden, S.A., Bagnall, D.M. *Appl. Phys. Lett.* 93: 133108 (2008).
Bohr, M.T. *IEEE Trans. Nanotechnol.* 1: 56-62 (2002).
Bolotin, K.J., Kuemmeth, F., Ralph, D.C. *Phys. Rev. Lett.* 97: 127202 (2006).
Boncheva, M. *et al. PNAS* (USA) 102: 3924-3929 (2005).
Borm, P.J.A., Berube, D. *Nanotoday* 3: 56-60 (2008).
Bosi, M., Pelosi, C. *Prog. Photovolt.* 15: 51-68 (2007).
Bosi, S. *et al. Eur. J. Med. Chem.* 38: 913-923 (2003).
Bosman, A.W., Sijbesma, R.P., Meijer, E.W. *Mater. Today* 7: 34-39 (2004).
Bosnick, K., Gabor, N.M., McEuen, P.L. *Appl. Phys. Lett.* 89: 163121 (2006).
Boudreau, D. *et al. Adv. Materials* 18: 2703 (2006).
Bourlinos, A.B. *et al. Eur. Phys. J.* E20: 109 (2006).
Bowden, N. *et al. Nature* 393: 146 (1998).
Bowler, D.R. *J. Phys. Cond. Matt.* 16: R721 (2004).
Brabec, C., Scherf, U., Dyakonov, V. (eds.). *Organic Photovoltaics: Materials, Device Physics, and Manufacturing Technologies*. Wiley-VCH, Weinheim (2008).
Breen, T.L. *et al. Science* 284: 948-951 (1999).
Brendel, R. *Sol. Energy* 77: 969-982 (2004).
Britto, P.J. *et al. Adv. Mater.* 11: 154-157 (1999).
Brown, A.S. *Aerospace America*, pp. 26-28 (Aug. 1998).
Browne, W.R., Feringa, B.L. *Nature Nanotechnol*. 1: 25-35 (2006).
Brust, M. *et al. J. Chem. Soc., Chem. Commun.* 801 (1994).
Buckley, P.R. *et al. IEEE Trans. Biomed. Eng.* 53: 2075 (2006).
Buehler, M.J. *PNAS* (USA) 103: 12285 (2006).
Buehler, M.J. *J. Comput. Theor. Nanosci.* 3: 603 (2006a).
Buehler, M.J., Gao, H. In: *Handbook of Theoretical and Computational Nanotechnology*, 2, 427-468. Rieth, M., Schommers, W. (eds.). (2005).
Bulusu, S., Wang, L.S., Zeng, X.C. *PNAS* (USA) 103: 8326-8330 (2006).
Burda, C. *et al. Chem. Rev.* 105: 1025 (2005).
Burdeniuc, J., Crabtree, R.H. *Science* 271: 340-341 (1996).
Buxton, G., Balazs, A.C. *Mol. Simul.* 30: 249 (2004).
Buxton, G., Balazs, A.C. *Phys. Rev. E Stat. Nonlin. Soft Matter Phys.* 67: 031802 (2003).
Buxton, G., Lee, J.Y., Balazs, A.C. *Macromolecules* 36: 9631 (2003).
Cai, Y. *et al. J. Phys. Chem.* 95: 7564-7568 (1991).
Calvo, E.J., Danilowicz, C., Wolosiuk, A. *J. Am. Chem. Soc.* 124: 2452-2453 (2002).
Calvo, E.J., Wolosiuk, A. *Chemophyschem.* 5: 235-239 (2004).
Calvo, E.J., Wolosiuk, A. *Chemphyschem* 6: 43-47 (2005).
Camera, S.G., Toma, H.E. *J. Photochem. Photobiol. A – Chem.* 151: 57-65 (2002).
Campbell, C.T. *Science* 294: 1471 (2001).
Cao, A. *et al. Science* 310: 1307 (2005).
Cao, J. *et al. Nature Mater.* 4: 745 (2005a).
Cao, Q., Rogers, J.A. *Adv. Mater.* 21: 29 (2009).
Cao, Y., Li, H. *Nature Mater.* 6: 109-114 (2007).

Capadona, J.R. *et al. Science* 319: 370-374 (2008).
Capello, J. *et al. Biotechnol. Prog.* 6: 198-202 (1990).
Carette, N. *et al. Nature Nanotechnol.* 2: 226-229 (2007).
Carless, P.A., Henry, D.A. *Br. J. Surg.* 93: 810-819 (2006).
Carta, R. *et al. Sens. Act.* A 156: 79 (2009).
Carter, J.M. *et al. J. Occup. Environ. Med.* 48: 1265-1278 (2006).
Casao-Rivera, E. *et al. J. Am. Chem. Soc.* 126: 4013 (2004).
Caspar, D.L. Biophys. *J.* 32: 103-138 (1980).
Cassagneau, T., Guerin, F., Fendler, J.H. *Langmuir* 16: 7318-7324 (2000).
Castelletto, V. *et al. Langmuir* 23: 6896 (2007).
Castleman, A.W., Jr., Jena, P. *PNAS* (USA) 103: 10552-10553 (2006).
Castleman, A.W., Jr., Wei, S. *Annu. Rev. Phys. Chem.* 45: 685-719 (1994).
Cedervall, T. *et al. PNAS* (USA) 104: 2050-2055 (2007).
Cen, C. *et al. Science* 323: 1026 (2009).
Cervini, R. *et al. Nanotechnology* 19: 175602 (2008).
Chaikof, E.L. Muscle mimic. *Nature* 465: 44-45 (2010).
Chakhalian, J. *et al. Nature Phys.* 2: 244 (2006).
Chaloupka, G., Khaliullin, G. *Phys. Rev. Lett.* 100: 016404 (2008).
Chambers, A. *et al. J. Phys. Chem. B* 102: 4253-4256 (1998).
Chan, C. *et al. Carbon* 43: 2431-2440 (2005).
Chang, K.Y., Kai, F. *GaAs High-Speed Devices.* Wiley, New York (1994).
Charlier, J.-C. *Acc. Chem. Res.* 35: 1063 (2002).
Charvin, G., Cross, F.R., Siggia, E.D. *PLoS ONE* 3: e1468 (2008).
Che, G.L. *et al. Nature* 393: 346-349 (1998a).
Cheetham, A.K. *et al. Stud. Surf. Sci. Catal.* 135: 788-795 (2001).
Chen, F. *et al. Proc. USDOE Hydrogen Prog. Rev.* (USA) (1998).
Chen, H.-T. *et al. Appl. Phys. Lett.* 93: 091117 (2008).
Chen, H.-T. *et al. Nature Photon.* 2: 295 (2008a).
Chen, I.-W., Rosenflanz, A. *Nature* 389: 701-704 (1997).
Chen, J. *et al. Science* 310: 1171 (2005).
Chen, J.-H., Kallenbach, N.R., Seeman, N.C. *J. Am. Chem. Soc.* 111: 6402-6407 (1989).
Chen, S. *et al. J. Am. Chem. Soc.* 125: 16186-16187 (2003).
Chen, X. *et al. Science* 295: 1698-1702 (2002).
Cheng-P.-Y., Baskin, J.S., Zewail, A.H. *PNAS* (USA) 103: 10570-10576 (2006).
Chien, P., Weissman, J.S., DePace, A.H. *Annu. Rev. Biochem.* 73: 617-656 (2004).
Chien, Y.C. *et al. Environ. Sci. Technol.* 35: 3259-3262 (2001).
Chiti, F., Dobson, C.M. *Annu. Rev. Biochem.* 75: 333-366 (2006).
Chiu, H.-Y. *et al. Nano Lett.* 8: 4342 (2008).
Cho, Y.K. *et al. Adv. Funct. Mater.* 17: 379-389 (2007).
Choi, H.J., Louie, S.G., Cohen, M.L. *Phys. Rev. Lett.* B 79: 094518 (2009).
Choi, M. *et al. Nature* 461: 246-249 (2009).
Chong, K.P., Garboczi, E.J. *Prog. Struct. Eng. Mater.* 4: 417 (2002).

Chopra, H.D. *et al. Nature Mater.* 4: 832-837 (2005).
Choresh, O., Bayarmagnai, B., Lewis, R.V. *Biomacromolecules* 10: 2852-2856 (2009).
Chou, S.Y., Krauss, P.R., Renstrom, P.J. *Science* 272: 85 (1996).
Chowdhury, D. *Curr. Sci.* 96: 923-932 (2009).
Chowdhury, D., Maoz, R., Sagiv, J. *Nano Lett.* 7: 1770-1778 (2007).
Chung, D.D.L. *J. Mater. Sci.* 39: 2645-2661 (2004).
Chung, J.Y., Chaudhury, M.K. *J. Roy.. Soc. Interface* 2: 55 (2005).
Chutinan, A., John, S., Toader, O. *Phys. Rev. Lett.* 90: 123901-123904 (2003).
Ciferri, A. (Ed.). *Supramolecular Polymers.* CRC, New York (2005).
Ciuparu, D. *et al. J. Phys. Chem.* B 108: 503-507 (2004).
Claessen, D. *et al. Phys. Rev. Lett.* 82: 1740 (1999).
Clark, T.D. *et al. J. Am. Chem. Soc.* 123: 7677-7682 (2001).
Cleary, J.P. *et al. Nature Neurosci.* 8: 79-84 (2005).
Clery, D., Lavine, M. *Science* 297: 961 (2002).
CMP Cientifica, CMP Cientifica, Madrid (2002).
CNCL. http://www.carbonnanoelectronics.com/applications.shtml (2008).
Colbert, D.T., Smalley, R.E. *Trends in Biotech.* 17: 46-50 (1999).
Coleman, J.N. *et al. Adv. Mater.* 18: 689 (2006).
Coleman, P. *Nature* 446: 379 (2007).
Collier, C.P. *et al. Science* 289: 1172 (2000).
Collins, M.L. *Nucleic acids Res.* 25: 2979 (1997).
Colombo, P. *J. Eur. Ceram. Soc.* 28: 1389 (2008a).
Colombo, P. *Science* 322: 381-383 (2008).
Coluci, V.R., Galvâo, D.S. Jorio, A. *Nanotechnology* 17: 617 (2006).
Colvin, V.L. *Nature Biotechnol.* 21: 1166-1170 (2003).
Colvin, V.L., Schlamp, M.C., Alivisatos, A.P. *Nature* 370: 354-357 (1994).
Compton, J. *Nature* 350: 91 (1991).
Conway, B.E. *Electrochemical Capacitors: Scientific Fundamentals and Technological Applications.* Kluwer, Dordrecht (1999).
Cordier, P. *et al. Nature* 451: 977-980 (2008).
Corma, A. *et al. J. Am. Chem. Soc.* 122: 2804-2809 (2000).
Corma, A., Botella, P., Mitchell, C. *Chem. Commun.* 2008-2010 (2004).
Corma, A. *et al. Nature* 443: 842-845 (2006).
Corma, A. *et al. Chem. Commun.* 6218-6220 (2008).
Corma, A. *Nature* 461: 182-183 (2009).
Correa-Duarte, M.A. *et al. Angew. Chem. Int. Ed.* 44: 4375 (2005).
Costa, C.A.R., Leite, C.A., Galembeck, F. *J. Phys. Chem.* B 107: 4747-4755 (2003).
Cox, B., Yang, Q. *Science* 314: 1102-1106 (2006).
Craighead, H.G. *Science* 290: 1532 (2000).
Cross *et al. Nano Lett.* 7: 10.1021/nl0624566 (2007).
Cruz, B.H. *et al. Electroanalysis* 12: 1130-1137 (2000).
Cubukcu, E. *et al. Nature* 423: 604-605 (2003).

Cui, H. *et al. Science* 317: 647 (2007).
Cui, W., Lieber, C.M. *Science* 291: 851 (2001).
Cushing, B.L., Kolesnichenko, V.L., O'Connor, C.J. *Chem. Rev.* 104: 3893-3946 (2004).
Custance, O., Morita, S. *Science* 319: 1051-1054 (2008).
Dai, H.J. *et al. Nature* 384: 147-150 (1996).
Dallas, P. *et al. Polymer* 48: 2007-2013 (2007).
Dalton, A.B. *et al. Nature* 423: 703 (2003).
Daniel, M.-C., Astruc, D. *Chem. Rev.* 104: 29 (2004).
Dao, M. *et al. Acta Mater.* 54: 5421-5432 (2006).
Das, B. *et al. Nanotechnology* 20: 125705 (2009).
Das, I., Ansari, S.A. *J. Sci. Industr. Res.* 68: 657-667 (2009).
Davis, I.J. *et al. Oral Microbiol. Immunol.* 20: 191 (2005).
Davis, M.E. *Nature* 417: 813-821 (2002).
De Greef, T.F., Meijer, E.W. *Nature* 453: 171-173 (2008).
Decher, G. *Science* 277: 1232 (1997).
Deepak, F.L., Govindaraj, A., Rao, C.N.R. *Chem. Phys. Lett.* 345: 5-10 (2001).
Degiron, A., Mock, J.J., Smith, D.R. *Opt. Express* 15: 1115 (2007).
Dekker, C. *Phys. Today* 52: 22-28 (1999).
Dekker, C. *Nature Nanotechnol.* 2: 209-215 (2007).
Delacourt, C. *et al. Science* 311: 977 (2006).
Demczyk, B.G. *et al. Mat. Sci. Eng.* A 334: 173 (2002).
Demers, L.M., Mirkin, C.A. *Angew. Chem. Int. Ed. Engl.* 40: 3069-3071 (2001).
Demus, D. *et al.* (eds.). *Handbook of Liquid Crystals.* Wiley-VCH, Weinheim (1998).
Denkov, N. *et al. Langmuir* 8: 3183 (1992).
Dennard, R.H. *et al. IEEE J. Solid-State Circuits* 9: 256 (1974).
Depree, C. *Water & Atmosphere* (NIWA, New Zealand) 16: 18-19 (2008).
Derycke, V. *et al. Nano Lett.* Published online 26 Aug. 2001. (http://pubs.acs.org/hotarticle/nalefd/nl015606!_revhtml).
Deshmukh, M.M *et al. Appl. Phys. Lett.* 75: 1631-1633 (1999).
Deshpande, A.V., Kumar, U. *Jour. Luminescence* 128: 1121 (2008).
Deshpande, A.V. *et al. Science* 323: 106 (2009).
Deville, S. *Adv. Eng. Mater.* 10: 155 (2008).
DeVries, G.A. *et al. Science* 315: 358-361 (2007).
Di Carlo, D., Wu, L.Y., Lee, L.P. *Lab Chip* 6: 1445-1449 (2006).
Di Ventra, M., Pershin, Y.V., Chua, L.O. *Proc. IEEE* 97: 1371 (2009).
Dickson, R.M., Lyon, L.A. (2000). *J. Phys. Chem.* B 104: 6095-6098 (2000).
Dietz, T.G. *et al. J. Chem. Phys.* 74: 6511-6512 (1981).
Dikin, D.A. *et al. Nature* 448: 457 (2007).
Ding, B., Seeman N.C. *Science* 314: 1583-1585 (2006).
Ding, L. *et al. J. Phys. Chem.* B109: 22337-22340 (2005).
Discher, D.E., Eisenberg, A. *Science* 297: 967-973 (2002).
Dobson, C.M. *Philos. Trans. R. Soc. London,* Ser. B *Biol. Sci.* 356: 133 (2001).

Dodabalapur, A. *Mater. Today* 9: 24 (2006).

Dogic, Z., Fraden, S. *Curr. Opin Colloid Interface Sci.* 11: 47-55 (2006).

Donald, A.M. *Nanotoday*, page 56 (May, 2005).

Douglas, J.F. *Nature* 463: 302-303 (2010).

Dowson, D. *History of Tribology*. Professional Engineering Publishing, London (1998).

Dresselhaus, M., Dresselhaus, G., Avouris, Ph. *Carbon Nanotubes: Synthesis, Structure, Properties and Applications*. Springer-Verlag, Berlin (2001).

Dresselhaus, M.S. *et al. Ann. Rev. Phys. Chem.* 58: 719 (2007).

Dresselhaus, M.S. *et al. Physical Properties of Carbon Nanotubes*. Imperial College Press, London (1998).

Dresselhaus, M.S., Dresselhaus, G. *Adv. Phys.* 51: 1-186 (2002).

Drexler, K.E. *PNAS* (USA) 78: 5275 (1981).

Drexler, K.E. *Sci. Am.* 285: 74-75 (2001).

Driscoll, T. *et al. Appl. Phys. Lett.* 91: 062511 (2007).

Driscoll, T. *et al. Appl. Phys. Lett.* 93: 024101 (2008).

Driscoll, T. *et al. Science* 325: 1518-1521 (2009).

Du, X.S. *et al. Synth. Met.* 143: 129-132 (2004).

Duan, X., Agarwal, Y., Lieber, C.M. *Nature* 421: 241-245 (2003).

Dubowski, J.J., Tanev, S. (eds.). *Photon-based Nanoscience and Nanobiotechnology*. Springer (2006).

Dudney, N.J. *et al. J. Electrochem. Soc.* 154: A805 (2007).

Duesberg, G.S. *et al. Diamond Rel. Mat.* 13: 354 (2004).

Duff, D.G., Baiker, A., Edwards, P.P. *Langmuir* 9: 2301-2309 (1993).

Duffy, D.C. *et al. Anal. Chem.* 71: 4669 (1999).

Dujardin, E. *et al. Nano Lett* 3: 413-417 (2003).

Dumitrica, T., Hua, M., Yakobson, B.I. *PNAS* (USA) 103: 6105-6109 (2006).

Duran, N. *et al. J. Biomed. Nanotech.* 3: 203-208 (2007).

Dusastre, V. *Nature* 451: 770-771 (2008).

Dutcher, J.R., Ediger, M.D. *Science* 319: 577-578 (2008).

Dzenis, Y. *Science* 319: 419-420 (2008).

Dzenis, Y., Reneker, D.H. *U.S. Patent* 6: 265, 333 (2001).

Ebron, V.H. *et al. Science* 311: 1580 (2006).

EC (European Commission). *European Technology Platform on NanoMedicine, Vision Paper and Basis for a Strategic Research Agenda for NanoMedicine*. Eur. Comm., Brussels (2005).

Edelstein, R.L. *et al. Biosensors Bioelectron.* 14: 805-813 (2000).

Egeblad, K. *et al. Chem. Mater.* 20: 946-960 (2008).

Eguchi, M. Philos. Mag. 49: 178-181 (1925).

Eigler, D.M., Schweizer, E.K. *Nature* 344: 524 (1990).

Eijkel, J.C.T., van den Berg, A. *Lab Chip* 6: 19-23 (2006).

Elechiguerra, J.L. *et al. J. Nanobiotechnol.* 3: doi: 10.1186/1477-3155-3-6 (2005).

Elias, D.C. *et al. Science* 323: 610-613 (2009).

El-Kady, I., Taha, M.M.R., Su, M.F. *Appl. Phys. Lett.* 88: 253109 (2006).

Ellenbogen, J.C., Love, J.C. *Proc. IEEE* 88: 386 (2000).

Ellingson, R.J. *et al. Nano Lett.* 5: 865 (2005).

Ellison, C.J., Torkelson, J.M. *Nature Mater.* 2: 695 (2003).
El-Sayed, M.A. *Acc. Chem. Res.* 37: 326-333 (2004).
Elvin, C.M. *et al. Nature* 437: 999-1002 (2005).
Emspak, J. *Sci. Amer.* 300:27 (Jan. 2009).
Endo, M. *et al. Carbon* 39: 1287-1297 (2001).
Engheta, N. *Science* 317: 1698 (2007).
Engheta, N., Ziolkowski, R. *Metamaterials: Physics and Engineering Explorations.* Wiley, Piscataway, NJ (2006).
Erdemir, A., Martin, J.-M. (eds.). *Superlubricity.* Elsevier (2007).
Ericson, L.M. *et al. Science* 305: 1447 (2004).
Eriksson, A. *et al. Nano Lett.* 8: 1224 (2008).
Fakhraai, Z., Forrest, J.A. *Science* 319: 600 (2008).
Fan T-X, Chow, S.K., Zhang, D. *Prog. Mater. Sci.* 54: 542 (2009).
Fan, Z. *et al. PNAS* (USA) 105: 11066-11070 (2008).
Fan, Z.Y. *et al. Adv. Mater.* 21: 3730 (2009).
Feinberg, A.W. *et al. Science* 317: 1366-1367 (2007).
Felker, P.M., Zewail, A.H. *Chem. Phys. Lett.* 94: 448-460 (1983).
Fendler, J.H. *Nanoparticles and Nanostructured Films: Preparation, Characterization and Applications.* Wiley VCH, Weinheim (1998).
Fendler, J.H., Meldrum, F. *Adv. Mater.* 7: 607-632 (1995).
Feng, L. *et al. Langmuir* 24: 4114 (2008).
Fernandez, J. *Material Architecture.* Architectural Press, Oxford (2006).
Fernandez, J.E. *Science* 315: 1807-1810 (2007).
Ferry, D.K. *Science* 319: 579-580 (2008).
Feyter, S.D., Schryver, F.C.D. *Chem. Soc. Rev.* 32: 139-150 (2003).
Fiebig, M.J. *J. Phys. D. Appl. Phys.* 38: R123 (2005).
Fire, A. *PNAS* (USA) 92: 4641 (1995).
Flandrois, S., Simon, B. *Carbon* 37: 165-180 (1999).
Fleischer, R.L., Price, P.B., Walker, R.M. *Nuclear Tracks in Solids. Principles and Applications.* Univ. California Press, Berkeley, CA (1975).
Flexner, C. *New Engl. J. Med.* 338: 1281-1292 (1998).
Flicker, J., Ready, J. *J. Appl. Phys.* 103: 113110 (2008).
Flores, M. *et al.* In: *Materials Res. Soc. Sympos. Proc.* Vol. 820. Materials Res. Soc., PA, USA (2004).
Florescu, M., Torquato, S., Steinhardt, P.J. *PNAS* (USA) 106: 20658-20663 (2009).
Florescu, M., Torquato, S., Steinhardt, P. *Phys. Rev.* B 80: 155112/1-7 (2009a).
Forester, T. *The Microelectronics Revolution.* MIT Press, Cambridge, MA (1982).
Fornasiero, F. *et al. PNAS* (USA) 105: 17250-17255 (2008).
Forrest, S.R., Thompson, M.E. *Chem. Rev.* 107: 923 (2007).
Fortin, D., Beveridge, T.J. In: *Biomineralization. From Biology to Biotechnology and Medical Applications.* Baeurien, E. (Ed.). Wiley-VCH, Weinheim (2000).
Frackowiak, E. *Phys. Chem. Phys.* 9: 1774-1785 (2007).
Frackowiak, E., Beguin, F. *Carbon* 39: 937-950 (2001).

Frankel, F.C., Whitesides, G.M. *No Small Matter: Science on the Nanoscale.* Belknap Press, Harvard University Press (2009).

Freestone, I. *et al. Gold Bull.* 40: 270 (2007).

Frenkel, D. *Nature* 460: 465-466 (2010).

Friedman, R.S. *et al. Nature* 434: 1085-1085 (2005).

Fu, G.F., Vary, P.S., Lin, C.T. *J. Phys. Chem.* 109: 8889-8898 (2005).

Fujikawa, S., Muto, E., Kunitake, T. *Langmuir* 25: 11563-11568 (2009).

Gabor, N.M. *et al. Science* 325: 1367-1371 (2009).

Galaev, I.Yu., Mattiasson, B. *Trends in Biotech.* 17: 335-340 (1999).

Galburt, E.A. *et al. Nature* 446: 820 (2007).

Galperin, M., Ratner, M.A., Nitzan, A. *J. Chem. Phys.* 121: 11965-11979 (2004).

Gandhi, S.K. In: *VLSI Fabrication Principles.* Wiley, New York, p. 728 (1994).

Ganteför, G. In: *Quantum Phenomena in Clusters and Nanostructures*, pp. 29-54. Khanna, S.N., Castleman, A.W., Jr. (eds.). Springer, Berlin (2003).

Gao, Y., Bulusu, S., Zeng, X.C. *J. Am. Chem. Soc.* 127: 15680-15681 (2005).

Gao, Y.F., Wang, Z.L. *Nano Lett.* 7: 2499-2505 (2007).

Garcia-Ruiz, J.U., Melero-Garcia, E., Hyde, S.T. *Science* 323: 362-365 (2009).

Garcia-Vidal, F.J., Moreno, E. *Nature* 461: 604-605 (2009).

Garnier, F. *et al. Science* 265: 1684 (1994).

Gates, B.D. *et al. Annu. Rev. Mater. Res.* 34: 339-372 (2004).

Gates, B.D. *et al. Chem. Rev.* 105: 1171-1196 (2005).

Gau, H. *et al. Science* 283: 46-49 (1999).

Ge, L. *et al. PNAS* (USA) 104: 10792 (2007).

Geim, A.K. *et al. Nat. Mater.* 2: 461 (2003).

Gelinck, G.H. *et al. Nature Mater.* 3: 106 (2004).

Genet, C., Ebbesen, T.W. *Nature* 445: 39 (2007).

Geng, Y. *et al. Nature Nanotech.* 2: 249 (2007).

Gessner, A. *et al. Int. J. Pharm.* 196: 245-249 (2000).

Gessner, A. *et al. J. Biomed. Mater. Res.* 65: 319-326 (2003).

Gewin, V. *Nature* 443: 137 (2006).

Ghosh, A.W. *et al. Nano Lett.* 4: 565-568 (2004).

Ghosh, S., Sood, A.K., Kumar, N. *Science* 299: 1042-1044 (2003).

Gibson, L.J., Ashby, M.F. *Cellular Solids, Structure and Properties.* Cambridge Univ. Press, Cambridge (1999).

Giessibl, F.J. *Rev. Mod. Phys.* 75: 949 (2003).

Gilbert, F. *et al. Science* 305: 651 (2004).

Girard, C., Dujardin, E. *J. Opt. A: Pure Appl. Opt.* 8: S73 (2006).

Glassmaker, N.J. *et al. PNAS* (USA) 104: 10786 (2007).

Glotzer, S.C., Solomon, M.J. *Nature Mater.* 6: 557-562 (2007).

Gnecco, E. *Nature* 461: 178-179 (2009).

Gnecco, E. *et al. Nanotechnology* 20: 025501 (2009a).

Goetzberger, A. *et al. Sol. Energy Mater. Sol. Cells* 92: 1570-1578 (2008).

Gogotsi, Y., Naguib, N., Libera, J.A. *Chem. Phys. Lett.* 365: 354 (2002).
Gogotsi, Y. *et al. Nature Mater.* 2: 591 (2003).
Gollub, J.N. *et al. Opt. Express* 17: 2122 (2009).
Gong, J.P. *Soft Matter* 2: 544 (2006).
Gong, T.Y., Marr, D.W.M. *Appl. Phys. Lett.* 85: 3846-3850 (2006).
Gong, X.J. *et al. Nature Nanotechnol.* 2: 709-712 (2007).
Goniakowski, J., Finocchi, F., Noguera, C. *Rep. Prog. Phys.* 71: 016501 (2008).
Göppert, T.M., Müller, R.H. *J. Drug Target.* 13: 179-187 (2005).
Gorb, S. *et al. J. R. Soc. Interface* 4: 271 (2007).
Gordon, P.A., Saeger, R.B. *Ind. Eng. Chem. Res.* 38: 4647-4653 (1999).
Gosal, W.J. *et al. Langmuir* 18: 7174 (2002).
Gosline, J. *et al. Phil. Trans. R. Soc. Lond.* B 357: 121-132 (2002).
Gozar, A. *et al. Nature* 455: 782 (2008).
Granasy, L. *et al. Nature Mater.* 3: 645 (2004).
Grass, R.N., Stark, W.J. *J. Mater. Chem.* 16: 1825-1830 (2006).
Graupner, R. *et al. J. Am. Chem. Soc.* 128: 6683-6689 (2006).
Greeden, J.E. *J. Mater. Chem.* 11: 37 (2001).
Green, M.A. *et al. Sol. Energy* 77: 857-863 (2004).
Green, M. *Prog. Photovoltaics* 13: 447 (2005).
Greer, A.L. *Nature Mater.* 5: 13-14 (2006).
Greer, A.L. *Nature* 464: 1137-1138 (2010).
Greiner, C., Arzt, E., Campo, A.D. *Adv. Mater.* 21: 479 (2009).
Griset, A.P. *et al. J. Am. Chem. Soc.* 131: 2469-2471 (2009).
Grobenly, J. *et al. Nanotechnology* 17: 5519-5524 (2006).
Groen, J.C. *et al. J. Am. Chem. Soc.* 129: 355-360 (2007).
Groisman, A. *et al. Nature Methods* 2: 685-689 (2005).
Grom, G.F. *et al. Nature* 407: 358-361 (2000).
Grossiord, N., Loos, J., Koning, C.E. *J. Mater. Chem.* 15: 2349-2352 (2005).
Grunlan, J.C., Liu, L., Kim, Y.S. *Nano Lett.* 6: 911-915 (2006).
Gu, H. *et al. Nature* 465: 202-205 (2010).
Gu, X. *et al. Phys. Rev. B Condens. Matter.* 70: 205401-01-205401-5 (2004).
Gudiksen, M.S. *et al. Nature* 415: 617-620 (2002).
Guldi, D.M. *Phys. Chem. Chem. Phys.* 9: 1400-1420 (2007).
Gumbsch, P. In: *Materials Science for the 21st Century*, vol. A, pp. 50-58. Soc. Materials Science, Kyoto (2001).
Gundiah, G. *et al. J. Mater. Chem.* 13: 209-213 (2003).
Guo, B.C., Kerns, K.P., Castleman, A.W., Jr. *Science* 255: 1411-1413 (1992).
Guo, T., Nikolaev, P., Thess, A., Colbert, D.T., Smalley, R.E. *Chem. Phys. Lett.* 243: 49-54 (1995).
Guo, X., Silva, S.R.P. *Science* 320: 618-619 (2008).
Guo, X. *et al. PNAS* (USA) 106: 20149-20154 (2009).
Gupta, A. *et al. Nature Med.* 5: 183 (1999).
Gupta, S. *et al. Nano Lett.* 6: 2066 (2006).

Gust, D., Moore, T.A., Moore, A.L. *Acc. Chem. Res.* 34: 40-48 (2001).

Haberland, H. (Ed.). *Clusters of Atoms and Molecules*. Springer, Berlin (1994).

Hadjichristidis, S., Pispas, S., Floudas, G. *Block Copolymers: Synthetic Strategies, Physical Properties, and Applications*. Wiley-Interscience, New York (2002).

Haensch, W. *et al. IBM J. Res. Dev.* 50: 339 (2006).

Hakkinen, H., Walter, M., Grönbeck, H. *J. Phys. Chem. B* 110: 9927 (2006).

Haldar, D. *Current Organic Synthesis* 5: 61-80 (2008).

Haldar, D., Drew, M.G.B., Banerjee, A. *Tetrahedron* 63: 5561-5566 (2007).

Haldar, J. *et al. PNAS* (USA) 103: 17667 (2006).

Hammami, A., Raymond, N., Armand, M. *Nature* 425: 635-636 (2003).

Hammond, P.T. *Adv. Mater.* 16: 1271-1293 (2004).

Han, J., Fu, J., Schoch, R.B. *Lab Chip* 8: 23-33 (2008).

Han, L. *et al. Biophys. J.* 92: 1384 (2007).

Han, W. *et al. Appl. Phys. Lett.* 69: 4111 (1996).

Hansen, E.D., Miller, T., Chiang, T.-C. *Phys. Rev. Lett.* 78: 2807 (1997).

Hanson, R. *et al. Rev. Mod. Phys.* 79: 1217 (2007).

Hassenzahl, W.V. *IEEE Trans. Appl. Supercond.* 11: 1447-1453 (2001).

Haung, Z. *et al. J. Photochem. Photobiol.* A130: 163 (2000).

He, H. *et al. Chem. Phys. Lett.* 287: 53-56 (1998).

Healy, K. *Nanomed.* 2: 459-481 (2007).

Hedin, L.H. In: *X-ray Spectroscopy*. Azaroff, L.V. (Ed.). p. 226. McGraw-Hill, New York (1974).

Hegner, M., Wagner, P., Semenza, G. *Surf. Sci.* 291: 39 (1993).

Heikal, A.A. *et al. Chem. Phys. Lett.* 242: 380-389 (1995).

Heinrich, A.J. *et al. Science* 298: 1381 (2002).

Hemminger, J. Directing matter and energy: Five challenges for science and the imagination. Report of Basic Energy Sciences Advisory Committee, US DOE (Dec. 20, 2007).

Henglein, A. *J. Phys. Chem.* 97: 5457-5471 (1993).

Hernandez, C.J., Mason, C.J. *J. Phys. Chem.* C111: 4477-4480 (2007).

Hess, O. *Nature* 455: 299-300 (2008).

Heyns, M., Tsai, W. *MRS Bull.* 34: 485-488 (2009).

Hill, J.P. *et al. Science* 304: 1481-1483 (2004).

Hill, M.T. *et al. Nature Photon* 1: 589-594 (2007).

Hillmyer, M.A. *Science* 317: 604-605 (2007).

Hinds, B.J. *et al. Science* 303: 62 (2004).

Hinnemann, B. *et al. J. Am. Chem. Soc.* 127: 5308 (2005).

Hirano, M., Shinjo, K. *Phys. Rev.* B41: 11837 (1990).

Hirata, M. *et al. Carbon* 42: 2929-2937 (2004).

Hirata, M., Gotou, T., Ohba, M. *Carbon* 43: 503-510 (2005).

Hirsch, L.R. *et al. PNAS* (USA) 100: 13549-13554 (2003).

Hirsch, P.B. (Ed.). In: *Defects*. Cambridge Univ. Press, Cambridge (1975).

Hla, S.W. *et al. Phys. Rev. Lett.* 85: 2777 (2000).

Hochella Jr., M.F. *et al. Science* 319: 1631-1635 (2008).

Hoeben, F.J.M. *et al. Chem. Rev.* 105: 1491-1546 (2005).
Hoeppener, S. *et al. Adv. Mater.* 14: 1036-1041 (2002).
Hoffman, A.J. *et al. Nature Mater.* 6: 946-950 (2007).
Hoffman, R.L., Norris, B.J., Wager, J.F. *Appl. Phys. Lett.* 82: 733 (2003).
Hogg, P.J. *Science* 314: 1100-1101 (2006).
Hohng, S.T. *et al. Science* 318: 279-283 (2007).
Holiday, B.J., Mirkin, C.A. *Angew. Chem. Int. Edn Engl.* 40: 2022-2043 (2002).
Holt, J.K. *et al. Science* 312: 1034-1037 (2006).
Hone, J., Deshpande, V.V. *Science* 325: 1084-1085 (2009).
Hong, R., Emrick, T., Rotello, V.M. *J. Am. Chem. Soc.* 126: 13572 (2004).
Hong, S., Mirkin, C.A. *Science* 288: 1836 (2002).
Hong, Y. *et al. J. Rheology* 43: 781 (1999).
Hörber, J.K.H., Miles, M.J. *Science* 302: 1002 (2003).
Hosek, J. *Veterinary Medicine* 51: 180 (2006).
Howarth, M. *et al. PNAS* (USA) 102: 7583 (2005).
HSE (Health and Safety Executive). *Health Effects of Particles Produced for Nanotechnologies.* Health and Safety Executive, London (2005).
Hsu, J.W.R. *et al. Nano Lett.* 5: 83-86 (2005).
Hu, P. *et al. J. Phys. Chem.* B 108: 936-938 (2004).
Hu, Y.Z., Granick, S. *Trib. Lett.* 5: 81 (1998).
Huang, X.H. *et al. J. Am. Chem. Soc.* 128: 2115 (2006).
Huang, Y. *et al. Science* 294: 1313 (2001).
Huang, Y., Lieber, C.M. *Pure Appl. Chem.* 76: 2051-2068 (2004).
Huey, L.G., Hanson, D.R., Howard, C.J. *J. Phys. Chem.* 99: 5001-5008 (1995).
Huey, LG. *Int. J. Mass Spectrom. Ion Proc.* 153: 145-150 (1996).
Huh, W.K. *et al. Nature* 425: 737-741 (2003).
Hui, C.Y. *et al. J. R. Soc. Interface* 1: 34 (2004).
Hummer, G., Rasaiah, J.C., Noworyta, J.P. *Nature* 414: 188-190 (2001).
Hunks, W.J., Ozin, G.A. *J. Mater. Chem.* 15: 3716 (2005).
Huo, F. *et al. Science* 321: 1658-1660 (2008).
Huo, F.W., Lytton-Jean, A.K.R., Mirkin, C.A. *Adv. Mater.* 18: 2304 (2006).
Husseman, M. *et al. Macromolecules* 32: 1424 (1999).
Hutchinson, J.L. *et al. Carbon* 39: 761-770 (2001).
Hüttel, A.K. *et al. Nano Lett.* 9: 2547 (2009).
Huvila, R. 3rd *Global Plastic Electronics Conference and Showcase,* Frankfurt (29 Oct. 2007); also see www.enfucell.com/Products.htm.
Huyghe, B. *et al. IEEE Trans. Adv. Packag.* 31: 802 (2008).
Huynh, W.U., Dittmer, J.J., Alivisatos, A.P. *Science* 298: 2425-2427 (2002).
Hwang, H.Y. *et al. Phys. Rev. Lett.* 77: 2041 (1996).
Hyde, S.T. *et al. Physica* A 339: 24 (2004).
Hyder, A.K. *et al.* In: *Spacecraft Power Technology.* 157: 223. Imperial College Press, London (2000).
Iijima, S. *Nature* 354: 56-58 (1991).

Iijima, S., Ichihashi, T. *Nature* 363: 603-605 (1993).
Ikesue, A. *et al. Annu. Rev. Mater.* 36: 397 (2006).
Imerito, T. *JOM* 57: 18 (Dec. 2005).
Ingram, A.L., Parker, A.R. *Phil. Trans. R. Soc.* B 363: 2465 (2008).
IPCS. In: *Handbook of Biodegradable Polymers*. Kost, J., Wiseman, D.M. (eds.). Harwood Academics, Amsterdam (1990).
Ishii, T. *et al. IEEE Trans. Elec. Dev.* 51: 1805 (2004).
Iyer, A.S., Lyon, L.A. *Angew. Chem. Int. Edn.* 48: 4562-4566 (2009).
IZA (International Zeolite Association). www.iza.org
Jackson, A.M., Myerson, J.W., Stellacci, F. *Nature Mater.* 3: 330 (2004).
Jadzinsky, P.D. *et al. Science* 318: 430 (2007).
Jaeger, N.I. *Science* 293: 1501 (2001).
Jaglinski, T. *et al. Science* 315: 620-622 (2007).
Jain, P.K. *et al. Acc. Chem. Res.* 41: 1578 (2008).
Jakse, N., Pasturel, A. *Phys. Rev. Lett.* 91: 195501 (2003).
Jakubowski, W., Matyjaszewski, K. *Angew. Chem. Int. Edn.* 45: 4482-4486 (2006).
James, D.K., Tour, J.M. *Top. Curr. Chem.* 257: 33-62 (2005).
Jang, D., Greer, J. *Nature Mater.* Doi: 10.1038/nmat2622 (2010).
Janssens, T.V.W. *et al. Top. Catal.* 44: 15 (2007).
Jaramillo, T.F. *et al. Science* 317: 100102 (2007).
Jaroniec, M. *Nature* 442: 638-640 (2006).
Javey, A. *et al. Nano Lett.* 7: 773-777 (2007).
Jayan, B.R. *et al. J. Phys. Chem.* C 112: 240-246 (2008).
Jena, P., Khanna, S.N., Rao, B.K. (eds.). *Physics and Chemistry of Small Metal Clusters*. Plenum, New York (1987).
Jensen, K., Kim, K., Zettl, A. *Nature Nanotechnol.* 3: 533 (2008).
Ji, M. *et al. Angew. Chem. Int. Ed.* 44: 7119-7123 (2005).
Jia, G. *et al. Environ. Sci. Technol.* 39: 1378-1383 (2005).
Jiang, K.L., Li, Q.Q., Fan, S.S. *Nature* 419: 801 (2002).
Jiang, W. *et al. Langmuir* 23: 8549-8553 (2007).
Jiang, X. *et al. Nano Lett.* 7: 3214-3218 (2007).
Joachim, C., Plevert, L. *Nanosciences: La Revolution Invisible*. Seuil: Paris (2008).
Joerger, R., Klaus, T., Granqvist, C.G. *Adv. Mater.* 12: 407-409 (2000).
Johansson, M.P., Sundholm, D., Vaara, J. *Angew. Chem. Int. Ed.* 43: 2678-2681 (2004).
John, B.S., Juhlin, C., Escobedo, F.A. *J. Chem. Phys.* 128: 044909 (2008).
John, S. *Phys. Rev. Lett.* 58: 2486-2489 (1987).
Jones, R.M. *Mechanics of Composite Materials*. Taylor and Francis, London (1999).
Joseph, S. *et al. Nano Lett.* 3: 1399 (2003).
Joseph, S., Aluru, N.R. *Nano Lett.* 8: 452-458 (2008).
Joseph, S.T.S. *et al. J. Phys. Chem.* B 110: 150 (2006).
Jo-Won Lee, M.S. Jhon. *KAST Review of Modern Science & Technol.* 4: 42-50 (2008).
Ju, S. *et al. Nano Lett.* Advance online publication doi: 10:1021/nl072538+ (2008).

Juhas, P. *et al. Nature* 440: 655-658 (2006).

Jung, B., Frey, W. *Nanotechnology* 19: 145303 (2008).

Jung, Y.J. *et al. J. Phys. Chem.* B 107: 6859-6864 (2003).

Kada, G., Kienberger, F., Hinterdorfer, P. *Nanotoday* 3: 12-19 (2008).

Kaetsu, I. *Adv. Polym. Sci.* 105: 81-97 (1993).

Kainuma, R. *et al. Nature* 439: 957 (2006).

Kajiwara, S., Kikuchi, T. *Acta Metall.* 38: 847 (1990).

Kakade, B.A., Pillai, V.K. *J. Phys. Chem.* C 112: 3183-3186 (2008).

Kalaitzidou, K. *et al. Carbon* 45: 1446 (2007).

Kalra, A. *et al. PNAS* (USA) 100: 10175 (2003).

Kamat, P.V. *J. Phys. Chem.* B 106: 7729 (2002).

Kang, S.J. *et al. Nature Nanotechnol.* 2: 230-236 (2007).

Kannan, A.M., Kanagala, P., Veedu, V. *Power Sources* 192: 297-303 (2009).

Kannan, R. *et al. Langmuir* 25: 8299-8305 (2009).

Kannan, R., Bhalchandra, A.K., Vijayamohanan, K.P. *Angew Chem. Int. Ed.* 47: 2653 (2008).

Kannan, R., Kakade, B.A., Pillai, V.K. *Angew Chem.* 120: 2693-2696 (2008).

Kannan, R., Pillai, V.K. *J. Indian Inst. Sci.* 89: 425-436 (2009).

Kaplan, D.L. *et al. Silk Polymers: Materials Science and Biotechnology.* Amer. Chem. Soc., Washington (1994).

Kaplan-Ashiri, I. *et al. PNAS* (USA) 103: 523-528 (2006).

Kapri, A. *et al. AIP Conf. Proc.* 1147: 469-474 (2009).

Kapri, A., Zaidi, M.G.H., Goel, R. *J. Microbiol. Biotechnol.* 20: 1032-1041 (2010).

Karen, K. *et al. Science* 302: 1380-1382 (2003).

Kasai, H., Wada, H., Matsumura, H. *Sol. Energy Mater. Sol. Cells* 65: 533-539 (2001).

Kastoff, R.N., Koytcheff, R.G., Lau, C.G.Y. *Curr. Sci.* 92: 1492-1498 (2007).

Katsnelson, M.I. *Mater. Today* 10: 20-27 (2007).

Katz, E., Willner, I. *ChemPhysChem* 5: 1084-1104 (2004).

Kaufmann, B.B. *et al. PLoS Biol.* 5: e239 (2007).

Kaul, S.N. *Curr. Sci.* 88: 78-82 (2005).

Kawasaki, M. *et al. Science* 266: 1540 (1994).

Kazes, M. *et al. Adv. Mater.* 14: 317-321 (2002).

Kegel, W.K., van Blaaderen, A. *Science* 287: 290-293 (2000).

Kelly, K.L. *et al. J. Phys. Chem.* B 107: 668 (2003).

Keskinen, H. *et al. Catal. Lett.* 111: 3-4 (2006).

Khang, D.Y. *et al. Nano Lett.* 8: 124 (2008).

Khanna, S.N., Castleman, A.W., Jr. (eds.). *Quantum Phenomena in Clusters and Nanostructures.* Springer, Berlin (2003).

Khanna, S.N., Jena, P. *Phys. Rev. Lett.* 69: 1664-1667 (1992).

Khanna, S.N., Jena, P. *Phys. Rev.* B51: 13705-13716 (1995).

Khanra, S. *et al. Inorg. Chem.* 47: 4605-4617 (2009).

Khanra, S. *et al. Angew. Chem. Int. Ed.* 46: 5564-5571 (2007).

Khanra, S. *et al. J. Mol. Struct.* 890: 157-162 (2008).

Khanra, S. *et al. Inorg. Chem.* 48: 5338-5349 (2009).

Khitrova, G. *et al. Nature Phys.* 2: 81-90 (2006).

Khodakovskya, M. *et al. ACS Nano* 3: 3221 (2009).

Kienberger, F. *et al. Biomaterials* 28: 2403 (2007).

Kienberger, F. *et al. Applied Scanning Probe Methods 2*; Springer Verlag, Berlin (2006).

Kijima, N. *et al. Chem. Lett.* 36: 568-569 (2007).

Kim, B. *et al. Small* 3: 1869-1872 (2007).

Kim, D.H. *et al. PNAS* (USA) 105: 15607 (2007).

Kim, D.-H. *et al. Science* 320: 507-511 (2008).

Kim, D.H. *et al. PNAS* (USA) 105: 18675 (2008a).

Kim, D.-H. *et al. Adv. Mater.* (published online 25 Jan. 2010: 10.1002/adma.2009902927) (2010).

Kim, E. *et al. J. Am. Chem. Soc.* 130: 4230-4231 (2008).

Kim, H.Y. *et al. Nano Lett.* doi: 10.1021/nl100824d (2010).

Kim, J. *et al. Appl. Phys. Lett.* 92: 031107 (2008).

Kim, J.Y. *et al. Science* 317: 222-225 (2007).

Kim, K.S. *et al. Nature* 457: 706 (2009).

Kim, K.T. *et al. Adv. Mater.* 21: 2787-2791 (2009).

Kim, S. *et al. Nano Lett.* 7: 2806 (2007).

Kim, S.H. *et al.* In: Proc. 2009 Sympos. on VLSI Technology, Kyoto, Japan, 15-17 June 2009, pp. 178-179 (2009).

Kim, S.K., Wang, J.K., Zewail, A.H. *Chem. Phys. Lett.* 228: 369-378 (1994).

Kim, W., Conticello, V.P. *J. Macromol. Sci.* C 47: 93-119 (2007).

Kim, Y.J. *et al. J. Electrochem. Soc.* 152: A710-A715 (2005).

Kinoshita, T. *et al. J. Magn. Magn. Mater.* 293: 106-110 (2005).

Klajn, R., Bishop, K.J.M., Grzybowski, B.A. *PNAS* (USA) 104: 10305-10309 (2007).

Klauk, H. *Nature* 451: 533-534 (2008).

Klaus-Joerger, T. *et al. Trends in Biotech.* 19: 15-20 (2001).

Klein, D.L. *et al. Nature* 389: 699-701 (1997).

Klein, J. *Proc. Inst. Mech. Eng. Part J.* 220: 691 (2006).

Klein, J. *et al. Nature* 370: 634 (1994).

Klein, J. *PNAS* (USA) 104: 2029-2030 (2007).

Knight, J.D., Miranker, A.D. *J. Mol. Biol.* 341: 1175-1187 (2004).

Knight, M.W. *et al. Nano Lett.* 7: 2346-2350 (2007).

Knoll, A. *et al. Microelec. Eng.* 83: 1692-1697 (2006).

Ko, H., Tsukruk, V.V. *Nano Lett.* 6: 1443-1448 (2006).

Kobayashi, M. *et al. Soft Matter* 3: 740 (2007).

Kodadek, T. *Biopolym. Peptide Sci.* 66: 134-140 (2002).

Kodadek, T. *Nature* 453: 861-862 (2008).

Koenig, D.R., Weig, E.M., Kotthaus, J.P. *Nature Nanotechnol.* 3: 482-485 (2008).

Köfinger, J., Hummer, G., Dellago, C. *PNAS* (USA) 105: 13218-13222 (2008).

Koida, T. *et al. Phys. Rev.* B 66: 144418 (2002).

Kojma, Y. *et al. J. Mater. Res.* 8: 1185 (1993).

Kokorin, V.V. *et al. Phys. Met. Metall.* 67: 202 (1989).

Kondo, Y., Takayanagi, K. *Phys. Rev. Lett.* 79: 3455-3458 (1997).

Korth, B.D. *et al. J. Am. Chem. Soc.* 128: 6562 (2006).

Kostarelos, K., Bianco, A., Prato, M. *Nature Nanotechnol.* 4: 627-633 (2009).

Kostoff, R.N. *et al.* Structure of the global nanoscience and nanotechnology research literature. DTIC Technical Report (http://www.dtic.mil/). Defense Technical Information Center, Fort Belvoir, VA, U.S.A. (2007).

Kostoff, R.N. *et al. J. Nanopart. Res.* 8: 193-213 (2006).

Kostoff, R.N. *et al. J. Nanopart. Res.* 8: 301-321 (2006a).

Kou, R. *et al. Electrochemistry Communications* 11: 954-957 (2009).

Kovtyukhova, N.I. *et al. Chem. Mater.* 11: 771-778 (1999).

Kowshik, M. *et al. Biotechnol. Bioeng.* 78: 583-588 (2002).

Kowshik, M. *et al. Adv. Mater.* 14: 815-818 (2002a).

Koyanagi, M. *et al. IEEE Trans. Electron. Dev.* 53: 2799 (2006).

Koziol, K. *et al. Science* 318: 1892-1895 (2007).

Kramer, R.M. *et al. Nature Mat.* 6: 533 (2007).

Kramer, S., Fuierer, R.R., Gorman, C.B. *Chem. Rev.* 103: 4369 (2003).

Krasheninnikov, A.V. *et al. Chem. Phys. Lett.* 418: 132 (2006).

Krasheninnikov, A.V., Banhart, F. *Nature Mater.* 6: 723-733 (2007).

Kratschmer, W. *et al. Nature* 347: 354-358 (1990).

Krebs, M.R.H. *et al. PNAS* (USA) 101: 14420 (2004).

Kreibig, U., Vollmer, M. *Optical Properties of Metal Clusters*. Springer-Verlag, New York (1995).

Krell, A. *et al. J. Am. Cer. Soc.* 86: 12 (2003).

Kresge, C.T. *et al. Nature* 359: 710-712 (1992).

Kreyling, W.G. *et al. J. Toxicol. Environ. Health* 65: 1513 (2002).

Krishna, C.S. (Ed.). *Patenting of Nanotech Inventions: A Debate*. The ICFAI University Press, Nagarjuna Hills, Hyderabad (2008).

Kroger, N., Deutzmann, R., Sumper, M. *Science* 286: 1129-1132 (1999).

Krogman, K.C., Druffel, T., Sunkara, M.K. *Nanotechnology* 16: S338 (2005).

Kroon, J.M. *et al. Progr. Photovolt.* 15: 1-18 (2007).

Kroto, H.W. *et al. Nature* 318: 162-163 (1985).

Krylov, S.Y. *et al. Phys. Rev.* E71: R65101 (2005).

Ku, J.-R., Stroeve, P. *Langmuir* 20: 2030-20323 (2004).

Kuiper, S. *et al. J. Membr. Sci.* 150: 1-8 (1998).

Kukar, T.L. *et al. Nature* 453: 925-929 (2008).

Kumacheva, E., Rubinstein, M. *Science* 329: 197-200 (2010).

Kumakura, M. *Process Biochem.* 32: 555-559 (1997).

Kumar, G., Tang, H.X., Schroers, J. *Nature* 457: 868 (2009).

Kumar, K.S., Van Swygenhoven, H., Suresh, S. *Acta Mater.* 51: 5743-5774 (2003).

Kumar, U., Deshpande, A.V. *Jour. Non-Crystalline Solids* 355: 501 (2009).

Künzel, H.M. *J. Thermal Envelope Building Sci.* 23: 95 (1999).

Kushmerick, J. *Nature* 462: 994-995 (2009).

Kwok, K., Ellenbogen, J. *Mater. Today* 5: 28-37 (2002).

Kymakis, G.A., Amaratunga, G.A. *Appl. Phys. Lett.* 80: 112 (2002).

Labarre, D. *et al. Biomaterials* 26: 5075-5084 (2005).

Labeit, S., Kolmerer, B. *Science* 270: 293-296 (1995).

Lacour, S.P. *et al. Appl. Phys. Lett.* 82: 2404 (2003).

Lacour, S.P. *et al. Proc. IEEE* 93: 1459 (2005).

Lahann, J. *et al. Science* 299: 371-374 (2003).

Lakes, R.S. *Philos. Mag. Lett.* 81: 95 (2001).

Lakes, R.S., Rosakis, P., Ruina, A. *J. Mater. Sci.* 28: 4667 (1993).

Lampert, C.M. *Mater. Today* 7: 28 (2004).

Land, M.F. *Prog. Biophys. Mol. Biol.* 24: 75 (1972).

Landy, N.I. *et al. Phys. Rev. Lett.* 100: 207402 (2008).

Langer, R., Tirrell, D.A. *Nature* 428: 487-492 (2004).

Langreth, D.C. *Phys. Rev. Lett.* 26:: 1229 (1971).

Langreth, D.C. In: *Collective Properties of Physical Systems*. Lundquist, B., Lundquist, S. (eds.). Academic, New York (1974).

Lantz, M.A., Wiesmann, D., Gotsmann, B. *Nature Nanotech.* Doi: 10.1038/nnano.2009.199 (2009).

Larbalestier, D. *et al. Nature* 414: 368-377 (2001).

Larsson, F., Svensson, L. *Comp.* A 33: 221 (2002).

Lassagne, B. *et al. Nano Lett.* 8: 3735 (2008).

Lassagne, B. *et al. Science* 325: 1107-1110 (2009).

Lavine, M. *Science* 314: 1099 (2006).

Law, M. *et al. Science* 305: 1269-1273 (2004).

Lazaris, A. *et al. Science* 295: 472-476 (2002).

Lee, C. *et al. Science* 321: 385-388 (2008).

Lee, C.-H. *et al. IEEE Int. Electron. Devices Meet. 2006* (295 (2006).

Lee, E.S., Na, K., Bae, Y.H. *Nano Lett.* 5: 325-329 (2005).

Lee, H. *et al. Nature* 434: 713 (2005).

Lee, H. *et al. Science* 318: 426-430 (2007).

Lee, H. *Nature* 465: 298-299 (2010).

Lee, I.R., Lee, W., Zewail, A.H. *PNAS* (USA) 103: 258-262 (2006).

Lee, J.U. *Appl. Phys. Lett.* 87: 073107 (2008).

Lee, J.Y. *et al. Phys. Rev. Lett.* 89: 155503 (2002).

Lee, J.Y., Shou, Z., Balazs, A.C. *Phys. Rev. Lett.* 91: 136103 (2003).

Lee, K.-B. *et al. Science* 295: 1702-1705 (2002).

Lee, P.J. *et al. Biotechniques* 44: 91-95 (2008).

Lee, S., Lee, Y. *Appl. Phys. Lett.* 76: 2877-2899 (2000).

Lee, S., Spencer, N.D. *Tribol. Int.* 38: 922 (2005).

Lee, S., Spencer, N.D. *Science* 319: 575-576 (2008).

Lee, S., Spencer, N.D. *Lubric. Sci.* 20: 21 (2008a).

Lee, W., Chiu, C.-S. *Optics Letters*. 26: 521-523 (2001).

Lee, W. *et al. Appl. Phys. Lett.* 85: 513-515 (2004).

Legnani, C. *et al. Appl. Phys. Lett.* 85: 10-12 (2004).

Lehn, J.M. *J. Inclusion Phenom.* 6: 351 (1988).

Lehn, J.-M. *Supramolecular Chemistry, Concepts and Perspectives*. VCH, Weinheim (1995).

Lele, A. Linking macromolecular dynamics to polymer processing: Some case studies. Abstract Ind. Acad. Sciences 21st Mid-Year Meeting, Bangalore, pp 15-16 (2010).

Lendentsov, N.N. *et al. J. Solid State Electron.* 40: 785-798 (1996).

Lendlein, A., Kelch, S. *Angew Chem. Int. Ed* 41: 2034-2057 (2002).

Leo, D. *Engineering Analysis of Smart Material Systems*. Wiley, N. York (2007).

Levitt, A.P. *Whisker Technology*. Interscience, New York (1970).

Lewis, J.A. *Adv. Funct. Mater.* 16: 2193 (2006).

Lewis, N., Nocera, D.G. *PNAS* (USA) 103: 15729 (2006).

Lewis, N.S. *Science* 315: 798-801 (2007).

Lewis, N.S., Crabtree, G. (eds.) *Basic Research Needs for Solar Energy Utilization*. Report of the Basic Energy Sciences Workshop on Solar Energy Utilization, U.S. Department of Energy, Washington, D.C. (2005).

Li, F. *et al. J. Am. Chem. Soc.* 125: 16166-16167 (2003).

Li, G. *et al. Nature Mater.* 4: 864-868 (2005).

Li, H. *et al. Nature* 418: 998-1002 (2002).

Li, J. *et al. Nature* 412: 166-169 (2001).

Li, J. *et al. Science* 299: 864-867 (2003).

Li, J.X., Banhart, F. *Nano Lett.* 4: 1143 (2004).

Li, P. *et al. Mater. Chem. Phys.* 106: 63-69 (2007).

Li, Q.Y., Cui, Z.F., Pepper, D.S. *J. Membr. Sci.* 136: 181-190 (1997).

Li, W. *et al. Langmuir* 21: 9386-9389 (2005).

Li, X. *et al. Angew. Chem. Int. Ed.* 41: 4786-4789(2002).

Li, X., Reinhoudt, D., Crego-Calama, M. *Chem. Soc. Rev.* 36: 1350 (2007).

Li, X.L. *J. Phys. D. Appl. Phys.* 41: 193001 (2008).

Li, Y. *et al. Mater. Today* 9: 18-27 (2006).

Li, Y.-L., Kinloch, I.A., Windle, A.H. *Science* 304: 276 (2004).

Li, Y., Tang, L., Li, J. *Electrochem. Commun.* 11: 846-849 (2009).

Li, Z., Hillmyer, M.A., Lodge, T.P. *Langmuir* 22: 9409 (2006).

Liang, W. *et al. Nature* 411: 665-669 (2001).

Liang, Z.C. *et al. Sol. Energy Mater. Sol. Cells* 80: 181-193 (2003).

Lieber, C.M., Wang, Z.L. *MRS Bull.* 23: 99-104 (2007).

Lima, M.M.S., Borsali, R. *Macromol. Rapid. Commun.* 25: 771-787 (2004).

Lin, D., Xing, B. *Environ. Pollut.* 150: 243 (2007).

Lin, J.F., Kamavaram, V., Kannan, A.M. *Power Sources* 195: 466-470 (2010).

Lin, J.H., Ko, T.H., Yen, M.Y. *Energy and Fuels* 23: 4042-4046 (2009).

Lin, Y. *et al. Science* 299: 226 (2003).

Lin, Y. *et al. J. Mater. Chem.* 14: 527-541 (2004).

Lin, Y. *et al. IEDM Technical Digest*, p. 687 (2004a).

Lin, Y. *et al. Nature* 434: 55 (2005).

Lin, Y.-T. *et al. Nanotechnology* 17: 5781 (2006).
Lin, Y. *et al. Science* 327: 662 (2010).
Ling, Y.H. *et al. Key Eng. Mater.* 280-283: 707-712 (2005).
Link, S., El-Sayed, M.A., Mohammed, M.B. *J. Phys. Chem.* B 109: 10531 (2005).
Linse, S. *et al. PNAS* (USA) 104: 8691-8696 (2007).
Lipinski, M. *et al. Sol. Energy Mater. Sol. Cells* 72: 271-276 (2002).
Litvinchuk, S. *et al. Nature Mater.* 6: 576-580 (2007).
Liu, J. *et al. J. Am. Chem. Soc.* 126: 6550 (2004).
Liu, Q. *et al. Nano Lett.* 9: 1007 (2009).
Liu, S. *et al. Nano Lett.* 2: 1055-1060 (2007).
Liu, S., Maoz, R., Sagiv, J. *Nano Lett.* 4: 845-851 (2004).
Liu, X. *et al. Adv. Mater.* 14: 231-234 (2002).
Liu, X.K., Jiang, M. *Angew. Chem. Int. Edit.* 45: 3846-3850 (2006).
Liu, Y., Galvin, S.P. *Acta Mater.* 45: 4431 (1997).
Liu, Y.H. *et al. Electrochemical and Solid-State Letters* 9: A356-A359 (2006).
Liu, Z. *et al. Science* 315: 1686 (2007).
Liz-Marzan, L.M. *Langmuir* 22: 32 (2006).
Logvenov, A. Gozar, A., Bozovic, I. *Science* 326: 699 (2009).
Loiseau, A. *et al.* In: *Understanding Carbon Nanotubes.* Lecture Notes in Physics Series. Springer (2006).
Long, K. *et al. IEEE Trans. Electr. Devices* 53: 1789-1796 (2006).
Love, J.C. *et al. Chem. Rev.* 105: 1105 (2005).
Loveless, D.M., Jeon, S.L., Craig, S.L. *J. Mater. Chem.* 17: 56-61 (2007).
Lowman, G.M. *et al. Langmuir* 20 9791-9795 (2004).
Lowman, G.M., Hammond, P.T. *Small* 1: 1070-1073 (2005).
Lozinsky, V.I. *Russ. Chem. Rev.* 71: 489-511 (2002).
Lozinsky, V.I. *et al. Trends in Biotech.* 21: 445-451 (2003).
Lozinsky, V.I., Plieva, F.M. *Enzyme Microb. Technol.* 23: 227-242 (1998).
Lu, J.P. *Phys. Rev. Lett.* 79: 1297 (1997).
Lu, J.R. *et al. J. Am. Chem. Soc.* 125: 3751-3757 (2003).
Lu, L. *et al. Science* 304: 422-426 (2004).
Lu, L. *et al. Acta Mater.* 53: 2169-2179 (2005).
Lu, W., Lieber, C.M. *Nature Mater.* 6: 841-850 (2007).
Lu, X., Xia, Y. *Nature Nanotech.* 1: 163 (2006).
Lucas, B.D. *et al. Adv. Mater.* 20: 1129-1134 (2008).
Lundqvist, M., Sethson, I., Jonsson, B.H. *Langmuir* 21: 5974-5979 (2005).
Luo, C. *et al. Biotechnol. Bioeng.* 101: 190-195 (2008).
Luque, A., Marti, A., Nozik, A.J. *MRS Bull.* 32: 236-241 (2007).
Luther, G.W., Rickard, D.T. *J. Nanopart. Res.* 7: 389 (2005).
Lutkenhaus, J.L., Hammond, P.T. *Soft Matter* 3: 804-816 (2007).
Luxembourg, D. *et al. Mater Sci. Eng.* B 108: 112 (2004).
Lv, S. *et al. Nature* 465: 69-73 (2010).
Lynch, I. *Physica* A 373: 511-520 (2007).

Lynch, I., Dawson K.A., Linse, S. *Science STKE* 10.1126/stke.3272006pe14 (2006).

Lynch, M.D., Patrick, D.L. *Nano Lett.* 2: 1197-1201 (2002).

Ma, J., Karaman, I. *Science* 327: 1468-1469 (2010).

Ma, M.L. *et al. Langmuir* 21: 5549 (2005).

Ma, Y.L. *et al. J. Electrochem. Soc.* 151: A8-A16 (2004).

Mace, C.R., Striemer, C.C., Miller, B.L. *Anal. Chem.* 78: 5578-5583 (2006).

Mack, S. *et al. Appl. Phys. Lett.* 88: 213101 (2006).

Mackay, M.E. *et al. Langmuir* 18: 1877 (2002).

Madou, M. *et al. Ann. Rev. Biomed. Engg.* 8: 601 (2006).

Madou, M.J. *Fundamentals of Microfabrications.* 2 ed. CRC Press, Boca Raton (2002).

Madou, M.J., Kellogg, C.J. In: *The LabCDTM: A centrifuge-based microfluidic platform for diagnostics, systems and technologies for clinical diagnostics and drug discovery.* Cohn, G.E., Katzir, A. (eds.). San Jose, California, SPIE, 80 (1998).

Maier, S.A. *IEEE J. Sel. Top Quant.* 12: 1671-1677 (2006).

Majumder, M. *et al. J. Am. Chem. Soc.* 127: 9062 (2005).

Majumder, M. *et al. Langmuir* 23: 8624 (2007).

Maki, T. *et al. Scr. Metall.* 18: 1105 (1984).

Makimoto, T. *Paradigm Shift in Electronic Industry.* GOSPL Conference, New Delhi (2004).

Mandal, S.K. *et al. J. Am. Chem. Soc.* 128: 1982-1994 (2006).

Mandal, S.K. *et al. Inorg. Chem.* 46: 7594-7600 (2007).

Mandal, S.K. *et al. Inorg. Chem.* 46: 10158-10168 (2007a).

Maniwa, Y. *et al. J. Phys. Soc. Jpn.* 71: 2863 (2002).

Maniwa, Y. *et al. Nature Mater.* 6: 135 (2007).

Mann, S. *Biomimetic Materials Chemistry.* VCH, New York (1996).

Manna, I. *Ann. Ind. Acad. Engg.* 3: 91-96 (2006).

Manna, I. In: Ind. Acad. Sci. Platinum Jubilee Meeting, Hyderabad (2-4 July 2009).

Mannhart, J., Schlom, D.G. *Science* 327: 1607-1611 (2010).

Mansur, H.S. *et al. J. Chem. Soc., Faraday Trans.* 91: 665-672 (1995).

Manz, A., Graber, N., Widmer, H. *Sensors Actuators* B 1: 244-248 (1990).

Mao, C.B. *et al. Science* 303: 213-217 (2004).

Marszalek, P.E. *et al. Nature* 396: 661 (1998).

Martin, F. *et al. J. Control. Release* 102: 123-133 (2005).

Martin, J.E. *Phys. Rev.* E 79: 011503 (2009).

Martin, J.E. *et al. J. Chem. Phys.* 118: 1557-1570 (2003).

Martin, J.E. *et al. Phys. Rev.* E 69: 021508 (2004).

Martin, J.E., Gulley, G. *J. Appl. Phys.* 106: 084301 (2009).

Martines, E. *et al. Nano Lett.* 5: 2097 (2005).

Martinez, L.-M., Angell, C.A. *Nature* 410: 663-667 (2001).

Mason, T.G., Weitz, D.A. *Phys. Rev. Lett.* 75: 2770-2773 (1995).

Maspoch, D., Ruiz-Molina, D., Veeiana, J. *Chem. Soc. Rev.* 36: 770 (2007).

Massey, J. *et al. Chem. Soc.* 120: 9533 (1998).

Matheson Jr., R.R. *Science* 297: 976 (2002).

Matsuo, N., Kihara, H., Takami, Y. *Solid State Electronics*, p. 1969 (2003).

Mattsson, J. *et al. Nature* 462: 83-86 (2009).

Matyjaszewski, K. *et al. PNAS* (USA) 103: 15309-15314 (2006).

May, B.D., Cartier, S.F., Castleman, A.W., Jr. *Chem. Phys. Lett.* 242: 265-272 (1995).

Maynard, A.D. *Nanotechnologies: Overview and Issues, Nanotechnology – Toxicol. Issues Environ. Safety.* Springer 1-14 (2007).

Maynard, A.D. *et al. Nature* 14: 11 (2006).

Medintz, I. *et al. Nature Mater.* 4: 435 (2005).

Mednikov, E.G., Jewell, M.C., Dahl, L.F. *J. Am. Chem. Soc.* 129: 11619 (2007).

Mehta, A., Zydney, A.L. *J. Membr. Sci.* 249: 245-249 (2005).

Meiwes-Broer, K. *Metal Clusters at Surfaces: Structure, Quantum Properties, Physical Chemistry.* Springer, Berlin (2000).

Menard, E. *et al. Chem. Rev.* 107: 1117-1160 (2007).

Merkle, R.C. *Nanotechnology* 11: 89-99 (2000).

Messing, G.L., Stevenson, A.J. *Science* 322: 383-384 (2008).

Metcalfe, A. *Biomaterials* 24: 491-497 (2003).

Michalet, X. *et al. Science* 307: 538-544 (2005).

Mie, G. *Ann. Phys.* 25: 377 (1908).

Miller, R. *et al. Science* 259: 1430-1433 (1993).

Miller, S.A. *et al. J. Am. Chem. Soc.* 126: 6226 (2004).

Minemoto, T. *et al. Sol. Energy Mater Sol. Cells* 90: 3009-3013 (2006).

Ming, T. *et al. Angew. Chem. Int. Ed.* 47: 9685 (2008).

Mirkin, C.A. *ACS Nano* 1: 79 (2007).

Mishra, L. *Scientific Basis for Ayurvedic Therapies.* CRC Press (2004).

Mitragotri, S. *Pharm. Res.* 26: 232-234 (2009).

Mitragotri, S., Lahann, S. *Nature Mater.* 8: 15-23 (2009).

Mitzi, D.B. *et al. Nature* 428: 299 (2004).

Miyamoto *et al. Am. Chem. Soc.* 128: 12636 (2006).

Miyazaki, H. *et al. Phys. Rev.* B67: 235109/1-5 (2003).

Mizuno, Y. *et al. J. Cer. Soc. Jpn.* 115: 181 (2007).

Mokkapati, S., Jagadish, C. *Mater. Today* 12: 22-32 (2009).

Molday, R.S., MacKenzie, D. *J. Immunol. Methods* 52: 353-367 (1982).

Möller, C. *et al. Biophys. J.* 77: 1150 (1999).

Mookerjea, S. *et al.* In: Proc. 2009 IEEE Internat. Electron Devices Meeting, Baltimore, MD, 7-9 Dec. 2009, pp. 949-951 (2009).

Moore, D.F. *The Friction and Lubrication of Elastomers.* Pergamon, Oxford (1972).

Moore, G.E. *Electronics*, p. 114 (1965).

Moore, G.E. Progress in Digital Integrated Electronics. IEEE, Electron device Meeting, Washington (1975).

Morones, J.R. *et al. Nanotechnology* 16: 2346-2353 (2005).

Motta, M. *et al. Nano Lett.* 5: 1529 (2005).

Mridha, S. *et al. Nanotechnology* 19: 275705-275710 (2008).

Muccini, M., Capelli, R. *Nature Mater.* doi: 10.1038/nmat2751 (2010).

Mufti, A.A. *Construct. Build. Mater.* 17: 379 (2003).

Mukherjee, A., Boskovic, C., Gable, R.W. *Acta Cryst.* C63: m71 (2007).

Mukherjee, A., Sadler, P.J. *Wiley Encycl. Chem. Biol.* Doi: 10.1002/ 9780470048672.wecb.333 (2009).

Mukhopadhyay, P.K. *et al. Physical Review* B 74: 064401 (2006).

Müller, M. *et al. Tribol. Lett.* 15: 395 (2003).

Mulvaney, P. *Nature* 450: 925 (2007).

Mulyana, Y. *et al. Inorg. Chem.* 48: 7765-7781 (2009).

Murphy, C.J. *et al. J. Phys. Chem.* B 109: 13857 (2005).

Mynar, J.L., Aida, T. *Nature* 451: 895-896 (2008).

Nadai, A. *Theory of Flow and Fracture of Solids*. McGraw Hill, New York (1950).

Nagpal, P. *et al. Science* 325: 594-597 (2009).

Naik, A. *et al. Nature* 443: 193 (2006).

Naik, V.M. *et al. Curr. Trends* in *Science*, Platinum Jubilee Special, Ind. Acad. Sci, Bangalore, pp. 129-148 (2009).

Nair, A.S., Pradeep, T. *Curr. Sci.* 84: 1560-1564 (2003).

Nair, P.R., Alam, M.A. *Nano Lett.* 8: 1281-1285 (2008).

Nakajima, K. *et al. Sol. Energy Mater Sol. Cells* 88: 323-329 (2005).

Nakajima, K., Fujiwara, K., Pan, W.G. *J. Electron. Mater.* 34: 1047-1052 (2005).

Nam, K.T. *et al. Science* 312: 885-888 (2006).

Nam, K.T. *et al. PNAS* (USA)105: 17727-17731 (2008).

Nam, K.T. *et al. Nano Lett.* 4: 23-27 (2004).

Nam, S. *et al. PNAS* (USA) 106: 21035-21038 (2009).

Nangrejo, M. *et al. Pharm. Dev. Technol.* 13: 425 (2008).

Nardin, C. *et al. Chem. Commun.* 1433-1434 (2000).

NASA. *Elastic Memory Composite Hinges Tested on Shuttle*, NASA Tech. Briefs 31(3): 20 (2007).

Natarajan, S., Gopalakrishnan, J. In: *Current Trends in Science* Platinum Jubilee Special, Ind. Acad. Sci., Bangalore, pp. 21-34 (2009).

Natarajan, S., Mahata, P. *J. Ind. Inst. Sci.* 88: 179 (2008).

Nath, M. *et al. Curr. Sci.* 85: 956-960 (2003).

Natsuki, T., Tantrakarn, K., Endo, M. *Carbon* 42: 39 (2004).

Nature Mater. Doi: 10.1038/nmat2033 (2007).

Nayfeh, A.H., Balachandran, B. *Applied Nonlinear Dynamics: Analytical, Computational, and Experimental Methods*. Wiley, New York (1995).

Nazin, G.V., Qiu, X.H., Ho, W. *Science* 302: 77 (2003).

Neal, A.L. *Ecotoxicology* 17: 362-371 (2008).

Negi, H. *et al. Int. Biodeter. Biodegrad.* 63: 553-558 (2009).

Nel, A. *et al. Science* 311: 622 (2006).

Nel, A.E. *et al. Nature Mater.* 8: 543-557 (2009).

Nelson, E.C., Braun, P.V. *Science* 318: 924 (2007).

Nembenna, S. *et al. J. Am. Chem. Soc.* 128: 13056-13057 (2006).

Newman, J.R. *et al. Nature* 441: 840-846 (2006).

Nicolis, G., Prigogine, I. *Self-organization in Non-Equilibrium Systems: From Dissipative Structure Formation to Order through Fluctuations.* Wiley, New York (1977).

Nicoll, G., Francisco, J.S. *Environ. Sci. Technol.* 33: 4102-4106 (1999).
Niemeyer, C.M. *Angew. Chem. Int. Edn. Engl.* 40: 4128-4158 (2001).
Niklaus, S. Traces of the future. *PRAXIS* (Bioengineering AG), Wald, Switzerland pp. 12-19 (2009).
Nirmal, M., Brus, L.E. *Acc. Chem. Res.* 32: 40 (1999).
Nishide, H., Oyaizu, K. *Science* 319: 737-738 (2008).
Nishimura, S. *et al. Am. J. Physiol. Heart Circ. Physiol.* 287: H196 (2004).
Nishino, T. *et al. Langmuir* 15: 4321 (1999).
Nitin, A.K., Joachim, M. *Adv. Mater.* 21: 2664-2680 (2009).
Niyogi, S. *et al. J. Am. Chem. Soc.* 128: 7720 (2006).
NNI (National Nanotechnology Initiative). *The Initiative and its Implementation Plan, National Science and Technology Council,* Committee on Technology, Subcommittee on Nanoscale, Science and Engineering, BNSET (2000).
Noginov, M.A. *et al. Phys. Rev. Lett.* 101: 226806 (2008).
Noginov, M.A. *et al. Nature* 460: 1110-1112 (2009).
Nosonovsky, M., Bhushan, B. *Microsyst. Technol.* 11: 535 (2005).
Novoselov, K.S. *et al. Science* 306: 666-669 (2004).
Novoselov, K.S. *et al. Nature* 438: 197 (2005).
Noy, A. *et al. Nanotoday* 2: 22-29 (2007).
Noy, A. *et al. PNAS* (USA) doi: 10.1073/pnas.0904850106 (2009).
Nunes, G.S. *et al. Eur. J. Inorg. Chem.* 1487-1495 (2005).
Nuxoll, E.E. *et al. Appl. Mater. Interfaces* 1: 888-893 (2009).
Oberdörster, G. *et al. Inhal. Toxicol.* 16: 437-445 (2004).
Oberdörster, G. *et al. Environ. Health Perspect.* 113: 201 (2005).
Oberdörster, G. *et al.* J. *Environ. Health Perspect.* 113: 823 (2005).
Obermeyer, P., Haase, C., Steibig, H. *Appl. Phys. Lett.* 92: 181102 (2008).
Ogata, S. *et al. Acta Mater.* 52: 233-238 (2004).
Oh, M., Mirkin, C.A. *Nature* 438: 651-654 (2005).
Oh, S., Oh, N., Applelford, M., Ong, J.L. *Am. J. Biochem. Biotech.* 2: 49 (2006).
Oh, S.H. *et al. Science* 310: 661-663 (2005).
Ohashi, T., Dai, L. *Carbon Nanotechnology.* Elsevier, Amsterdam (2006).
Ohnishi, H., Kondo, Y., Takayanagi, K. *Nature* 395: 780-783 (1998).
Oka, M. *et al. Metal Based Drugs* 1: 511 (1994).
Oloffs, A.C. *et al. Biomaterials* 15: 753-758 (1994).
Onoe, H., Matsumoto, K., Shimoyama, I. *Small* 3: 1383-3189 (2007).
Orhan, Y., Buyukgungor, H. *Int. Biodeter. Biodegrad.* 45: 49-55 (2000).
Ormonde, A.D.H. *et al. Langmuir* 20: 6927-6931 (2004).
Osborne, I., Lavine, M., Coontz, R. *Science* 327: 1595 (2010).
Osterman, N. *et al. Phys. Rev. Lett.* 103: 228301 (2009).
Otsuka, K., Ren, X. *Prog. Mater. Sci.* 50: 511 (2005).
Otsuka, K., Wayman, C.M. (eds.). *Shape Memory Materials.* Cambridge Univ. Press, Cambridge (1998).
Oulton, R.F. *et al. Nature* 461: 629-630 (2009).
Ouyang, M., Awscholom, D.D. *Science* 301: 1074-1078 (2003).

Oyabu, N. *et al. Nanotechnology* 16: S112 (2005).

Ozbay, E. *Science* 311: 189-193 (2006).

Paik, D.H., Kim, N.J., Zewail, A.H. *J. Chem. Phys.* 118: 6930-6929 (2003).

Paik, D.H. *et al. Science* 306: 672-675 (2004).

Pal, B. *et al. Phys. Rev.* B 75: 125322 (2007).

Pal, B. *et al. Appl. Phys. Lett.* 93: 073105 (2008).

Pal, B., Masumoto, Y. *Phys. Rev.* B 80: 125334 (2009).

Palmer, D.J. *Nanotoday* 3(3-4): (2008).

Pan, Z.Y. *et al. J. Phys. Chem.* B 101: 8703-9709 (1997).

Pankhurst, Q.A., Connolly, J., Dobson, J. *J. Phys. D Appl. Phys.* 36: R167-R181 (2003).

Paradiso, J.A., Starner, T. *Pervasive Comput.* 4: 18-27 (2005).

Parak, W.J. *et al. Adv. Mater.* 14: 882-885 (2002).

Park, J., *et al. J. Phys. Chem.* 108: 13594-13598 (2004).

Park, J., Hammond, P.T. *Adv. Mater.* 16: 520-525 (2004).

Park, J.H. *et al. Nanotechnology* 17: 895 (2006).

Park, K.-S. *et al. IEEE Trans. Nanotechnol.* 4.2, p. 242 (2005).

Park, M. *et al. Science* 276: 1401-1404 (1997).

Park, S.I. *et al. Adv. Funct. Mater.* 18: 2673 (2008).

Park, S.J. *et al. Science* 295: 1503 (2002).

Park, T.-J. *et al. J. Mater. Chem.* 16: 141-154 (2006).

Parker, A.R., Lawrence, C.R. *Nature* 414: 33 (2001).

Patolsky, F. *et al. Science* 313: 1100 (2006).

Pavel, C.C., Schmidt, W. *Chem. Commun.* 8: 882-884 (2006).

Pease, A.R., Stoddart, J.F. *Molecular Machines and Motors, in Structure and Bonding*. Sauvage, J.P. (Ed.), pp. 189-236. Springer (2001).

Pendry, J.B. *Phys. Rev. Lett.* 85: 3966-3969 (2000).

Pendry, J. *Nat. Mater.* 5: 599 (2006).

Pendry, J.B., Schurig, D., Smith, D.R. *Science* 312: 1780 (2006).

Peng, X. *et al. Nature Nanotechnol.* 4: 353-357 (2009).

Penza, M., Antolini, F., Antisari, M.V. *Sensors Actuators B: Chem.* 100: 47-59 (2004).

Percec, V. *et al. J. Am. Chem. Soc.* 128: 14156-14165 (2006).

Perez, L. *et al. Mat. Res. Soc. Symp. Proc.* Vol. 737. Materials Research Society (2003).

Perkowitz, S. *Universal Foam: From Cappuccino to the Cosmos*. Walker, New York (2000).

Perutz, M.F. *et al. PNAS* (USA) 99: 5591-5595 (2002).

Pettersson, A., Magnusson, P., Lundberg, P., Nygren, M. *Int. J. Imp. Eng.* 32: 387 (2005).

Pfeifer, R. *et al. Science* 318: 1088 (2007).

Phillips, D. *100 Years of Physical Chemistry, Royal Society of Chemistry*, pp. 105-106. Black Bear Press, Cambridge, U.K. (2003).

Phillips, J.C. *PNAS* (USA) 107: 1307-1310 (2010).

Philp, D., Stoddart, J.F. *Angew. Chem. Int. Edn. Engl.* 35: 1154-1196 (1996).

Pierson, H.O. *Handbook of Carbon, Graphite, Diamonds and Fullerenes*. Noyes Publications, NJ (1993).

Pile, D.F.P. *et al. Appl. Phys. Lett.* 87: 061106 (2005).

Pileni, M.-P. *Nature Mater.* 2: 145-150 (2003).
Pilgrim, J.S., Duncan, M.A. *J. Am. Chem. Soc.* 115: 6958-6961 (1993).
Pine, D.J. *Nature* 448: 876-877 (2007).
Piner, R.D., Zhu, J., Xu, F., Hong, S., Mirkin, C.A. *Science* 283: 661 (1999).
Podsiadlo, P. *et al. Science* 318: 80-83 (2007).
Polarz, S., Smarsly, B. *J. Nanosc. Nanotech.* 2: 581 (2002).
Postma, H. *Nano Lett.* Doi: 10.1021/nl9029237 (2010).
Poulton, S.W., Raiswell, R. *Chem. Geol.* 218: 203 (2005).
Powers, E.T., Kelly, J.W. *J. Am. Chem. Soc.* 123: 775-776 (2001).
Pradeep, T. (pp. 7-8) (Abst.). Ind. Acad. Sci. Platinum Jubilee Meeting, Hyderabad (2-4 July 2009).
Praharaj, S. *et al. Langmuir* 20: 9889 (2004).
Pramod, P., Thomas, K.G. *Adv. Mater.* 20: 4300 (2008).
Prasad, K.E. *et al. PNAS* (USA) 106: 13186-13189 (2009).
Prasad, N.K. *et al. J. Mater. Chem.* 17: 5042-5051 (2007).
Prato, M. *Top. Curr. Chem.* 199: 173-187 (1999).
Prato, M. *Nature* 465: 172173 (2010).
Press, D. *et al. Nature* 465: 218-221 (2008).
Proc. of IEEE Photovoltaic Conference. IEEE 2005.
Prucha, M. *Shock* 22: 29 (2004).
Pryor, C.E., Flatte, M.E. *Appl. Phys. Lett.* 88: 233108 (2006).
Pugno, N.M., Ruoff, R.S. *Philos. Mag.* 84: 2829-2845 (2004).
Pum, D., Sleytr, U.B. *Trends Biotechnol.* 17: 8-12 (1999).
Purohit, P.K. *et al. Biophys. J.* 88: 851-866 (2005).
Pushparaj, V.L. *et al. PNAS* 104: 13574-13577 (2007).
Putnam, D. *Nature. Mater.* 5: 439-451 (2006).
Py, C. *et al. Phys. Rev. Lett.* 98: 156103 (2007).
Pyayt, A.L. *et al. Nature Nanotechnol.* 3: 660-665 (2008).
Pyzik, A.J., Li, C.G. *Int. J. Appl. Ceram. Tech.* 2: 440 (2006).
Qazilbash, M.M. *et al. Appl. Phys. Lett.* 92: 241906 (2008).
Qi, M. *et al. Nature* 429: 538 (2004).
Qiao, R., Aluru, N.R. *J. Chem. Phys.* 118: 4692 (1997).
Qin, Y., Wang, X., Wang, Z.L. *Nature* 451: 809-813 (2008).
Quintino, M.S.M., Araki, K., Toma, H.E., Angnes, L. *Electroanalysis* 14: 1629-1634 (2002).
Rabinovitch, K., Pagis, A. *Appl. Optics* 14: 1326 (1975).
Rabotyagova, O.S., Cebe, P., Kaplan, D.L. *Biomacromolecules* 10: 229-236 (2009).
Radomski, A. *Br. J. Pharmacol.* 146: 882 (2005).
Radomski, A. *et al. Br. J. Pharmacol.* 146: 882-893 (2005).
Radström, P. *et al.* Pre-PCR processing. *Mol. Biotech.* 26: 133 (2004).
Raether, H. *Surface Plasmons.* Springer, Berlin (1988).
Rai, P.K. *et al. Am. Chem. Soc.* 128: 591-595 (2006).
Rajalakshmi, N. *et al. Electrochim. Acta* 45: 4511-4515 (2000).
Ralph, J.M., Schoeler, A.C., Krumpelt, M. *J. Mater. Sci.* 36: 1161 (2001).

Raman, A., Melcher, J., Tung, R. *Nanotoday* 3: 20-26 (2008).

Ramanathan, T. *et al. Nature Nanotech.* 3: 327-331 (2008).

Ramesh, S. *et al. J. Phys. Chem.* B 108: 8794-8798 (2004).

Ramos-Nino, M.E. *et al. Cancer Res.* 63: 3539-3545 (2003).

Rana, S., Misra, R.D.K. *J. Miner. Met. Mater. Soc.* 57: 65-69 (2005).

Ranjan, R. *et al. Physical Review* B74: 224443 (2006).

Rao, C.N.R., Gopalakrishnan, J. *New Directions in Solid State Chemistry.* Cambridge University Press, Cambridge (1997).

Rao, C.N.R. *et al. Chem. Phys. Chem.* 2: 78-105 (2001).

Rao, C.N.R., Govindaraj, A. *Acc. Chem. Res.* 35: 998 (2002).

Rao, C.N.R. *et al. Chem. Eur. J.* 8: 29-35 (2002).

Rao, C.N.R. *et al. Curr. Sci.* 85: 1041-1045 (2003).

Rao, C.N.R., Natarajan, S., Vaidhyanathan, R. *Angew. Chem. Int. Ed.* 42: 1466 (2004).

Rao, C.N.R., Govindaraj, A. *Nanotubes and Nanowires.* The Royal Society of Chemistry, London (2005).

Rao, C.N.R., Muller, A., Cheetham, A.K. (eds.). *Nanomaterials Chemistry: Recent Development and New Directions.* Wiley-VCH, Weinheim (2007).

Rao, C.N.R. *et al. J. Mater Chem.* 19: 2457-2469 (2009).

Reed, M.A. *et al. Appl. Phys. Lett.* 78: 3735 (2001).

Regal, C.A., Teufel, J.D., Lehnert, K.W. *Nature Phys.* 4: 555 (2008).

Rehr, J.J., Albers, R. *Rev. Mod. Phys.* 72: 621-654 (2000).

Rehr, J.R. *Nature* 440: 618-619 (2006).

Reinhold, C. *Nanotoday* 1: 16 (2006).

Renner, L. *et al. Langmuir* 21: 4571-4577 (2005).

Reufer, M. *et al. Appl. Phys. Lett.* 91: 171904/1-171904/3 (2007).

Reuss, R.H. *et al. Proc. IEEE* 93: 1239 (2005).

Reyren, N. *et al. Science* 317: 1196 (2007).

Rice, D.C. *et al. J. Surg. Oncol.* 73: 224-227 (2000).

Rich, J.B. *et al. Neurology* 45: 51-55 (1995).

Rief, M. *et al. Science* 276: 1109-1112 (1997).

Rijnders, G.J.H.M. *et al. Appl. Phys. Lett.* 70: 1888 (1997).

Rim, S.B. *et al. Appl. Phys. Lett.* 91: 243501 (2007).

Riul, A. *et al. Sensors Actuators B – Chem.* 98: 77-82 (2004).

Rockenberger, J., Scher, E.J., Alivisatos, A.P. *J. Am. Chem. Soc.* 121: 11595-11596 (1999).

Roco, M.C., Bainbridge, W.S. Societal Implications of Nanoscience and Nanotechnology – Improving Benefits to Humanity (NSET and National Science Foundation), Arlington, Virginia (2005).

Rogers, J.A., Someya, T., Huang, Y. *Science* 327: 1603-1607 (2010).

Rossi, M.J. *Chem. Rev.* 103: 4823-4882 (2003).

Roth, C.B., Dutcher, J.R. In: *Soft Materials: Structure and Dynamics.* Dutcher, J.R., Marangoni, A.G. (eds.), pp. 1-38 . CRC, New York (2004).

Rothemund, P.W.K. *Nature* 440: 297-302 (2006).

Rowat, A.C. *et al. PNAS* (USA) 106: 18149-18154 (2009).

Rowsell, J.LC., Yaghi, O.M. *Angew. Chem. Int. Ed.* 44: 4670 (2005).

Russell, T. *Curr. Opin. Colloid Interface Sci.* 1: 107-115 (1996).

Russell, T.P. *Science* 297: 964-967 (2002).

Rustagi, S.C. *et al. IEEE Electr. Device Lett.* 28: 1021-1024 (2007).

Ryan, A.J. *Nature* 456: 334-336 (2008).

Ryan, K.M. *et al. Nano Lett.* 6: 1479 (2006).

Rycenga, M., McLellan, J.M., Xia, Y. *Adv. Mater.* 20: 2416 (2008).

Sagahara, S., Tanaka, M. *Applied Physics Letters* 84.13, 206-214 (2005).

Sah, A. *et al. J. Microbiol. Biotechnol.* 20: 908-916 (2010).

Sahai, N., Schoonen, M.A.A. (eds.). *Medical Mineralogy and Geochemistry*. Mineral. Soc. America, Chantilly, VA (2006).

Sahni, V., Blackledge, T.A., Dhinojwala, A. *Nature Commun.* 1. doi: 10.1038/ ncomms1019 (2010).

Saitoh, T. *et al. J. Lightwave Tech.* 3: 288 (1985).

Salahuddin, S., Datta, S. *Nano Lett.* 8: 405 (2008).

Salaita, K. *et al. Angew. Chem. Int. Ed.* 45: 7220 (2006).

Samokhvalov, A. *et al. J. Phys. Chem.* B 107: 4245-4252 (2003).

Sanders, A.W. *et al. Nano Lett.* 6: 1822-1826 (2006).

Sandler, R., Bosso, C.J. *Issues Sci. Technol.*, pp. 28-30 (Summer 2007).

Sapsford, K.E., Berti, L., Medintz, I.L. *Angew. Chem. Int. Ed.* 45: 4562 (2006).

Sarawate, N.N., Dapino, M.J. *Appl. Phys. Lett.* 93: 062501 (2008).

Sarikaya, M. *et al. Nature Mater.* 2: 577-585 (2003).

Sarva, S. *et al. Int. J. Imp. Eng.* Online 15 Sep. 2005. 10.1016/j.ijimpeng.2005.07.006 (2005).

Sasaki, T. *et al. J. Antibiot.* 41: 835-842 (1988).

Sastry, M. *et al. Curr. Sci.* 85: 162-170 (2003).

Sastry, S., Debenedetti, P.G., Stillinger, F.H. *Nature* 393: 554-557 (1998).

Satishkumar, B.C. *et al. Appl. Phys. Lett.* 77: 2530-2532 (2000).

Satlewal, A. *et al. J. Microbiol. Biotechnol.* 18: 477-482 (2008).

Sato, A. *et al. Acta Metall.* 30: 1171 (1982).

Sauvage, J.P. *Acc. Chem. Res.* 31: 611-619 (2000).

Savariar, E.N., Krishnamoorthy, K., Thayumanavan, S. *Nature Nanotechnol.* 3: 112-117 (2008).

Scalari, G. *et al. Laser Photon Rev.* 3: 45-66 (2009).

Scardi, V. *Methods Enzymol.* 135B: 293-299 (1987).

Schafer, A.I., Fane, A.G., Waite, I.D. (eds.). *Nanofiltration: Principles and Applications*. Elsevier, New York (2005).

Scheffler, M., Colombo, P. (eds.). *Cellular Ceramics: Structure, Manufacturing, Properties and Applications.* Wiley-VCH, Weinheim (2005).

Scheibel, T. *et al. PNAS* (USA) 100: 4527 (2003).

Schirmeisen, A. *et al. Phys. Rev.* B71: 245403 (2005).

Schlichting, I. *et al. Science* 287: 1615 (2000).

Schlom, D.G. *et al. Appl. Phys. Lett.* 53: 1660 (1988).

Schmid, G. *Chem. Rev.* 92: 1709-1727 (1992).

Schmidt, J., Montemagno, C. In: *Introduction to Nanoscale Science and Technology*, pp. 549-574. Di Ventra, M., Evoy, S., Heflin, J.R. Jr. (eds.). Springer, New York (2004).

Schön, J.H., Meng, H., Bao, Z. *Nature* 413: 713 (2001).

Schülli, T.U. *et al. Nature* 464: 1174-1177 (2010).

Schurig, D., Pendry, J.B., Smith, D.R. *Opt. Express* 15: 14772 (2007).

Schwab, K.C., Roukes, M.L. *Phys. Today* 58: 36 (2005).

Schwierz, F., Liou, J.J. *Solid-State Electron.* 51: 1079-1091 (2007).

Scrosati, B. *Nature Nanotechnol.* 2: 598 (2007).

Seaton, A., Donaldson, K. *Lancet* 365: 923-924 (2005).

Seeman, N.C., Belcher, A.M. *PNAS* (USA) 99: 6451-6455 (2002).

Seeman, N.C., Lukeman, P.S. *Rep. Prog. Phys.* 68: 237-270 (2005).

Seinfeld, J.H., Pandis, S.N. *Atmospheric Chemistry and Physics.* Wiley, New York (1998).

Sekitani, T. *et al. Science* 321: 1468 (2008).

Sekitani, T. *et al. Nature Mater.* 8: 494 (2009).

Sekitani, T., Someya, T. *Adv. Mater.* (published online 12 Mar. 2010: 10.1002/adma.200904054) (2010).

SIA (Semiconductor International Association). International Technology Roadmap for Semiconductors (2001).

Service, R.F. *Science* 301: 909-911 (2003).

Service, R. *Science* 314: 45 (2006).

Service, R.F. *Science* 320: 1584-1585 (2008).

Service, R.F. *Science* 327: 1596-1597 (2010).

Setoyama, T. *Catal. Today* 116: 250-262 (2006).

Shalaev, V.M. *Nature Photonics* 1: 41-48 (2007).

Shalaev, V.M. *Science* 322: 384-385 (2008).

Shan, Z.W. *et al. Nature Mater.* 7: 115-119 (2008).

Shannon, J.M., Balon, F. *IEEE Trans. Elec. Dev.* 54: 354 (2007).

Shannon, J.M., Gerstner, E.G. *IEEE Elec. Dev. Lett.* 24: 405 (2003).

Shannon, M.A. *et al. Nature* 452: 301-310 (2008).

Shao, Z., Vollrath, F. *Nature* 418: 741 (2002).

Shaw, G.A. *et al. Adv. Mater.* 17: 1123-1127 (2005).

Shchukin, D.G., Radtchenko, I.L., Sukhorukov, G.B. *J. Phys. Chem.* B 107: 86-90 (2003).

Shen, Z., Zhao, Z., Peng, H., Nygren, M. *Nature* 417: 266-269 (2002).

Sheng, H.W. *et al. Nature* 439: 419-425 (2006).

Shevchenko, E.V. *et al. Nature* 439: 55 (2006).

Shin, H.J. *et al. J. Phys. Chem.* B 107: 4699-4704 (2003).

Shiu, J.-Y. *et al. Chem. Mater.* 16: 561 (2004).

Shull, R.D., McMichael, R.D., Ritter, J.J. *Nanostruct. Mat.* 2: 205-211 (1993).

Si, M. *et al. Macromolecules* 39: 4793 (2006).

Siegel, R.W., Hu, E., Roco, M.C. *Nanostructure Science and Technology.* Kluwer Academic Publishers, Dordrecht (1999).

Siegrist, J., Peytavi, R., Madou, M. *Curr. Trends in Sci.* Platinum Jubilee Special. Ind. Acad. Sci., Bangalore 111-127 (2009).

Sivaraman, S.K. *et al. Curr. Sci.* 97: 1055-1059 (2009).

Skaff, H. *et al. J. Am. Chem. Soc.* 124: 5729 (2002).

Skoulidas, A.I. *et al. Phys. Rev. Lett.* 89: 185901 (2002).

Sloan, E.D. *Nature* 426: 353 (2003).

Smalley, R.E. *Sci. Am.* 285: 76-77 (2001).
Smestad, G.P., Lampert, C.M. *Sol. Energy Mater Sol. Cells* 91: 440-444 (2007).
Smith, D.R., Pendry, J.B., Wiltshire, M.C.K. *Science* 305: 788 (2004).
Smogunov, A. *et al. Nature Nanotechnol.* 3: 22-25 (2008).
Socoliuc, A. *et al. Science* 313: 207-210 (2006).
Socoliuc, A., Bennewitz, R., Gnecco, E., Meyer, E. *Phys. Rev. Lett.* 92: 134301 (2004).
Sofo, J.O., Chaudhari, A.S., Barber, G.D. *Phys. Rev.* 75: 153401-153404 (2007).
Sokolov, A. *et al. Nature Nanotechnol.* 2: 171-175 (2007).
Somorjai, G.A. *Introduction to Surface Chemistry and Catalysis.* Wiley, New York (1994).
Song, H. *et al. Nature* 462: 1039-1043 (2009).
Song, J.H., Zhou, J., Wang, Z.L. *Nano Lett.* 6: 1656-1662 (2006).
Song, W., Windle, A.H. *Macromolecules* 38: 6181-6188 (2005).
Song, W., Windle, A.H. *Adv. Mater.* 20: 3149-3154 (2008).
Song, Y., Murray, R.W. *J. Am. Chem. Soc.* 124: 7096 (2002).
Song, Y. *et al. J. Am. Chem. Soc.* 126: 635-645 (2004).
Soni, R. *et al. J. Polym. Environ.* 17: 233-239 (2009).
Sorin, E.J., Pande, V.S. *J. Am. Chem. Soc.* 128: 6316 (2006).
Sousa, M.E. *et al. Appl. Phys. Lett.* 87: 173115 (2005).
Sousa, S.R. *et al. Langmuir* 20: 9745-9754 (2004).
Southam, G., Beveridge, T.J. *Geochim. Cosmochim. Acta* 60: 4369-4376 (1996).
Spittle, J.A. *Int. Mater. Rev.* 51: 247-269 (2006).
Srinivasan C. *Curr. Sci.* 92: 1338-1339 (2007).
Srinivasan, C., Saraswathi, R. *Curr. Sci.* 97: 302-303 (2009).
Srivastava, M. *et al. Jour. Sci. Indust. Res.* 68: 93-96 (2009).
Srivastava, M. *et al.* In: Bahl, O.P., Dhami, T.L., Manocha, L.M. (eds), *Proc. INDO Carbon 2001*, pp. 236-242 (2001).
Srivastava, S. *et al. Science* 327: 1355-1359 (2010).
St. Louis, P. *Clin. Biochemi.* 33: 427 (2000).
Stankovich, S. *et al. Nature* 442: 282-286 (2006).
Stankovich, S. *et al. J. Mater. Chem.* 16: 155-158 (2006a).
Stankovich, S. *et al. Carbon* 44: 3342 (2006b).
Stankovich, S. *et al. Carbon* 45: 1558 (2007).
Starke, E.M. *Ann. New York Acad. Sci.* 1098: 345 (2007).
Stauth, S.A., Parviz, B.A. *PNAS* (USA) 103: 13922 (2006).
Stebe, K.J., Lewandowski, E., Ghosh, M. *Science* 325: 159-160 (2009).
Steele, B.C.H., Heinzel, A. *Nature* 414: 345-352 (2001).
Steele, G.A. *et al. Science* 325: 1103-1107 (2009).
Steele, G.A., Gotz, G., Kouwenhoven, L.P. *Nature Nano* 4: 363 (2009).
Steen, E., Prinsloo, E.F. *Catal. Today* 71: 327-334 (2002).
Stein, A., Melde, B.J., Schroden, R.C. *Adv. Mater.* 12: 1403-1419 (2000).
Stein, A., Schroden, R.C. *Curr. Opin. Solid State Mater. Sci.* 5: 553 (2001).
Stoddart, J.F. *Acc. Chem. Res.* 34: 410-411 (2001).

Stojkovic, D. *et al. Phys. Rev.* B 68: 195406-195410 (2003).
Storm, A.J. *et al. Nature Mater.* 2: 537-540 (2003).
Striemer, C.C. *et al. Nature* 445: 749-753 (2007).
Studart, A.R. *et al. J. Am. Ceram. Soc.* 89: 1771 (2006).
Stupp, S.I. *et al. MRS Bull.* 30: 864-873 (2005).
Stupp, S. *et al. Nature Mater.* Doi: 10.1038/NMAT2778 (2010).
Su, M. *et al. J. Am. Chem. Soc.* 124: 1560-1561 (2002).
Subhramannia, M. *et al. Chem. Mater.* 20: 601 (2008).
Sudeep, P.K., Joseph, S.T.S., Thomas, K.G. *J. Am. Chem. Soc.* 127: 6516 (2005).
Suga, T. *et al. Chem. Commun.* 1730 (2007).
Sugimoto, Y. *et al. Nature Mater.* 4: 156 (2005).
Sui, H. *et al. Nature* 414: 872 (2001).
Sun, C.H., Jiang, P., Jiang, B. *Appl. Phys. Lett.* 92: 061112 (2008).
Sun, D.-H. *et al. J. Biomed. Mater. Res.* A65: 290-298 (2003).
Sun, J. *et al. Nature* 458: 1154-1157 (2009).
Sun, L. *et al. Science* 312: 1199-1202 (2006).
Sun, L. *et al. Phys. Rev. Lett.* 101: 156101 (2008).
Sun, L., Crooks, R.M. *Langmuir* 15: 738 (1999).
Sun, Q. *et al. Phys. Rev. Lett.* 93: 186803 (2004).
Sun, S., Zeng, H. *J. Am. Chem. Soc.* 124: 8204-8205 (2002).
Sun, X. *et al. Chemical Physics Letters* 379: 99-104 (2003).
Sun, Y., Rogers, J.A. *Adv. Mater.* 19: 1897 (2007).
Sun, Y., Rogers, J.A. *Mater. Chem.* 17: 832 (2007a).
Sun, Y.P. *et al. Acc. Chem. Res.* 35: 1096-1104 (2002).
Sungjin, P., Ruoff, R.S. *Nature Nanotechnol.* 4: 217-224 (2009).
Sur, U.K. *Curr. Sci.* 98: 1562-1563 (2010).
Suresh, S. and Li J. *Nature* 456: 716-717 (2008).
Sutton, A.P., Balluffi, R.W. *Interfaces in Crystalline Materials.* Clarendon, Oxford (1995).
Suzuki, K., Iijima, T., Wakihara, M. *Electrochim. Acta* 44: 2185-2191 (1999).
Suzuki, K., Ikari, K., Imai, H. *J. Am. Chem. Soc.* 126: 462-463 (2004).
Symonds, J.L. *Phys. Today* 48: 26-32 (1995).
Syrgiannis, Z. *et al. Angew. Chem. Int. Edn.* 49: 3322-3325 (2010).
Taberna, P.L. *et al. Nature Mater.* 5: 567 (2006).
Takagi, H. *et al. Mater. Sci. Engg.* B 108: 143-147 (2004).
Takagi, H., Hwang, H.Y. *Science* 327: 1601 (2010).
Tan, E. *et al. Analytical Chemistry* 77: 7984 (2005).
Tan, J.L. *et al. Tissue Eng.* 10: 865 (2004).
Tan, M.X. *et al. Progr. Inorg. Chem.* 41: 21-144 (1994).
Tan, V.B.C., Tay, T.E., Teo, W.K. *Int. J. Solids Struct.* 42: 1561 (2005).
Tan, W.H., Takeuchi, S. *PNAS* (USA) 104: 1146-1151 (2007).
Tanaka *et al. Science* 319: 1083 (2008).
Tanaka, H. *et al. Nature* 415: 192 (2002).

Tanaka, Y. *et al. Lab Chip* 7: 207 (2007).
Tanaka, Y. *et al. Science* 327: 1488-1490 (2010).
Tang, C. *et al. Science* 322: 429-432 (2008).
Tang, C.-Y., O'Brien, M.J., Hawkins, G.F. *JOM* 57: 32 (Mar. 2005).
Tang, D.-M. *et al. PNAS* (USA) 107: 9055-9059 (2010).
Tang, Z., Kotov, N.A., Giersig, M. *Science* 297: 237-240 (2002).
Tans, S.J., Vershueren, A.R.M., Dekker, C. *Nature* 393: 49 (1998).
Tao, H. *et al. Opt. Lett.* 16: 7181-7188 (2008).
Tarjus, G. *Nature* 448: 758-759 (2007).
Tasis, D. *et al. Chem. Rev.* 106: 1105-1136 (2006).
Taskova, R.M. *et al. J. Biosci.* 61: 347 (2006).
Taton, T.A. *Trends* in *Biotech.* 20: 277-279 (2002).
Taton, T.A. *et al. J. Am. Chem. Soc.* 122: 6305-6306 (2000).
Taur, Y. *et al. Proc. IEEE*, p. 486 (1997).
Templeton, A.C. *et al. J. Phys. Chem.* B 104: 564 (2000).
Templeton, A.C., Wuelfing, W.P., Murray, R.W. *Acc. Chem. Res.* 33: 27 (2000).
Tenne, R. *Nature Nanotechnol.* 1: 103-111 (2006).
Terfort, A., Bowden, N., Whitesides, G.M. *Nature* 386: 162-164 (1997).
Ternes, M. *et al. Science* 319: 1066 (2008).
Terrones, M. *Annu. Rev. Mater. Res.* 33: 419 (2003).
Terrones, M. *et al. Phys. Rev. Lett.* 89: 075505 (2002).
Tessler, N. *et al. Science* 295: 1506-1508 (2002).
Theis, T.N., Solomon, P.M. *Science* 327: 1600-1601 (2010).
Thelander, C. *et al. Mater. Today* 9: 28-35 (2006).
Thess, A. *et al. Science* 273: 483-487 (1996).
Thomas, J.M. *Angew. Chem. Ind. Ed.* 33: 913 (1994).
Thomas, J.M., Raja, R., Lewis, D.W. *Angew. Chem. Int. Ed.* 44: 6456 (2005).
Thomas, K.G. *et al. J. Phys. Chem.* B 108: 13066 (2004).
Thomas, K.G. In: *Current Trends in Science.* Platinum Jubilee Special, Ind. Acad. Sciences, Bangalore, pp. 53-66 (2009).
Thomas, K.G., Kamat, P.V. *Acc. Chem. Res.* 36: 888 (2003).
Thomassin, J.M. *et al. Membrane Sci.* 303: 252-257 (2007).
Thompson, B.C., Frechet, J.M.J. *Angew Chem. Int. Ed.* 47: 58-77 (2008).
Thompson, D.P. *Nature* 417: 237 (2002).
Thompson, J.M.T. *Philos. Trans. R. Soc. London Ser. A* 292: 1 (1979).
Thompson, R. *et al. Science* 292: 2469 (2001).
Thompson, S., Packan, P., Bohr, M. *Intel Technology Journal* (1998).
Thomson, D.A., Best, J.S. *IBM J. Res. Dev.* 3: 311-321 (2000).
Thurmond, F.A., Trotter, J.A. *J. Exp. Biol.* 199: 1817-1828 (1996)
Tian, B. *et al. Nature* 449: 885-890 (2007).
Titelman, G., Bron, S. *Method of Manufacturing of Graphite Oxide (GO) P1391-P1392.* Research Disclosure 500. IMI TAMI Institute for Research and Development Ltd., Haifa (2005).

Toma, H.E. *Curr. Sci.* 95: 1202-1225 (2008).

Toma, H.E., Araki, K. *Coord. Chem. Rev.* 196: 307-329 (2000).

Toma, S.H., Toma, H.E. *Electrochem. Commun.* 8: 1628-1632 (2006).

Tong, H.D. *et al. Nano Lett.* 4: 283-287 (2004).

Toohey, K.S. *et al. Nature Mat.* 6: 581 (2007).

Topol, A.W. *et al. IBM J. Res. Dev.* 50: 491-506 (2006).

Torquato, S., Jiao, Y. *Nature* 460: 876-879 (2009).

Tosheva, L., Valtchev, V.. *Chem. Mater.* 17: 2494-2513 (2005).

Treacy, M.M.J., Ebbesen, T.W., Gibson, J.M. *Nature* 381: 678 (1996).

Troisi, A. *et al. PNAS* (USA) 104: 14255-14259 (2007).

Tromp, R.M. *et al. Nano Lett.* 8: 469-472 (2008).

Trotter, J.A. *et al. Biochem. Soc. Trans.* 28: 357-362 (2000).

Tsakalakos, L. *Mater Sci. Eng.* R 62: 175-189 (2008).

Tsakmakidis, K.L., Boardman, A.D., Hess, O. *Nature* 450: 397-401 (2007).

Tsang, P.H. *et al. PNAS* (USA) 103: 5764-5768 (2006).

Tseng, G., Ellenbogen, J.C. *Science* 294: 1293-1294 (2001).

Tskhovrebova, L., Trinick, J. *Nature Rev. Mol. Cell Biol.* 4: 679-689 (2003).

Tu, Y. *et al. PNAS* (USA) 106: 18120-18124 (2009).

Uehara, H. *et al. ACS Nano* 3: 924-932 (2009).

Urbakh, M., Klafter, J., Gourdon, D., Israelachvili, J. *Nature* 430: 525 (2004).

Urry, D.W. *et al. Phil. Trans. R. Soc. Lond.* B357: 169-184 (2002).

Usuki, A. *et al. J. Mater. Res.* 8: 1179 (1993).

Utsunomiya, S. *et al. Environ. Sci. Technol.* 38: 2289 (2004).

Vaia, R., Baur, J. *Science* 319: 420-421 (2008).

Vaia, R.A., Wagner, H.D. *Mater. Today* 7: 32-37 (2004).

Valentine, J. *et al. Nature* 455: 376-379 (2008).

Van de Jagt, O.C., Beukers, A. *Polymer* 40: 1035 (1999).

Van den Berg, A., Wessling, M. *Nature* 445: 726 (2007).

Van Hest, J.C.M. *Nature* 461: 45-47 (2009).

Venema, L. *Nature* 442: 994 (2006).

Venkataramanan, M. *et al. Adv. Func. Mater.* 17: 3462 (2007).

Venkataramanan, M. *et al. J. Phys. Chem.* C 112: 17745 (2008).

Venkataramanan, M. *et al. Small* 4: 105 (2008a).

Venkataramanan, M. *et al. Adv. Mater.* 21: 4025 (2009).

Venkataramanan, M. *et al. J. Mater. Chem.* 19: 3149 (2009a).

Veselago, V.G., Nasimanov, E.E. *Nat. Mater.* 5: 759 (2006).

Vettiger, P. *et al. IEEE Trans. Nanotechnol.* 1: 39-55 (2002).

Vijayamohanan, K.P. *Abstracts of Lectures*. Nineteenth Mid-Year Meeting, 4-5 July 2008, Indian Academy of Sciences, Bangalore (2008).

Villers, S.H. *et al. J. Phys. Chem.* B 110: 25916-25925 (2006).

Vivekchand, S.R.C., Ramamurthy, U., Rao, C.N.R. *Nanotechnology* 17: S344-S350 (2006).

Vlassiouk, I. *et al. PNAS* (USA) 106: 21039-21044 (2009).

Vogel, S. *J. Biosci.* 31: 309 (2006).
Vogelaar, L. *et al. Small* 1: 645-655 (2005).
Vollrath, F. *et al. Nature* 345: 526-527 (1990).
Wada, Y. *Proc. IEEE* 89: 1147 (2001).
Waite, J.H., Tanzer, M.L. *Science* 212: 1038-1040 (1981).
Wales, D.J., Scheraga, H.A. *Science* 285: 1368-1372 (1999).
Walker, A.J. *et al. IEEE Trans. Elec. Dev.* 51: 1856 (2004).
Wan, C. *et al. J. Chem. Phys.* 106: 4353-4356 (1997).
Wang *et al., Langmuir*, 21(4): 1153 (2005).
Wang, H., Pinnavaia, T.J. *Angew. Chem. Int. Edn. Engl.* 45: 7603-7606 (2006).
Wang, S. *et al. Nano Lett.* 2: 817-822 (2002).
Wang, X., Waje, M., Yan, Y. *Electrochemical and Solid State Letters* 8: A42-A44 (2005).
Wang, X. *et al. J. Power Sources* 158: 154-159 (2006).
Wang, X., Song, J., Liu, J., Wang, Z.L. *Science* 316: 102-105 (2007).
Wang, X. *et al. Science* 317: 644 (2007a).
Wang, X.D. *et al. Nano Lett.* 7: 2475-2479 (2007a).
Wang, Y. *et al. Phys. Rev.* B 79: 212408 (2009).
Wang, Z.L., Song, J.H. *Science* 312: 242-246 (2006).
Wanlass, M.W. *et al. Sol. Cells* 27: 191 (1989).
Waychunas, G.A., Kim, C.S., Banfield, J.F. *J. Nanopart. Res.* 7: 409 (2005).
Weeks, E.R. *et al. Science* 287: 627-631 (2000).
Weiner, J., Ho, P.-T. *Light-matter Interaction: Fundamentals and Applications.* Wiley and Sons (2003).
Weiss, V., Thiruvengadathan, R., Regev, O. *Langmuir* 22: 854-856 (2006).
Weller, H. *Angew. Chem. Int. Ed. Engl.* 37: 1958-1959 (1998).
Welton, T. *Chem. Rev.* 99: 2071-2083 (1999).
West, A.R. *Solid State Chemistry and its Applications.* Wiley, Chichester (2003).
Westermark, P. *et al. Amyloid* 12: 1-4 (2005).
Whaley, S.R. *et al. Nature* 405: 665-668 (2000).
Whetten, R.L., Price, R. *Science* 318: 407-408 (2007).
Whitby, M., Quirke, N. *Nature Nanotechnol.* 2: 87-87-94 (2007).
White, S.R. *et al. Nature* 409: 794-797 (2001).
Whitesides, G.M. *Nature Biotechnol.* 1161-1165 (2003).
Whitesides, G.M. *Nature* 442: 368-373 (2006).
Whitesides, G.M. *et al. Annu. Rev. Biomed. Eng.* 3: 335 (2001).
Whitesides, G.M., Boncheva, M. *PNAS* (USA) 99: 4769-4774 (2002).
Whitesides, G.M., Grzybowski, B. *Science* 295: 2418-2421 (2002).
Whitten, R.C. (Ed.). *The Stratospheric Aerosol Layer.* Springer, Berlin (1982).
Wigginton, N.S., Haus, K.L., Hochella Jr., M.F. *J. Environ. Monit.* 9: 1306 (2007).
Wiley, B. *et al. MRS Bulletin* 30: 356 (2005).
Wilkie, I.C. *J. Exp. Biol.* 205: 159-165 (2002).
Willert, M. *et al. Chem. Mater.* 13: 4681-4685 (2001).
Williams, D.N. *et al. J. Nanobiotechnol.* 4: 3 (2006).

Wilson, M. *Superconducting Magnets*. Clarendon, Oxford (1983).
Wilson, S.T. *et al. J. Am. Chem. Soc.* 104: 1146 (1982).
Witkamp, B., Poot, M., van der Zant, H.S.J. *Nano Lett.* 6: 2904 (2006).
Wochner, P. *et al. PNAS* (USA) 106: 11511-11514 (2009).
Wolf, S.A. *et al. Science* 294: 1488 (2001).
Wong, E.W., Sheehan, P.E., Lieber, C.M. *Science* 277: 1971-1975 (1997).
Wong, S.S. *et al. J. Am. Chem. Soc.* 120: 603-604 (1998).
Wong, S.S. *et al. Nature* 394: 52-55 (1998a).
Wong, S.Y., Pelet, J.M., Putnam, D. *Prog. Polym. Sci.* 32: 799-837 (2007).
Wouters, D. *et al. Adv. Funct. Mater.* 15: 938-944 (2005).
Wu, J. *et al. Small* 1: 210-221 (2005).
Wyart, M. *Ann. Phys. Fr.* 30: 1-96 (2005).
Xi, J.Z., Schmidt, J.J., Montemagno, C.D. *Nature Mater.* 4: 180 (2005).
Xia, X., Hu, Z. *Langmuir* 20: 2094-2098 (2004).
Xiang, J. *et al. Nature* 441: 489-488 (2009).
Xiao, F.S. *et al. Angew. Chem. Int. Edn. Engl.* 45: 3090-3093 (2006).
Xie, X.-L., Mai, Y.-W., Zhou, X.-P. *Material Science and Engineering Reports* 49: 89-112 (2005).
Xu, B., Tao, N.J. *Science* 301: 1221-1223 (2003).
Xu, H. *et al. Small* 4: 1972 (2008).
Xu, H.Q. *Nature Mater.* 4: 649-650 (2005).
Yager, P. *Ann. Rev. Biomed. Engg.* 10: 107 (2008).
Yair, H. *et al. Angew. Chem. Int. Ed.* 48: 5443 (2009).
Yakayama, Y. *et al.* 10.1021/nl801649v. .
Yamada, T. *et al. Nature Nanotechnol.* 1: 131 (2006).
Yamaguchi, A. *et al. Nature Mater.* 3: 337-341 (2004).
Yamamoto, Y. *et al. PNAS* (USA) 106: 21051-21056 (2009).
Yan, F., Goedel, W.A. *Adv. Mater.* 16: 911-915 (2004).
Yan, H. *et al. Science* 301: 1882-1884 (2003).
Yan, R. *et al. PNAS* (USA) 106: 21045-21050 (2009).
Yang, C. *et al. Nano Lett.* 6: 2929 (2006).
Yang, J.G. *et al. Chem. Lett.* 35: 648-649 (2006).
Yang, L., Watts, D.J. *Toxicol. Lett.* 158: 122 (2005).
Yang, R.S., Wang, Z.L. *J. Am. Chem. Soc.* 128: 1466 (2006).
Yang, Z.P. *et al. Nano Lett.* 8: 446-451 (2008).
Yao, J. *et al. Science* 321: 930 (2008).
Yeh, S.W. *et al. Macromolecules* 38: 6559 (2005).
Yen, R.T., Fung, Y.C. *Am. J. Physiol.* 235: H251-H257 (1978).
Yin, Y., Alivisatos, A.P. *Nature* 437: 664-666 (2005).
Yin, Y.D. *et al. J. Am. Chem. Soc.* 123: 8718 (2001).
Yoffe, A.D. *Adv. Phys.* 42: 173-266 (1993).
Yoo *et al. PNAS* (USA) 107: 10.1072/pnas.0913879107 (2010).
Yoo, J.-W., Mitragotri, S. *PNAS* (USA) 107: 11205-11210 (2010).

Yoo, P.J. *et al. Nature Mater.* 5: 234-240 (2006).
Yoon, J. *et al. Nature* 465: 329-333 (2010).
Yoon, J.S. *et al. Nature Mater.* 7: 907-915 (2008).
Yoshida, J., Kobayashi, T. *J. Magn. Magn. Mater.* 194: 176-184 (1999).
Yoshimoto, K. *et al. J. Chem. Phys.* 122: 144712 (2005).
Yu, G. *et al. Science* 270: 1789-1791 (1995).
Yu, G. Cao, A., Lieber, C.M. *Nature Nanotech.* 2: 372-327 (2007).
Yu, G.H., Gao, A.Y., Lieber, C.M. *Nature Nanotechnol.* 2: 372-377 (2007).
Yu, K. *et al. Small* 1: 332 (2005).
Yu, L.H., Natelson, D. *Nano Lett.* 4: 79-83 (2004).
Yu, M.-F. *et al. Science* 287: 637 (2000).
Yu, S. *et al. J. Am. Chem. Soc.* 131: 10557-10566 (2009).
Yu, Z. *et al. J. Neurotrauma* 26: 1135 (2009).
Yuan, D. *et al. Electrochemistry Communications* 9: 2473-2478 (2007).
Yuan, H.-C. *et al. J. Appl. Phys.* 100: 013708 (2006).
Yuan, H.-C., Ma, Z. *Appl. Phys. Lett.* 89: 212105 (2006).
Yumura, T. *et al. Phys. Chem. Chem. Phys.* 11: 8275-8284 (2009).
Yurdumakan, B. *et al. Chem. Comm.* 30: 3799 (2005).
Zachariassen, K.E., Kristiansen, E. *Cryobiology* 41: 257-279 (2000).
Zade, S.S. In: Scientific Brochure 2009, p. 83. Indian Institute of Science Education and Research, IISER, Kolkata (2010).
Zade, S.S., Bendikov, M. *Chem. Eur. J.* 13: 3688 (2007).
Zade, S.S., Bendikov, M. *Chem. Eur. J.* 14: 6734 (2008).
Zade, S.S., Zamoshchik, N., Bendikov, M. *Chem. Eur. J.* 15: 8613 (2009).
Zakri, C. *Liq. Cryst. Today* 16: 1-11 (2007).
Zana, R., Kahler, E.W. (eds.). *Giant Micelles: Properties and Applications.* CRC Press, Boca Raton, FL (2007).
Zawaneh, P.N. *et al. Biomacromolecules* 7: 3245-3251 (2006).
Zawaneh, P.N. *et al. PNAS* (USA) 107: 11014-11019 (2010).
Zewail, A.H. *Femtochemistry – Ultrafast Dynamics of the Chemical Bond.* World Scientific, Singapore (1994).
Zewail, A.H. In: *Les Prix Nobel: The Nobel Prizes 1999*, pp. 103-203. Frängsmyr, T. (Ed.). Almquist & Wiksell, Stockholm (2000).
Zhai, H.J., Li, J., Wang, L.S. *J. Chem. Phys.* 121: 8369-8374 (2004).
Zhang *et al. Science* 327: 1634 (2010).
Zhang, D. *et al. Chem. Lett.* 34: 240-241 (2005).
Zhang, J. *et al. Nature* 466: 91-95 (2010).
Zhang, M. *et al. Science* 309: 1215 (2005).
Zhang, S., Kumar, S. *Small* 4: 1270-1283 (2008).
Zhang, S.G. *Nature Biotechnol.* 21: 1171-1178 (2003).
Zhang, X. *et al. J. Phys. Chem.* 110: 1158-1165 (2006).
Zhang, Y., Wang, L.W., Mascarenhas, A. *Nano Lett.* 7: 1264-1269 (2007).
Zhang, Z. *et al. Nano Lett.* 7: 1670 (2007).

Zhang, Z., Ly, T., Kodadek, T. *Chem. Biol.* 8: 391-397 (2001).
Zhao, D.Y. *et al. Science* 279: 548-552 (1998).
Zhao, K., Mason, T.G. *Phys. Rev. Lett.* 99: 268301 (2007).
Zheludev, N.I. *et al. Nature Photon* 2: 351-354 (2008).
Zheng, L. *et al. Biol. Trace Elem. Res.* 104: 83 (2005).
Zheng, M. *et al. Atmos. Environ.* 39: 3967 (2005).
Zheng, Z. *et al. J. Am. Chem. Soc.* 128: 7730 (2006).
Zhi, L. *et al. Angew. Chem. Int. Ed.* 44: 2120-2123 (2005).
Zhong, Q. *et al. Mater. Lett.* 62: 188-190 (2008).
Zhong, X. *et al. J. Am. Chem. Soc.* 125: 8589-8594 (2003).
Zhou, F. *et al. Chem. Lett.* 34: 1114-1115 (2004).
Zhou, G.Q. *et al. Prog. Biochem. Biophys.* 35: 998-1006 (2008).
Zhou, L. *et al. Appl. Phys. Lett.* 88: 083502 (2006).
Zhou, M., Chen, S., Zhao, S. *Chem. Lett.* 35: 332-333 (2006).
Zhou, Q., Wood, J.R., Wagner, H.D. *Appl. Phys. Lett.* 78: 1748-1750 (2001).
Zhou, W.D. *et al. J. Appl. Phys.* 102: 103105 (2007).
Zhu, J., Brink, M., McEuen, P.L. *Appl. Phys. Lett.* 87: 242102 (2005).
Zhu, T. *et al. PNAS* (USA) 104: 3031-3036 (2007).
Zhu, Y.C. *et al. J. Am. Chem. Soc.* 125: 16196-16197 (2003).
Zhu, Y.Q. *et al. Adv. Mater.* 17: 1500 (2005).
Zhu, Y.T.T., Liao, X.Z. *Nature Mater.* 3: 351-352 (2004).
Zimmermann, W.H. *et al. Cardiovasc. Res.* 71: 419 (2006).
Zimmler, M.A. *et al. Appl. Phys. Lett.* 93: 051101 (2008).
Zoetendal, E.G. *et al. System. Appl. Microbiol.* 24: 405 (2001).
Zuttel, A. *Naturwissch.* 91: 157 (2004).
Zuttel, A. *et al. Int. J. Hydrogen Energy* 27: 20'3-212 (2002).

Index